The Frontiers Collection

The books in this collection are devoted to challenging and open problems at the forefront of modern science and scholarship, including related philosophical debates. In contrast to typical research monographs, however, they strive to present their topics in a manner accessible also to scientifically literate non-specialists wishing to gain insight into the deeper implications and fascinating questions involved. Taken as a whole, the series reflects the need for a fundamental and interdisciplinary approach to modern science and research. Furthermore, it is intended to encourage active academics in all fields to ponder over important and perhaps controversial issues beyond their own speciality. Extending from quantum physics and relativity to entropy, consciousness, language and complex systems—the Frontiers Collection will inspire readers to push back the frontiers of their own knowledge.

More information about this series at https://link.springer.com/bookseries/5342

Thomas Dittrich

Information Dynamics

In Classical and Quantum Systems

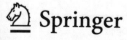 Springer

Thomas Dittrich
Departamento de Física
Universidad Nacional de Colombia
Bogotá D.C., Colombia

ISSN 1612-3018   ISSN 2197-6619   (electronic)
The Frontiers Collection
ISBN 978-3-030-96747-5   ISBN 978-3-030-96745-1   (eBook)
https://doi.org/10.1007/978-3-030-96745-1

This Springer imprint is published by the registered company Springer Nature Switzerland AG
The registered company address is: Gewerbestrasse 11, 6330 Cham, Switzerland

Preface

"Information" is one of the most common technical terms in our times, the market abounds with literature on the subject—do we need yet another book on this subject? The present work grew out of the demand for a text accompanying a topical lecture on the concept of information in physics, at the graduate to postgraduate levels. The course, in turn, was and still is an experiment in teaching physics, an attempt to introduce a novel kind of interdisciplinary subject in a traditional science curriculum.

The lectures offered by a university or college physics department are usually characterized by a high degree of specialization and compartmentalization: "Theoretical Solid-State Physics", "Quantum Optics", "Experimental High-Energy Physics" are standard titles reflecting an organization of knowledge motivated by phenomenological criteria or by mere historical development. Correspondingly, the formation of a professional physicist is perhaps intended to be broad, but not primarily to be comprehensive. At the same time, there is an increasing conscience that a wider perspective, the ability to think "across the Two Cultures", to work in interdisciplinary teams, is a highly desired competence in the academy as well as in administration and industry. The demand is growing for courses that contribute to this facet of science education, oriented towards transversal concepts and broad syntheses. However, they mostly appear as separate parts of the curriculum, say under the label *studium generale*, hardly interlocked with the technical courses proper, and do not form a bridge between specialized and holistic teaching. Achieving it requires a reorganization of traditional contents, presenting them as components of a larger endeavour, and stimulating students to construct a coherent whole of the diverse pieces of the puzzle they are confronted with during their studies.

The teaching project underlying this book [Dit14] is intended as a contribution to this general objective, focussing on a subject that appears particularly suitable for the purpose: The concept of information, even in its most specific, quantitative meaning, proves applicable not only in physics, but also in the adjacent disciplines, in mathematics, including foundations and logics, and in biology, from genetics through development through evolution. Beyond science, the concept is fruitful in the humanities, e.g., in philosophy, linguistics, and history, as well as in the social sciences, in particular in Sociology and in economy (Fig. 1).

Fig. 1 Information
dynamics overlaps
significantly with the
neighbouring sciences,
mathematics, physics, and
biology, as well with
humanities such as
philosophy and
hermeneutics, and of course
with computer science and
other branches of
engineering (not shown)

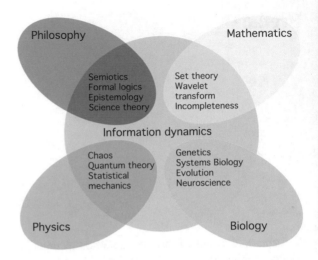

The concept of information is an extraordinarily powerful and versatile analytical tool. While in the sciences, information theoretical approaches are already well established and often even institutionalized as chairs, departments or entire faculties, they open a fresh novel access to many fields of the humanities, enabling a quantitative analysis where it was hitherto inconceivable and providing a firm common ground for a dialogue with the natural sciences.

A hypothesis that holds much of this book together is that these diverse disciplines do not merely use similar concepts referred to with the same term "information", but that they are actually dealing with manifestations of one and the same quantity, exchanged and transformed between different systems, places, and scales.

It cannot be the objective of this monograph to cover such a vast range of subjects. Rather, I would like to provide a conceptual basis, precise and solid enough to allow for mathematical reasoning in many cases, and at the same time sufficiently general to open the view towards a broad range of applications. This means that, unlike traditional physics lectures, a comprehensive idea plays the guiding role, in a wider sense than just a formalism or a set of axioms like Newton's laws or Maxwell's equations. The central aim is to demonstrate the unifying power of the concept of information. More specific topics are addressed selectively as case studies to illustrate and substantiate the general argumentation. The style of reasoning, adapted to an interdisciplinary content and a wider audience, also has to leave the traditional scheme of theoretical-physics lectures behind, occasionally replacing detailed derivations by qualitative arguments, often based on analogies.

With its level of rigour and completeness, the book occupies an intermediate position between semi-popular science [Llo06, Sei06, Flo10, Gle11, Sto15] and specialized monographs pretending completeness and technical accuracy [Bri56, Pie80, Ash90, Roe00, Mac03, CT06]. While detailed knowledge from molecular biology through mathematical logics obviously cannot be expected from a typical reader, some tolerance in the face of advanced mathematics and the willingness to switch frequently between very diverse topics will be helpful.

The title "Information Dynamics" has also been chosen to indicate a particular emphasis on the flow of information in space and time, its exchange among parts of a system, and its ascent and descent between different scales. Indeed, this work is rooted in the author's academic background in chaos and quantum chaos and is even inspired by a specific article: In a seminal paper [Sha80], Robert Shaw interprets deterministic chaos in terms of a directed flow of information from invisible microscopic scales to macroscopic visibility. As an immediate consequence, this idea leads to the question of how information is processed at the smallest scales, that is, in Quantum Mechanics. There is increasing agreement that the fundamental limits quantum physics imposes on information provide the most appropriate approach even to quantum phenomena we perceive as enigmatic, such as entanglement and nonlocality. Towards the largest scales, basic questions of cosmology invite to being interpreted in terms of entropy. In this sense, the book might well be titled "The information cycle", referring by analogy to the hydrological cycle.

The theory of deterministic chaos has been appreciated as a step towards replacing randomness as an irreducible element in the foundations of physics. Precisely, quantum mechanics, otherwise considered as the last stronghold of insurmountable unpredictability, however, prevents truly chaotic behaviour. The analysis of this surprising outcome inspired a number of profound insights into the character and origin of random processes in physics. Indeed, clarifying concepts and phenomena around randomness, bearing on causality vs. Chance, predictability vs. randomness, symmetry vs. disorder, ..., is another aim of this work: A third possible title could read "Chance: classical stochasticity vs. quantum determinism".

The first eight chapters of this volume are dedicated to interpreting natural phenomena in terms of information processing. They span an arc from basic definitions and simple applications through epistemological aspects to the central part on information processing in physical systems. Following an introduction from a historical point of view, the subsequent two sections cover elementary material such as general definitions and ramifications of the concept of information. Chapter 2 applies them to illustrative examples from quite diverse fields. The three applications presented there, the genetic code, formal logics, and Fourier transformation, not only demonstrate the use, but also the enormous versatility of the concept. They also serve as a reference for later chapters, concerning, e.g., the notion of gates, to recur in the context of computation, and uncertainty relations, so central for quantum mechanics.

Chapters 3 and 4 examine more fundamental issues around information that share their epistemological nature. They analyze the concepts of causality and predictability and oppose them to the creation of novelty, epitomized by randomness. With Chaps. 5–7, attention returns to physics in a more direct sense and thus to a safer ground as concerns mathematical tools. They present a detailed account of how information is processed in classical physical systems, beginning in Chap. 5 on the microscopic level. Hamiltonian dynamics provides the most detailed description of how information is exchanged between scales and among different degrees of freedom. The central tenet here is its strict conservation in all processes. From a rather more macroscopic viewpoint, Chap. 6 considers dissipative systems as sinks and chaotic systems as sources of entropy. Chapter 7 attempts a synthesis of these contrasting views, both relevant for systems far from equilibrium. It reveals their intimate relation, manifests, in particular, in fluctuation-dissipation theorems, and culminates in an interpretation of the time arrow implicit in the Second Law of Thermodynamics in terms of information flows.

This reasoning directs attention towards the bottom level of the information cycle, where the inconsistencies of classical physics in this respect become inescapable. Chapter 8 turns to the radical way quantum mechanics redraws our image of information processing in physical systems. The progress achieved in recent decades in quantum optics and related areas suggests a contemporary view on quantum mechanics based on information, which may eventually supersede the traditional Copenhagen interpretation. Forming a cornerstone of this book, Chapter 8 covers a wide range of topics, from information-related postulates through quantum measurement and quantum chaos.

The last two chapters take up the same metaphor of natural systems as information processors but read it backwards, considering computers as physical systems. They are dedicated to the conditions the laws of physics impose on information processing, in particular, in artificial systems. Also, here, quantum mechanics implies a radically different view. Chapter 9 introduces the concept of computation as a particular feature attributable to man-made systems, but by no means restricted to them. How far and in which sense computation in quantum systems is different from classical computation is explored in Chap. 10. Chapters 9 and 10 are definitely not nearly suited to replace monographs on conventional or quantum computation, such as Refs. [BEZ00, NC00, BCS04]. Quantum information and computation are among the buzzwords associated with the information era. However, rather than praising quantum high technology, these sections should provide a conceptual basis solid enough to enable a more balanced view towards these developments. Readers waiting for Alice and Bob to enter the stage will be disappointed.

Even with the broad scope pretended in this book, it is inevitable that some topics, intimately akin to its subject, are not given the space they deserve. This is the case for two areas in particular: Nonequilibrium statistical mechanics is at the heart of natural information processing. The fundamental role of the concept of entropy for statistical mechanics has been made popular notably by the seminal work of E. T. Jaynes [Jay57]. In the present book, nonequilibrium phenomena show up in various contexts. However, hardly a chapter—parts of Chap. 7—could be dedicated specifically to this vast subject. Readers are instead referred to the bibliography.

Talking of the particular way computers process information, it would have been tempting to compare it with Nature's solution to the problem: neural computing, specifically in the human brain. In brain research, neuro- and cognitive science are huge subjects in rapid development where information theory attains an increasingly crucial role. It could not be done any justice in this book. Again, readers are invited to follow citation paths.

This book does not conclude a mature subject. It is a snapshot of an active research area in rapid development. Like this, I hope that the reader will close it with the impression of a buzzing construction site, not of a proud venerable edifice. The intention is to inspire ideas that hopefully will contribute to the intellectual endeavour outlined in this work. Should the book provoke more questions than it can answer, it would have achieved its purpose.

The present work would not have been possible without the help and support from numerous sides. Gert-Ludwig Ingold gave the decisive initial impulse by asking me, "Why do you not consider complementing your topical lecture on the concept of information in physics by a textbook?". He accompanied the progress of the book patiently throughout the years and advised me, in particular, on issues related to quantum mechanics. Philippe Binder not only encouraged me untiringly from the very beginning, but became a real mentor of the book project. He was always eager to answer my questions on all topics around classical information and complexity, and he accompanied the writing with suggestions and constructive criticism. Angela Lahee, editor at Springer Nature, gave me the confidence and safety that was indispensable for me to enter into serious work after writing a few sketches of the first sections. Behind her editorial decisions and her patience with my repeated pleas to extend the deadline for completion, I always enjoyed her enthusiasm for my book project. I am owing a very different kind of encouragement to the students of my lectures on information dynamics at Universidad Nacional de Colombia, who motivated me permanently with their curiosity, their sincere interest in the subject, and countless bright, exploratory, and often enough critical questions. In the final phase of the project, Frank Großmann, Frank Jülicher, Leonardo Pachón, Holger Schanz, Walter Strunz, and Carlos Viviescas agreed to review different sections of the book with their particular expertise in the corresponding subjects. I am deeply indebted to all these colleagues and friends.

Besides persons, it was two institutions that made this book possible by their support. The first sketches of a pedagogical project covering the dynamics of information, in particular, in systems with complex dynamics, arose during my time as a postdoctoral researcher at the Max Planck Institute for the Physics of Complex Systems in Dresden. I am sure that the particularly stimulating atmosphere, the countless opportunities inviting to freely discuss even highly sophisticated questions with colleagues working on related subjects, contributed to those first ideas. Numerous short stays ever since at the MPIPKS consolidated my sense of belonging to this institute. As an indispensable practical aid, without which this book could never have completed, the institute granted me access to its electronic journal library throughout the writing process. Universidad Nacional de Colombia, the university I am affiliated with for more than twenty years, gave me the required freedom to dedicate a substantial part of my working time to this book. Above all, the Physics Department accepted and encouraged my proposal of a topical lecture on the concept of information and gave me the opportunity to hold it every few semesters as part of my teaching duty. I am sure that the liberal and innovative spirit at this university has facilitated these generous concessions.

Bogotá, spring 2022
Thomas Dittrich

Contents

Part I
Natural Systems as Information Processors

Chapter 1
The Concept of Information

1.1 Some History

Just as trilobites and hand axes characterize their respective paleontological and archaeological strata, there are concepts that can serve as "leitfossils" of certain eras of cultural history, in particular of the history of science and technology. The notion of information certainly has reached this level of prominence, as could easily be evidenced, say, by a word count including all classes of documents produced in our times.

In this respect, information has much in common with a related category, energy: Physical by nature, it applies to a huge variety of situations also outside the scientific realm. In fact, both concepts also share the dichotomy between a precise meaning in physics, including a quantitative definition, and a rather broad and sloppy usage in everyday language. Their adoption from and penetration back into common vocabulary can be traced back to similar historical processes: The history of science and technology on the one hand and the general political, social, and economic development on the other are in permanent interaction [Ber71]. Obviously, the evolution of science depends on economic and social boundary conditions, it reacts in its specific way, for instance, to periods of peace and of war. At the same time, major progress achieved in science and technology can have significant repercussions on common life, even on political decisions. While small steps usually go unnoticed by the general public, it is well possible to correlate grand epochs in the history of science with concomitant global changes in the society, often but not necessarily coinciding with scientific revolutions in the sense of Thomas Kuhn [Kuh70].

Table 1.1 is a crude attempt to identify such long-term parallel developments during the last half millennium, not attempting precision nor completeness or uniqueness. It also indicates how these last three eras have been accompanied by simultaneous technological progress, reflected in novel devices being invented and entering daily life. The central concepts like energy and information have played the role of universal metaphors, applied not only in their original field but to just about every subject that invites for scientific understanding and interpretation. While Laplace and

© Springer Nature Switzerland AG 2022
T. Dittrich, *Information Dynamics*, The Frontiers Collection,
https://doi.org/10.1007/978-3-030-96745-1_1

Table 1.1 Subdividing the last half millennium into three major epochs demonstrates strong correlations between scientific and technological progress, paradigms dominating science and philosophy, and social and economic development.

Centuries	Era	Characteristic quantity	Characteristic devices	Paradigms	Prevailing socio -eco - nomic model
16 - 17	Mechanistic	Force	Wedge, lever, pulley, tooth wheel, clock, printing press	The universe as clockwork, celestial me- chanics, Laplace's demon	Agriculture, manufacture
18 - 19	Thermo - dynamic	Energy	Watermill, windmill, steam engi- ne, combus- tion motor	Energy conser- vation, thermo- dynamic systems, Maxwell's demon	Industrial
20 - 21	Informatic	Entropy, information	Telegraph, telephone, computer	The world as computer/ network	Postindustrial knowledge society

his contemporaries gained deep insight interpreting nature as a deterministic mechanism, as "cosmic clockwork", we enjoy "Programming the Universe" [Llo06], seeing it as an all-embracing quantum computer.

In a similar way as Laplace and his contemporaries interpreted all of nature as a deterministic mechanism, as "cosmic clockwork", we tend to associate organs and their functions to engines or parts thereof, to compare our brains with computers and genetic control with algorithms.

The category of information penetrates our life from the TV news till the purchase of a computer or a cell phone. The present book is intended to contribute to a conscious critical use of the concept, based on solid knowledge of its meaning and its limitations.

The way this notion gained shape in the course of academic discussion is a particularly fascinating chapter of the recent history of science and technology. Suffice it to mention but a few of the most important protagonists [Sei06, Gle11]:

The title of founding father is, without doubt, due to Ludwig Boltzmann. The essence of his conception of entropy in the early nineteenth century has essentially survived all modifications and ramifications the notion suffered ever since. Before spreading further outside physics, the idea revolutionized statistical mechanics and inspired his great contemporaries: Maxwell's demon continues stimulating discussion and research till today. Gibbs' paradox may be considered as the first crack in the edifice of classical physics, paving the way towards quantum mechanics. Planck's conception of the quantum of action was inspired by profound reflections on the nature of entropy [Pla49]. While nineteenth century engineering was focused on machinery converting energy from one form into another (see Table 1.1), such as the Steam engine, it has already seen the first sophisticated apparatus dedicated to information processing, including the control unit of the Jacquard weaving loom (Fig. 1.1) and Babbage's "difference engine " and " analytical engine".

Fig. 1.1 The Jacquard weaving loom (1805) is probably the first man-made tool generating patterns from a digital Memory medium, astonishingly similar to the punch cards of the 20th cty. It continues to serve as a metaphor to explain reproducible pattern formation controlled by digital data, for example the genetic control of embryonic development [Nus04].

It was mathematicians who saw the enormous potential of the concept, gave it a rigorous form, sufficiently general to be applied all over science, and explored its implications till the most mind-boggling consequences. Shannon, in the course of quite practical work for the Bell Telephone Co., proposed the basic definition (see Eq. 1.3.9 below), valid till today. Charles Babbage, Ada Lovelace, and Alan Turing, three personalities with a somewhat tragic biography, pioneered the idea of a computer and the foundations of its mathematical description. Kurt Gödel did not refer to information explicitly, but it is undisputed by now that it is the key to understanding his Incompleteness Theorem, arguably the greatest intellectual achievement of the twentieth century [Cha75, Cha82].

Information gained impact again within physics with the work of theoreticians who recognized its relevance in quantum mechanics. John von Neumann not only "quantized" the concept, as the Von-Neumann entropy, but also applied it to discuss fundamental problems of quantum mechanics, such as the measurement process [Neu18]. Leon Brillouin, besides his direct contributions to the field. was the first to conceive a systematic account of the role of information in physics, with his classical textbook on the subject [Bri56]. Leo Szilard reduced Maxwell's demon to its very essence, a single-bit decision, bringing out for the first time the equivalence of entropy and information [Szi29].

From the 1960s onwards, the concept already became so commonplace that selecting the most relevant contributors is inevitably arbitrary. In physics, names come to mind such as Richard Feynman who was the first to contemplate quantum computing [Fey82]. Ilya Prigogine and Hermann Haken introduced concepts closely related to information into statistical mechanics far from equilibrium: dissipative structures [PN77] and synergetics [Hak83], resp. Edwin T. Jaynes recast statistical mechanics, basing it on the principle of maximum entropy [Jay57]. Philip W. Anderson, in the shadow of his epochal work on solid-state physics, contributed important reflections on the concept of complexity [And72]. Rolf Landauer was the first to address systematically the question of fundamental physical limits to computation and coined the statement "Information is physical" [Lan91]. The last sections of this book are largely based on his work.

Among mathematicians of these generations excels, above all, Andrey Kolmogorov, who advanced mathematical physics in a broad spectrum of objects related to probability and information. Aleksandr Khintchin, closer to pure mathematics and probability theory, contributed in a similar direction to this field. With a rather unique profile merging mathematical physics with epistemology, Ray Solomonoff is known as the father of algorithmic complexity [Sol64], his work forming the nucleus of Sect. 4.2.2.

The contributions of contemporary philosophers, mathematicians, physicists will be accounted for in the sequel, in the context of the respective subject.

1.2   The "Three Dimensions" of Information

Before delving into physics and mathematics ,an approach to information should be addressed that originates in philosophy, anthropology, and linguistics. It is in many respects broader than and complements the scientific point of view, emphasizes other facets and allows us to embed the concept of information in a more general context.

Under the headline "semiotics", a theory of Communication has been pioneered by Ferdinand de Saussure [Sau77] and Charles Sanders Peirce [Pei34] and developed further by Thomas A. Sebeok [Seb76] and others. Its most well-known feature is the distinction of three categories, three "dimensions" of information, which however are not to be understood as a division into disjoint sectors, but rather as nested sets:

syntactic information

syntactic information refers to the quantity and formal structure of messages. It counts signs, circumscribes their repertoire and their mutual relations. Alluding to the general meaning of "syntax", it defines rules and restrictions how to compose legitimate sequences of signs, often in a hierarchical fashion. The motion of Messages in space and time, reduced to a mere Flow of information, pertains to the syntactic realm. The notions of sender and receiver reduce to mere sources and sinks. In the context of languages, it is orthography and grammar ("syntax") that stand for syntactic information.

The syntactic aspect of information is relevant for physics and chemistry, for structural sciences such as mathematics and informatics, and for technology related to communication and computation.

It is important to keep in mind that meaning is excluded in syntactic information. The meaning of signs, and more generally, communication as an exchange of messages, enter the scene only with:

semantic information

Semantics is dedicated to the relationship between symbols and their meaning. Obviously, to apply, semantic information requires a language to exist, be it in the most rudimentary form. It includes the relation between symbols and the elements of reality they are referring to (encyclopedias) as well as between equivalent symbols of different codes and languages (dictionaries). Semantic information is concerned with synonyms (different words within the same language with identical or overlapping meanings) and homonyms (words with two or more distinct meanings).

In the context of semantics, the concepts of sender and receiver acquire the additional aspect of *understanding*, not contained in syntactic information. Only on this level, it makes sense talking of *truth* (as opposed to falsehood, to error, lie, etc.). Truth has to do with the relationship between two levels of reality, objects and their symbols, and cannot be grasped within the symbolic level alone, as numerous futile attempts to "define" truth confirm (such as "The proposition 'It rains' is true if and only if it rains."). Representing reality in images, words, formulas, …, is a collective endeavour that cannot be reduced to a formal criterion.

Within the natural sciences, only biology deals with codes and languages and thus with sematic information. It becomes the central aspect in the humanities, in particular in linguistics, in hermeneutics and literature, and in the arts. Both the fine arts (painting and sculpture and their more recent ramifications) and the performing arts (music and theatre) typically do not use established codes but create meaning by inventing new symbols "in the making". The concept of truth is of obvious relevance for journalism.

Pragmatic information

Pragmatic information finally takes into account that senders may have intentions emitting messages, and receivers may or may not react as desired by these intentions. With pragmatics, norms come into the play, the dichotomy of true and false is complemented by that of good and evil.

Contexts where we can meaningfully talk of Pragmatic information are even more restricted than in the case of semantic information. It only applies to the communication of intelligent beings we can attribute intentions to. The disciplines where pragmatic information becomes relevant range from psychology through the Social sciences (particularly history, political science, and economy), through law.

Syntactic, semantic, and pragmatic information require increasingly higher degrees of organization and complexity in the systems they are ascribed to. This suggests a hierarchical relation between them, as sketched in Fig. 1.2.

It will become clear in the following sections that binding the concept of information too tightly to the triad "sender – message – receiver" as its fundamental elements, as is implicit in the semiotic approach and is frequently reproduced in recent literature on information, is in fact inappropriate and leads to an overly narrow outline of

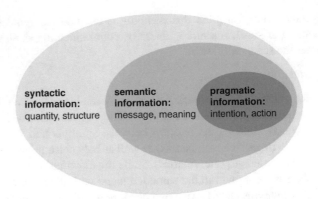

Fig. 1.2 Semiotics distinguishes three "dimensions" of the concept of information: Syntactic information deals with the sheer quantity of information and the internal structure of symbol sequences. With semantics, messages transmit meanings from senders to receivers, while pragmatic information analyzes intentions of senders and the actions their messages evoke in the receivers.

Fig. 1.3 René Magritte's painting "La trahison des images" (1928–1929) illustrates, from an artist's point of view, the distinction between an object and the symbol representing it: An image of a pipe is not a pipe; one cannot smoke it. The perplexing effect of the painting is due to Magritte's mixing three levels of meaning into one image, the object (the pipe), the symbol referring to it (the picture of the pipe) and a metalanguage statement "(Ceci n'est pas une pipe)".

the notion. There is a host of subjects in physics and mathematics that invite to be discussed in terms of information but do not allow nor require identifying anything as sender or receiver.

It is a fascinating question, to be addressed en passant, exactly where in the history of nature a phenomenon to be called "meaning" emerged for the first time. As far as we know till now, symbols ,that is, objects encoding for others, did not exist in any sense before the advent of life on Earth. With the genetic code, a systematic relationship—a dictionary—arose between one class of long-lived molecules (RNA or DNA), serving as memory, and another class of very reactive but short-lived molecules (proteins), capable of affecting and altering their environment, see Sect. 2.2 . The genetic code is located on the borderline between natural phenomena that could still be described without reference to meaning and the onset of semantic information.

Further on in biological evolution, languages of very diverse type developed several times, from chemical cell signalling through olfactory, acoustical and optical signals in multicellular organisms through the formal object language in humans.

It is particularly illuminating to consider how objects, images, smells, sounds, gestures, … assume the role of symbols, that is, come to be understood as pointers towards specific phenomena as their meaning. An important factor in this process is certainly the existence of correlations, which at least initially relate symbol and meaning. As an illustrative example, take conditioning, epitomized in the paradigmatic experiment known as Pavlov's Dog [CM10]: Being confronted in a systematically correlated manner with the sound of a bell and the serving of meat, in this order, the dog learns to associate the bell with the food and drools already upon hearing the bell alone. For the dog, the sound of the bell has become an acoustic signal anticipating the meat, a symbol that represents the upcoming feeding. Indeed, in many languages we can still observe the origin of words in an onomatopoetic imitation of sounds correlated to their meaning, and in primordial scripts the origin of graphical symbols in pictogram-like representations of the objects they refer to is evident [Jac81].

In this context, a terminology should be introduced that will be become indispensable in the discussion of self-reference below: Symbols, irrespective of their acquired function in communication, continue forming part of objective reality, and as such can be referred to by other symbols (Fig. 1.3). Symbols whose meaning is itself a symbol can be associated to a distinct, higher layer of language, called *metalanguage*, each one forming a *metasymbol*. Symbols referring to *metasymbols*, in turn, are *metametasymbols* and belong to a *metametalanguage* etc., in an obvious sequence of nested recursive definitions: For example, ✄ is a pictogram, "Ceci n'est pas une pipe" is a sentence in French, the quotation marks indicating that this sentence belongs to the *object language*, relative to this English metalanguage sentence embedding it, while "'Word' n'est pas un mot français" adds one more level to the hierarchy, this time indicated by single apostrophes.

A comprehensive feature associated with information, both in its common and in its scientific usage, is its independence of the medium carrying it. It is an everyday experience that the audio signal of a telephone call is transferred back and forth between acoustic waves, electric impulses in cables, radio-frequency carrier waves, light pulses in optical fibres and other media, without being severely distorted. Marshall McLuhan's proverbial phrase "The medium is the message" is an ironic inversion of what in general we consider as a fundamental contrast: A message can be transmitted via any medium, a medium can carry whatever message.

More specifically, in physics, the same amount of information, now in its syntactic sense, can be stored in the shape of a solid object, in the magnetization pattern imprinted in a magnetizable material, or the absorbence of photographic emulsion, or even in the probability density distribution of an ensemble in phase space. It can be transmitted by acoustic waves in air or surface waves in water, by electromagnetic and even gravitational waves, and as a matter wave according to quantum mechanics.

Information shares this universality with the concept of energy, in its multiple manifestations in every context in physics, including sheer mass as implied by Special Relativity Theory.

The contrast of message and medium recurs here as the dichotomy of information vs. mass or energy, resembling the opposite pair form and matter, which pervades the philosophical discussion since the antiquity. In a radical view, every structure in space and time ultimately reduces to information, leaving only a featureless mass for the component of matter. It is taken to the extreme, complemented by the idea of the two-state system as minimum manifestation of a quantum system, in John Archibald Wheeler's succinct dictum "It from bit" [Whe90a,Whe90b], and has been worked out by Carl Friedrich von Weizsäcker as a metaphysical system, the theory of the "ur-alternative" [Wei71, Wei85]. The following subsection resumes proposals to define information quantitatively and their consequences.

1.3   From Boltzmann's Entropy to Shannon's Information

In the context of physics, information is distinct from all other physical quantities in that an operational definition, clear instructions how to measure it, can hardly be given. That it belongs to a separate category is also evident from the fact that it applies to the very concept of measurement, as a transfer of information. Basically, information is a mathematical concept, it pertains to the realm of structural sciences, not to the natural sciences. Yet it is so ubiquitous in physics and in particular, as thermodynamical entropy, becomes indirectly measurable, that the statement "Information is physical" [Lan91] appears well justified.

With this intermediate position between physics and mathematics, not unlike geometrical concepts, information is not only comparable with dimension but is even closely related to it in its meaning, see Sect. 6.4: another instance of a quantitative notion that is crucial in physics but cannot be measured with a "dimensiometer".

We owe it to Ludwig Boltzmann's ingenious intuition when the concept of entropy is introduced as counting the number N of "complexions" of a physical system, a huge step forward from previous attempts to define it as a thermodynamic quantity.

In conceiving this term, Boltzmann thought of microscopically distinct states of matter, states that can be distinguished by some microscopic physical quantity, but are not necessarily discernible by macroscopic observation. Anticipating later generalizations, in particular by Shannon, we can define complexions as *distinguishable states of a system*, thus avoiding a reference to physical terms. All the subtleties of this definition are hidden in the word "distinguishable". It implies a high degree of subjectivity: distinguishable by whom, according to which criterion, with respect to which quantity, measured to which accuracy? We shall see below that by analyzing this subjectivity, a rich structure emerges which in fact contributes to the versatility of the concept. A quantity that allows to interpret Boltzmann's complexions in modern terms is the probability density in phase space, see Sect. 5.2. It is remarkable that Boltzmann himself was already quite aware of this problem and in fact based his original argument [Bol09] on the assumption of a *finite number of discrete states*, corresponding, e.g., to energies $E_n = n\Delta E/N$. He considered this, however, as an intermediate formal step, to be complemented by a limit process $N \to \infty$, $\Delta E \to 0$.

Quantifying the repertoire of states a system can assume was the radically new aspect Boltzmann introduced with his entropy. Otherwise the quantity should comply with all general conditions other thermodynamic potentials fulfill. In particular, he required it to be additive: Consider a physical system composed of two separate subsystems, 1 and 2, i.e., two parts not interacting with one another, not interchanging matter or energy. In this case, the total entropy should just be the sum of the entropies of the two subsystems.

If systems 1 and 2 comprise N_1 and N_2 states, resp., the total number of states is the product, $N = N_1 N_2$. The function replacing products by sums is the logarithm, hence the decision to define entropy S as the logarithm of this number,

$$S = c\log(N),\tag{1.3.1}$$

c denoting an arbitrary constant that fixes the unit of entropy. In the context of statistical mechanics, the total number of distinguishable states is proportional to the volume Ω of the accessible phase space, the quotient being given by the size $\Delta\Omega$ of the smallest discernible phase-space cell,

$$S = c\ln\left(\frac{\Omega}{\Delta\Omega}\right).\tag{1.3.2}$$

The denominator $\Delta\Omega$ is also required to render the argument of the logarithm dimensionless. This equation is equivalent to the emblematic formula $S = k\log W$ engraved on Boltzmann's tombstone.

As elsewhere in physics, appropriate units depend on context. In thermodynamics and statistical mechanics, the standard application is to macroscopic ensembles. In this case, manageable numbers for the entropy are achieved defining

$$S = k_B\ln(N),\tag{1.3.3}$$

where $k_B = 1.38 \times 10^{-16}$erg/K is the Boltzmann constant, with dimension energy over absolute temperature, in units of erg over degrees Kelvin. In the context of symbol strings, the established procedure refers to the shortest sequence of yes–no questions that defines the sequence at hand. This amounts to

$$I = \text{lb}(N) = c\ln(N),\tag{1.3.4}$$

with $c = \text{lb}(e) = 1/\ln(2)$ ("lb" denoting "binary logarithm"), defining the bit as unit of information.

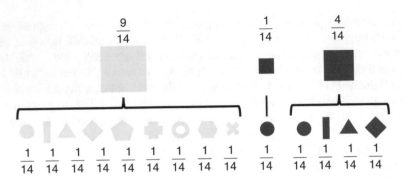

Fig. 1.4 Boltzmann's original entropy definition assigns the same weight to all distinct microstates. In later generalizations, sets of microstates, considered indistinguishable on a higher level, are lumped into single items, weighted according to the number of microstates they contain. In the figure, the corresponding weights would be 9/14, 1/14, and 4/14, respectively.

Boltzmann's original version, Eqs. (1.3.1) or (1.3.2), assumes that there are no a priori preferences for anyone of the N microstates, attributing democratically the same weight to all of them. In almost all applications of entropy, however, the states or symbols counted in measuring an information content already correspond to sets of microstates that are considered as equivalent or indistinguishable (Fig. 1.4). Depending on the size of these sets, the states no longer occur with the same frequency but can be assigned different probabilities. This freedom is indispensable for example in the context of abbreviations and substitution rules that replace sequences of more elementary signs by higher-level symbols.

An important instance and application of varied frequencies, quite unrelated to statistical physics, is the design of codes. In this context, the role of macrostates is played by high-level symbol sets or alphabets, with a large number of symbols, for example the Latin alphabet with around 26 characters, to be encoded by a more elementary symbol set, say the Morse code with only two signals (dot · and dash —), analogous to the microstates. Fixing the number of signals per Morse code symbol representing a single Latin character, this would require $5 = \lceil \log_2(26) \rceil$ signals per letter of the alphabet. However, admitting also shorter lengths, symbols of four signals per letter or less are sufficient. The code can then be optimized, concerning the total number of signals per average message, by assigning the shortest symbols (one signal) to the most frequent Latin letters (in a language of reference, say English) and the longest symbols (four signals) to the rarest letters. This is how the Morse code has been devised (Fig. 1.5). The appropriate criterion for an optimal symbol length is the Shannon information per symbol.

If M is the total number of microstates, M_j of which belong to the jth distinguishable class, $j = 1, \ldots, J, \sum_{j=1}^{J} M_j = M$, then probabilities

$$p_j = \frac{M_j}{M} \qquad (1.3.5)$$

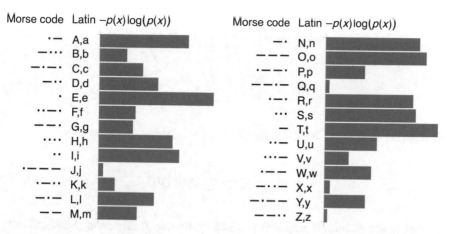

Fig. 1.5 The Morse code assigns a combination of dots · and dashes – to each letter of the Latin alphabet. In order to optimize the code regarding the number of single signals per transmitted information, the length of each Morse code sign, from one to four signals, is adapted to the relative frequency $p(x)$ of each letter (in English). Frequent letters such as "e" and "t" transmit least information and are encoded by a single signal, rare letters such as "q" and "z" are encoded by four signals. An appropriate criterion to determine the length of each code symbol is the weighted logarithm $p(x)\log(p(x))$, see Eq. (1.3.9) (bar chart).

can be attributed to the class j, with $\sum_{j=1}^{J} p_j = 1$. The total number of distinct sequences consisting of M of these symbols is not $M!$, but has to be divided by all the numbers $M_j!$ of combinations of microstates that can be formed within each class j,

$$N = \frac{M!}{\prod_{j=1}^{J} M_j!}. \tag{1.3.6}$$

The corresponding total information is [Bri56]

$$I_M = c \ln(N) = c \left(\ln(M!) - \sum_{j=1}^{J} \ln(M_j!) \right). \tag{1.3.7}$$

If the number of symbols is large, $M_j \gg 1$ for all classes j, this expression can be simplified using Stirling's formula, $\ln(x!) \approx x \ln(x) - x$, valid for $x \gg 1$:

$$I_M \approx c \left(M \ln(M) - M - \sum_{j=1}^{J} M_j \ln(M_j) + \sum_{j=1}^{J} M_j \right)$$

$$= cM \left(\ln(M) - \sum_{j=1}^{J} \frac{M_j}{M} \ln(M_j) \right)$$

$$= -cM \sum_{j=1}^{J} \frac{M_j}{M} \ln\left(\frac{M_j}{M} \right). \tag{1.3.8}$$

Comparing with Eq. (1.3.5), this implies for the information per symbol,

$$I_1 = \frac{I_M}{M} = -c \sum_{j=1}^{J} p_j \ln(p_j). \tag{1.3.9}$$

This is the definition proposed in 1948 by Claude E. Shannon [Sha48,SW49], following a similar expression suggested by J. W. Gibbs in the context of statistical mechanics [Gib02,Gib93]. It generalizes Eq. (1.3.1) but contains it as a special case; for equal probabilities $p_j = 1/J$, it reduces to Boltzmann's expression,

$$I = -c \sum_{j=1}^{J} \frac{1}{J} \ln\left(\frac{1}{J} \right) = c \ln(J). \tag{1.3.10}$$

The minus sign compensates for the fact that, with $0 \leq p_j \leq 1$, $j = 1, ..., J$, the logarithms in Eq. (1.3.9) are all negative. At the other extreme, opposite to equal probabilities, Shannon's definition also contains the deterministic case that only a single choice j_0 applies,

$$p_j = \delta_{j-j_0} = \begin{cases} 1 & j = j_0, \\ 0 & \text{else}, \end{cases} \tag{1.3.11}$$

so that

$$I = -c \sum_{j=1}^{J} \delta_{j-j_0} \ln(\delta_{j-j_0}) = 0. \tag{1.3.12}$$

At the same time, the two cases (1.3.10) and (1.3.12) mark the absolute maximum and minimum values, resp., the entropy can assume, a fact to be analyzed further in the sequel.

1.4 Sign: Entropy and Negentropy: Actual Versus Potential Information

Till here, the concepts of entropy and of information have been used indiscriminately, despite their highly divergent origins in thermodynamics and in communication, resp. Indeed, it is one of the objectives of this book to demonstrate that this is a legitimate manifestation of the enormous versatility of the concept. Notwithstanding, where it appears appropriate, the term "entropy" will be preferred in thermodynamic contexts and "information" where symbol strings and other discrete entities are concerned, always keeping in mind that this distinction is ambiguous and in fact unnecessary.

Calling the quantity defined in Eq. (1.3.9) "information" would appear inadequate also on the background of the colloquial use of the word. As the two limiting cases concluding the last subsection show, I reaches its maximum precisely when all states involved are equally probable, that is, if nothing is known about the system, while the minimum 0 is assumed when the state is exactly fixed. It could hardly be more counterintuitive.

Erwin Schrödinger and Leon Brillouin, contemporaries of Shannon, were quite aware of this problem and suggested the term "negative entropy" or "negentropy" instead, for $-S$ as the more legitimate candidate for a measure of information. As a result, however, an incurable confusion concerning the sign of this quantity has been created that continues to cause misunderstandings in the literature.

For the sake of consistency of the terminology to be used in this volume, I introduce another distinction that helps keeping these meanings apart. What is in fact measured by entropy is the magnitude of state space accessible to the system or the number of signs available to compose a message of. It is therefore adequate to denominate it *potential* information. Furthermore, to account also for ambiguities with respect to the state of knowledge, resolution, measurement accuracy etc., implicit in the term "distinguishable", it makes sense to define a *relative* information, with respect to some reference value. Together, this leads to the definition

$$\Delta I_{pot} = c\left(\ln(N_{final}) - \ln(N_{initial})\right),\qquad(1.4.1)$$

where "initial" and "final" refer to a dynamical time evolution, to a measurement, an observation, or whatever process causes a change in the number of available states. Potential information is equivalent to the lack of knowledge, to ignorance, to uncertainty about the state of the system.

As complementary quantity, introduce *actual* information as measuring what is already known about the system. This would amount to

$$\Delta I_{act} = c(\ln(N_{initial}) - \ln(N_{final})) = -\Delta I_{pot},\qquad(1.4.2)$$

Actual information stands for what is known, for the knowledge an observer, a measurement, a theory, ..., has on the state of the system. It measures the fraction

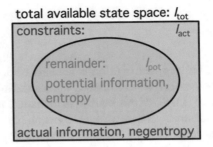

Fig. 1.6 The dichotomy entropy vs. negentropy can be made more explicit by relating the two terms to the set of states that remain accessible to the system, the potential information I_{pot}, under the constraints imposed by what is already known about it, to be called the actual information I_{act}. The sum of the two quantities, $I_{pot} + I_{act} = I_{tot}$, is the total information, measuring the size of the total state space of the system.

of its state space that is not excluded by these constraints and is analogous to the concept of negentropy, thus coincides with the familiar sense of information. The identity $\Delta I_{pot} = -\Delta I_{act}$ means that with each increase of actual Information, some white spot on the map is filled, some part of what can be known about the system is discovered, reducing the potential information by the same amount.

It is tempting to conclude for the absolute values that also $I_{act} = -I_{pot}$. However, they cannot have opposite sign (both are positive), but their sum, the total information, is a constant,

$$I_{pot} + I_{act} = I_{tot} = \text{const},\qquad(1.4.3)$$

so that the simultaneous changes of both quantities are indeed opposed, $\Delta I_{pot} = -\Delta I_{act}$. The relation between these three quantities is illustrated in Fig. 1.6.

The total information, I_{tot}, coincides with the maximum value of I_{act} ("everything that can possibly be known about the system", corresponding to the zero of I_{pot}) and, conversely, also with the maximum of I_{pot} ("nothing is known", marking the zero of I_{act}). It is a measure of the entire repertoire of states the system disposes of, its "universe". Not even this quantity is free of a subjective component, depending, e.g., on the best resolution available to determine the state of the system. It may vary even within a given discipline, according to which level of description is considered a "complete account" of the system, as explained in the next subsection. The total information will remain a fictitious quantity in almost all contexts, unattainable in practical applications. The one remarkable exception is quantum mechanics, where the total information available on a system is well defined: Knowledge of the full state vector of the system amounts to zero entropy, $I_{pot} = 0$, see Sect. 1.8 below.

Throughout the rest of this book, the terms "information" and "entropy" are to be understood as "potential information", if not explicitly indicated otherwise.

1.5 Hierarchical Structures

Two physical systems, data sets, bodies of knowledge, ..., can be related "horizontally" if they coexist separately side by side. The extreme case of two independent systems has been addressed above, where the respective information contents just add. We shall see that a similar rule applies if systems are related "vertically", that is, if sets of elements of one system can be identified with elements of the other. An emblematic example is the tree structure of the organization of archives provided by most of the current computer operation systems like Unix.

The idea, however, is much older. The first well-known instance is in fact the ontology of "levels of reality", developed by the German philosopher of the nineteenth century, Nicolai Hartmann [Har40]. He subdivides phenomena into subsequent layers, each of which contains the phenomena of the next lower level, but grouped into more encompassing units. Similar hierarchies are known since Greek philosophy.

Contemporary physics, for example, analyzes a piece of metal as a macroscopic object, in terms of crystal domains, unit cells, ions, atoms, elementary particles, ending presently at quarks and leptons (Fig. 1.7). Biology sets in at the level of super-organisms (e.g., ant states), followed by individual organisms, organs, tissues, cells, cell organelles, macromolecules, before merging into physics at the level of molecules and ions. Similar levels of resolution can be discerned in many other contexts: A novel can be subdivided into chapters which in turn consist of paragraphs, of sentences, ... words, ... letters, down even to pixels of print.

Number systems show in a nutshell how information theory allows to analyze the information contained in a hierarchy of levels, each one corresponding to a specific scale of resolution. In the decimal system for numbers less than unity, digits can alternatively represent units, tenths, hundredths, etc. The binary system, where

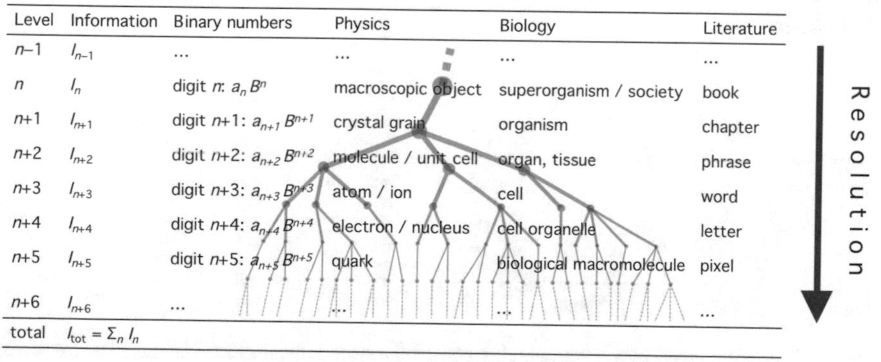

Level	Information	Binary numbers	Physics	Biology	Literature
$n{-}1$	I_{n-1}	...	...	...	...
n	I_n	digit n: $a_n B^n$	macroscopic object	superorganism / society	book
$n{+}1$	I_{n+1}	digit $n{+}1$: $a_{n+1} B^{n+1}$	crystal grain	organism	chapter
$n{+}2$	I_{n+2}	digit $n{+}2$: $a_{n+2} B^{n+2}$	molecule / unit cell	organ, tissue	phrase
$n{+}3$	I_{n+3}	digit $n{+}3$: $a_{n+3} B^{n+3}$	atom / ion	cell	word
$n{+}4$	I_{n+4}	digit $n{+}4$: $a_{n+4} B^{n+4}$	electron / nucleus	cell organelle	letter
$n{+}5$	I_{n+5}	digit $n{+}5$: $a_{n+5} B^{n+5}$	quark	biological macromolecule	pixel
$n{+}6$	I_{n+6}	...	...	...	...
total	$I_{tot} = \Sigma_n I_n$				

Resolution →

Fig. 1.7 Hierarchies of structures are abundant in science as well as in daily life, where elementary items at the nth level resolve into a set of smaller parts at the subsequent, $(n{+}1)$th level, providing a finer resolution. Disregarding correlations, the information contents of each level just add up to the total information of the structure at hand.

digits represent units, halves, quarters, ..., correspondingly towards positive powers of 2, directly demonstrates that the information contained in each digit can simply be summed: If a number x is specified by N binary digits,

$$x = \sum_{n=n_0}^{n_0+N-1} a_n 2^n, a_n \in \{0, 1\}, \tag{1.5.1}$$

it contains N bits of information. As will be shown below, this rule applies generally to hierarchies of levels n, $n = 1, ..., N$, each of which contributing an information content I_n (independent of and in addition to the previous levels, see Fig. 1.8). The total information is then

$$I_{\text{tot}} = \sum_{n=1}^{N} I_n, \tag{1.5.2}$$

independently of how the tree is organized in detail. The bottom and top of a hierarchy, its lowest and uppermost levels, depend on context and convention. They are of course limited by the state of knowledge at any given time; some hundred years ago physics considered atoms as the smallest units and galaxies as the largest known structures. In biology, progress has extended the range to molecular cell biology below and to ecosystems above.

The idea of hierarchical structures in science immediately evokes the controversy around reductionism vs. holism. An information theoretical view can contribute to it, increasing precision and objectivity of the arguments. In particular, as will become clear in the sequel, the contributions of different tiers of the hierarchy do no longer simply add (as would be implied by reductionism), as soon as the components at a given level are correlated, "the whole is more than the sum of its parts" (as insinuated by holism).

Fig. 1.8 Different levels of a hierarchical structure can store independent contents. The messages it contains at distinct levels of resolution need not even be consistent with each other.

1.6 Properties of Shannon's Definition of Information

In this subsection, three fundamental mathematical properties of Shannon's definition, Eq. (1.3.9), are discussed and proven step by step, resuming features that have already been alluded to above. Readers not interested in mathematical detail may well skip the proofs.

1.6.1 An Extremum Property

From a mathematical point of view, a conspicuous feature of Eq. (1.3.9) is that the prefactors p_j of the logarithms under the sum coincide with their arguments. What if we lift this restriction, considering the two sets of quantities independent, but still defined as probabilities? That is, if we assume two n-tuples,

$$p_j, q_j, \quad j = 1, \ldots, J, \quad 0 \le p_j, q_j \le 1, \ \sum_{j=1}^{J} p_j = \sum_{j=1}^{J} q_j = 1, \qquad (1.6.1)$$

the sum

$$R = \sum_{j=1}^{J} p_j \ln(q_j) \qquad (1.6.2)$$

reaches its absolute maximum (or equivalently, the generalized Shannon information $S' = -c \sum_{j=1}^{J} p_j \ln(q_j)$ reaches its minimum) with respect to variation of the q_j if the two sets are chosen identical, $p_j = q_j$, $j = 1, \ldots, J$.

Proof [Bri56] Define the discrepancies between the two sets of probabilities,

$$u_j := q_j - p_j, \quad j = 1, \ldots, J, \qquad (1.6.3)$$

so that

$$q_j = p_j + u_j = p_j \left(1 + \frac{u_j}{p_j} \right). \qquad (1.6.4)$$

By normalization of the probabilities, the differences u_j cannot be either all positive or all negative, and

$$\sum_{j=1}^{J} u_j = 0. \qquad (1.6.5)$$

Substituting in Eq. (1.6.2),

$$R = \sum_{j=1}^{J} p_j \ln\left(p_j\left(1 + \frac{u_j}{p_j}\right)\right) = \sum_{j=1}^{J} p_j \ln(p_j) + \sum_{j=1}^{J} p_j \ln\left(1 + \frac{u_j}{p_j}\right). \quad (1.6.6)$$

In this form, it suggests itself to refer to the convexity of the logarithm, $\ln(1 + x) \le x$, $x \in \mathbb{R}$, to apply the inequality

$$\ln\left(1 + \frac{u_j}{p_j}\right) \le \frac{u_j}{p_j}, \, j = 1, \dots, J, \quad (1.6.7)$$

to Eq. (1.5.6),

$$R \le \sum_{j=1}^{J} p_j \ln(p_j) + \sum_{j=1}^{J} p_j \frac{u_j}{p_j} = \sum_{j=1}^{J} p_j \ln(p_j) + \sum_{j=1}^{J} u_j = \sum_{j=1}^{J} p_j \ln(p_j),$$
$$(1.6.8)$$

the last step making use of Eq. (1.6.5). As a consequence,

$$S' = -c \sum_{j=1}^{J} p_j \ln(q_j) \ge -c \sum_{j=1}^{J} p_j \ln(p_j) = S. \quad (1.6.9)$$

An important application of generalized entropies of type S' in Eq. (1.6.9) is the *Kulback entropy*, also referred to as *Kulback-Leibler divergence* [Kul68,CT06], where the probabilities q_j are interpreted as expectations for the actual values p_j arising from some hypothesis or approximation. The Kullback entropy will be contrasted with conditional information in Sect. 1.7 below.

1.6.2 Equal Probabilities Imply Maximum Entropy

As has been announced above, the Shannon information reaches its maximum if all probabilities are equal, $p_j = 1/J$, $j = 1, \dots, J$.

Proof [Bri56] It is to be shown (i) that equal probabilities correspond to an extremum of the information, and (ii) that this extremum is a maximum.

As to (i), consider $I(p_1, \dots p_J)$ as a function of actually only $J - 1$ free variables, say $p_1, \dots, p_{J-1}$, while p_J is fixed by normalization, $p_J = 1 - \sum_{j=1}^{J-1} p_j$ so that

$$I = -c \left(\sum_{j=1}^{J-1} p_j \ln(p_j) + \left(1 - \sum_{j=1}^{J-1} p_j \ln(p_j) \right) \ln\left(1 - \sum_{j=1}^{J-1} p_j \ln(p_j) \right) \right).$$

$$(1.6.10)$$

The condition for this to be an extremum is then that

$$\frac{\partial I}{\partial p_j} = -c \left(\ln(p_j) + 1 - \ln\left(1 - \sum_{j=1}^{J-1} p_j \right) - 1 \right) = -c \ln\left(\frac{p_j}{p_J} \right)^! = 0,$$

$$j = 1, \ldots, J - 1.$$

$$(1.6.11)$$

Given the general condition of positivity of the p_j, it can only be satisfied if $p_j = 1/J$, $j = 1, \ldots, J$.

Equation (1.6.11) is complemented to become a sufficient condition for a maximum by requiring the curvature of $I(p_1, \ldots p_J)$ to be negative at this point. To verify it, find the second derivatives of this function,

$$\frac{\partial^2 I}{\partial p_j \partial p_k} = -c \frac{\partial}{\partial p_j} \bigg|_{p_1 = \cdots = p_J = 1/J} \left(\ln(p_k) - \ln\left(1 - \sum_{l=1}^{J-1} p_l \right) \right)$$

$$= -Jc(1 + \delta_{k-j}), \ j, k = 1, \ldots, J - 1.$$

$$(1.6.12)$$

The determinant of this matrix of second derivatives at $p_j = 1/J$, $j = 1, \ldots, J$ is

$$\det\left(\frac{\partial^2 I}{\partial p_j \partial p_k} \right) = \underbrace{\begin{vmatrix} -2Jc & -Jc & \cdots & -Jc \\ -Jc & -2Jc & & \vdots \\ \vdots & & \ddots & -Jc \\ -Jc & \cdots & -Jc & -2Jc \end{vmatrix}}_{J-1 \text{ columns}} \left. \begin{matrix} \\ \\ \\ \\ \end{matrix} \right\} J - 1 \text{ rows} = J(-JC)^{J-1}.$$

$$(1.6.13)$$

Indeed, for its sign we find

$$\text{sgn}\left(\det\left(\frac{\partial^2 I}{\partial p_j \partial p_k} \right) \right) = (-1)^n,$$

$$(1.6.14)$$

which coincides with the condition for this point to be a maximum.

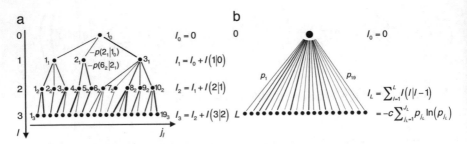

Fig. 1.9 In a hierarchical structure, the total information content I_L at each level L is given by the sum of the relative contributions $I(l|l-1)$, Eq. (1.6.16), of all the previous levels (**a**), equivalent to a structure without any intermediate tiers of the hierarchy, Eq. (1.6.17) (**b**).

1.6.3 Information Content of Tree Structures

Information-containing systems structured as a tree, with a succession of levels where items branch out into sets of sub-items, have been discussed in Sect. 1.5, culminating in the statement that their full information content is just the sum of the information contained at each level, independently of any details of the tree. This will be demonstrated here.

Consider a decision tree with L levels (Fig. 1.9). At each level l of the tree, there is a choice of J_l options (branches), each realized with a conditional probability (see Sect. 1.7 below)

$$p(j_l|j_{l-1}) = \frac{p_{j_l}}{p_{j_{l-1}}}, \; j_{l-1} = 1, \ldots, J_{l-1}, \; j_l = 1, \ldots, J_l, \qquad (1.6.15)$$

relative to the option j_{l-1} at the previous (next higher) level from which it branches off. All conditional probabilities relative to the same root j_{l-1} sum up to unity. With the conditional information (see Sect. 1.7) at level l, relative to level $l-1$,

$$I(l|l-1) = -c \sum_{j_l=1}^{J_l} p_{j_l} \ln(p(j_l|j_{l-1})), \qquad (1.6.16)$$

the total information at the lowest level L of the hierarchy is given by

$$I_L = -c \sum_{j_L=1}^{J_L} p_{j_L} \ln(p_{j_L}) = \sum_{l=0}^{L} I(l|l-1), \qquad (1.6.17)$$

i.e., as if the nested branch structure were replaced by a single alternative of J_L options.

Proof [Bri56] In view of the central role of the index l for the hypothesis, it suggests itself to demonstrate it by complete induction with respect to this index.

$L = 1$:

The probability at the root level 0 is $p_0 = 1$, at the only branch level 1 they are $p_{j_1} =: p_j$. The information at this level is

$$I_1 = I_0 + I(1|0) = 0 - c \sum_{j_1=1}^{J_1} p_{j_1} \ln(p_{j_1}|0)$$

$$= -c \sum_{j_1=1}^{J_1} p_{j_1} \ln\left(\frac{p_{j_1}}{p_0}\right) = -c \sum_{j_1=1}^{J_1} p_{j_1} \ln(p_{j_1}). \tag{1.6.18}$$

$L-1$ to L:

By definition, $I_L = c \sum_{j_L=1}^{J_L} p_{j_L} \ln(p_{j_L})$. Subdivide the summation over j_L into J_{L-1} sections indexed by j_{L-1}, each of which contributes a sum over j_L running only from the leftmost branch $j_{l,\min}(j_{L-1})$ to the rightmost one $j_{l,\max}(j_{L-1})$ rooted in this same item j_{L-1},

$$I_L = -c \sum_{j_{L-1}=1}^{J_{L-1}} \sum_{j_L=j_{L,\min}(j_{L-1})}^{j_{L,\max}(j_{L-1})} p_{j_L} \ln(p_{j_L})$$

$$= -c \sum_{j_{L-1}=1}^{J_{L-1}} \sum_{j_L=j_{L,\min}(j_{L-1})}^{j_{L,\max}(j_{L-1})} p_{j_L} \ln\left(\left(\frac{p_{j_L}}{p_{j_{L-1}}}\right) - \ln(p_{j_{L-1}})\right)$$

$$= -c \sum_{j_{L-1}=1}^{J_{L-1}} \sum_{j_L=j_{L,\min}(j_{L-1})}^{j_{L,\max}(j_{L-1})} p_{j_L} \ln\left(\frac{p_{j_L}}{p_{j_{L-1}}}\right) - c \sum_{j_{L-1}=1}^{J_{L-1}} \ln(p_{j_{L-1}}) \sum_{j_L=j_{L,\min}(j_{L-1})}^{j_{L,\max}(j_{L-1})} p_{j_L}. \tag{1.6.19}$$

In the second line, first term, the argument of the logarithm can be interpreted as a conditional probability $p(j_L|j_{L-1})$. In the right term, the inner sum over probabilities at the Lth level amounts to the total probability $p_{j_{L-1}}$ of the branch j_{L-1} at level $L-1$:

$$I_L = -c \sum_{j_L=1}^{J_L} p_{j_L} \ln(p(j_L|j_{L-1})) - c \sum_{j_{L-1}=1}^{J_{L-1}} p_{j_{L-1}} \ln(p_{j_{L-1}}) = I(L|L-1) + I_{L-1}. \tag{1.6.20}$$

Applying this procedure at level $L-1$ and repeating it further along the hierarchy till $L = 0$, one arrives at Eq. (1.6.17),

$$I_L = I_0 + \sum_{l=0}^{L-1} I(l|l-1) + I(L|L-1) = \sum_{l=0}^{L} I(l|l-1). \tag{1.6.21}$$

1.7   Joint, Conditional, Mutual Information, Bayes' Law, Correlations and Redundancy

In Sect. 1.3, the additivity of entropy has been introduced referring to the case of two perfectly separate and statistically independent subsystems. This is an idealized extreme situation. In most cases where we perceive a total system as subdivided into parts that can be delimited against each other, these subsystems are neither absolutely autonomous nor completely identical with each other. An intermediate situation characterized by a partial coincidence is the rule.

For this case, probability theory provides a number of tools that allow to quantify the interdependence of two sets of events or elements, which can be translated directly into corresponding information measures, referring to two systems to be compared to one another: joint, conditional, and mutual information, quantities conceived to measure the overlap between correlated systems in terms of information.

In view of Shannon's definition (1.3.9), assign probabilities to the elements of two state spaces, A and B,

$$p_j^A, j = 1, \ldots, J, \sum_{j=1}^{J} p_j^A = 1, p_k^B, k = 1, \ldots, K, \sum_{k=1}^{K} p_k^B = 1. \qquad (1.7.1)$$

Merging them conceptually into a single total system, the combined states can be identified by the respective indices j, k of the subsystem states and are attributed to a corresponding *joint probability* [Fel68],

$$p_{jk}^{AB}, j = 1, \ldots, J, k = 1, \ldots, K, \sum_{j=1}^{J} \sum_{k=1}^{K} p_{jk}^{AB} = 1. \qquad (1.7.2)$$

Individual probabilities in each system are extracted by summing over the indices of the other system, that is, by projecting over it,

$$p_j^A = \sum_{k=1}^{K} p_{jk}^{AB}, \quad p_k^B = \sum_{j=1}^{J} p_{jk}^{AB}. \qquad (1.7.3)$$

The Shannon information for the combined states jk,

$$I^{AB} = -c \sum_{j=1}^{J} \sum_{k=1}^{K} p_{jk}^{AB} \ln\left(p_{jk}^{AB}\right), \qquad (1.7.4)$$

is the *joint information* of the two systems. It suggests itself to compare it to the two individual information contents I^A and I^B,

$$I^A = -c \sum_{j=1}^{J} p_j^A \ln\left(p_j^A\right) = -c \sum_{j=1}^{J} \sum_{k=1}^{K} p_{jk}^{AB} \ln\left(p_j^A\right)$$

$$I^B = -c \sum_{k=1}^{K} p_k^B \ln\left(p_k^B\right) = -c \sum_{k=1}^{K} \sum_{j=1}^{J} p_{jk}^{AB} \ln\left(p_k^B\right), \tag{1.7.5}$$

and to their sum $I^A + I^B$. Given the positivity of probabilities, it is clear that $p_{jk}^{AB} \leq \sum_{k=1}^{K} p_{jk}^{AB} = p_j^A$, $p_{jk}^{AB} \leq \sum_{j=1}^{J} p_{jk}^{AB} = p_k^B$ and thus $\left(p_{jk}^{AB}\right) \leq \min\left(\ln(p_j^A), \ln(p_k^B)\right)$. This allows to estimate

$$I^{AB} = -c \sum_{j=1}^{J} \sum_{k=1}^{K} p_{jk}^{AB} \ln\left(p_{jk}^{AB}\right) \geq -c \sum_{j=1}^{J} \sum_{k=1}^{K} p_{jk}^{AB} \ln\left(p_j^A\right)$$

$$= -c \sum_{j=1}^{J} \ln\left(p_j^A\right) \sum_{k=1}^{K} p_{jk}^{AB} = -c \sum_{j=1}^{J} p_j^A \ln\left(p_j^A\right) = I^A, \tag{1.7.6}$$

and likewise for system B. For the sum, we have from Eq. (1.7.5),

$$I^A + I^B = -c \sum_{j=1}^{J} \sum_{k=1}^{K} p_{jk}^{AB} \left(\ln\left(p_j^A\right) + \ln\left(p_k^B\right)\right) = -c \sum_{j=1}^{J} \sum_{k=1}^{K} p_{jk}^{AB} \ln\left(p_j^A p_k^B\right).$$

$$\tag{1.7.7}$$

Here we can make use of the extremum property (1.6.9), considering the arguments of the logarithms in Eq. (1.7.7) as a set of probabilities $q_{jk} := p_j^A p_k^B$, to find

$$I^A + I^B = -c \sum_{j=1}^{J} \sum_{k=1}^{K} p_{jk}^{AB} \ln(q_{jk}) \geq -c \sum_{j=1}^{J} \sum_{k=1}^{K} p_{jk}^{AB} \ln\left(p_{jk}^{AB}\right) = I^{AB}. \tag{1.7.8}$$

Combining Eqs. (1.7.6) and (1.7.8), we obtain as lower and upper bounds for the joint information,

$$\max\left(I^A, I^B\right) \leq I^{AB} \leq I^A + I^B. \tag{1.7.9}$$

Also here, identities result if we go to the opposite extremes of probabilities that are either uncorrelated or identical between the two systems. In the former case, joint probabilities factorize, $p_{jk}^{AB} = p_j^A p_k^B = q_{jk}$, so that $I^{AB} = I^A + I^B$. In the latter, $p_{jk}^{AB} = p_j^A \delta_{k-j}$ and $I^{AB} = I^A = I^B$. The interpretation is obvious: The information content of the combined system ranges between the sum of the two individual values in the absence of correlations and reduces to either one if the subsystems are identical (irrespective of the sign of the correlations!). The inequality

(1.7.9) readily generalizes to sets of more than two subsystems. In the broader context of composed systems in science, it can be read as an exact quantitative manifestation of "the whole is more than the sum of its parts" (in terms of potential information as in Eq. (1.7.9), "more knowledge" means "less information").

The non-negative difference

$$I^{A \cap B} := I^A + I^B - I^{AB} = -c \sum_{j=1}^{J} \sum_{k=1}^{K} p_{jk}^{AB} \ln\left(\frac{p_j^A p_k^B}{p_{jk}^{AB}}\right) \qquad (1.7.10)$$

measures the *common, shared,* or *mutual* information contained in I^A as well as in I^B. In the context of transmission channels, $I^{A \cap B}$ can also be interpreted as the information of the original message A that remains in the received version B, and therefore can be used to measure channel capacities [Pie80].

While the mutual entropy is manifestly bounded from below, $I^{A \cap B} \geq 0$, it also satisfies an upper bound. Equations (1.7.9) and (1.7.10) imply that

$$I^{A \cap B} \leq I^A + I^B - \max(I^A, I^B) = \min(I^A, I^B), \qquad (1.7.11)$$

as is evident also directly from the Venn diagram, Fig. 1.10.

In the preceding paragraphs, the two subsystems A and B have been treated in an unbiased manner. A different perspective arises if we take the knowledge of one of them, say A, as granted and ask, how much additional information—if any—is gained as we include knowledge of B? As above, we can refer to a corresponding probability, the *conditional probability* [Fel68]

$$p_{jk}^{AB} = p^{B|A}(k|j) p_j^A \Leftrightarrow p^{B|A}(k|j) := \frac{p_{jk}^{AB}}{p_j^A}, \qquad (1.7.12)$$

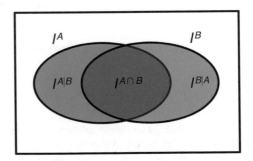

Fig. 1.10 A simple Venn diagram suffices to visualize the concepts of conditional and of mutual or common information. If the correlated systems A and B are associated to sets, then the conditional information $I^{A|B}$ (for states in A, given the state in B) corresponds to the set $A\backslash B$, $I^{B|A}$ to $B\backslash A$, and the mutual information $I^{A \cap B}$ (contained in both systems) to the intersection $A \cap B$.

(read $k|j$ as "k, provided j" or "k, given j"), accordingly for $p^{A|B}(j|k)$, to define the *conditional information* for B, given A,

$$I^{B|A} := -c \sum_{j=1}^{J} \sum_{k=1}^{K} p_{jk}^{AB} \ln\left(p^{B|A}(k|j)\right) = -c \sum_{j=1}^{J} \sum_{k=1}^{K} p_{jk}^{AB} \ln\left(\frac{p_{jk}^{AB}}{p_j^A}\right) \qquad (1.7.13)$$

It differs from all previous information definitions in that here, the argument of the logarithm does not coincide with its prefactor. Invoking Eq. (1.7.12), we arrive at an expression comparing conditional to unconditional information,

$$I^{B|A} = -c \sum_{j=1}^{J} \sum_{k=1}^{K} p_{jk}^{AB} \left(\ln\left(p_{jk}^{AB}\right) - \ln\left(p_j^A\right)\right)$$

$$= -c \left(\sum_{j=1}^{J} \sum_{k=1}^{K} p_{jk}^{AB} \ln\left(p_{jk}^{AB}\right) - \sum_{j=1}^{J} p_j^A \ln\left(p_j^A\right)\right) = I^{AB} - I^A \qquad (1.7.14)$$

which suggests interpreting $I^{B|A}$ as information surplus of A relative to B. Owing to the estimate (1.7.9), conditional information is non-negative. Note that the extremum property (1.6.9), which for the two sets of probabilities p_{jk}^{AB} and p_j^A would imply that $I^A \geq I^{AB}$, does not hold for the second terms in both members of Eq. (1.7.14), because the probabilities p_j^A are not normalized to unity with respect to the full system, $\sum_{j=1}^{J} \sum_{k=1}^{K} p_j^A = K$.

Conditional information should not be confused with Kulback entropy or Kulback-Leibler divergence [Kul68,CT06], alluded above to in the discussion of Eq. (1.6.9). Using a similar notation as in Eq. (1.7.12), it is defined as

$$I_{KL}^{A|B} := -c \sum_{j=1}^{J} p_j^A \ln\left(\frac{p_j^B}{p_j^A}\right) = -c \left(\sum_{j=1}^{J} p_j^A \ln\left(p_j^B\right) - \sum_{j=1}^{J} p_j^A \ln\left(p_j^A\right)\right). \qquad (1.7.15)$$

In this case, Eq. (1.6.9) does apply and guarantees that the Kulback entropy is positive. The difference is that here, p_j^A and p_j^B apply to the same system with J states indexed by j, but the p_j^B represent an alternative (expected, averaged, hypothetical, ...) estimate of the likelihoods of these states. They can also be interpreted as defining an alternative code for a set of J symbols. The Kulback entropy then measures the surplus information in the system pretended by this suboptimal code as compared to the optimal code defined by p_j^A. In conditional information, Eq. (1.7.13), unlike Kulback entropy, the alternative distribution p_j^B corresponds to a subset, compared to a composed set AB, and comprises accordingly a smaller number of states. Therefore, conditional information cannot be understood as a special case of Kulback entropy. The identities

$$I^{AB} = 1^{B|A} + I^A = I^{A|B} + I^B \Leftrightarrow 1^{B|A} = I^B - I^A + 1^{A|B}, \tag{1.7.16}$$

combined with the inequality $I^{AB} \le I^A + I^B$, imply that

$$I^{B|A} \le I^B, \quad I^{A|B} \le I^A. \tag{1.7.17}$$

The kinship of the identities (1.7.16) with Eq. (1.7.10) is made explicit in the relation

$$\left. \begin{array}{l} I^{B|A} = I^{AB} - I^A \\ I^{A|B} = I^{AB} - I^B \end{array} \right\} \Rightarrow I^{B|A} + I^{A|B}$$

$$= 2I^{AB} - I^A - I^B = I^A + I^B - 2I^{A\cap B} = I^{AB} - I^{A\cap B}. \tag{1.7.18}$$

Conditional information is also closely related to shared information. Recalling Eq. (1.7.10), $I^{A\cap B} = I^{AB} - \left(I^A + I^B\right)$, the shared information can also be written as

$$I^{A\cap B} = I^{AB} - I^{B|A} - I^{A|B}. \tag{1.7.19}$$

(cf. Fig. 1.10). Resolving for the conditional information, this means

$$I^{A|B} = I^A - I^{A\cap B}, \quad I^{B|A} = I^B - I^{A\cap B}, \tag{1.7.20}$$

which implies a third possible definition of the mutual information, namely as

$$I^{A\cap B} = I^A - I^{A|B} = I^B - I^{B|A} = I^{B\cap A}. \tag{1.7.21}$$

Again it is instructive to anchor these relations in the limiting cases of vanishing versus maximum correlations. For statistically independent systems,

$$p^{B|A}(k|j) = \frac{p_j^A p_k^B}{p_j^A} = p_k^B, \, p^{A|B}(j|k) = \frac{p_k^B p_j^A}{p_k^B} = p_j^A, \tag{1.7.22}$$

(knowing A, nothing is gained in order to predict B), the conditional information reduces to

$$I^{B|A} = -c \sum_{j=1}^{J} \sum_{k=1}^{K} p_{jk}^{AB} \ln\left(p_k^B\right) = -c \sum_{j=1}^{J} p_j^A \sum_{k=1}^{K} p_k^B \ln\left(p_k^B\right) = I^B \tag{1.7.23}$$

(likewise for $I^{A|B}$). For identical probabilities $p^{B|A}(k|j) = p^{A|B}(j|k) = \delta_{k-j}$, the surplus information vanishes,

$$I^{B|A} = I^{A|B} = -c \sum_{j=1}^{J} \sum_{k=1}^{K} p_j^A \delta_{k-j} \ln(\delta_{k-j}) = 0. \tag{1.7.24}$$

A third limiting case of special interest is that events in system B are necessary conditions for events in system A, or equivalently, events in A are sufficient conditions for those in B, be it in the sense of a logical tautology $A \to B$ (see Sect. 2.1.4) or of a cause–effect relation (see Sect. 3.1.2). This means that no actual information is gained in going from A to B, or in terms of conditional information

$$I^{A|B} = -c \sum_{j=1}^{J} \sum_{k=1}^{K} p_{jk}^{AB} \ln\left(\frac{p_{jk}^{AB}}{p_k^B}\right) = 0, \tag{1.7.25}$$

which requires that for all $j = 1, ..., J$, $p_{jk}^{AB} = p_k^B$ and therefore $I^{AB} = I^B$. For the mutual information, shared by A and B, Eq. (1.7.21) allows to conclude that

$$I^{A\cap B} = I^A. \tag{1.7.26}$$

At the same time, $A \to B$ does by no means require that $I^{B|A} = 0$. If indeed $I^{B|A} > 0$ while $I^{A|B} = 0$, this implies that $I^B > I^A$. Conversely, however, $I^B > I^A$ is not a sufficient criterion for an implication $A \supset B$. The condition $I^{A|B} = 0$ is crucial, as will be discussed further in Sects. 2.1.4 for logical implication and 3.1.2 for causality.

Important insights are gained as soon as we involve a third system, C. Combining Eq. (1.7.18) with the corresponding identity for C,

$$I^{B|A} = I^B - I^A + I^{A|B},$$
$$I^{C|A} = I^C - I^A + I^{A|C}, \tag{1.7.27}$$

and subtracting the second from the first line leads to the conclusion that

$$I^{B|A} - I^{C|A} = I^B - I^C + I^{A|B} - I^{A|C}. \tag{1.7.28}$$

It could adequately be called "Bayes' theorem for conditional information", in analogy to Bayes' theorem for conditional probabilities [Fel68], a lemma of Eq. (1.7.12),

$$\frac{p^{B|A}(k|j)}{p^{C|A}(l|j)} = \frac{p_k^B \, p^{A|B}(j|k)}{p_l^C \, p^{A|C}(j|l)}. \tag{1.7.29}$$

For tripartite systems, the analogous relations are

$$I^{A|BC} = I^A - I^{A\cap B\cap C} - I^{(A\cap B)|C} - I^{(C\cap A)|B} \tag{1.7.30}$$

(etc., with *A*, *B*, *C* permuted cyclically). For larger systems, the number of terms to be included increases according to combinatorics, see Fig. 1.11, so that defining and isolating the individual share of a subsystem in the information content of a multipartite system becomes practically unmanageable, an important obstacle if in many-body physics, information is to be attributed to subsystems or even individual degrees of freedom.

The fact that in the presence of correlations, the total information contained in a system may be less than the sum of the partial information contents of its subsystems, Eq. (1.7.9), is also known as *redundancy*. Redundancy comes with negative connotations like inefficiency and needlessness. In fact, it is an indispensable tool in all kinds of communication and computation, as a fundamental strategy against loss, misunderstanding, and errors (Fig. 1.12). A standard example from everyday technology is the checksum, see Sect. 10.5.2, essentially a number of redundant additional bits,

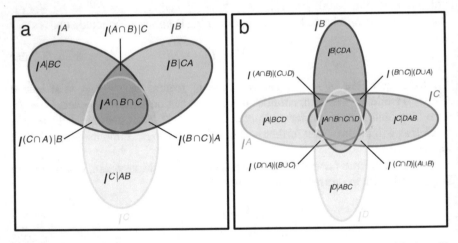

Fig. 1.11 While for a bipartite system, individual information can still be distinguished readily from shared information (cf. Fig. 1.10), the number of terms to be included increases rapidly with the system size, as shows their representation as Venn diagrams, here for systems comprising three (**a**) and four (**b**) parts.

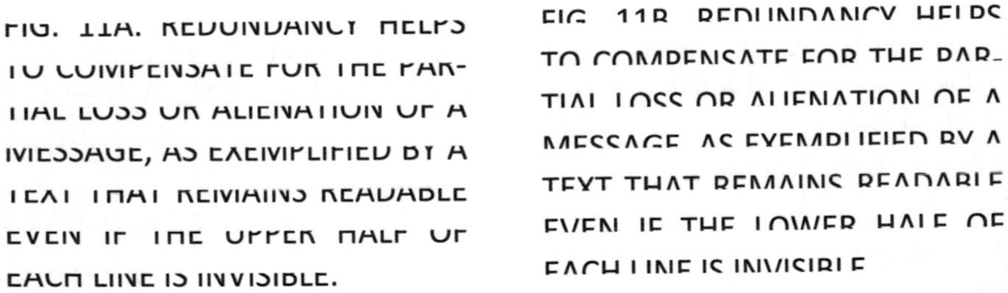

FIG. 11A. REDUNDANCY HELPS TO COMPENSATE FOR THE PARTIAL LOSS OR ALIENATION OF A MESSAGE, AS EXEMPLIFIED BY A TEXT THAT REMAINS READABLE EVEN IF THE UPPER HALF OF EACH LINE IS INVISIBLE.

FIG. 11B. REDUNDANCY HELPS TO COMPENSATE FOR THE PARTIAL LOSS OR ALIENATION OF A MESSAGE AS EXEMPLIFIED BY A TEXT THAT REMAINS READABLE EVEN IF THE LOWER HALF OF EACH LINE IS INVISIBLE

Fig. 1.12 Rdndnc hlps t cmpnst fr th prtl lss r lntn f mssg, s xmplfd b txt tht rmns rdbl vn f ll vwls r rmvd.

which allow to check the correctness of a transmitted message—not with absolute certainty but up to a probability that increases with the number of redundant bits. In oral and written natural language, a high degree of redundancy guarantees the fluidity and security of everyday human communication. Indeed, were it not for redundancy, we could not identify misprints: A text containing typographical errors would just be a different text, and a piece of music interpreted by an imperfect musician would be a new composition. The fact that we are able to predict and control an increasing part of natural phenomena is evidence that also here, redundancy is ubiquitous, as will be discussed in Chap. 3 below.

A necessary and sufficient condition for redundancy is correlations among states or symbols. However, there is no general functional relationship between the two quantities. This is already evident from the fact mentioned above, that positive as well as negative correlations imply the existence of mutual information. Nevertheless, the parallelism between the two concepts goes further than demonstrated till now. This applies in particular if we go to systems with more than two parts, messages comprising more than two symbols. In this case, higher (order) correlations come into play. For English (and in all other natural languages) it is well known that there exist correlations on all scales, in terms of symbol separation or block size: from letter pairs, where only a small fraction of all possible combinations actually occur, to the scale of entire novels, where the beginning of a love story already allows us to anticipate the happy end. This is evidenced for example by compiling random sequences of letters. Taking increasingly higher-order correlations into account in the generation of random strings, the outcome asymptotically approaches fictitious, but intelligible text passages [Gle11].

Higher-order correlations can readily be included in the above formal treatment by considering strings of two or more symbols as individual signs. Blocks of N symbols are defined recursively, adding symbols one by one, as

$$\mathbf{x}^{(N)} := \left(\mathbf{x}^{(N-1)}, x_N\right) = (x_1, \ldots, x_{N-1}, x_N). \tag{1.7.31}$$

The conditional probability to go from $\mathbf{x}^{(N-1)}$ to $\mathbf{x}^{(N)}$ is then, see Eq. (1.7.12),

$$p\left(\mathbf{x}^{(N)}\right) = p\left(\mathbf{x}^{(N-1)}\right) p\left(x_N | \mathbf{x}^{(N-1)}\right), \tag{1.7.32}$$

corresponding to a conditional information, Eq. (1.7.13), contained in the Nth symbol, of

$$I^{N|N-1} = -c \sum_{\mathbf{x}^{(N-1)}, x_N} p\left(\mathbf{x}^{(N)}\right) \ln\left(p\left(x_N | \mathbf{x}^{(N-1)}\right)\right) \leq I^1 \tag{1.7.33}$$

where I^1 is the information contained in a single symbol, and the inequality is implied by Eq. (1.7.17). As a consequence, the information gain per added symbol diminishes monotonously with increasing length N of the sequence, as long as there exist correlations of order up to N. Taking them all into account, we can define an asymptotic information content of very long strings as

$$I^{\infty} = \lim_{N \to \infty} I^N \qquad (1.7.34)$$

and the redundancy of the language at hand as

$$R = \frac{I^1 - I^{\infty}}{I^{\infty}}, \qquad (1.7.35)$$

that is, as the average reduction of information content, comparing infinitely long sequences with isolated single symbols.

A paradigm of redundancy is symmetry, understood in the most general sense as a coincidence of different parts of an image, a message, a time series etc. What immediately comes to mind are spatial symmetries. They figure in physics, e.g., in the classification of lattice systems in crystallography, and in chemistry (molecular structures). Examples in biology range from the multitude of radial symmetries in radiolaria [Hae05] to the imperfect bilateral symmetry in most vertebrates, in animals. In plants, prime examples are the elaborate radial and chiral symmetries of blossoms. Above all, symmetries play a fundamental role in the arts, in the visual arts as well as in music. In all these diverse areas, symmetries also have a strong aesthetic appeal. Our perception tends to attribute beauty to this particular type of correlations. Hermann Weyl's classic monograph "Symmetry" gives masterly overview over all these aspects of symmetries, from the point of view of a mathematician [Wey52].

As redundancy in general, symmetries amount to a substantial reduction of the effective information content. Wherever the number of identical copies of a minimum part is exactly determined by the type of symmetry group involved, the compression of information can be expressed by group theoretical characteristics [Mul07]. Pure geometrical symmetries can readily be complemented by symmetries relating to other than spatial parameters, for example an inversion of the density value in visual data or an inversion of the note value (retrograde) or the time order in music. A number of examples of this kind of discrete symmetry will be worked out in the context of Fourier transformation (Sect. 2.3.1), where it will become clear that they lead to the same reduction of information content, given by the number of replicas related by the symmetry (e.g., by a factor 2 if the relation is a binary symmetry).

In most practical cases, in particular in analogue data, spatial and comparable symmetries apply only approximately, deviations breaking the symmetry are unavoidable. In reverse conclusion, we can say that these deviations will contribute substantially more to the information content than the basic pattern that otherwise would repeat identically. In exaggerated words, the information I in the asymmetry [Mul07]. In fact, such imperfections are not necessarily a nuisance. In the arts, it is exactly the fine deviations from exact symmetry that account for the aesthetic value of an image or an interpretation of a musical composition. We perceive exact repetition, as e.g., in digital production, as monotonous and boring. It is the extreme sensitivity of our perception for small deviations that seeks to be stimulated by nontrivial input. Concluding, we may say that symmetry breaking generates information and is tantamount to creativity.

1.8 Information in Continuous Physical Quantities

As argued in Sect. 1.3, an "entropymeter" that measures information does not exist. At the same time, quantifying a common feature of observations, of data, of communications, information *applies* to physical quantities and their measurement, hence can be classified as a metaconcept.

The simplest information-theoretic question to be asked about a physical observable is, "How much information contains a single value x of this quantity?" (assuming, to begin with, that it is a scalar). The answer requires to reflect circumstances not directly visible in a statement like "$E = 0.5873$ kg m^2/s^2". The numbers entering Shannon's definition are probabilities p_j, $j = 1, \ldots, J$, which have to be interpreted and evaluated for the case at hand. Above all, the discrete index does not apply readily to continuous quantities, typically classified in mathematical terms as real numbers. In some exceptional cases, the measurement itself already implies a natural discretization, for example if a digital device is used or if upon reading the instrument, the value is approximated by the closest tick on the scale. In most situations, however, the most adequate approach is to invoke a typical uncertainty (resolution) Δx of the measurement, providing an appropriate discretization step size (Fig. 1.13).

In order to calculate probabilities, a largest scale X has to be specified besides Δx. It is given by the range of the measurement, determined by technical features of the instrument, fundamental physical bounds, etc., as they may apply to the instrument at

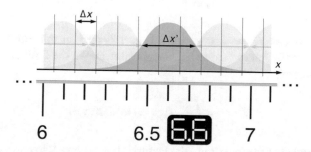

Fig. 1.13 An important parameter estimating the information contained in a measurement is its resolution. The instrument itself may have a basic built-in step size Δx, as e.g., in a digital device. The actual accuracy of the measurement, reflected, e.g., in the width $\Delta x'$ of a Gaussian distribution, may in fact be much larger.

hand. Then, if as the simplest option equal probabilities are assumed over the entire accessible range, the individual probabilities are $p_j = \Delta x/X$, and the information content per measurement is

$$I = -c \sum_{j=1}^{J} p_j \ln(p_j) = -c \sum_{j=1}^{\lceil X/\Delta x \rceil} \frac{\Delta x}{X} \ln\left(\frac{\Delta x}{X}\right) = c \ln\left(\frac{X}{\Delta x}\right). \qquad (1.8.1)$$

It is intuitively plausible that the information content increases with the relative resolution $X/\Delta x$. An important insight for later discussions is that no meaningful result is obtained without specifying a finite minimum deviation $\Delta x > 0$: In the limit $\Delta x \to 0$, the information according to Eq. (1.8.1) diverges as $-\ln(\Delta x)$. This implies in particular that the mathematical concept of real, even that of rational numbers, is at least problematic in this context, since specifying a real number in general amounts to an infinite information content (for remarkable exceptions, see Sect. 1.3 below). Equation (1.8.1) refers to the potential information gained by a measurement of x with the given optimal resolution. The actual information associated to a measurement with this outcome and the same precision is 0, that is, the limiting resolution Δx determines the zero of entropy. The typical case of measurements with suboptimal accuracy $\Delta x' > \Delta x$ yields intermediate positive values for the information.

Generally, the probability of observing a value of x even within an interval Δx depends itself on x, as a continuous variable. In that case, the discrete probabilities p_j in Eq. (1.8.1) have to be replaced by continuous probability density distributions $\rho(x)$, related to probabilities $p(x)$ by

$$p(x) = \rho(x)dx. \qquad (1.8.2)$$

A version of Eq. (1.8.1), augmented accordingly,

$$I = -c \sum_{j=1}^{\lceil X/\Delta x \rceil} \Delta x \rho(x_j) \ln(\Delta x \rho(x_j)), \qquad (1.8.3)$$

with discrete values $x_j = j\Delta x$, $j = 1, ..., \lceil X/\Delta x \rceil$ in this form, the sum in Eq. (1.8.3) is readily interpreted as a Riemann sum, and in the limit $\Delta x \to 0$ would become an integral,

$$I = -c \int_{x_{min}}^{x_{max}} dx \rho(x) \ln(d_x \rho(x)). \qquad (1.8.4)$$

In the argument of the logarithm, the factor d_x has to be kept to render the product with the probability density dimensionless, and above all, to define the reference resolution that determines the zero of entropy. It measures the minimal space (length, area, volume, ...) a state must occupy in state space, a *minimum separabile*, to be counted as a distinguishable state of its own. In fact, reducing d_x without bound leads to a divergence of information. This simple consideration already indicates that leaving the resolution undefined has far-reaching consequences: It is at the heart of the conceptual problems of classical mechanics, solved only by quantum mechanics introducing a fundamental bound for the resolution of certain measurements, see Chap. 8.

In the context of physical quantities and their observation, traditional accounts of information theory introduce the distinction of discrete vs. continuous quantities as a fundamental alternative. The previous discussion should already have indicated that with a finite insurmountable resolution, this distinction is at least blurred, as now also continuous quantities are specified with a finite amount of information. It proves completely obsolete as soon as arbitrary probability distributions $p_j, j = 1, ..., J$, are admitted. In terms of distribution functions, a single measurement typically corresponds to a box function (for example in the case of a digital instrument) or a Gaussian (in the case of a visual meter reading). More exotic distributions are perfectly conceivable.

In particular, dealing with an arbitrary normalizable function $p(x)$, it suggests itself to recur to Fourier series or to expansions in other sets of orthogonal functions, such as Chebyshev polynomials. In all these cases, the finite resolution Δx implies that the resulting expansions are truncated at a finite number J of terms. What is more, as will be demonstrated in Sect. 2.3 for harmonic functions (Fourier series), *this value is always the same*, irrespective of the function set chosen. Hence it constitutes an important invariant quantity, equivalent to the effective dimension of the function space at the given resolution.

Very often in physics, one is dealing with vectors, continuous functions, or even fields, that is, with quantities of a large or even infinite number of components. They can be included in an information-theoretical analysis by mapping them to the case of systems composed of subsystems, discussed in the previous subsection. For most vectors and functions, we can even assume that each component constitutes an independent quantity, not correlated to the other components. Therefore, the additivity of entropy applies in its simplest form, and the respective information contents just add. For a vector with d dimensions, $\mathbf{x} = (x_1, ..., x_d)$, where Eq. (1.8.1) is valid for each component $x_n, n = 1, ..., d$, this means

$$I^d = -c \sum_{n=1}^{d} \sum_{j_n}^{J_n} p_{j_n} \ln\left(p_{j_n}\right) = -c \sum_{n=1}^{d} \sum_{j_n=1}^{X_n \bmod \Delta x_n} \frac{\Delta X_n}{X_n} \ln\left(\frac{\Delta x_n}{X_n}\right) = d I^1, \quad (1.8.5)$$

where I^1 is the information contributed by each component. Equation (1.8.5) is a first example of the close relationship between the concepts of information and dimension, to be discussed further in Sect. 6.4.

Confronted with functions $f(x)$ of a continuous variable $x \in \mathbb{R}$, the same situation arises as addressed above in the context of smooth probability densities of a single variable, now for the function values themselves. Indeed, the same strategy as above helps to reduce this problem to a finite number of dimensions: A maximum resolution Δx of the argument implies that any expansion of the function in a suitable orthonormal basis of the function space will be truncated after a finite number d of terms,

$$f(x) = \sum_{n=1}^{d} f_n a_n(x). \tag{1.8.6}$$

This is then the effective dimension of this space, and Eq. (1.8.5) applies to the expansion coefficients a_n, considered as components of a d-dimensional vector. In the discussion of Fourier transformation in the subsequent section, this method proves of crucial importance to measure the information content of continuous functions.

The rather technical considerations of the preceding subsections served to clear out some common misunderstandings around the concept of information and to present a few of its most important features and ramifications. In Chap. 2, they will be filled with life, applying them to subjects as diverse as formal logics, genetics, and data analysis. In the discussion of Fourier transformation in Sect. 2.3, for instance, Eq. (1.8.3) will prove of crucial importance to measure the information content of continuous functions.

Chapter 2
Simple Applications

The conceptual framework outlined in Chap. 1 will be illustrated here applying it to three examples, propositional logics, the genetic code, and Fourier transform. In each of these areas, an information-theoretic point of view opens unusual or even unknown perspectives. Insights to be gained are a deep relationship between logical interference and information flow, Nature's invention of a highly optimized digital code for the structure of molecules, as well as the accumulation of actual information on the environment in the genome and other channels of heritage, and a quantitative analysis of the information contained in data sets that resolves the apparent contrast of digital versus analogue.

2.1 Logics

Traditional proposition logics with its *tertium non datur* invites to be interpreted in terms of binary codes, and thus qualifies as a subject of information theory. A brief exploration of this idea not only allows us to introduce logical gates in passing as elements of digital information processing, but also opens the view to a remarkable chain of isomorphisms from propositional logics through Boolean algebra and set theory [Gre98] through information theory, culminating in an information theoretic account of logical inferences.

© Springer Nature Switzerland AG 2022
T. Dittrich, *Information Dynamics*, The Frontiers Collection,
https://doi.org/10.1007/978-3-030-96745-1_2

2.1.1 Propositional Logics

Logics was introduced by philosophers in classical Greece as a branch of rhetoric, a set of rules of the game that should facilitate valid reasoning and at the same time allow to identify misleading arguments. They observed that arguments frequently used in speech follow certain patterns that are independent of the specific subject addressed. The nucleus of the field is logical inference, passing from a set of hypotheses to a conclusion without any reference to reality in the transition. The most elementary part of it is propositional logics, the area within logics concerned with statements composed of parts that are connected by conjunctions such as "and", "or", "provided that" etc. Each of them implies specific rules of the game the speaker has to accept, which of the component propositions she or he has to be verify in order to gain acceptance for the entire statement. A central achievement of propositional logics is the formalization of these rules, reducing them to clear schemes how the truth of the composed statement is related to the truth of its components.

In what follows, I shall take the *tertium non datur* (law of the excluded middle) for granted, the assumption that the truth value of a proposition can be mapped to a binary choice, "true" or "false", abbreviated as t and f, resp., in the following. To be sure, alternatives to this rule are as old as logics itself and culminate in contemporary developments around continuous truth values, known as *fuzzy logic* [NPM99]. However, for the present discussion they constitute unnecessary complications. Within propositional logics, "true" and "false" lack any semantic connotations, they are a disjoint pair of attributes satisfying certain formal rules. There is not even any inherent feature of these attributes that would allow to distinguish them without reference to the respective opposite: They share this symmetry with other pairs such as, for example, "left" and "right".

In mathematical terms, conjunctions of everyday language can be represented as maps from the set $\{t,f\}$ or its square $\{t,f\} \times \{t,f\}$ onto itself. There are two *unary* operations,

$$\text{identity}: \begin{matrix} t \mapsto t \\ f \mapsto f \end{matrix} \text{ and negation } (\neg): \begin{matrix} t \mapsto f \\ f \mapsto t \end{matrix} \tag{2.1.1}$$

Among the 16 possible *binary* operations $\{t,f\} \times \{t,f\} \to \{t,f\}$, the most common ones are [KB07].

AND (∧)		
a	b	a∧b
t	t	t
t	f	f
f	t	f
f	f	f

OR (∨)		
a	b	a∨b
t	t	t
t	f	t
f	t	t
f	f	f

EQUIVALENCE or

XNOR (≡)		
a	b	a≡b
t	t	t
t	f	f
f	t	f
f	f	t

IMPLICATION (⊃)		
a	b	a⊃b
t	t	t
t	f	f
f	t	t
f	f	t

NAND (∧̄)		
a	b	a∧̄b
t	t	f
t	f	t
f	t	t
f	f	t

NOR (∨̄)		
a	b	a∨̄b
t	t	f
t	f	f
f	t	f
f	f	t

XOR (≠)		
a	b	a≠b
t	t	f
t	f	t
f	t	t
f	f	f

IDENTITY a		
a	b	a
t	t	t
t	f	t
f	t	f
f	f	f

Binary operators can be classified according to properties such as symmetric (invariant under exchange of the two arguments, $a \circ b = b \circ a$, "$\circ$" representing any binary operator, e.g., $a \wedge b = b \wedge a$, $a \vee b = b \vee a$) balanced (same number of outcomes t and f, e.g., EQUIVALENCE, IDENTITY), and independence of one argument (IDENTITY). A property referring to the symmetry of "true" and "false" is behaviour under "conjugation", interchange of t and f in factors as well as product, replacing $a \circ b$ by $\neg(\neg a \circ \neg b)$. It maps AND and OR to one another and XOR and XNOR. IDENTITY a is invariant under this operation. A feature all these gates have in common is that not both inputs can be reconstructed from the output bit. In a few cases, one of the two inputs is conserved (e.g., for IDENTITY a), or both can be reconstructed for one of the outputs (e.g., for $a \supset b = f$). This fact may appear as a mere curiosity but will prove crucial in the context of classical and quantum computation, see Chaps. 9 and 10.

The subject of formal logics is identifying properties of composite propositions that depend only on the way they are composed, not on the truth values of the elementary statements, in analogy to theorems of algebra. A standard finding of obvious practical relevance is the equivalence of two expressions $A(a,b,\dots)$ and $B(a,b,\dots)$: $A(a, b, \dots) \leftrightarrow B(a, b, \dots)$ if and only if the truth values of A and B coincide for all truth values of the arguments a, b, ..., for example

$$\neg(a \wedge b) \leftrightarrow \neg a \vee \neg b,$$
$$(a \vee b) \wedge (\neg a \vee c) \leftrightarrow (a \wedge c) \vee (\neg a \wedge b) \tag{2.1.2}$$
$$(a \vee \neg b) \wedge (\neg a \vee c) \wedge (b \vee \neg c) \leftrightarrow (a \wedge c) \vee (\neg a \wedge b)$$

Statements on the equivalence of expressions are metastatements in the sense of Sect. 1.2. They can be reduced to propositions, taking into account that the definition of the equivalence of expressions is analogous to the operator XNOR (EQUIVA-LENCE) introduced above. In this way, the three example statements can be replaced by

$$\left(\neg(a \wedge b) = (\neg a \vee \neg b)\right) = t$$
$$\left((a \vee b) \wedge (\neg a \vee c) = (a \wedge c) \vee (\neg a \wedge b)\right) = t \qquad (2.1.3)$$
$$\left((a \vee \neg b) \wedge (\neg a \vee c) \wedge (b \vee \neg c) = (a \wedge c) \vee (\neg a \wedge b)\right) = t$$

Of particular interest are theorems concerning the reducibility of the full set of two unary and 16 binary operations to smaller subsets of them, e.g., to three operators $\{\neg, \vee, \wedge\}$ or two $\{-, \supset\}$.. It is even possible to express all unary and binary operations by $\overline{\wedge}$ (NAND) or $\overline{\vee}$ (NOR) alone, so that the functional dependence of an expression $A(a,b,\ldots)$ is encoded entirely in the bracketing structure of the equivalent expression reduced to a single operator or in the topology of an equivalent circuit.

Similarly to the definition of the metastatement $A \leftrightarrow B$ and its equivalence to the operator $\equiv$, one defines a logical inference $A(a, b, \ldots) \to B(a, b, \ldots)$ by requiring that all interpretations of the arguments a, b, ..., that make A (the *antecedent*) true, also make B (the *consequent*) true [KB07], for example,

$$a \wedge b \to a \vee b,$$
$$\neg a \supset \neg b \to b \supset a \qquad (2.1.4)$$
$$\neg a \to a \supset b$$

In logical inference, the antecedent is a *sufficient* condition for the consequent, the consequent a *necessary* condition for the antecedent.

As before, the inference $A \to B$ as a metastatement can be replaced by the requirement that the implication $A(a, b, \ldots) \supset B(a, b, \ldots)$ be true for all truth values of the arguments a, b, ... That means, for the above examples,

$$\left((a \wedge b) \supset (a \vee b)\right) = t,$$
$$\left((\neg a \supset \neg b) \supset (b \supset a)\right) = t \qquad (2.1.5)$$
$$\left(\neg a \supset (a \supset b)\right) = t$$

Statements that become true for every possible combination of truth values of their arguments, that is, irrespective of their interpretation, are called *tautologies*. They form constant functions over proposition space, analogous to constants of motion in classical mechanics. In this sense, we may define that

The logical inference $A(a, b, \ldots) \to B(a, b, \ldots)$ is valid if and only if the implication $A(a, b, \ldots) \supset B(a, b, \ldots)$ is a tautology.

Logical inference and inference chains should be clearly distinguished from arguments referring to ordered sequences of contingent facts, expected on basis of empirical evidence, such as in particular causal chains, see Sect. 3.1.

2.1.2 Boolean Algebra and Electronic Implementations

How close the relationship is between logics and information theory emerges as soon as the dichotomy true–false is interpreted as a binary choice, readily mapped to the pair 1–0. The unary and binary operators of propositional logics are thus identified with the simplest operations of information processing [Gre98], that is, of a succession of symbol strings that undergoes changes according to deterministic rules. In information processing, such steps are called *gates*.

The standard way of implementing the elementary alternative 0–1 in electronics is as two distinct levels (low–high) of the potential. Like this, gates become elements of electronic circuitry, represented as black boxes with one or two entrances and one exit lead, analogous to the operators listed above [Whi61, Gre98]:

AND				OR				XOR				Implication (no standard symbol)		
$\wedge$	0	1		$\vee$	0	1		$\not\equiv$	0	1		$\supset$	0	1
0	0	0		0	0	1		0	0	1		0	1	1
1	0	1		1	1	1		1	1	0		1	0	1

NAND				NOR				XNOR				NOT	
$\overline{\wedge}$	0	1		$\overline{\vee}$	0	1		$\equiv$	0	1		$\neg$	
0	1	1		0	1	0		0	1	0		0	1
1	1	0		1	0	0		1	0	1		1	0

Combining logical gates in series and in parallel, any kind of composed proposition, such as e.g., in Eqs. (2.1.2), can be implemented. They can be realized even with the simple tools from the electrician's shelf like ordinary switches, for example

AND

a   b

XNOR

a   b

OR

a

b

XOR

a   b

The interpretation of truth values as binary digits allows for another relevant isomorphism, to *Boolean algebra*, or algebra *modulo 2* [Whi61]. The binary numbers with addition and multiplication mod 2 form a *finite field* or *Galois field* of order 2 [Jac09]. In particular, we can associate

$$f \leftrightarrow 0$$
$$t \leftrightarrow 1$$
$$\neg a \leftrightarrow 1 - a$$
$$a \neq b \leftrightarrow a + b \pmod 2$$
$$a \wedge b \leftrightarrow a \times b \pmod 2 \tag{2.1.6}$$

The last two identities become evident comparing the truth tables for XOR and AND with addition and multiplication mod 2, resp.,

$$
\begin{array}{c|cc}
+ & 0 & 1 \\
\hline
0 & 0 & 1 \\
1 & 1 & 0
\end{array}
\qquad
\begin{array}{c|cc}
\times & 0 & 1 \\
\hline
0 & 0 & 0 \\
1 & 0 & 1
\end{array}
\tag{2.1.7}
$$

In this way, the replacement rules of propositional logics are reinterpreted as ordinary identities of Boolean algebra, for example

$$\big((a \vee \neg b) \wedge (\neg a \vee c) \wedge (b \vee \neg c)\big) \equiv \big((a \wedge c) \vee (\neg a \wedge b)\big) \leftrightarrow$$
$$\big(1 - (1 - a) \times b\big) \times \big(1 - a \times (1 - c)\big) \times \big(1 - (1 - b) \times c\big) \tag{2.1.8}$$
$$= 1 - (1 - a \times c) \times \big(1 - (1 - a) \times b\big)$$

2.1.1 Set Theory

There exists yet another isomorphism between propositional logics and Boolean algebra and a third important field in mathematics: set theory [Gre98]. Here, the starting point is a general rule how to associate propositions to sets:

$x \in A$ if and only if the proposition $a(x)$ is true, that is, if and only if x has the property a.

It induces an association of logical constants and operations with those of set theory, as follows

$$t \leftrightarrow U \text{ (universal set)}$$
$$f \leftrightarrow \emptyset \text{ (empty set)}$$
$$\neg a \leftrightarrow \overline{A} \text{ (complement of } A)$$
$$a \wedge b \leftrightarrow A \cap B$$
$$a \vee b \leftrightarrow A \cup B$$
$$a \supset b \leftrightarrow \overline{A} \cup B \tag{2.1.9}$$

They become plausible representing them as Venn diagrams as in Fig. 2.1.

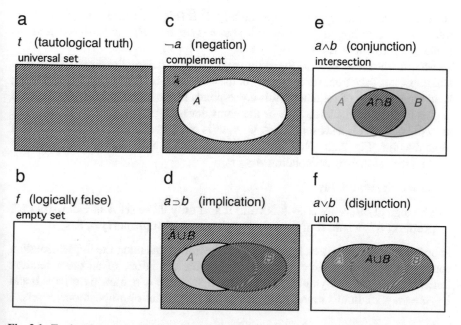

Fig. 2.1 Truth values and operations of popositional logics relate to constants and operations of set theory, as visualized by Venn diagrams: **a** true, **b** false, **c** negation, **d** implication, **e** conjunction, and **f** disjunction.

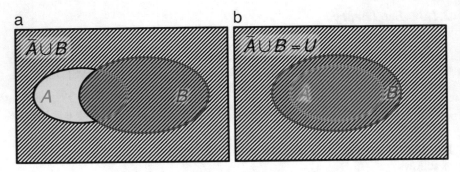

Fig. 2.2 The binary logical operation "$a \supset b$" (implication) is equivalent to the set-theoretical expression $\overline{A} \cup B$, represented as a Venn diagram in part **a**. Logically valid inferences correspond to implications that are always true, independently of contingent facts, $a \supset b = t$. In terms of sets, this amounts to $\overline{A} \cup B = U$, the universal set, or $A \subseteq B$, as in part **b**.

Of particular interest, in order to work out the intimate relationship between logics and information theory, is logical inference. In general, the binary operator "$\supset$", rewritten as

$$a \supset b = \neg a \vee b = \neg(a \wedge \neg b), \tag{2.1.10}$$

corresponds to the set $\overline{A} \cup B$ or equivalently $\overline{A \setminus B}$ (Fig. 2.2a): It is only false if a is true ($x \in A$, according to the above rule) but b is not ($x \notin B$)—to be true, x must be in the complement of $A \setminus B$.

An inference $a \to b$, to be logically valid, requires the implication $a \supset b$ to be a tautology, i.e., to be true independently of the truth values of the propositions entering it. In the language of set theory, this corresponds to the requirement that the union $\overline{A} \cup B$ be the universal set. This or the equivalent expression $A \setminus B = \emptyset$, however, are valid only if A is a subset of B, $A \subseteq B$, as is obvious from its representation as a Venn diagram, Fig. 2.2b. (Note the discrepancy between the logical notation $a \supset b$ $\equiv t$ and the set-theoretical notation $A \subseteq B$,)

Summarizing, we can say

The logical inference $a \to b$ is valid if and only if the set A of elements with property a is a subset of the set B of elements with property b, $A \subseteq B$.

In both representations, as a logical or a set theoretical operator, the implementation constitutes an asymmetric relation, satisfying the conditions of an order relation (reflexivity, $a \leq a$, antisymmetry, $a \leq b$ and $b \leq a \Rightarrow a = a$, transitivity, $a \leq b$ and $b \leq c \Rightarrow a \leq c$). In this way it gives rise to ordered chains of propositions or sets.

2.1.4 Inference Chains

To perform the last step, from logical inference in the language of set theory to its interpretation in terms of information, we invoke the concept of cardinality (number of elements) of sets as a link between the two levels. On the side of set theory, there is a simple necessary condition for the inclusion $A \subset B$ of two sets and their cardinalities $|A|$, $|B|$: $A \subseteq B$ implies $|A| \leq |B|$. This is not a sufficient condition, since the cardinality only refers to the number, not to the identity of elements.

At the same time, we can identify the elements of a set with symbols or states of a system, making contact with information content, as alluded to in Sect. 1.4 and illustrated in Fig. 1.5. In this sense, actual information measures the number of true statements a on a system and thus is related to the cardinality $|A|$ of the set A of elements to which these statements apply. $|A| \leq |B|$ therefore requires that $I_{pot}(a) \leq I_{pot}(b)$, or $I_{act}(a) \geq I_{act}(b)$. We can complement the above statement on inference in terms of set theory by

The logical inference $a \rightarrow b$ is valid only if the actual information does not increase from the premises a to the conclusion b, $I_{act}(b) \leq I_{act}(a)$.

This is not surprising if we take into account that logical inferences are true independently of contingent facts, they do not require checking against real facts and therefore do not allow us to learn anything new on the real world. As elementary instances, consider the inferences $a \wedge b \rightarrow a$, where the actual information contained in b is lost, and $a \rightarrow a \vee b$, where the additional option b augments the potential information.

Logics emerged from rhetoric, developing arguments that can become quite long. They form *inference chains*, where each single step can be justified and verified, referring to established rules, but the relation between premises and final conclusion may get out of sight in the course of the reasoning. If we assume that from the premise a_0 to the conclusion a_N each step $a_n \rightarrow a_{n+1}, n = 0, \ldots, N-1$, is a logical inference, we arrive at the chain of inequalities

$$I_{act}(a_0) \geq \cdots \geq I_{act}(a_n) \geq I_{act}(a_{n+1}) \geq \cdots \geq I_{act}(a_N) \quad (2.1.11)$$

that is, the actual information content is a monotonically decreasing function of the statement index n, while the potential information increases monotonically (Fig. 2.3). This allows to cast the above statement alternatively as

In logical inference chains $\ldots \rightarrow a_0 \rightarrow \ldots \rightarrow a_n \rightarrow \ldots$, potential information remains constant or increases from sufficient to necessary conditions.

In the extreme, only an empty statement (a tautology) may remain as ultimate conclusion. Metaphorically said, logical inference chains can be constructed from the logician's loft in the ivory tower, without looking down to the vicissitudes of real life.

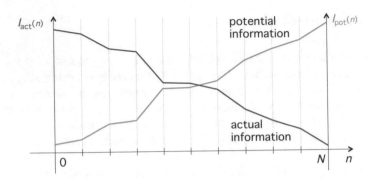

Fig. 2.3 An inference chain can be characterized as a sequence of propositions $a_0 \rightarrow \ldots \rightarrow a_n \rightarrow a_{n+1} \rightarrow \ldots \rightarrow a_N$ where the entropy increases monotonically at each step and no information on reality is gained: $\mathrm{Inct}(a_0) \leq \ldots \leq I_{pot}(a_n) \leq I_{pot}(a_{n+1}) \leq \ldots \leq I_{por}(a_N)$.

This information-theoretical property applies not only to inference chains composed of logical implications in a rigorous sense, but likewise to mathematical proofs and to "derivations", their analogues in theoretical physics and other quantitative sciences. Also there, one has to be aware for example of steps that lose actual information and therefore are not invertible, such as the step from a pair of equations $a = b, c = d$, to their sum $a + c = b + d$.

Interpreting the index n as a discrete time, the monotonous decrease of actual information finds a remarkable analogy in physical systems that define an increasing volume in phase space, as will be discussed in the contexts of causality (Sect. 3.1), chaotic systems (Sect. 6.3) and the Second Law of Thermodynamics (Sect. 7.3).

An extreme instance of this principle, complementary to the case of $I_{act}(a_N) = 0$ (tautology) is the case $I_{pot}(a_0) = 0$, commonly known as *ex falso quodlibet* (false implies anything). In this case, *any* statement a_N qualifies as logical consequence of a_0: $I_{pot}(a_0) = 0$ means that the hypothesis a_0 cannot be true, irrespective of contingent facts. that is, it must be logically false, $a_0 = f$ (the negation of a tautology). Equation (2.1.9) is then always fulfilled, any conclusion a_N whatsoever can be logically inferred from a_0. In the mathematical context, this would result in disaster. This is the reason why the self-consistency of sets of axioms is an issue of such crucial importance for the foundations of mathematics, see Sect. 4.3.1.

2.2   The Genetic Code

The genetic code excels among the examples presented in this section in that it stands for the invention of semantics by nature—as far as present-day science can judge. It marks the transition from dead matter, as the legitimate subject of physics, to life. Sitting right at the interface, it is tempting and has been pretended more than once to describe and explain the emergence of the genetic code in exclusively physical terms [Kup90], in particular as a phenomenon of non-equilibrium statistical mechanics.

Life appears on a planet under the conditions of a steep gradient of entropy, receiving intense short-wavelength, "hot" radiation from Sun and releasing longer wavelength, "cool" thermal radiation back into space [Mic10, Kle10, Kle12, Kle16]. The appearance of molecules that can store a virtually unlimited amount of information in their structure, stable far beyond the common timescales of molecular physics and readily accessible for readout, is a unique event confronting non-equilibrium statistical mechanics with an unprecedented situation, a condition that requires to rethink all its established concepts and that is at the heart of the question of the origin of life [Kup90].

2.2.1 Syntax

One of the pioneers of a physical account of living matter was Erwin Schrödinger, in particular as author of his 1944 trailblazing work "What is life?" [Sch44], where he anticipates essential features of DNA and coins the concept of "aperiodic crystal" for it. His speculation was confirmed and substantiated by the discovery of the double helix in 1953 by Francis Crick and James D. Watson. DNA (DesoxyriboNucleic Acid) is not literally a crystal, but a linear polymer, which combines the two features, anticipated by Schrödinger: Its basic structure is periodic, the repeating element being a sugar molecule. The helical shape is in fact irrelevant as far as the information content is concerned. What matters is that the repeating elements, the nucleotides or monomers of the chain molecule, come in four slightly different versions. Adenine (abbreviated A), Cytosine (C), Guanine (G), and Thymine (T) differ by a single group that represents the genetic information proper, while the rest of the monomer, which provides the physical stability of the chain, is identical. The decisive deviation from a crystal is that the order in which they appear does not result in significant differences in the total energy of the polymer, so that there is no energetic preference for any particular sequence, in particular not for periodic sequences as in a proper crystal: DNA constitutes "frozen chemical disorder", it conserves what on the face of it are random sequences of monomers, in a highly stable manner on chemical time scales and with an extreme density of information storage. The high-dimensional potential energy landscape of the total molecule, in the space of all possible sequences, does not have a single minimum at a periodic crystal structure, as in a metal, but is nearly flat, with roughly the same stability for each of the 4^N sequences, if the molecule comprises N monomers. Counting only the information encoded in the nucleotide sequence, disregarding all other physical and chemical details of the molecule, a polynucleotide of N units stores $\mathrm{lb}(4^N) = 2N$ bits, according to Eq. (1.3.4).

An additional characteristic comes into play in most of the known forms of DNA, in that it forms a *double* helix. It consists of two parallel strands linked along the middle axis, nucleotide by nucleotide, by hydrogen bonds between the opposing information-bearing groups, as in a zipper. This structure does not contribute to the information content; the genetic information encoded in the two strands is exactly the same. However, it is crucial for the physical stability of the molecule, and moreover

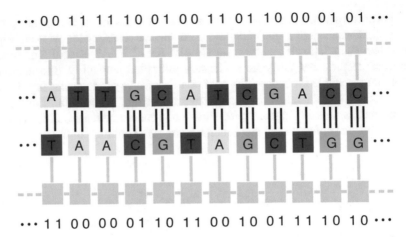

Fig. 2.4 DNA is a polymer, periodic up to the variation of a single information bearing group that comes in four variants, A, C, G, and T (nucleobases, see text). Its chemical structure is symmetric with respect to the central axis along which opposing nucleobases are linked by two (A = T) or three (C ≡ G) hydrogen bonds. The 2 bits of information per monomer could be encoded as bitstrings, with two digits per monomer.

for its ability to replicate as identical copies. In fact, the two strands are invariant under a symmetry operation combining parity, i.e., here, reflection along the middle axis of the molecule, i.e., $x \rightarrow -x$, $y \rightarrow -y$, $z \rightarrow z$, if z is the direction along the axis, and "conjunction", a one-to-one interchange of the nucleobases, A ↔ T and C ↔ G (Fig. 2.4).

The specificity of this correspondence owes itself to the structure of the hydrogen bonds connecting the two chains. In the case of the A–T affinity, it consists of two hydrogen bonds, while C and G are linked by three such bonds. This difference is sufficiently specific so as to guarantee, together with a number of elaborate correction and repair mechanisms, the astonishing fidelity of DNA replication in the living cell. The process consists in opening the zipper from one side and complementing the two resulting single strands, monomer by monomer, with free nucleotides present in the cellular medium, so that two identical double helices form as the process reaches the opposite end of the chain.

Besides DNA, a closely related molecule, RNA (RiboNucleic Acid) is as important for life [WH&87]. It is distinct from DNA by a slight difference in the sugar backbone (bearing an additional hydroxyl group per nucleotide) and by the fact that instead of Thymine, it contains Uracil (abbreviated U). Above all, RNA does not form double strands. Its single chains are more reactive than DNA, therefore serve predominantly to transfer information, not to store it, as does DNA. RNA even possesses catalytic activity itself, if not as strong and versatile as that of proteins. With this twofold capacity, RNA could combine the roles of DNA and proteins in a single class of molecules and thus precede DNA as a primordial precursor of life [Joy02, CSM07].

2.2.2 Semantics, The Central Dogma

Typically, languages develop, codes are conceived, to communicate real facts. They exist in function of reality. Therefore, one might expect that also DNA and RNA have emerged to enable the reproduction of certain amino acid chains (proteins), which, by their action on their environment, are in some sense successful. In fact, all evidence known to date indicates that the historical order was in fact inverse. It was probably free RNA (not DNA) molecules, natural memory chips of huge capacity but without anything to memorize, that existed first and then developed a functional partnership with proteins. The question how this interaction arose and why RNA molecules coding for chemically more active proteins proliferated stronger has not yet found a conclusive answer. It is all the more fascinating since at this preliminary stage, not involving any manifestation of live, a rule assumed in evolution theory as self-evident is not yet applicable: that successful reproduction is a positive value in itself, necessarily aimed at by nature. A plausible alternative explanation in this case is that the emergence of semantics—RNA molecules encoding for proteins—mutu-ally rendered both sides more stable in otherwise uncorrelated processes generating random sequences of nucleotides and amino acids, resp. A division of labour thus emerged, DNA and RNA conferring memory to Proteins, proteins imparting catalytic activity to DNA and RNA.

"Semantics" means, in the present case, that there exists a mechanism that controls the synthesis of linear amino acid chains in an order determined by the nucleotide sequence of the likewise linear DNA molecule, in an exactly reproducible manner. It is not a "code" in any anthropomorphic sense but a physico-chemical process so extraordinary specific and precise that in the vast majority of cases, the amino acid sequence is uniquely determined by DNA.

The "target language" of this translation process is formed by the twenty amino acids found in living matter:

ala	alanine	gln	glutamine	leu	leucine	ser	serine
arg	arginine	glu	glutamic acid	lys	lysine	thr	threonine
asn	asparagine	gly	glycine	met	methionine	trp	tryptophan
asp	aspartic acid	his	histidine	phe	phenylalanine	tyr	tyrosine
cys	cysteine	ile	isoleucine	pro	proline	val	valine

As there are 20 items in the list but only four different nucleotides available as symbols, the minimum number of symbols required to encode the entire list is

$$N_{min} = \lceil \log_4(20) \rceil = \lceil 2.160964 \rceil = 3 \tag{2.2.1}$$

This is exactly the value that has been consolidated in the genetic code. Three nucleotides together form a *codon*, the minimum unit of genetic information, compa-rable to a byte in computer science (the codon composed of the complementary nucleotides is called *anticodon*). At the same time, the total available number of distinct codons, 64, exceeds the number of amino acids. The surplus is used for two "purposes". It provides redundancy: There are typically more than one, and up to six, codons corresponding to the same amino acid. Moreover, it allows to define direction and reference frame for the reading process, that is, a reference point fixing which triples of nucleotides are to be combined to codons in the otherwise unstructured sequence of nucleotides. Both functions are fulfilled by stopping codons that mark the end of a gene, the next larger unit, see Fig. 2.4.

The association of amino acids to codons, the genetic dictionary, materializes in a specialized class of molecules. *Transfer-RNA* (tRNA) consists of a binding site specific for a single amino acid, tightly linked to an anticodon corresponding to one of the codons associated to that amino acid according to the genetic code. The redundancy of the code is implemented on the level of tRNA by a "wildcard nucleotide", inosine, that binds unspecifically to all the four nucleobases and replaces, e.g., the base at an irrelevant third position, see Fig. 2.5. The genetic information is not read off directly from DNA double strands but from single strands of RNA that have been formed by *transcription* as a copy of the original DNA. Hence the term messenger-RNA (mRNA) for them [WH&87].

In the living cell, the process of translation, that is, the assembly of amino acids following the genetic information encoded in RNA, occurs at organelles fulfilling only this function. Ribosomes are assembly lines for proteins, controlling the temporary binding of the anticodons and the addition of amino acids in the correct order to the growing chain, as well as the final detachment of the protein and its well-shaped folding (see below).

The entire procedure, from DNA through RNA through amino acid chain, has a unique inherent direction. By difference to many technical codes, it is not invertible

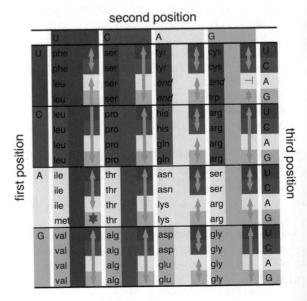

Fig. 2.5 The genetic code associates at least one codon, consisting of three nucleotides of DNA or RNA, to each amino acid. Since there are 64 different codons available for only 20 amino acids, considerable redundancy ensues. Typically, the third nucleotide is irrelevant, resulting in four codons corresponding to the same amino acid, but this number ranges from 1 (methionine, *) to 6. Moreover, codons are reserved to mark the end of a gene (—|) and to define the reading frame, i.e., how the nucleotides are to be grouped in triplets to form codons.

but only proceeds from DNA to protein, as is evident for example from the fact that the genetic code is a many-to-one (surjective) map from codons to aminoacids, hence has no unique inverse [WH&87]:

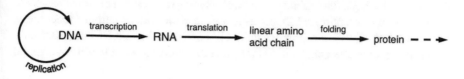

This is the first instance in this book of a directed information flow.

The traditional term "central dogma" for this marked directionality in genetics has connotations of ideology and rigidity. This is more than a coincidence. The central dogma played an important role in the early discussions of evolution, as genetic basis of Darwinism and thus as a main argument against Lamarck's hypothesis that acquired traits can be inherited, thus transmitting information from the organism to the genome. Lamarck's idea was later taken up and revived as Lysenkoism, and served as scientific support for the hope that under conditions of a socialist society, a "new man" might develop, less egoistic, more altruistic than ourselves—indeed a dogmatic view that may have motivated calling also the opposite hypothesis a "dogma".

Only recently, the central dogma sees itself challenged by serious scientific arguments. Experimental evidence indicates that certain acquired characteristics can indeed be inherited at least over one or a few generations, in plants and even in man. However, it is already clear that this heritage pathway does not affect the genetic information proper, but only the way it is expressed by the cell. *Epigenetics* studies the mechanisms controlling the frequency of transcription and translation of genes [AJR07]. For example, DNA can be packed into a dense form inaccessible to transcription and unpacked again, similar to data compression in computer science, and genes can be selectively neutralized by attaching methyl groups to G-C nucleotide pairs (DNA methylation), analogous to the possibility of "commenting out" passages of a computer code that is implemented in common programming languages. These conditions can be transferred through the reproduction process. However, they do not delete nor add genetic information, hence do not affect the basic contents of genetic heritage.

2.2.3 Pragmatics and Discussion

With the assembly of a linear chain of amino acids, the process of gene expression is not nearly complete. In fact, the final product of the entire genome is only the fully-fledged organism. Obviously, various stages of processing of genetic information are to follow, most of them far more difficult to analyze in information theoretic terms than the initial steps sketched above. Since the subsequent stages have to do increasingly with biological functions of proteins, cell organelles, organs, etc., we may subsume them, in a certain abuse of the term, as the pragmatic facet of genetic information.

A pertinent example is already the next step that immediately follows the assembly of amino acid strings, the folding of the linear chain to a three-dimensional molecule [FG04]. As highly optimized as proteins are, in their role as cellular machinery, they depend crucially on a precise, uniquely determined result of this procedure, concerning the average spatial configuration of the molecule as well as its dynamical properties.

Proteins possess spatial structure on at least three different length scales, besides the amino acid sequence as primary structure: The second level consists of folding the chain into helical (alpha helices) or plane (beta sheets) shapes, stable configurations preferred by the amino-acid chain. They are combined to form the tertiary structure, an almost arbitrary spatial composition of helices and sheets, but with a relatively compact, often approximately spherical, total outline. These parts are assembled in turn, not unlike a mechanical device, into a quaternary structure, typically with a few effective mechanical degrees of freedom describing the global configuration, such as rotation axes and centres.

Exactly how the initial linear sequence, directly incorporating the genetic information, is guided through an almost deterministic folding process is still not fully understood [Kar97, DC97, Dur98]. It is clear that this process adds spatial information, rooted in the governing physico-chemical laws, to the genetic information proper, thus augments the total information content. Indeed, besides the mere amino acid sequence, the result depends very sensitively on parameters characterizing the ambient medium, such as temperature, acidity, local electric fields, and even the presence or absence of specific molecules, in particular other proteins. The astonishing precision and speed that characterize the folding process exceed all expectations based on near-equilibrium thermodynamics, a phenomenon known as Levinthal's paradox [Kar97, DC97, Dur98].

A particularly promising hypothesis assumes that the free energy landscape of the molecules resemble a funnel [DC97, DO&08], so that relaxation leads the molecule through metastable relative minima directly to its native configuration, forming the global minimum. How precise this process indeed is, is evident from the fact that there exist only very few cases of failure. A prominent example is prions, proteins folded incorrectly and possibly catalyzing the inappropriate folding of further like molecules [Pru98]. They tend to accumulate in the brains of patients and constitute a very probable cause of neurodegenerative diseases such as scrapie in sheep and the Creutzfeldt-Jakob syndrome in humans.

The subsequent step at the next larger scale is the formation of cells. By difference to protein folding, at the present stage of evolution, it never occurs from scratch in vidrio, that is, even stem cells form in the presence of mature cells providing basic infrastructure for the process. Pattern formation without a mould occurs again in multicellular organisms where under the orchestration by the genome, large-scale spatial structures are formed in a self-organized process. Resembling protein folding, the genetic information determines a complex pattern of time- and space-dependent gene expression, complemented by physico-chemical laws, to accomplish the development of the organism [Nus06].

A common overall feature of the entire sequence of information-processing steps, from genome to organism, is the increase of total (actual) information in the result. This is evident already from the fact that a multicellular organism contains countless copies of its own genetic blueprint in its cells, therefore, as a whole, must comprise more information than that presented by the genome alone. A detailed account of this augmentation, identifying the source and quantifying the amount of information added at each stage of development, is an open question.

An aspect of the genetic control of multicellular organisms that cannot be ignored in this context is behaviour. The long-standing debate, in how far it involves genetic information or is even determined by it, is known as the nature vs. nurture controversy. Apart from all disputed details, it is clear by now that, similar to embryonic development, behavioural patterns emerge from a permanent interplay between external factors and genetically predisposed responses, so intricate that separating and quantifying the respective impact of either side appears futile.

2.2.4 *Genetic Information in Phylogenesis*

Returning to the early stages of the genesis of life on Earth, one may well ask about the origin of the information stored in primordial RNA strands that encoded for certain proteins. While RNA sequences without semantic role can be considered random, the coupling to amino acid chains with a catalyzing effect on their environment constitutes a first instance of selection, still without any valuating connotation of fitness. In this way, primordial RNA already reflected the conditions of its chemical ambience, that is, bore information on it (actual information/negentropy as opposed to potential information, see Sect. 1.4). This situation prevails throughout evolution till the present. As Konrad Lorenz put it [Lor78], the genome of a fish, determining its shape, skeleton, and musculature and optimized for high performance in swimming, diving, etc., contains a huge amount of information about hydrodynamics. The same idea, applied to cognition as a complementary channel of heritage, has given rise to evolutionary epistemology, individual and collective knowledge as a result of variation (trial and error) and selection [Lor78, Vol05, BH20], see the discussion in Sect. 3.1.5.

Comparing modern multicellular organisms, above all vertebrates, with early microbes, it is fair to say that the information on their ambient world, stored in the genome of these organisms, has steadily increased in amount and specificity. This applies all the more if we include culture as an additional pathway of heritable collective experience that emerged in primates, as well as individual learning and even learning on the cellular and subcellular levels by the immune system. Simultaneously, we observe a global increase in the complexity of life, in terms of the degree of organization and the number of levels of the structural hierarchy exhibited by organisms (Fig. 1.7) [SS95]. In how far these two tendencies are coupled or arise independently, and whether one of them or both can be considered inherent trends or even objectives of evolution, are open questions of enormous relevance within and beyond biology.

According to the central dogma, selection cannot inject information directly into the genome, it can only work by restricting a larger initial "offer" of alternatives to choose from. It cuts down initially broad distributions of random proposals, to be interpreted as a supply of potential information. The role of providing the random input is played by mutation. Selection then imparts structure to this input, much as a mould forms dough. In this way, information on the environment does get into the genome, albeit on much longer timescales than those of individual selection. This principle of operation, implemented intentionally also as an optimization strategy in engineering, in machine learning, and in artificial intelligence [Rec73,Bac96], will be analyzed further in Sect. 3.1.5.

Also mutations are prone to an analysis in terms of information: A relevant problem on the smallest scales of evolution is therefore, whence the information that becomes manifest in random modifications of the genome, caused by thermal fluctuations affecting the involved reactions, by other physical or chemical mutagens, or even expressing quantum indeterminacy [Low63].

2.3 Fourier Transform

Fourier transformation is a standard tool of data analysis and processing. It is therefore not surprising that it abounds with cross-references to information theory and constitutes an excellent example for its application. This subsection addresses four specific aspects, the effect of discrete symmetries on the information content in 2.3.1, the sampling theorem relating analogue to digital data in 2.3.2, the uncertainty relation as a mathematical, as opposed to physical, theorem in 2.3.3, and Fast Fourier Transform, a paradigmatic example of a sophisticated algorithm, in 2.3.4.

2.3.1 Discrete Symmetries in Fourier Transformation

Fourier transformations are applied in order to represent datasets into a form that exposes certain features that are hard to detect in the original data, rendering them directly visible. As this process should not add or delete anything, fidelity in the sense of one-to-one conservation of the information content, despite the drastic change in representation, is a crucial issue in Fourier transformations.

Fourier transformations order data according to their inherent scales, instead of their position in a time series or a spatial arrangement. They allow us, for example, to perceive a symmetry present in a pattern, an image, a sound and how it becomes manifest as redundancy, as a reduction of the effective information content, by a factor directly related to the size of the symmetry group.

Beginning with the most common version of Fourier transformation, this can be demonstrated in the case of Fourier series. It applies to periodic signals, i.e., to functions $f : \mathbb{R} \to \mathbb{C}, \mathbb{R} \ni t \mapsto f(t) \in \mathbb{C}$, that are periodic with period T, i.e., $f(t + T) = f(t)$. The function can then be decomposed as a series of harmonics,

$$f(t) = \sum_{n=-\infty}^{\infty} c_n \exp(in\omega t), c_n \in \mathbb{C}, \omega = \frac{2\pi}{T}. \qquad (2.3.1)$$

The complex numbers c_n, in turn, are determined by the function f through

$$c_n = \frac{1}{T} \int_0^T dt f(t) \exp(-in\omega t) = a_n + ib_n, a_n, b_n \in \mathbb{R}. \qquad (2.3.2)$$

They form an infinite-dimensional vector. According to Eqs. (1.8.2) and (1.8.3), its information content is then likewise infinite. Notwithstanding, quotients can be specified between the cardinalities of related subsets, such as in comparing even with odd numbers or with all integers.

In typical applications, the function f, besides its periodicity, is often characterized by one or more of the following twofold symmetries:

a f is purely real, $f(t) \in \mathbb{R}$

Using the identity $f^* = f$, one finds for the coefficients c_n,

$$c_n^* = \frac{1}{T} \int_0^T dt f^*(t) \exp(in\omega t) = \frac{1}{T} \int_0^T dt f(t) \exp(-i(-n)\omega t) = c_{-n}.$$
$$(2.3.3)$$

Series of complex coefficients with this property are called *Hermitean*. For their real and imaginary components, this means

$$a_n - ib_n = a_{-n} + ib_{-n} \Leftrightarrow a_n = a_{-n}, b_n = -b_{-n}, b_0 = 0, \qquad (2.3.4)$$

that is, real parts with negative index are identical, imaginary parts identical up to sign, with those with positive index, reducing the information content by a factor 2.

b f is reflection symmetric, $f(-t) = f(t)$ (Cosine Transform)

For a similar reason as above, the condition $f(-t) = f(t)$ implies for the coefficients c_n,

$$c_n = \frac{1}{T} \int_0^T dt f(t) \exp(-in\omega t) = \frac{1}{T} \int_0^T dt f(-t) \exp(in\omega(-t)), \qquad (2.3.5)$$

and, transforming the integration variable $t \to -t$ and using periodicity,

$$
\begin{aligned}
c_n &= \frac{-1}{T} \int_{-T}^{0} d(-t) f(-t) \exp(in\omega(-t)) \\
&= \frac{1}{T} \int_{0}^{T} d(t+T) f(t+T) \exp(in\omega(t+T)) \\
&= \frac{1}{T} \int_{0}^{T} dt f(t) \exp(in\omega t) = C_{-n}.
\end{aligned}
\tag{2.3.6}
$$

This reflection symmetry with respect to the index n applies to real and imaginary parts alike,

$$
a_n + ib_n = a_{-n} + ib_{-n} \Leftrightarrow a_n = a_{-n}, b_n = b_{-n}.
\tag{2.3.7}
$$

Again, these identities reduce the number of independent real coefficients by 2.

c f is antisymmetric under reflection, $f(t) = -f(-t)$ (Sine Transform)

A derivation analogous to that for symmetric functions now leads to

$$
c_n = \frac{-1}{T} \int_{0}^{T} dt f(-t) \exp(in\omega(-t)) = -c_{-n},
\tag{2.3.8}
$$

hence

$$
a_n + ib_n = -a_{-n} - ib_{-n} \Leftrightarrow a_n = -a_{-n}, -b_n = b_{-n}, a_0 = b_0 = 0,
\tag{2.3.9}
$$

reducing the number of independent real coefficients by 2.

The previous symmetries can be combined, for example:

d f is real and symmetric, $f(t) = f(-t) \in \mathbb{R}$
Taking Eqs. (2.3.4) and (2.3.6) simultaneously into account,

$$
a_n = a_{-n}, b_n = b_{-n} = -b_n = 0,
\tag{2.3.10}
$$

resulting in a reduction of the information content by 4.

e f is real and antisymmetric, $f(t) = -f(-t) \in \mathbb{R}$

With Eqs. (2.3.4) and (2.3.9),

$$
a_n = a_{-n} = -a_n = 0, b_n = b_{-n}.
\tag{2.3.11}
$$

Again, the information content diminishes by 4.

It is possible that the "nominal" period T is not actually the primitive period, for example if the function in fact repeats twice within each interval of length T:

f f has half the nominal period, $f(t + T/2) = f(t)$

This allows subdividing the integration range in Eq. (2.3.2) in two equal halves,

$$c_n = \frac{1}{T} \int_0^{T/2} dt f(t)(\exp(-in\omega t) + \exp(-in\omega(t + T/2)))$$

$$= \frac{1}{T} \int_0^{T/2} dt f(t)\left(1 + (-1)^n\right) \exp(-in\omega t) = \begin{cases} c_n & n \text{ is even} \\ 0 & n \text{ is odd} \end{cases}. \quad (2.3.12)$$

The deletion of every second coefficient reduces the information content by 2.

g f repeats negatively within the nominal period, $f(t + T/2) = -f(t)$

A similar calculation as above yields

$$c_n = \frac{1}{T} \int_0^{T/2} dt f(t)\left(1 - (-1)^n\right) \exp(-in\omega t) = \begin{cases} 0 & n \text{ is even} \\ c_n & n \text{ is odd} \end{cases}. \quad (2.3.13)$$

As before, the information content reduces by 2.

Figure 2.6 summarizes and illustrates these symmetries.

The benefits of Fourier transformation are by no means restricted to periodic functions. Considering a periodic function as defined on a compact interval of length T, it is plausible that aperiodic functions can be included by letting T approach infinity. Indeed, if each term in the Fourier expansion (2.3.1) is weighted by a factor inversely proportional to the period, the summation over the index n can be interpreted as a Riemann sum, approaching an integral as $T \to \infty$ and $\omega_0 := 2\pi/T \to 0$,

$$f(t) = \sum_{n=-\infty}^{\infty} c_n \exp(in\omega t) = \omega_0 \sum_{n=-\infty}^{\infty} \frac{c_n}{\omega_0} \exp(in\omega t)$$

$$\xrightarrow{\omega_0 \to 0, T \to \infty} \int_{-\infty}^{\infty} d\omega \tilde{f}(\omega) \exp(i\omega t). \quad (2.3.14)$$

It expresses the aperiodic *time-domain* function by a *Fourier integral* over a continuous spectral density $\tilde{f}(\omega)$ in the *frequency domain*. It is related to the discrete spectral coefficients $\tilde{f}_n$ by

$$\tilde{f}(\omega) = \frac{\tilde{f}}{[\omega_0/\omega_0]} = \frac{T}{2\pi} \frac{1}{T} \int_0^T dt f(t) \exp(-in\omega_0 t)$$

$$\xrightarrow{\omega_0 \to 0, T \to \infty} \frac{1}{2\pi} \int_{-\infty}^{\infty} dt f(t) \exp(-i\omega t). \quad (2.3.15)$$

Up to the prefactor $1/2\pi$ and the negative phase in Eq. (2.3.15), this inverse transformation is now formally identical to the forward transform (2.3.14). In many contexts, this prefactor is replaced, for the sake of symmetry, by $1/\sqrt{2\pi}$ in both directions.

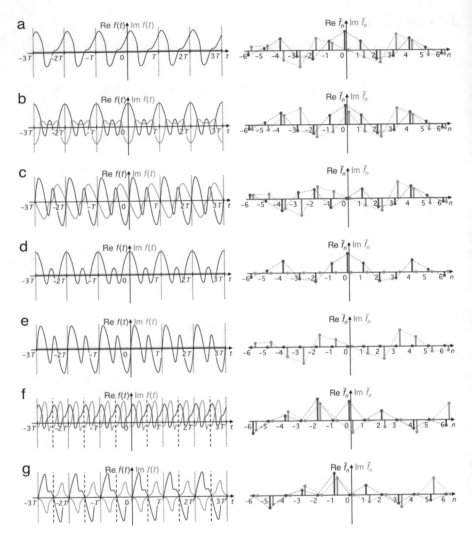

Fig. 2.6 In Fourier series, discrete symmetries of the original periodic function are manifest in a corresponding reduction of the number of real independent coefficients $\tilde{f}_n = \mathrm{Re}\,\tilde{f}_n + \mathrm{i}\,\mathrm{Im}\,\tilde{f}_n$ (one for the real part, red, one for the imaginary part, green, per index n) of the series expansion. Binary symmetries result in a reduction of their number, and thus of the information content of the series, by 2, as in purely real (**a**), reflection symmetric (**b**), and antisymmetric functions (**c**). Combining two invariances, as in symmetric (**d**) and antisymmetric (**e**) real functions, implies coefficients that are purely real and symmetric or purely imaginary and antisymmetric, resp., reducing the information content by 4. If the original function repeats identically (**f**) or negatively (**g**) within the nominal period T, all coefficients with odd or even index, resp., vanish, again resulting in a reduction by 2. Apart from the symmetries mentioned, the data shown are fictitious.

Applying Eq. (2.3.15) twice to the same function leads to

$$\tilde{\tilde{f}}(t) = \frac{1}{2\pi} \int_{-\infty}^{\infty} d\omega \left(\int_{-\infty}^{\infty} dt' f(t') \exp(-i\omega t') \right) \exp(i\omega t) = f(-t). \quad (2.3.16)$$

It results in a mere inversion of the argument, so that a fourfold application restores the original function identically. These transformations can be interpreted as a sequence of four rotations in time–frequency space by $\pi/2$ each. Interpolating this discrete set constitutes a continuous family of transformations, known as fractional Fourier transforms [OZK01], parameterized by an angle ϕ, which takes the values $\pi/2$, π, and $3\pi/2$ for common forward and backward Fourier transforms. The analogy to rotation in phase space by ϕ, generated by the Hamiltonian of the harmonic oscillator [GPS01], is evident. Combined with the relation between Fourier transformation and differentiation,

$$\frac{1}{2\pi} \int_{-\infty}^{\infty} dt f^{(n)}(t) \exp(-i\omega t) = \frac{(i\omega)^n}{2\pi} \int_{-\infty}^{\infty} dt f(t) \exp(-i\omega t) = (i\omega)^n \tilde{f}(\omega),$$

$$(2.3.17)$$

replacing the integer exponent n by a continuous real parameter, it gives rise to a definition of fractional derivatives.

The consistency of this construction can be checked by returning to the case of time-domain functions with a finite period T, Eq. (2.3.2),

$$\tilde{f}(\omega) = \frac{1}{2\pi} \int_{-\infty}^{\infty} dt \sum_{n=-\infty}^{\infty} \tilde{f}_n \exp(in\omega_0 t) \exp(-i\omega t)$$

$$= \sum_{n=-\infty}^{\infty} \tilde{f}_n \frac{1}{2\pi} \int_{-\infty}^{\infty} dt \exp(i(n\omega_0 - \omega)t) = \sum_{n=-\infty}^{\infty} \tilde{f}_n \delta(\omega - n\omega_0). \quad (2.3.18)$$

It results in a train of delta functions, a "comb", weighted by coefficients $\tilde{f}_n$, embedding the original discrete Fourier sum in continuous frequency space. An important application is the generation of frequency comb in laser light, an indispensable tool for example for the measurement of high frequencies [Hal06, Han06, PH19]. They are obtained from pulsed lasers (with a period in the range of femtoseconds) with mode locking, a precisely stabilized relative phase between the modes of the lasing cavity.

Conversely, contracting the time domain function to a delta train, say equidistant with a separation T and weighted with discrete coefficients f_n, one finds in the frequency domain,

$$\tilde{f}(\omega) = \frac{1}{2\pi} \int_{-\infty}^{\infty} dt \sum_{n=-\infty}^{\infty} f_n \delta(t - nT) \exp(-i\omega t) = \frac{1}{2\pi} \sum_{n=-\infty}^{\infty} f_n \exp(-in\omega T),$$

$$(2.3.19)$$

where $\tilde{f}(\omega)$ is a periodic function of the frequency with period ω_0,

$$\tilde{f}(\omega + \omega_0) = \frac{1}{2\pi} \sum_{n=-\infty}^{\infty} f_n \exp(-in(\omega + \omega_0)T)$$

$$= \frac{1}{2\pi} \sum_{n=-\infty}^{\infty} f_n \exp(-in(\omega + 2\pi/T)T) = \tilde{f}(\omega). \qquad (2.3.20)$$

The situation is therefore converse to Fourier series: An aperiodic discrete chain of coefficients, the amplitudes of isolated peaks in the time domain, separated by a distance T, transforms into a continuous function with finite period $\omega_0 = 2\pi/T$ in the frequency domain.

Finally, based on the expressions (2.3.14) and (2.3.15) for the Fourier integral, the case of a periodic delta comb, repeating after N peaks, is readily construed. In the time domain, it takes the form

$$f(t) = \sum_{m=-\infty}^{\infty} T \sum_{n=0}^{N-1} f_n \delta(t - (n + mN)T), \qquad (2.3.21)$$

introducing a factor T to render the expression dimensionless. The corresponding spectral function is

$$\tilde{f}(\omega) = \int_{-\infty}^{\infty} dt \sum_{m=-\infty}^{\infty} T \sum_{n=0}^{N-1} f_n \delta(t - (n + mN)T) \exp(-i\omega t), \qquad (2.3.22)$$

Rearranging the two sums and applying the Poisson summation formula $\sum_{n=-\infty}^{\infty} \exp(in\omega T) = 2\pi \delta(\omega T \bmod 2\pi)$ [Lan68] allows evaluating Eq. (2.3.22) as

$$\tilde{f}(\omega) = \sum_{m=-\infty}^{\infty} T \sum_{n=0}^{N-1} \tilde{f}_n \delta\left(\omega - (n + mN)\frac{\omega_0}{N}\right). \qquad (2.3.23)$$

The result turns out to be another periodic delta chain, now in the frequency domain, a frequency comb with spectral weights

$$\tilde{f}_n = \frac{1}{N} \sum_{n'-0}^{N-1} f_{n'} \exp\left(-2\pi i \frac{nn'}{N}\right) = \tilde{f}_{n+N}. \qquad (2.3.24)$$

At the same time, the weights of the delta functions in the time domain, Eq. (2.3.21), are related to the spectrum by

$$f_n = \sum_{n'=0}^{N-1} \tilde{f}_{n'} \exp\left(2\pi i \frac{nn'}{N}\right) = f_{n+N^*}.$$ (2.3.25)

As in the case of the Fourier integral, comparing Eqs. (2.3.24) and (2.3.25), we have achieved symmetry between time and frequency domain, again up to a prefactor and an inversion of the phase. A decisive difference, however, is that both transformations now only involve a *finite number of discrete coefficients* and therefore a well-defined information content, without the need to introduce a finite resolution as an extra parameter. This privileges the discrete Fourier transform as a mathematically robust procedure that is particularly suitable as basis for numerical algorithms like the Fast Fourier Transform, see Sect. 2.3.4 below. Moreover, it proves identical to the transformation between position and momentum space in quantum dynamics on a finite-dimensional Hilbert space. The four categories of Fourier transform discussed in this subsection are summarized and compared in Table 2.1. A remarkable conclusion is that continuous functions on a compact interval are equivalent, as concerns their information content, to an enumerably infinite set of discrete coefficients.

2.3.2 Sampling

An instructive application of Fourier series is the *Sampling theorem*. It allows to analyze the apparent dichotomy between analogue and discrete data in a quantitative manner. Recall smooth continuous T-periodic functions, $f(t + T) = f(t)$. In order to avoid an unbounded information content, its Fourier expansion has to be restricted to an efficient total number of, say, N terms, by imposing a suitable envelope of more or less smooth shape (a Gaussian, a Fermi function, a box ...). The simplest choice, a sharp frequency cutoff at $N\omega$, $\omega = 2\pi/T$, amounts to coarse-graining the original function to

$$f_N(t) = \frac{1}{N} \sum_{n=0}^{N-1} \tilde{f}_n \exp(in\omega t).$$ (2.3.26)

The coefficients $\tilde{f}_n$ are usually not accessible to direct measurement. Instead, a natural access is through a sufficient number, namely N, of sampling points in time, equidistant within the period T, $t_n = n\Delta t$, $\Delta t = T/N$, each one yielding a sample function value $f_n = f(t_n)$. Sampling is an elementary example of *analog-to-digital conversion* (ADC). The continuous function $f_N(t)$ is then reconstructed from these finite sample points by interpolation, replacing sharp peaks at the t_n by smooth but sharply localized functions $g_N(t)$,

Table 2.1 Fourier transforms come in four categories: From continuous periodic functions to infinite sequences of discrete coefficients (Fourier series), from infinite sequences back to periodic functions, between two aperiodic functions on a continuous set (Fourier integral) and between two finite sets of discrete coefficients (discrete Fourier transform). In all cases, the smallest scale resolved in the frequency domain is the inverse (up to a factor 2π) of the largest scale covered in the time domain, and vice versa. From the two asymmetric cases it is evident that a continuous function on a compact interval and an enumerably infinite sequence of coefficients are equivalent, as concerns their information content.

Time domain	Smallest scale	Largest scale	Frequency domain	Smallest scale	Largest scale
Periodic with period T: finite continuous interval	0	T	Discrete with separation $\Delta\omega = 2\pi/T$ enumerably infinite sequence	$2\,ji\,IT$	00
Discrete with separation $\Delta t = T$: enumerably infinite sequence	T	∞	Periodic with period $\omega = 2\pi/T$. finite continuous interval	0	$2\pi/T$
Aperiodic continuous: infinite continuous set	0	∞	Aperiodic continuous: infinite continuous set	0	00
Discrete with separation T/N and period T: discrete finite set with N elements	T/N	T	Discrete with separation $2\pi/NT$ and period $2\pi/T$: discrete finite set with N elements	$2\pi/NT$	$2\pi/T$

$$f(t) = \frac{1}{N} \sum_{n=0}^{N-1} f_n g_N(t - t_n) \tag{2.3.27}$$

that have to be chosen in an appropriate manner.

In order to work as a smooth version of a periodic Kronecker delta function $\delta_{n(\bmod N)}$, $g_N(t)$ must fulfill the conditions (pink dots in Fig. 2.7)

$$g_N(t) = \begin{cases} 1 & t = 0(\bmod T) \\ 0 & t = n\Delta t, n \neq 0(\bmod N) \end{cases}. \tag{2.3.28}$$

This is the closest approximation to a periodic Dirac delta that can be achieved as a superposition of N harmonics. It amounts to passing the periodic delta function through a low-pass filter with a cutoff frequency $N\omega$ and can be constructed as the discrete Fourier transform of a constant, Eq. (2.26) with coefficients $\tilde{f}_n = 1, n = 0, \ldots, N - 1$ (Fig. 2.7),

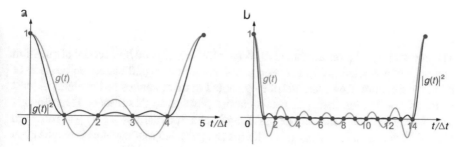

Fig. 2.7 Smooth approximations $g_N(t)$ of a periodic delta function $\delta(t(\bmod T))$, composed of N harmonics of the fundamental frequency $\omega = 2\pi/T$, Eq. (2.3.29) (green), and their absolute squares $|g_N(t)|^2$(blue), for $N = 5$ **(a)** and $N = 15$ **(b)**. Pink dots indicate points where these functions are constrained to coincide identically with a discrete delta comb, Eq. (2.3.28).

$$g_N(t) = \frac{1}{N}\sum_{n=0}^{N-1}\exp(in\omega t) = \frac{1}{N}\exp\left(i(N-1)\frac{\omega t}{2}\right)\frac{\sin(N\omega t/2)}{\sin(\omega t/2)}. \qquad (2.3.29)$$

Substituting $g_N(t)$ back into Eq. (2.3.26), the original function is thus recovered as

$$f_N(t) = \frac{1}{N}\sum_{n,r'-0}^{N-1}\exp\left(in'\omega\left(t-\frac{n}{N}T\right)\right) = \frac{1}{N}\sum_{n,n'-0}^{N-1}\exp\left(in'\left(\omega t - 2\pi\frac{n}{N}\right)\right).$$

$$(2.3.30)$$

Comparing with Eq. (2.3.26), we find that the discrete coefficients f_n in the time domain and $\tilde{f}_n$ in the frequency domain are related by

$$\tilde{f}_n = \frac{1}{N}\sum_{n'=0}^{N-1}f_{n'}\exp\left(-2\pi i\frac{nn'}{N}\right), \quad f_n = \sum_{n'=0}^{N-1}\tilde{f}_{n'}\exp\left(2\pi i\frac{nn'}{N}\right). \qquad (2.3.31)$$

This pair of transformations coincides with the discrete Fourier transform, Eqs. (2.3.24) and (2.3.25). The conclusion is that a continuous function representing analogue data, reduced to a finite resolution, can be represented in various alternative ways as a set of discrete coefficients, the only common invariable quantity being the number N of coefficients, to be interpreted as the dimension of the underlying function space. The step from the original function $f(t)$ to the approximating $f_N(t)$ is an instance of analogue-to-digital conversion. Once reduced to a finite frequency range, there is no fundamental difference any longer between the approximate "analogue" form (2.3.26) and its "discretization" as a set of coefficients f_n or $\tilde{f}_n$ Eq. (2.3.31)—the information content is the same.

2.3.3 Uncertainty Relations

The uncertainty relation, attributed to Werner Heisenberg and his account of quantum mechanics, is a relationship between physical quantities and hence, on the face of it, a law of nature. However, it is deeply rooted in mathematics and contains at least as much information theory as it talks about physics. In this respect, Heisenberg's uncertainty relations resemble Gauss' Law in electrodynamics [Jac75], as rooted in Gauss' Theorem in vector integral calculus [Kap03], but mathematics is maybe even more prevalent in the case of uncertainty.

Comparing the four cases summarized in Table 2.1, a duality between the two sides, the time and the frequency domain, is evident, not unlike Noether's theorem relating symmetries to conservation laws in classical mechanics [JS98]. Indeed, in all cases the largest scale covered in one domain is the inverse of the smallest resolved scale in the other domain. This is particularly clear in discrete Fourier transform. Considering a two-dimensional space spanned simultaneously by time and frequency, the two alternative representations of the same function appear as tessellations of a square of side lengths $N \times N$ (the total number of coefficients), subdivided into equal rectangles of side lengths 1 and N, resp., oriented either parallel to the frequency axis (vertically, time domain) or the time axis (horizontally, frequency domain), but always of area N, cf. Fig. 2.6.

Looking for a more general rule that would include and explain the four cases, it is instructive to consider specific functions with their Fourier transform, as listed in Table 2.2. For each function pair it gives the variances in the time and in the frequency domain

$$\langle (t - \langle t \rangle)^2 \rangle = \int_{-\infty}^{\infty} dt \, (t - \langle t \rangle)^2 |f(t)|^2,$$

$$\langle (\omega - \langle \omega \rangle)^2 \rangle = \int_{-\infty}^{\infty} d\omega \, (\omega - \langle \omega \rangle)^2 |\tilde{f}(\omega)|^2. \tag{2.3.32}$$

(Obviously, for these definitions to make sense, the functions $f(t)$ and $\tilde{f}(\omega)$ must be square-integrable, see Eq. (2.3.33) below.) The last column states the dimensionless product $\langle \Delta t \rangle \times \langle \Delta \omega \rangle = \langle (t - \langle t \rangle)^2 \rangle^{1/2} \langle (\omega - \langle \omega \rangle)^2 \rangle^{1/2}$. For the cases shown, it ranges from 1 (Gaussians) through positive values larger than unity, through ∞ (not defined), but all of them independent of the respective parameters controlling the widths of $f(t)$ and $\tilde{f}(\omega)$. The fact that Gaussians are a special case, mapped to Gaussians up to scaling under Fourier transformation, suggests that the value 1 might mark a lower bound, valid generally for all Fourier transform pairs.

This can indeed be shown [Bri56]. To avoid irrelevant technicalities, assume $f(t)$ and $\tilde{f}(\omega)$ to have unit norm and zero mean,

Table 2.2 Four square-integrable (see Eq. (2.3.33)) functions (first column), widely used in physics and statistics, with their Fourier transforms (third column). The table also gives the mean square deviations in the time domain (second column) and in the frequency domain (fourth column) as well as the dimensionless uncertainty product (fifth column). The fourth row is dedicated to the eigenfunction of order n of the quantum mechanical harmonic oscillator, $H_n(x)$ denotes the nth Hermite polynomial. They are invariant (up to scaling) under Fourier transform. The Gaussians featured in the first row are equivalent to the harmonic-oscillator eigenfunction of order 0.

$f(t)$	$\langle \Delta t \rangle^2$	$\tilde{f}(\omega)$	$\langle \Delta \omega \rangle^2$	$\langle \Delta t \rangle \langle \Delta \omega \rangle$
Gaussian $\left(\frac{1}{2\pi\sigma}\right)^{1/4} \exp\left(-\frac{t^2}{4\sigma}\right)$	σ	Gaussian $\left(\frac{2\sigma}{\pi}\right)^{1/4} \exp\left(-\sigma\omega^2\right)$	$\frac{1}{\sigma}$	1
box $\frac{1}{\sqrt{l}}\left(\Theta\left(t+\frac{l}{2}\right) - \Theta\left(t+\frac{l}{2}\right)\right)$	l^2	sine $\left(\frac{l}{2\pi}\right)^{1/2} \frac{\sin(\omega l/2)}{\omega l/2}$	∞	∞
tent $\left(\frac{3}{2l^3}\right)^{1/2} \begin{cases} l - \|t\| & \|t\| \le 1 \\ 0 & \text{else} \end{cases}$	$\frac{l^2}{10}$	sine squared $\frac{1}{2\pi}\left(\frac{3}{2l}\right)^{1/2}\left(\frac{\sin(\omega l/2)}{\omega l/2}\right)^2$	$\frac{3}{l^2}$	$\sqrt{\frac{3}{10}}$
nth eigenfunction, harmonic oscillator $\frac{1}{\sqrt{2^n \pi^{1/2} \sigma n!}} e^{-\frac{t^2}{2\sigma^2}} H_n\left(\frac{t}{\sigma}\right)$	$\sigma\left(n+\frac{1}{2}\right)$	nth eigenfunction, harmonic oscillator $\left(\frac{\sigma}{2^n \pi^{1/2} n!}\right)^{1/2} e^{\frac{\sigma^2\omega^2}{2}} H_n(\sigma\omega)$	$\frac{n+1/2}{\sigma}$	$n+\frac{1}{2}$

$$\int\limits_{-\infty}^{\infty} dt\, |f(t)|^2 = 1, \quad \int\limits_{-\infty}^{\infty} d\omega\, |\tilde{f}(\omega)|^2 = 1,$$

$$\langle t \rangle = \int\limits_{-\infty}^{\infty} dt\, t |f(t)|^2 = 0, \quad \langle \omega \rangle = \int\limits_{-\infty}^{\infty} d\omega\, \omega |\tilde{f}(\omega)|^2 = 0. \tag{2.3.33}$$

If they are related by Fourier transformations with common symmetrized prefactor $1/\sqrt{2\pi}$,

$$f(t) = \frac{1}{\sqrt{2\pi}} \int\limits_{-\infty}^{\infty} d\omega\, \tilde{f}(\omega) e^{-i\omega t}, \quad \tilde{f}(\omega) = \frac{1}{\sqrt{2\pi}} \int\limits_{-\infty}^{\infty} dt\, f(t) e^{i\omega t}, \tag{2.3.34}$$

we can express the variance of, say, the frequency by a Fourier transform of the original time-domain function and its second derivative,

$$\langle \omega^2 \rangle = \int_{-\infty}^{\infty} d\omega \omega^2 \, \tilde{f}(\omega) \tilde{f}^*(\omega) = \frac{1}{\sqrt{2\pi}} \int_{-\infty}^{\infty} d\omega \omega^2 \, \tilde{f}(\omega) \int_{-\infty}^{\infty} dt f^*(t) e^{-i\omega t}. \quad (2.3.35)$$

Considering the factor ω^2 in the integrand as the result of a double time derivative of the phase factor $e^{-i\omega t}$, this takes the form

$$\langle \omega^2 \rangle = - \int_{-\infty}^{\infty} dt f^*(t) \frac{d^2 f(t)}{dt^2}. \quad (2.3.36)$$

The inequality sought for the uncertainty product can be derived as a consequence of *Schwarz' inequality* [Kap03]. A fundamental feature of linear vector spaces H endowed with a scalar product, it states that for any pair of vectors $\mathbf{a}, \mathbf{b} \in$ H,

$$|\mathbf{a}|^2 |\mathbf{b}|^2 \geq |\mathbf{a} \cdot \mathbf{b}|^2. \quad (2.3.37)$$

Interpreting the integral $\mathbf{a} \cdot \mathbf{b} = \int_{-\infty}^{\infty} dx a(x) b^*(x)$ as a scalar product on H, Schwarz' inequality takes the form

$$\int_{-\infty}^{\infty} dx |a(x)|^2 \int_{-\infty}^{\infty} dx |b(x)|^2 \geq \left| \int_{-\infty}^{\infty} dx \frac{1}{2} \left(a^*(x) b(x) + a(x) b^*(x) \right) \right|^2 \quad (2.3.38)$$

(the form of the integrand on the right-hand-side has been chosen so as to symmetrize it in cases of complex functions $a(x), b(x)$). Identifying $a(x): = tf(t)$ and $b(x): = df(t)/dt$, Eq. (2.3.38) implies

$$\int_{-\infty}^{\infty} dt t^2 |f(t)|^2 \int_{-\infty}^{\infty} dt \left| \frac{df(t)}{dt} \right|^2 \geq \left| \int_{-\infty}^{\infty} dx \frac{1}{2} \left(tf^*(t) \frac{df(t)}{dt} + tf(t) \frac{df^*(t)}{dt} \right) \right|^2. \quad (2.3.39)$$

The first integral on the left-hand side of Eq. (2.3.38) gives the variance in the time domain, $\langle \Delta t \rangle^2$. The second integral can be integrated by parts,

$$\int_{-\infty}^{\infty} dt \left| \frac{df(t)}{dt} \right|^2 = \int_{-\infty}^{\infty} dt \frac{df^*(t)}{dt} \frac{df(t)}{dt} = - \int_{-\infty}^{\infty} dt f^*(t) \frac{d^2 f(t)}{dt^2} = \langle \omega^2 \rangle. \quad (2.3.40)$$

The boundary term emerging from the partial integration vanishes if the function $f(t)$ is square-integrable, see Eq. (2.3.33). The last member follows from Eq. (2.3.36).

The right-hand-side of Eq. (2.3.38) also invites for integration by parts, invoking Eq. (2.3.36),

$$\int_{-\infty}^{\infty} dt \frac{1}{2} \left(t f^*(t) \frac{df(t)}{dt} + t f(t) \frac{df^*(t)}{dt} \right) = \frac{1}{2} \int_{-\infty}^{\infty} dt t \frac{d}{dt} |f(t)|^2$$

$$= t |f(t)|^2 \Big|_{-\infty}^{\infty} - \frac{1}{2} \int_{-\infty}^{\infty} dt |f(t)|^2 = -\frac{1}{2}. \qquad (2.3.41)$$

Again, the boundary term vanishes as a consequence of the square-integrability of $f(t)$, and the last step uses its normalization, Eq. (2.3.33).

Substituting Eqs. (2.3.40) and (2.3.41) in the Schwarz inequality (2.3.39),

$$\langle t^2 \rangle \langle \omega^2 \rangle = \int_{-\infty}^{\infty} dt t^2 |f(t)|^2 \int_{-\infty}^{\infty} dt \left| \frac{df(t)}{dt} \right|^2$$

$$\geq \left| \int_{-\infty}^{\infty} dt \frac{1}{2} \left(t f^*(t) \frac{df(t)}{dt} + t f(t) \frac{df^*(t)}{dt} \right) \right|^2 = \left| -\frac{1}{2} \right|^2 = \frac{1}{4}. \qquad (2.3.42)$$

Till here, what we have obtained is a relation of purely mathematical nature between duration and bandwidth of an oscillation, related by Fourier transform. Defining uncertainties Δx as related to variances $\langle x^2 \rangle$ (for $\langle x \rangle = 0$) by $\Delta x = 2\sqrt{\pi \langle x^2 \rangle}$, Eq. (2.3.41) takes the form of an uncertainty relation [Wey31, Hir57, Bec75]

$$\Delta t \Delta \omega \geq 2\pi. \qquad (2.3.43)$$

It is converted in fundamental inequalities for quantum mechanics by invoking basic quantum relationships, such as Einstein's $E = \hbar \omega$. Multiplying both sides of Eq. (2.3.43) by $E = \hbar$, this leads to the well-known energy-time uncertainty relation

$$\Delta t \Delta E \geq 2\pi \hbar. \qquad (2.3.44)$$

If ω is interpreted as the spatial frequency (wavenumber) k of matter waves, related to their momentum by de Broglie's relation $p = \hbar k$ and t with the position q, gives the position-momentum uncertainty relation,

$$\Delta q \Delta p \geq 2\pi \hbar. \qquad (2.3.45)$$

Representing Eqs. (2.3.44) and (2.3.45) in the common space spanned by time and frequency, on the mathematical level, or by energy and time, momentum and position (phase space), angular momentum and angle, ..., in the physical context, suggests to interpret them as statements about a lower bound for the area occupied by

Fourier transform pairs in these spaces. As a consequence, the total area A covered, for example, by a given data set in t-ω space, cannot accommodate more than $N = \lceil A/2\pi \rceil$ states weighted by independent coefficients, or $N = \lceil S/2\pi\hbar \rceil$ quantum eigenstates within a phase-space area S. This simple geometrical estimate has given rise to the concepts of *Gabor cells* of area 2π, in signal processing, and *Planck cells* of size $2\pi\hbar$, in quantum mechanics (Fig. 2.8). In the latter context, it turns out that the image of a rectangle of area $2\pi\hbar$, within which the system is found with certainty but never outside, is by far too restrictive. As detailed in Chap. 8, quantum mechanics allows for smooth probability density distributions in phase space, with contours of any shape. Nevertheless, Gabor cells and Planck cells are closely akin, both based on the same fundamental restriction of the information content per unit area of time–frequency space.

The reversible character of Fourier transformations implies that both representations of a data set, in the time domain and in the frequency domain, are already complete in themselves. A combined space therefore is unnecessarily large to represent the same data, resulting in redundancy. That is a plausible explanation why in time–frequency space, the resolution cannot reach the maximum given by a single "pixel" $\Delta t \, \Delta\omega$ at specific values ω and t of frequency and time. Indeed, analyzing the information contributed by the two representations corroborates this suspicion.

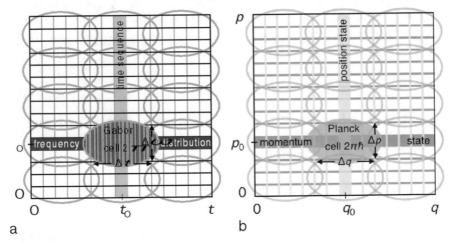

a   b

Fig. 2.8 The time–frequency uncertainty relation for Fourier transformation (**a**) and Heisenberg's quantum mechanical uncertainty relations (**b**) are consequences of the same underlying information theoretical rule. In data analysis, the same information content can be represented in the time domain (**a**, vertical rectangles), as a frequency distribution (horizontal rectangles) with N sampling points (here, $N = 15$) or as a mixed representation with intermediate resolutions in frequency and time, forming N Gabor cells of size 2π. Similarly, a quantum mechanical state vector can be represented, e.g., as a wave function in position space (**b**, vertical rectangles), in momentum space (horizontal rectangles) or determined by simultaneous position and momentum values, with a resolution restricted by the position-momentum uncertainty relation. In phase space, the analogue of a Gabor cell is a Planck cell of size $2\pi\hbar$, in units of action. Their number N now gives the Hilbert-space dimension.

Thus required to measure the information contained in $f(t)$ and $\tilde{f}(\omega)$ in a way different from that used in the previous subsections, where the sets of coefficients pertaining to either domain were treated as high-dimensional, possibly infinite dimensional, vectors, each of their components distributed according to some probability density. Instead, we here consider $f(t)$ and $\tilde{f}(\omega)$ as functions which themselves give rise to distributions, interpreting, as in quantum mechanics, their absolute squares as probability densities that describe the distribution of the single quantity in their argument,

$$p_t(t) := |f(t)|^2, \; p_\omega(\omega) := |\tilde{f}(\omega)|^2. \tag{2.3.46}$$

Their normalization, as implied by Eq. (2.3.33), is taken for granted. In this way, the information contained in either domain is not that of N vector components but that of a single scalar distributed according to these densities, Eq. (1.8.4),

$$I_t = -c \int_{-\infty}^{\infty} dt\, p_t(t) \ln(d_t\, p_t(t)) = -c \int_{-\infty}^{\infty} dt\, |f(t)|^2 \ln(d_t |f(t)|^2),$$

$$I_\omega = -c \int_{-\infty}^{\infty} d\omega\, p_\omega(\omega) \ln(d_\omega\, p_\omega(\omega)) = -c \int_{-\infty}^{\infty} d\omega\, |\tilde{f}(\omega)|^2 \ln(d_\omega |\tilde{f}(\omega)|^2), \tag{2.3.47}$$

where the parameters d_t and d_ω control the resolution in the two respective domains. As explained in Sect. 1.8, non-zero values of d_t and d_ω are required to keep the total information finite, see Eq. (1.8.4).

How do these information contents change if we scale time and frequency by inverse factors, say $t \rightarrow \alpha t$ and $\omega \rightarrow \omega/\alpha$, that is, if we squeeze the corresponding contours in (t, ω)-space, keeping shapes and total areas of the distributions unaltered? Denoting the above unscaled distributions $p_{t,1}(t)$ and $p_{\omega,1}(\omega)$, resp., we define

$$p_{t,\alpha}(t) = \alpha p_{t,1}(\alpha t), \quad p_{\omega,\alpha}(\omega) = \frac{1}{\alpha} p_{\omega,1}\left(\frac{\omega}{\alpha}\right). \tag{2.3.48}$$

Prefactors α and $1/\alpha$, resp., are included to maintain these distributions normalized to unity. With this modification, the information in the time domain becomes

$$I_{t,\alpha} = -c \int_{-\infty}^{\infty} dt\, p_{t,\alpha}(t) \ln(d_t\, p_{t,\alpha}(t)) = -c \int_{-\infty}^{\infty} dt\, \alpha p_{t,1}(\alpha t) \ln(d_t \alpha p_{t,1}(\alpha t))$$

$$= -c \ln(\alpha) \int_{-\infty}^{\infty} dt'\, p_{t,1}(t') - c \int_{-\infty}^{\infty} dt'\, p_{t,1}(t') \ln(d_t\, p_{t,1}(t')) = I_{t,1} - c \ln(\alpha),$$

$$\tag{2.3.49}$$

where in the second line, the integration variable has been scaled $t \to t' = \alpha t$. By an analogous calculation,

$$I_{\omega,\alpha} = I_{\omega,1} + c \ln(\alpha) \tag{2.3.50}$$

The scaling changes the information contents by the same amount in opposite directions. For example, if $\alpha > 1$, that is, compressing the duration in the time domain and expanding the bandwidth in frequency, the potential information in the time domain decreases while it increases by the same amount in the frequency domain. In this way,

the total information remains constant,

$$I_{\text{tot},\,\alpha} = I_{t,\alpha} + I_{\omega,\alpha} = I_{t,1} - c \ln(\alpha) + I_{\omega,1} + c \ln(\alpha) = I_{\text{tot},1}. \tag{2.3.51}$$

This conservation law is independent of the choice of d_t and d_ω, as long as they are finite. It includes in particular the limiting cases $\alpha \to \infty$ where $p_t(t)$ approaches a sharp pulse in time of infinitesimal width d_t, and $\alpha \to 0$, corresponding to a sharp spectral peak of infinitesimal width d_ω. It is not to be confounded with the invariance under Fourier transformation of the information contained in set of coefficients, discussed in Sect. 2.1, where every coefficient carries a fixed amount of information.

The constant $I_{\text{tot},1}$ appearing on the right-hand-side of Eq. (2.3.51) characterizes the shape of $p_{t,1}(t)$ and $p_{\omega,1}(\omega)$. It constitutes a measure, independent of any scaling of time and frequency, of the compactness of these distributions. For example, for Gaussians,

$$I_{\text{tot},1} = I_{t,1} + I_{\omega,1} = c, \tag{2.3.52}$$

where c depends only on the unit of information, see Eq. (1.3.4). Gaussians, which are invariant up to scaling under Fourier transformation, mark the absolute minimum of the total information contained in both domains.

The reasoning leading to Eq. (2.3.52) applies only to a narrow class of transformations, linear dilations and compressions, of the two Fourier-related distributions $p_t(t) = |f(t)|^2$ and $p_\omega(\omega) = |\tilde{f}(\omega)|^2$. In fact, it can be shown that generally for such pairs [Hir57, Bec75],

$$I_t + I_\omega \geq ck, \tag{2.3.53}$$

where k is a numerical constant of order 1. It depends on the specific form of the Fourier transformation relating $f(t)$ to $\tilde{f}(\omega)$ but otherwise is independent of these functions and thus constitutes an absolute lower bound for the total information I_{tot} appearing in Eq. (2.3.51). Equation (2.3.53) can be considered as an information-theoretical version of the mathematical uncertainty relation (2.3.43). An analogous version of the quantum–mechanical uncertainty relations (2.3.44, 2.3.45) will be given in Sect. 8.1.2.

2.3.4 Fast Fourier Transformation

Logical gates and circuits composed of gates represent examples of organized information flow and processing. They receive a binary input and generate from it, in a controlled and deterministic manner, an output that depends exclusively on the input and the structure and elements of the circuit. As such, they can also be considered as rudimentary instances of *algorithms*, see Sects. 9.3.1 and 10.3. Far more advanced, concerning the complexity of the data and their processing, are the standard algorithms of numerical mathematics. An illustrative example, still with a transparent structure, is *Fast Fourier Transformation* (FFT) [SB10], to be outlined in this subsection.

The only version of the Fourier transformation which consistently involves only finite amounts of information and therefore can be implemented numerically as it stands, is the discrete Fourier transform, Eqs. (2.3.24) and (2.3.25),

$$f_k = \frac{1}{\sqrt{N}} \sum_{l=0}^{N-1} \tilde{f}_l \exp\left(2\pi i \frac{kl}{N}\right), \quad \tilde{f}_l = \frac{1}{\sqrt{N}} \sum_{k=0}^{N-1} f_k \exp\left(-2\pi i \frac{kl}{N}\right) \qquad (2.3.54)$$

repeated here in a form symmetrized with respect to the prefactors $1/\sqrt{N}$. These two transformations can be considered as matrix multiplications,

$$f = F_N \tilde{f}, \quad (F_N)_{ik} = \frac{1}{\sqrt{N}} \exp\left(2\pi i \frac{kl}{N}\right), \quad \tilde{f} = \tilde{F}_N f,$$

$$\left(\tilde{F}_N\right)_{kl} = \frac{1}{\sqrt{N}} \exp\left(-2\pi i \frac{kl}{N}\right). \qquad (2.3.55)$$

The matrices F_N and $\tilde{F}_N$ are unitary and inverses of one another. Implemented directly as matrix multiplications, these operations require a number of floating-point operations (FLOPs) of the order of N for each of the N vector components, resulting in a total of

$$N_{\text{FLOP}} = O(N^2) \qquad (2.3.56)$$

FLOPs per transformation.

The simple form of the matrix elements (2.3.54) already suggests that there may exist possibilities for shortcuts. A first step towards exploiting this potential is a recursive construction of the matrices F_N and $\tilde{F}_N$. Here and in what follows, it proves advantageous to restrict the total size N to powers of 2, $N = 2^n$. Generalizations to arbitrary values of N are straightforward but technically cumbersome. For the case $N = 2$, define the seed

$$F_2 = \frac{1}{\sqrt{2}} \begin{pmatrix} 1 & 1 \\ 1 & -1 \end{pmatrix}. \tag{2.3.57}$$

To go to $N = 4$, introduce in addition the auxiliary matrices

$$F_4^{(1)} = \frac{1}{\sqrt{2}} \begin{pmatrix} F_2 & 0 \\ 0 & F_2 \end{pmatrix}, \quad F_4^{(2)} = \begin{pmatrix} I_2 & I_2 \\ I_2' & -I_2' \end{pmatrix}, \tag{2.3.58}$$

where I_2 is the 2×2 unit matrix and $\left(I_2'\right)_{nm} = \exp(2\pi i n/N)\delta_{n-m}$. They are unitary as well. F_4 is then construed from their product

$$F_4' = F_4^{(1)} F_4^{(2)} = \frac{1}{2} \begin{pmatrix} 1 & 1 & 1 & 1 \\ 1 & -1 & 1 & -1 \\ 1 & i & -1 & -i \\ 1 & -i & -1 & i \end{pmatrix} \tag{2.3.59}$$

by permuting rows and columns,

$$F_4 = J_4 F_4' = \begin{pmatrix} 1 & 0 & 0 & 0 \\ 0 & 0 & 1 & 0 \\ 0 & 1 & 0 & 0 \\ 0 & 0 & 0 & 1 \end{pmatrix} F_4' = \frac{1}{2} \begin{pmatrix} 1 & 1 & 1 & 1 \\ 1 & i & -1 & -i \\ 1 & -1 & 1 & -1 \\ 1 & -i & -1 & i \end{pmatrix}. \tag{2.3.60}$$

The matrices for $N = 8, 16, 32$ are obtained by the same procedure. Based on this scheme, a detailed algorithm for the efficient calculation of the coefficients

$$\tilde{f}_i = \frac{1}{\sqrt{N}} \sum_{k=0}^{N-1} f_k \exp\left(2\pi i \frac{kl}{N}\right), \quad l = 0, \ldots, N-1 \tag{2.3.61}$$

has been worked out, known as the *Sande-Tukey algorithm* [SB10]:
   Introduce the phase factors

$$\varepsilon_n := \exp\left(-\frac{2\pi i}{N}\right) = \exp\left(-\frac{2\pi i}{2^n}\right) \tag{2.3.62}$$

with the properties

$$\varepsilon_n^2 = \varepsilon_{n-1}, \quad \varepsilon_n^{2^{n-1}} = \varepsilon_n^{N/2} = -1, \quad \varepsilon_n^N = \exp(-2\pi i) = 1. \tag{2.3.63}$$

Coefficients $\tilde{f}_1$ with even vs. odd indices, $l = 2j$ or $l = 2j + 1$, resp., can thus be written as

$$\sqrt{N} f_{2j} = \sum_{k=0}^{N-1} f_k \varepsilon_n^{2jk} = \sum_{k=0}^{N/2-1} (f_k + f_{k+N/2}) \varepsilon_{n-1}^{jk} = \sum_{k=0}^{N/2-1} f_k' \varepsilon_{n-1}^{jk},$$

$$\sqrt{N} \tilde{f}_{2j+1} = \sum_{k=0}^{N-1} f_k \varepsilon_n^{(2j+1)jk} = \sum_{k=0}^{N/2-1} \left((f_k - f_{k+N/2}) \varepsilon_n^k \right) \varepsilon_{n-1}^{jk} = \sum_{k=0}^{N/2-1} f_k^* \varepsilon_{n-1}^{jk},$$

$$(2.3.64)$$

with new time-domain coefficients

$$f_k' := f_k + f_{k+N/2}, \quad f_k^N := (f_k - f_{k+N/2}) \varepsilon_n^k. \tag{2.3.65}$$

This procedure reduces the number of terms in the Fourier sum by $1/2$, on the expense of one more addition or subtraction per term. Applying it once, one is left with essentially the same task, to be applied to two vectors of dimension $N/2$, so that it suggests itself repeating it $n = \mathrm{lb}(N)$ times till only a single term is left (Fig. 2.9a). Counting the ordinal numbers of these steps backwards, from $m = n = \mathrm{lb}(N)$ (initial) to $m = 0$ (final), the procedure at stage m takes the general form

$$\sqrt{N} f_{j+j} = \sum_{k=0}^{M-1} f_{j,k}^{(m)} \varepsilon_m^{kl}, j = 0, \ldots, J-1, I = 0, \ldots, M-1, m = n, \ldots, 0,$$

$$(2.3.66)$$

with $M = 2^m$ and $J = 2^{n-m} = N/M$. In particular, the initialization reduces to $\sqrt{N} \tilde{f}_j = f_{j,0}^{(0)}, j = 0, \ldots, N-1$. We define the intermediate coefficients recursively as

initialization $m = n$	$f_{0,l}^{(n)} \leftarrow f_l$	$l = 0, \ldots, N-1$
subsequent steps	$f_{jJ}^{(m-1)} \leftarrow f_{j,1}^{(m)} + f_{j,j+m/2}^{(m)} = f_{j,l}^{\prime(m)}$	$m = n, \ldots, 1$
$m \to m-1$	$f_{j+J,}^{(m-1)} \leftarrow \left(f_{j,l}^{(m)} - f_{j,l+m/2}^{(m)} \right) \varepsilon_m^j = f_{j,l}^{(m)}$	$\begin{matrix} j = 0, \ldots, J-1 \\ l = 0, \ldots, M/2 - 1 \end{matrix}$

$$(2.3.67)$$

At each stage, the algorithm requires memory space for 2^{n-m} vectors of 2^m components each, or a constant number of $N = 2^n$ complex coefficients, reflecting the conservation of information under Fourier transform. Within this space, each step $m \to m - 1$ replaces coefficients according to

$$f_{j,l}^{(m)} \to f_{j,l}^{(m-1)}, f_{j,l+M/2}^{(m)} \to f_{j+J,l}^{(m-1)}. \tag{2.3.68}$$

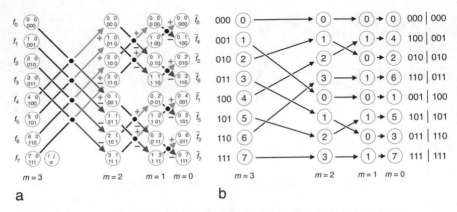

Fig. 2.9 Fast Fourier Transform (FFT) replaces multiplication with an $N \times N$ matrix by a sequence of Ib(N) steps (**a**, for $N = 8$), each one involving N operations as in Eq. (2.3.67) (green and red arrows, resp.). The memory allocation scheme (**b**) implies that, upon exit, the index l of the Fourier coefficients is obtained as the memory position, with binary digits in inverse order.

The memory allocation is appropriately organized as follows (Fig. 2.9b): Store the coefficient $f_{j,l}^{(m)}$ at stage m at the element number $s(m, j, l)$ of the total array. The operation (2.3.66) then requires defining

$$
\begin{array}{lll}
\text{initialization} & s(n, 0, l) := l & l = 0, \ldots, N - 1 \\
m = n & & \\
\text{subsequent} & s(m - 1, j, l) := s(m, j, l) & m = n, \ldots, 1 \\
\text{steps} & & \\
m \to m - 1 & s(m - 1, j + J, l) := s(m, j, l - M/2) & \begin{array}{l} j = 0, \ldots, J - 1 \\ l = 0, \ldots, M/2 - 1 \end{array}
\end{array}
$$

$$(2.3.69)$$

It is useful to write the indices j and l in binary notation. Their ranges $J = 2^{n-m}$ and $M = 2^m$, resp., have a constant product $JM = N = 2^n$, so that the total number of binary digits $\alpha_k \in \{0, 1\}$, $-k = 0, \ldots, n - 1$, is also constant, n. Dedicate the first $n - m$ of them to j and the remaining m to l,

$$
\begin{aligned}
j &= \alpha_{n-1}2^0 + \cdots + \alpha_m 2^{n-m-1}, \\
l &= \alpha_0 2^0 + \cdots + \alpha_{m-1}2^{m-1}.
\end{aligned}
$$

$$(2.3.70)$$

Arranging the digits for j in inverse order, as compared to those for l, is equivalent to the reshuffling of matrix elements by the matrices J_N, see Eq. (2.3.60). This rule allows calculating the addresses $s(m, j, l)$ explicitly,

$$
\begin{aligned}
&s\left(m, \alpha_{n-1}2^0 + \cdots + \alpha_m 2^{n-m-1}, \alpha_0 2^0 + \cdots + \alpha_{m-1}2^{m-1}\right) \\
&= \alpha_0 2^0 + \cdots + \alpha_{n-1}2^{n-1}.
\end{aligned}
$$

$$(2.3.71)$$

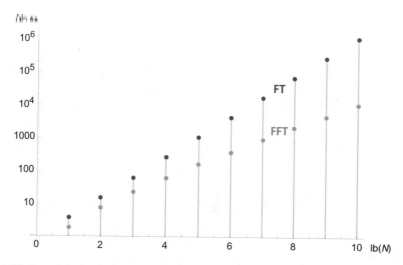

Fig. 2.10 A straight-forward implementation of the Fourier transform, for a data set of size N, requires of the order of N^2 floating point operations (pink). The Fast Fourier Transform, implemented as the Sande-Tukey algorithm, needs only of the order of $N \operatorname{lb}(N)$ of them (green).

Upon exit, the Fourier coefficients are found at the memory loci $\tilde{f}(s(0, j, 0))$ with

$$s\left(0, \alpha_{n-1}2^0 + \cdots + \alpha_0 2^{n-1}, 0\right) = \alpha_0 2^0 + \cdots + \alpha_{n-1}2^{n-1}, \qquad (2.3.72)$$

i.e., the binary number j, "read backwards", coincides with the index l of $\tilde{f}_l$ (Fig. 2.9b).

For $N = 2^n$, n steps of type (2.3.66, 2.3.67) are necessary to complete the task, each of them comprising N floating-point operations as in Eq. (2.3.64). The total number of FLOPs per Fourier transformation thus reduces to

$$N_{\text{FLOP}} = O\left(N \ln(N)\right). \qquad (2.3.73)$$

compared to $N_{\text{FT}} = O(N^2)$ for a straightforward implementation.

For the large data sets processed in professional applications, this is an enormous gain, see Fig. 2.10, which saves huge amounts of computer capacity, every day all over the world. In a more recent development, the Sande-Tukey algorithm served as a model for the construction of a quantum Fourier transformation. Combining the same sequence of $\operatorname{lb}(N)$ major steps with quantum parallelism, it achieves a further reduction to $N_{\text{QMOP}} = O\left[(\operatorname{lb}(N))^2\right]$ qubit operations, see Sect. 10.3.4.

Chapter 3
Epistemological Aspects

In previous sections, information has been treated as a static phenomenon. Time, as a parameter that counts and organizes changes, appeared there in a rudimentary form and indirectly at most, behind transformations of states, reflected in their information content. In the sequel, it becomes the central aspect, opening the way to information dynamics proper and allowing us to introduce concepts such as information flows. Conversely, information turns out to be a crucial element in our understanding of time. Observing our subjective time perception, we find that it depends strongly on "how much happens": Uneventful periods, such as standing in a queue, seem to last endlessly, but in hindsight shrink to almost nothing, while we even forget how time moves on if we are bombarded with new experiences, in particular if they come unexpected and surprise. This applies both to personal and to historical time. As Richard Feynman put it in his "Lectures on Physics" [Fey63], "Time is what happens when nothing else happens".

Explanation as well as prediction are particular forms of how we perceive and interpret ordered sequences of events. They deal with the flow of information, its density and direction, they interpolate and extrapolate observations. But they go much further in that they provide an extraordinarily effective way of structuring observed data. Extracting regularities, patterns and laws, explanation and prediction require even more sophisticated analytical tools than, for example, correlations and spectral analysis. From sensory perception to scientific theory, they represent instances of high-end information processing. This chapter and Chap. 4 are dedicated to the tight links that exist between information and epistemology.

Information is immediately reflected in particular to the direction of time [Rei56]. We know the past, future is open. The "time arrow", in turn, is intimately related to the notion of causality, it separates cause and effect [Rei56, Skl93]. Causal explanations respond to "why" questions in the widest sense. The present section discusses causality in terms of information flows as a conceptual analysis, leaving intricacies, implied for example by quantum mechanics, for later (Chap. 8). Causal chains are defined as ordered and directed networks of events, the immediate relation of causality to signal transmission, in particular in Special Relativity, is pointed out.

© Springer Nature Switzerland AG 2022
T. Dittrich, *Information Dynamics*, The Frontiers Collection,
https://doi.org/10.1007/978-3-030-96745-1_3

It demonstrates that finality, answering "what for" questions and often considered as antipode to causality, nevertheless is closely related and can be reduced to it. A more thorough scrutiny of causality and determinism in the framework of classical mechanics will be provided in Chaps. 5, 6 and 7.

Prediction is more than explanation turned into the future. The deep questions that arise from basic limits of information content, notably if predictions are made "from within the system", will be considered in Sect. 3.2. Here, in contrast to explanation, the forecast comes into focus as a causal agent of its own right. The ability to anticipate, based on experience, is constitutive for learning and adaptation. The role acquired knowledge and anticipation play for living systems, as a decisive factor in their survival and evolution, will be sketched in Sect. 3.3.

An indispensable condition for sophisticated prediction is the capacity to discern complex patterns in images and other percepts. It goes far beyond merely detecting correlations and results in a highly efficient compression of data. What prediction based on theory and laws achieves is abstracting universal regular patterns from contingency. In this way, it pushes the frontier of foreseeability further ahead into the realm of randomness. Chapter 4 focuses on the far side of this border and illuminates chance as incompressibility of data. It culminates interpreting Gödel's incompleteness theorem as a fundamental principle referring to finite systems that emulate other finite systems, thus to the limits of information compressibility.

Perception, induction, prediction, cognition occur in individuals embedded in a society. They are social phenomena. The construction of truth is a collective task. It cannot be appreciated taking onto account only the information arriving at an individual observer. The echo its perceptions and conclusions have in fellow subjects is crucial. Sections 3.3.2 and 3.3.3 address the information exchange among subjects and the effect the response of the society may have in terms of concrete physical parameters, for example in the context of climate change.

3.1 Causality

In our subjective imagination, time is related to causality as a stream of events, which forms ordered chains connecting the past with the future. They can run in parallel, cross, merge, and split (Fig. 3.1). Behind this perception of order is an intuitive notion of hierarchy: We consider the former as "responsible" for the latter, in a vague sense: The past is determined, the future is open and amenable to change. Questions related to causality range from personal life ("If I suffer a car accident during a solar eclipse, can I blame the eclipse for my bad luck?") through controversial technical questions of political relevance ("Is glyphosate carcinogenic?", "Does CO_2 emission cause global warming?) to profound problems of social and other sciences ("Do major religious transformations pave the way for economic changes, or does economic progress induce confessional reorientation?"). The following subsections suggest a systematic approach, based on information, towards these intuitive ideas.

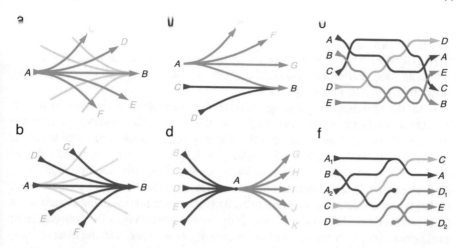

Fig. 3.1 A causal relation from cause A to effect B (**a**) implies that A be a *sufficient* condition for B to occur. However, generally A will not be a *necessary* condition for B, i.e., there may be other events in addition to A that can cause B, too. At the same time, A can cause side effects $C, D, E, F,...$, besides B. Conversely, if A is only a necessary condition for B (**b**), there may be additional conditions $C, D, E, F,...$, besides A, that must be satisfied to make B happen. B is then a *multicausal event*. Divergent arrows in a, convergent arrows in b can be interpreted as loss or gain, respectively, of actual information (gain or loss of potential information) on the initial state A. For an analogous graphical representation, but in terms of potential information, see Fig. 3.8. In general, a correlated pair of events cannot be associated uniquely to one of these categories but combines features of both (**c**). Generally, from on a given event A (**d**), relations to alternative causes fan out into the past, relations to possible effects branch out into the future (see the time structure in Special Relativity, Fig. 3.11b). Determinism, by contrast, implies strict one-to-one relations between events at different time sections (**e**). In particular, it excludes (**f**): branching (as in $D \rightarrow D_1, D_2$)), merging (as in $A_1, A_2 \rightarrow A$), disappearing (as in B), or emerging (as in E) world lines or trajectories.

Philosophers have reflected on the concept of causality at least since antiquity. Ideas that never lost their relevance have been proposed in particular by protagonists of enlightenment, such as David Hume [Hum90] and Immanuel Kant [Kan04, Kan98]. Notably Kant considered the abstract concept of causality (not specific cause-effect relations) as "synthetic a priori knowledge", a necessary condition for scientific thought that has to be presupposed in order to extract ordered sequences of events from the bulk of perceptions.

In physics, as the principal source of experience on how to make use of this concept, it permanently lurks in the background as a general guideline, a criterion for scientific method, but is only exceptionally referred to explicitly. For instance, anticausal (or advanced) solutions of Maxwell's equations are dismissed as unphysical, as soon as many-body phenomena such as dissipation come into play [Jac75], and the apparent non-causal character of variational principles, such as Hamilton's principle, as they are invoked in classical mechanics [GPS01, JS98], arouses concern. By contrast, in other disciplines, in particular more applied sciences that have to make sense of large data sets, identifying causal relations can be a serious challenge of everyday research, for example in geophysics [SM09, KCG11], neuroscience

[DRD08, SBB15], epidemiology [RGL08], and economy [Gra69]. Again, information proves pivotal. Using this concept to cast causality in sufficiently precise terms, so as to apply it to quantitative data, is an active research area at the crossroads of natural and social sciences, mathematics, cognitive science, and philosophy [Sch00, KS04, PV07, AP08, Pea09, HP11].

Causality must be clearly distinguished from a concept that is closely related to it but differs in subtle nuances: determinism. In the philosophical debate, determinism does not refer to the directly observable, that may or may not be amenable to causal explanation. It rather alludes to an underlying level of rigorous lawfulness that is not necessarily accessible to observation or insight. Deterministic laws exclude in particular a merely statistical account; they form the converse of probabilistic rules. At the same time, determinism is not related to time order. Instead, it implies events at different times to be linked one-to-one along endless fibres (Fig. 3.1e): trajectories or world lines running in parallel without branching or merging, without disappearing into or emerging from nowhere (Fig. 3.1f). Time-reversal invariant evolution equations, such as in classical and quantum mechanics, are paradigms of deterministic laws. As a consequence, determinism often coincides with but is neither sufficient nor necessary for causality. A patent counterexample is deterministic chaos (see Sect. 5.3.3). It reveals practically unpredictable, thus acausal, processes as solutions of deterministic evolution equations.

An elementary benefit of invoking causality to organize sequences of events is that it induces a directed linear order. "Linear" is here used not in its full geometrical

Fig. 3.2 Time travel between two sections of time centred at t_1 and t_2, resp., creates a shortcut, allowing for causal relations in both directions (**a**). In effect, it forces the otherwise linear time order into a closed loop (**b**), a *closed time-like curve* (CTC). Within this loop, any instant can be causally related to any other, time order is suspended, and causes and effects become interchangeable (**c**). Time travel is therefore incompatible with a global linear causal order.

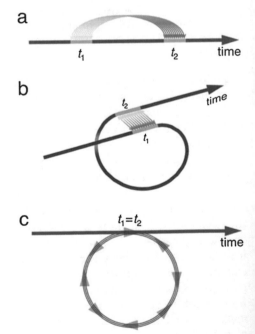

ᴵ ⁽⁾ⁱⁱⁱⁱⁱⁱⁱⁱ⁰ⁱⁱⁱ ⁱⁱⁱⁱ ⁱⁱ ⁰ᵘᶜᵉⁱⁱ ⁱⁱⁱ ⁱⁱ⁰ᵖᴬˡᵒᵍʸ. Given two events A and B, B either precedes or follows A, and a third event C, if it does not come before or after the other two, must be in between [Rei56]. Linear order excludes, specifically, loops in time, "time travel" or *closed time-like curves* (CTCs) (Fig. 3.2), that have inspired so many adventures in science fiction.

To contrive a fictitious counterexample: Had a present-day radio station conducted an interview with Geheimrat Johann Wolfgang von Goethe, say on the occasion of his 50th birthday in his own time, on August 28, 1799, commemorating his 270th anniversary in 2019, the poet would have left at least some notice in his diary on this remarkable experience, and this notice would still be known in 2019. It could thus be referred to in retrospect in this same interview, which in turn would have found its way into Goethe's notes, etc. Moreover, via this same communication with a twenty-first century journalist, as highly interested in science as Goethe was, he might have learned about major advances achieved during the nineteenth and twentieth century, such as the theory of evolution, relativity theory, and quantum mechanics. Through Goethe's notes, in turn, Darwin, Einstein, and other protagonists of modern science would have known about their respective discoveries, even long before they actually occurred to them, casting a new light on these inventions. The implications of this causal bypass are even more drastic: Goethe's record of the interview, known through his notes to the interviewers, could have been addressed by them during the conversation, thus entering the same record in turn, ... A regression results that forces time to run in circles of a period given by the time span bridged, in the present example 220 years. The only possibility to render this circle self-consistent is to assume exact periodicity: Time travel rolls up time in a coil, it inhibits any further progress of time. With the immediate protagonist, Geheimrat Goethe, and his contemporaries, history as a whole is caught in a closed loop. A single time loop cuts a ladder into the fabric of time. It is incompatible with causality: The ordered flow of information renders time travel unconceivable.

A second feature constitutive for cause-effect relations, which can be a major challenge when identifying and verifying them, is their directedness. Order is not enough; an independent criterion is required to distinguish cause from effect, past from future [Rei56]. A causal chain is a linear sequence of events where in addition, its two ends are uniquely distinguishable. In more formal terms, a relation that is reflexive, antisymmetric, and transitive, such as "$\leq$" between numbers, generates an order. However, the opposite relation "$\geq$" generates the same sequence—just backwards. A complementary independent criterion, say, to tell positive from negative numbers could be that for the former, $\mathrm{sgn}(a^2) = \mathrm{sgn}(a)$ while for the latter, $\mathrm{sgn}(a^2) \neq \mathrm{sgn}(a)$.

Causal chains share this and other features with inference chains, see Sect. 2.1.4. However, causal explanation is to be clearly distinguished from logical tautology. For example, a syllogism such as "All surgeons are doctors". "Why is Alex a doctor?"— "Because he is a surgeon." does not qualify as an explanation. It gains its certainty from its independence of material facts. By contrast, causal explanation must be based on observations in the real world. Nonetheless, an instructive first approach to cause-effect relations can be found in the concepts of necessary vs. sufficient conditions, introduced in the context of logical inference. The interpretation that approximates most closely our intuitive understanding of causality is [Wri71],

Causes are sufficient conditions for their effects to occur, effects are necessary conditions for their causes.

More precisely, the cause (the *explanans*) must comprise at least all conditions that are required to make the effect (the *explanandum*) happen. As a consequence, the cause may comprise a bundle of more restrictive factors, each of them a necessary condition, and even include other factors that have only minor impact or are even unessential. At the same time, there may exist other, alternative, causes, which are likewise sufficient for the same effect. In terms of propositional logic, several necessary conditions A_1, A_2, ... combine by conjunction, $(A_1 \cap A_2 \cap \cdots) = A \rightarrow E$, alternative causes A, B, ... by disjunction, $A \cup B \cup \cdots \rightarrow E$. Conversely, a cause can lead to simultaneous, secondary effects E_1, E_2, ... ("side effects" in medicine, "collateral damage" in military jargon), which combine as a logical conjunction, $A \rightarrow (E_1 \cap E_2 \cap \cdots)$, (Fig. 3.1a).

The alternative interpretation, the cause as a necessary condition for the effects, is more consistent with the intuitive idea of a partial cause. To achieve the effect, a partial cause A_1 has to coincide with additional factors $A_2, A_3, \ldots$, so that together they form a sufficient cause $A = (A_1 \cap A_2 \cap A_3 \cap \cdots)$, as indicated above. Effects depending on several necessary conditions are considered as multicausal events (Fig. 3.1b). In logical terms, they become sufficient conditions for all their partial causes.

Summarizing these logical properties, we may say that the cause must be more specific than the effect it explains. This suggests to carry the analogy with logical inference even further, to interpret causation in terms of a directed information flow, see Sect. 2.1.4,

potential information (entropy) remains constant or increases from cause to effect,

or equivalently,

actual information (negentropy) on the initial state remains constant or is lost from cause to effect.

Defining information in terms of probability distributions as in Eq. (1.3.9), this means that from cause to effect, probabilities approach an equidistribution, variations get evened out. It is consistent with interpreting cause-effect relations as transmitting signals. They operate as a direction-sensitive elements, preferring transmission in one direction while preventing or at least suppressing it in the opposite direction, analogous to a check valve in a piping system or a diode in an electronic circuit:

Signals are transmitted preferentially from cause to effect but suppressed from effect to cause.

Working out details in the sequel, this principle will prove valuable as a guideline, but requires further specification and refinement.

Turning to classical Hamiltonian mechanics as a physical model, phase-space points as elementary events are connected uniquely by trajectories, one-dimensional manifolds that imprint a linear order to the time evolution (Fig. 3.2e, cf. Chap. 5). However, if the underlying equations of motion are symmetric with respect to time reversal, they cannot define a time arrow. Therefore, cause and effect are indistinguishable; in deterministic, time-reversal invariant systems such as in Hamiltonian mechanics, the concept of causality loses its meaning (which however does not prevent assigning the roles of cause and effect temporarily to parts of a Hamiltonian system, see the case of coupled harmonic oscillators, Sect. 5.6.1).

This symmetry, inherent in the fundamental evolution equations of classical as well as of quantum mechanics, is broken only when going to macroscopic equations by some kind of coarse-graining process that distinguishes small from large spatial scales, as in statistical mechanics (see Sect. 7.3.2 for details). This imparts a certain degree of subjectivity to the time arrow, shared consequently by the concept of causality—it is perceivable only from a macroscopic viewpoint. Could intelligence be based exclusively on reversible microscopic processes, it would develop only a rudimentary concept of time, if any, not based on any asymmetry between the known on one side of the present and the unknown on the other, and notably lacking causality.

Often enough, a causal relation has to be attributed to (pairs of) individual unique events, for example in justice and in medicine. So-called token-level causality is however difficult to verify and only exceptionally allows for a quantitative account [Eel91]. In those sciences where events are represented by statistical time series, conditions are much more propitious for a quantitative description. Based on statistical data, the interdependence of events can be measured probabilistically as correlations, or more generally, as mutual information (defined in Eq. (1.7.10)).

Correlations between events A and B, however, do not imply causation. They can be explained in three different, not mutually exclusive ways: (i) A causes B, (ii) B causes A, or (iii), they are both effects of a common cause C. In order to distinguish them, an asymmetry between A and B is required, separating cause from effect. Interpreting causality as a directed transfer of information from A to B indicates a more systematic approach, starting with series of discrete events organized in networks.

3.1.1 Causality from Topology: Markov Chains and Bayesian Networks

The most elementary framework to discuss sequences of events is that of Markov processes [KS04, Pea09], as a part of dynamical systems theory, dedicated to the probabilistic evolution of discrete states in discrete time. It includes in particular systems where no deterministic evolution laws are known, but only statistical relations between its possible states. In that case, the evolution laws can be expressed as transition probabilities $p_{m \to n}$ from state m to state n in a Markov chain (Fig. 3.3a) or, equivalently, organized in a matrix [Llo91, Pea09] (Fig. 3.3b).

In order to qualify for a Markov process, a system has to satisfy an essential condition: The transition to the next (future) state must only depend on the present, but not on previous (past) states. It may then be called memory-free or *Markovian* [Pea09]. In fact, this condition can be weakened to include also a limited number of past states, representing an extended present [KS04] of finite duration: If a sufficient number, say k, of subsequent states, the memory, are lumped into such a "present", forming a k-dimensional *delay vector*, it is again the present, and only the present, that determines the future, in a *Markov process* or *Markov chain of order k* [Sch00,

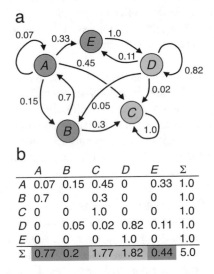

a

b

	A	B	C	D	E	Σ
A	0.07	0.15	0.45	0	0.33	1.0
B	0.7	0	0.3	0	0	1.0
C	0	0	1.0	0	0	1.0
D	0	0.05	0.02	0.82	0.11	1.0
E	0	0	0	1.0	0	1.0
Σ	0.77	0.2	1.77	1.82	0.44	5.0

Fig. 3.3 A Markov chain formalizes the evolution of a discrete dynamical system in discrete time, following probabilistic evolution laws. It can be represented graphically as a network of system states, in this example A, B, C, D, E, connected by links weighted by transition probabilities (**a**) or algebraically as a matrix (**b**). In the matrix representation, the normalization of all exit probabilities per state is equivalent to a sum rule for the rows of the matrix, $\Sigma_n p_{m \to n} = 1$. A similar sum rule for columns does not exist, but the sum of each column indicates whether the corresponding state is stable, hence an attractor ($\Sigma_m p_{m \to n} > 1$, highlighted green) or unstable, hence a repeller ($\Sigma_m p_{m \to n} < 1$, pink). Loops from a node back to itself indicate that the evolution may stall at that state with the given probability. Deterministic evolution laws are reached if the probabilities reduce to the alternative 0 or 1.

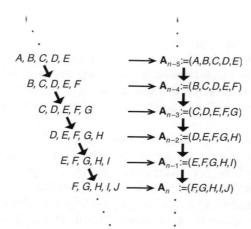

Fig. 3.4 On the face of it, a dynamical system where the next state depends not only on the present, but also on k (here, 5) past states, is not a Markov process. It can be rendered Markovian, though, by merging sequences of k sequential states each into vectors (words) of dimension k. If each of them is considered a single state, the future, here state $\mathbf{A}_n$, is again determined by the previous state $\mathbf{A}_{n-1}$ alone.

KS04, PV07, HP11] (Fig. 3.4).

$$\mathbf{A}_n := (A_n, A_{n-1}, \ldots, A_{n-k+1}). \tag{3.1.1}$$

This strategy is closely related to finding an embedding dimension that allows identifying a dynamical system as deterministic [KS04].

For Markov processes of order k, an entropy rate can be defined as the additional or conditional information contained in state A_{n+1}, if the k preceding states $\mathbf{A}_n$ are known, generalizing Eq. (1.7.12) [Sch00],

$$h_{n+1}^{A,k} := I^{\mathbf{A}_{n+1}} - I^{\mathbf{A}_n} = -c \sum_{j_{n+1}, j_n, \ldots, j_{n-k+1}} p\big(A_{n+1, j_{n+1}}, \mathbf{A}_{n, \mathbf{j}_n}\big) \ln\big(p\big(A_{n+1, j_{n+1}} \big| \mathbf{A}_{n, \mathbf{j}_n}\big)\big),$$

$$\tag{3.1.2}$$

(indices j_n in $A_{n+1, j_{n+1}}$ and the index vector $\mathbf{j}_n := (j_n, \ldots, j_{n-k+1})$ in $\mathbf{A}_{n, \mathbf{j}_n}$ refer to different realizations of the same processes A_n and $\mathbf{A}_n$, resp.). The argument of the logarithm is the conditional probability $p\big(A_{n+1, j_{n+1}} \big| \mathbf{A}_{n, \mathbf{j}_n}\big) = p\big(A_{n+1, j_{n+1}}\big) / p\big(\mathbf{A}_{n, \mathbf{j}_n}\big)$. The entropy rate is a first step towards measuring the transmission of information along a sequence of events such as a causal chain. It already introduces an implicit direction of (discrete) time in that it considers the state A_{n+1} as successor, not as precursor, of the ordered sequence of states $\mathbf{A}_n$. This theoretical reference direction will later be used to define a direction of information flow in causal processes, which in turn gives rise to an empirical time arrow.

Markov processes are, however, lacking a crucial ingredient: They do not allow distinguishing events at different sites or in different subsystems. It could be incorporated implicitly by giving the states additional depth so as to include the spatial position where it takes place, similar to the strategy implemented for time to define processes of order k. However, that would conceal the spatial structure in components of a state vector and not enable defining information currents between states (Fig. 3.5).

In many contexts, information is considered to flow along discrete channels (Fig. 3.5): Telephone signals are transmitted along telephone lines, voltage modulations move along cables, mechanical pulses transmit through Bowden cables, pressure pulses propagate through hydraulic tubes: Signal transmission epitomizes the essence of causal chains (Fig. 3.6). Generally, the triplet "sender, message, receiver", so fundamental in communication theory [Flo10], implicitly refers to a discrete channel. In fact, channels are an idealization of the more general case of continuous information flows in space, to be described as a continuous velocity field. We shall assume this more exact point of view in Chaps. 5, 6 and 7, where information currents in physical, specifically mechanical, systems will be considered, focussing on concepts such as vorticity, convergence vs. divergence, sinks vs. sources of the flow.

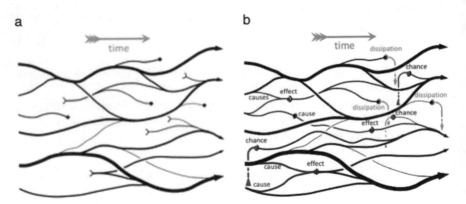

Fig. 3.5 A naïve rendering of a network of causal chains (**a**) consists of a set of "world lines" (see Sect. 3.1.4), each of them representing a continuous sequence of events, with a common direction defined by a global parameter "time". Some lines terminate in blind ends (bullets), corresponding to events without consequences, some enter apparently from nowhere (semicircles), symbolizing effects lacking a discernible cause or simply "chance". Interpreting world lines as information channels, their thickness symbolizing the strength of the flow, and complementing them with fictitious additional currents (**b**), allows to visualize elementary concepts related to information flow, such as causal explanations (blue lines), including multicausal events with two or more causes, randomness as originating in invisibly small scales and microscopic degrees of freedom (red bullets), see Chaps. 4 and 7, and dissipation as information loss into invisible scales (green semicircles), see Chap. 6. This construction requires modification and refinement, as by the more sophisticated concept of Bayesian networks, see Fig. 3.7.

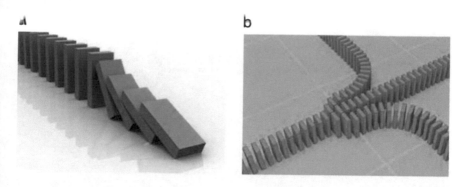

Fig. 3.6 Chains of toppling dominos are perfect exemplifications of causal chains as transmission lines, sending a signal from cause (trailing end of the domino chain) to effect (leading end). They allow to visualize many basic features of causality, such as the inherent time arrow (**a**), side effects as ramifications (**b**), and various others.

The concentration of continuous flow fields into approximately discrete channels is not only conceptually convenient, it frequently occurs in nature. A patent example are rivers, where geological and hydrological forces produce an instability that contracts the backflow of liquid atmospheric water to the oceans into narrow riverbeds, converging further from numerous small rivulets to a few mighty streams. Similarly, the physiology of vertebrates organizes blood circulation as an arterial tree, reaching from main arteries down to capillaries. If the flow of information is discretized not only laterally, but also in the longitudinal direction, we arrive at ordered chains of sequential events, at causal chains. Causal chains and information channels are nearly synonymous, the latter pointing directly to a quantitative account.

In close analogy to the hydrological model, the concept of *Bayesian networks* [Llo91, AP08, Pea09] has been conceived as a graphical criterion for causality. The appropriate general mathematical notion to define Bayesian networks is indeed graphs: A graph consists of a set of vertices (nodes), connected by edges (links), see Fig. 3.7a. Like this, it applies to a huge variety of situations, from quantum mechanics to the Internet. If we assign a direction to every link, defining the two nodes it connects as an ordered pair, we obtain a *directed graph* (Fig. 3.7b).

However, even if the network attaches an arrow to each individual edge, the total graph need not have a global directionality. To achieve it, we have to further exclude all cycles or loops—"time travel", see above—that is, sequences $V_n \rightarrow V_{n+1} \rightarrow V_{n+2} \rightarrow \ldots \rightarrow V_{n+m} \rightarrow V_n$ of vertices and directed links that lead back to the starting point. This restriction defines *directed acyclic graphs* (DAGs, Fig. 3.7c), the category equivalent to Bayesian networks if applied to events linked by transition probabilities. Note that this does not exclude feedback loops, as a central concept of cybernetics [Wie61]. Feedback is equivalent to loops between sites or subunits of a system separate in space, it does not imply loops in time, hence does not lead to a conflict with linear time order.

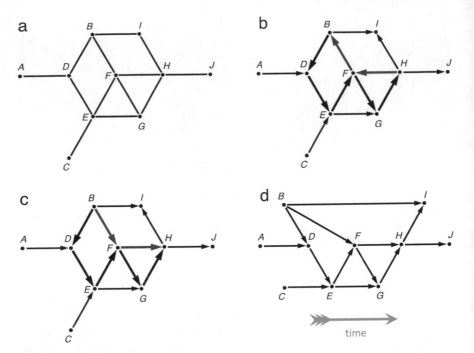

Fig. 3.7 A graph with 10 nodes and 13 links (**a**) becomes a directed graph by assigning directions to each link (**b**). It may contain closed cycles or loops, such as *BDEF* and *FGH* (bold arrows). Eliminating them by inverting the orientation of some arrows, here *FB* and *HF* (red), leads to a *directed acyclic graph* (DAG) or Bayesian network (**c**). It can be consistently ordered, thus defining an inherent global direction, as can be visualized by straightening out the network (**d**), emphasizing roots (nodes with only outbound links), here *A*, *B*, and *C*, and sinks (nodes with only inbound links), here *I* and *J*.

In order to anchor the construction of a Bayesian network in reality, it has to be based on data, appropriately on statistics. Returning to the question of distilling causality from correlations, should arrows be drawn from *A* to *B*, from *B* to *A*, or rather only through a third node *C*? At this point, it is tempting to fall back on an empirically observed temporal order, introducing time lags or phase lags, detected in the time or in the frequency domain, between correlated events at *A*, *B*, and *C* as a criterion for causality [Gra69]. We shall avoid this option, though: If the objective is to anchor the global time arrow in causal order in terms of information flow, referring to time in basic definitions would imply the risk of a circle [Rei56]. For an elementary instance of phase lags indicating cause-effect relations, though, see the example of a coupled pair of harmonic oscillators, Sect. 5.6.1.

If we want to track the ramifications of a real network, a blood circuit, the sewerage, or a stream system, we inject a tracer, for instance a dye or a radioactive substance, at a suitable source point and follow its course in the network. Applied to information flow in causality, the corresponding strategy is *screening* [Rei56, Llo91, AP08, Pea09]. Mere correlation between *A* and *B* allows to infer *A* from *B* as well as *B* from *A*.

a directed causal relation $A \rightarrow B$ would require that actively affecting (injecting information at) A entails changes of B but not vice versa. In order to scrutinize the possible role of a third event C, screening it here means blocking this source upstream in the Bayesian network by fixing its state, to observe which effects, if any, this has at A and B. This provides clear evidence as to their causal relation: If cutting off the input from C decorrelates A from B, there cannot be any direct causal relation between them. If C interrupts (blocks) the exchange of information between A and B, correlations can only be attributed to C as the common cause. Inversely, if C is a common effect of A and B, fixing its state will impose mutual correlations on the two causes. In this way, an order between the pair A and B and a third node or set of nodes, disjoint from A and B, can be unambiguously defined. These alternatives can be distinguished by analyzing the common probability distribution $p_{jkl...}^{ABC}$ at the nodes involved, as will be detailed below. A Bayesian network is then a directed acyclic graph with a topology of links consistent with the screening relations implied by $p_{jkl...}^{ABC}$. It defines a global direction in that it has at least one *root* (node with only outbound edges, A in Fig. 3.7d) and at least one *sink* (node with only inbound edges, J in Fig. 3.7d) and induces at least a partial ordering of the nodes $A, B, C, ...,$ as in the step from Fig. 3.7c, d.

However, constructing Bayesian networks from screening relations is not a complete solution, since this step is not unique and therefore does not fix causal relations based on information flows. Specifically, Bayesian nets leave it open how information incoming through two convergent links is combined at their common vertex. For information encoded in a binary sequence, the node could be represented by any one of the binary gates addressed in Sect. 2.1.1. For example, nodes in a Bayesian network could be interpreted either as OR or as AND gates in the sense that they transmit the information from one of the inputs identically to the output if the other input has the value 0 (OR) or 1 (AND). This means that a Bayesian network cannot distinguish between causes as sufficient and causes as necessary conditions for their effects.

3.1.2 Causality from Information Flow: Transfer Information

That causality is intimately related to the transfer of information from cause to effect is suggested by an intuitively evident analogy already alluded to: Causal chains can be interpreted as transmission lines in the widest sense (Fig. 3.6). The final effect is a message sent by the original cause. The analogy is instructive but requires refinement and clarification. Besides referring to the sheer quantity of information, conditions for causality have to include a certain continuity of the transfer of contents, which is the major challenge.

In order to achieve a more precise criterion for causation, it is necessary to depart from a view focussing on individual events and consider classes of events and their statistics, replacing binary logic with probabilities, determinism with probabilistic evolution [Eel91], as alluded to in the context of Markov chains. The idea addressed above, that cause-effect relations operate as one-way valves in the information flow, is most appropriately expressed in terms of conditional probabilities $p^{B|A}$, see Sect. 1.7. In essence, a causal relation from A to B requires that varying A as the condition should be reflected in a marked increase (or decrease) of the probability of B, (signal A to B goes through) but not vice versa (signal B to A is blocked). Instead of implying the effect B in a deterministic sense, the cause A now only raises (or lowers) the probability for B to occur, as compared to its average disregarding A, that is, $p^{B|A} \neq p^B$. Specifying this further, however, depends on the precise logical interpretation of causality. For example, considering the cause as a necessary condition for the effect, reading the truth table of the logical implication $A \subset B$ as a matrix of probabilities, results in attenuating the alternative $A = (false, true)$ to conditional probabilities $p^{B|A} = 0.5$ and $p^{B|\neg A} = 0$ ($A = f$ requires $B = f$, while $A = t$ leaves B undetermined). This is consistent with the more general condition [Eel91]

$$p^{B|A} > p^{B|\neg A}. \qquad (3.1.3)$$

("given A, it is more probable that B occurs than if not A is given").

If conversely the cause is interpreted as sufficient condition for the effect, the truth table of $A \supset B$ amounts to conditional probabilities $p^{B|A} = 1$ and $p^{B|\neg A} = 0.5$ ($A = t$ implies $B = t$, while $A = f$—*ex f also quodlibet*—leaves B undetermined). This is also consistent with Eq. (3.1.3) but is more appropriately expressed as

$$p^{B|A} > p^{\neg B|A} \qquad (3.1.4)$$

("given A, it is more probable that B occurs than that it does not occur"). Equation (3.1.3) appears as plausible as Eq. (3.1.4). It excludes the possibility though that B could not only be caused by A but quite as well in the absence of A by other, alternative causes, which is only consistent with the cause as sufficient condition for the effect, i.e., with Eq. (3.1.4).

Both forms, Eq. (3.1.3) as well as Eq. (3.1.4), can be interpreted in the sense that knowing whether A is the case or not allows to predict with more certainty whether B will come true. Compared to prediction unaware of A, it reduces the ignorance, hence the potential information, regarding B. However, since alternative causes would also increase p^B, knowing B provides less actual information on A than knowing that A itself was the case, which in turn is consistent with the condition, surmised above, that potential information should grow from cause to effect. The advantage of this viewpoint is that it treats cause and effect on equal footing, as is appropriate for consecutive links in a causal chain. Seen like this, detaching the previous link in the chain, A, by going from $p^{B|A}$ to p^B, replaces complete knowledge of A—yes–no alternative, $p^A = 1$ ($p^{\neg A} = 0$) vs. $p^A = 0$ ($p^{\neg A} = 1$), by partial knowledge $p^B \in [0, 1]$, hence increases the potential information.

To be acceptable as criteria for causal relations, however, Eqs. (3.1.3) and (3.1.4), are still too crude. Both refer to an event and to its negation, of the effect in the former, of the cause in the latter, which is hard to define in general terms and difficult to apply in practice. In particular, if the positive statement has the form of an existential statement ("there exists …"), its negation is a universal quantification ("for each …") [KB07]. As such, it cannot be rigorously verified in empirical science. Moreover, criteria based on probabilities of only the two events in question cannot comply with a basic requirement, discerning direct causation from correlations induced by a common cause. And they do not include negative causation: Anecdote has it that millers wake up in the night if the noise of their running mill stops, because that would indicate a disturbance in the operation or a damage of the mill: Also negative correlations or a reduction in the frequency of an effect can be a signal and thus manifest causality. Criteria based on information readily include them by construction, probability conditions such as Eqs. (3.1.3 and 3.1.4) would have to be modified.

Returning to the simple picture of a sender connected by a channel to a receiver, we would associate the sender with the cause, the receiver with the effect, and the channel with causation. Physical intuition then tells us that the actual information (see Sect. 1.4) transmitted by the sender will arrive at the receiver in general only partially, in ideal cases completely, but will never increase during transmission. As a first criterion for a causal relationship, we might therefore propose, as anticipated above, that entropy (potential information, see Sect. 1.4) should increase from cause A to effect B:

$$I^B \geq I^A. \tag{3.1.5}$$

Referring to the above relationship of causal chains to transmission lines, the decrease of actual information (negentropy) implied by Eq. (3.1.5) is directly analogous to the information loss along a channel, the increase of entropy due to noise accumulating along the line. The fact that Eq. (3.1.5) agrees with two very distinct principles, the logical inference criterion (2.1.8) and the Second Law of Thermodynamics, is more than a coincidence. It will become clear in the sequel that Eq. (3.1.5) does not apply unconditionally to all causal chains, but as a general tendency, this asymmetric relation, inducing an order among events, is at the heart of the time arrow.

It is not yet a sufficient and not even a necessary condition, since it does not say anything about the identity of the signals taken into account: The mere quantity of information arriving at B may satisfy Eq. (3.1.3), yet its content may be completely different from that emitted by A. At the same time, other signals, "crosstalk" coming from a different source than the sender, may enter the line, to be considered as noise (see Chap. 7). Therefore, the received information may contain the complete signal from A but even exceed its size due to added noise. The sheer quantity of information needs to be complemented by another criterion.

Coincidence of the emitted and the received signals can be quantified as mutual information $I^{A \cap B}$, defined in Eq. (1.7.10). This suggests requiring

$$I^{A \cap B} = I^{AB} - \left(I^A + I^B\right) \geq 0. \tag{3.1.6}$$

Equation (3.1.6), in turn, fails to establish an order, because it is symmetric with respect to interchanging A and B, a notorious problem for criteria of causality: It can neither tell cause from effect nor exclude correlations owing to a common cause. It is more promising to refer to conditional information (see Eq. (1.7.12))

$$I^{B|A} = -c \sum_j \sum_k p_{jk}^{AB} \ln\left(p^{B|A}\left(k|j\right)\right), \tag{3.1.7}$$

since the conditional probability $p^{B|A}\left(k|j\right) = p_{jk}^{AB}/p_j^A$ breaks the symmetry between cause and effect. Indeed, as argued in Sect. 1.7, a positive $I^{B|A} > 0$ while $I^{A|B} = 0$ or at least $I^{A|B} < I^{B|A}$, i.e.,

$$0 \leq I^{A|B} < I^{B|A}, \tag{3.1.8}$$

is a suitable criterion for a directed flow of information from cause to effect.

However, even Eq. (3.1.8) is not yet sufficient to scrutinize the possible role of a third event C as a common cause of A and B. Here screening, alluded to above as a strategy to analyze the pattern of information flow between three nodes in a Bayesian network, comes in handy: The consequences of blocking C as a node correlating A with B become visible in the conditional probabilities of A and B, given C. Comparing their individual probabilities with their joint probability (see Eq. (1.7.2)), all conditioned on C, the intervention of the third event can be assessed directly,

$$p^{AB|C} \begin{cases} = p^{A|C} p^{B|C} & \text{if } C \text{ is a common cause of } A \text{ and } B \\ \neq p^{A|C} p^{B|C} & \text{generally otherwise} \end{cases} \tag{3.1.9}$$

that is, fixing C makes A statistically independent of B. This suggests to introduce the *conditional mutual information* of A and B, given C [PV07]

$$I^{A \cap B|C} := I^{A|C} + I^{B|C} - I^{AB|C} = I^{A \cap B \cap C} - I^{A \cap C} - I^{B \cap C}, \tag{3.1.10}$$

as a measure of the interdependence of A and B that is not mediated by C. If $p^{AB|C} = p^{A|C} p^{B|C}$, it follows that $I^{AB|C} = I^{A|C} + I^{B|C}$ and the conditional mutual information (3.1.10) vanishes.

The concept of *transfer entropy* takes up the idea of screening and applies it not only to pairs, but to entire chains of events, representing causal chains. Conditional probability breaks the symmetry between cause and effect, but in order to measure an information flow, it has to be complemented by a rudimentary order in time. It can be included by referring to chains of events, as introduced above with the concept of Markov processes, and combine conditional probability with the entropy rate, defined in Eq. (3.1.2), which measures the increase of entropy along discrete chain of events. A suitable quantity to compare the entropies for two different distributions conditioned on the same premise C, say $p_{i,j}^{D|C}$ and $q_{i,j}^{D|C}$, is the Kullback entropy, see Eq. (1.7.14), for conditional probabilities

$$I_{KL}^{D|C} := I_q^{D|C} - I_p^{D|C} = c \sum_{i,j} p_{i,j}^{CD} \ln\left(\frac{p_{i,j}^{D|C}}{q_{i,j}^{D|C}}\right), \qquad (3.1.11)$$

with I_q and I_p denoting the information measures based on q and p, resp. In order to apply Eq. (3.1.10) to entropy rates, substitute the k predecessor states $\mathbf{A}_n :=$ $(A_n, \ldots, A_{n-k+1})$ and $\mathbf{B}_n := (B_n, \ldots, B_{n-k+1})$ of the Markov processes A and B for the condition C in Eq. (3.1.11), the subsequent state B_{n+1} for the conditioned event D, and the probabilities conditioned only on B itself, not on A, for the alternative distribution $q^{D|C}$: The transfer entropy [Sch00]

$$h^{A \to B} := I^{B_{n+1}|\mathbf{B}} - I^{B_{n+1}|\mathbf{A},\mathbf{B}}$$

$$= c \sum_{i_{n+1},i,j} p\left(B_{n+1,j_{n+1}}, \mathbf{A}_{n,i}, \mathbf{B}_{n,j}\right) \ln\left(\frac{p\left(B_{n+1,j_{n+1}}, \mathbf{B}_{n,j} \middle| \mathbf{A}_{n,i}, \mathbf{B}_{n,j}\right)}{p\left(B_{n+1,j_{n+1}} \middle| \mathbf{B}_{n,j}\right)}\right). \qquad (3.1.12)$$

(notations j_{n+1} in $B_{n+1,j_{n+1}}$ and $\mathbf{j}_n := (j_n, \ldots, j_{n-k+1})$ in $\mathbf{B}_{n,\mathbf{j}_n}$ refer to different realizations of the same process $(B_{n+1}, \mathbf{B}_n)$, as in Eq. (3.1.2)) measures the information carried over from process A to process B, and thus the degree of dependence of B on A, beyond the impact of the prehistory $\mathbf{B}_n$ of B itself. Constructed as a Kullback entropy, it is a positive quantity. If strictly $h^{A \to B} > 0$, there exists a causal dependence of B on A. In general, $h^{A \to B} \neq h^{B \to A}$, and the larger one of the two amounts indicates the dominant direction of transfer [HP11]. Like this, the transfer entropy already defines an information flow in discrete time.

3.1.3 Causality in Continuous Time: Kolmogorov-Sinai Entropy

In numerous applications in physics, in economy, in epidemiology and other fields, causes and effects not only connect to an ordered sequence, a causal chain, but form a continuous process evolving in continuous time, as epitomized by a spreading wave or a propagating pulse. This suggests treating the discrete sequences of events $A_0, A_1, A_{2,...}, A_n, \ldots$, considered above as parameterized by time, $A_0 = A(t), A_1 = A(t + \Delta t), \ldots, A_n = A(t + n\Delta t)$ and letting the increment Δt approach zero. In this way, criteria for causality based on information measures can be rephrased in terms of entropy production rates.

A positive transfer entropy between two events only tells us that "something is being transmitted", it does not contain any clue as to the general quality of signal transmission along a chain, as alluded to in Eq. (3.1.5). A more suitable quantity to measure the increase or decrease of entropy along a Markov process or a Bayesian network is the entropy rate, Eq. (3.1.2) [KS04, CT06]. Alternatively, it can be written as

$$h_{n+1}^{(k)} = I^{\mathbf{A}_{n+1}} - I^{\mathbf{A}_n}, \tag{3.1.13}$$

that is, as the increment in information from $\mathbf{A}_n$ to $\mathbf{A}_{n+1}$. Reading Eq. (3.1.9) as a difference equation, it can be interpreted as the discrete version of a differential equation for systems with a discrete state space, resulting from a partition into a finite set of cells.

Analyzing causality in the framework of discrete events in discrete time is problematic, if only because most relevant processes in the natural as well as in the social and human sciences evolve in continuous time. Processes cannot be contracted to a single instant and may overlap, so that a unique linear order can hardly be defined. They have to be described in continuous time, combined with a correspondingly continuous state space. In the limit of infinitesimal time steps Δt and infinitesimal resolution Δx in state space, the entropy rate becomes the *Kolmogorov-Sinai entropy*

$$h_{\mathrm{KS}} := \lim_{\Delta t \to 0} \lim_{\Delta x \to 0} \lim_{N \to \infty} \frac{1}{N} \sum_{n=0}^{N-1} \frac{I^{(n+1)} - I^{(n)}}{\Delta t} \tag{3.1.14}$$

(a misnomer since it is a rate with dimension time^{-1}) [Sch84, LL92, BS93, KS04]. In effect, it measures the average overall rate of information increment required to predict the state of the system at time $t + \Delta t$ if its state at t is known. The significance of this quantity becomes particularly evident if applied to systems evolving along trajectories, representing causal chains in some D-dimensional geometry. A particularly suggestive example is the $2f$-dimensional phase space (f denoting the number of degrees of freedom) of classical Hamiltonian mechanics (Chap. 5). In this context, the loss or gain of actual information translates to the divergence or convergence of trajectories. More details of such processes will be considered in physical terms in Chap. 6.

Suffice it to discuss causality in this framework for two prototypical situations representing opposite extremes (Fig. 3.8):

Convergent trajectories—dissipative systems

Where a dynamical system possesses an attractor (a stationary state, a limit cycle, or a strange attractor) of a dimension (0, 1, or fractal, resp.) below that of the state space, trajectories converge onto these manifolds, typically in an exponential manner. As a consequence, initially close trajectories $\mathbf{x}_1(t)$, $\mathbf{x}_2(t)$ further approach one another exponentially, at least in some of the directions of state space, such that

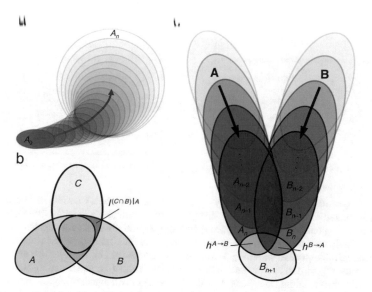

Fig. 3.8 Venn diagrams allow for an expressive graphical representation of information criteria for causality. Panel (a) depicts the general scheme of a causal chain satisfying the increase of conditional probability, Eq. (3.1.8). The basic set configuration to visualize transfer information is the diagram for a tripartite system as in Fig. 1.9a. In panel **b**, the relevant subset representing the conditional mutual information $I^{C \cap B | A}$, cf. Eq. (3.1.10), is marked in orange. Based on this scheme, the information transfer between chains of events A and B (**c**) is the information a final event B_{n+1} shares with one of the chains, independently of the information inherited from the other.

their separation diminishes in time as

$$|\mathbf{x}_2(t) - \mathbf{x}_1(t)| = |\mathbf{x}_2(0) - \mathbf{x}_1(0)| \exp(-\lambda_- t), \tag{3.1.15}$$

where λ_- is a contraction rate, often with the physical significance of a damping constant (see Sect. 6.2). On the background of a fixed partition of state space into disjoint cells, the number of cells covered by the initial distribution (state A in Fig. 3.9a) will therefore initially reduce by integer steps, on average by the same rate. As soon as it shrinks below the bin size Δx of the last occupied cell, however, the number of covered cells no longer decreases but stalls at 1 (state B), and the KS entropy reduces to zero. In terms of knowledge about the system's state, this means that once the system has settled down to the attractor, its future state is completely known within the given resolution, so that observing the state at t does not increase the information concerning its state at $t + \Delta t$. At the same time, referring to the initial condition instead of the evolving present state, actual information about it is completely lost [Sha80] while the potential information diverges, in agreement with the general rule for causal processes pointed out in the previous subsection.

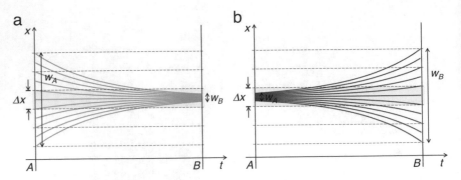

Fig. 3.9 In systems evolving continuously in a continuous state space, causality and the time arrow are closely linked to the convergence or divergence of trajectories. Given a partition with finite resolution Δx, a contraction of state space (panel a) will initially lead to a negative entropy rate given by a dissipation coefficient, cf. Eq. (3.1.15), but only for a finite time and if the initial width exceeds the cell size (state A, width $W_A > \Delta x$), till the state shrinks into a single cell of the partition (state B, width $W_B < \Delta x$) and the rate stalls at zero. By contrast, in an expanding state space (**b**), an arbitrarily narrow initial state (state A, width $W_A < \Delta x$) will eventually grow beyond the cell size (state B, width $W_B > \Delta x$) and then contribute a positive entropy rate, cf. Eq. (3.1.16), given by the Lyapunov exponent. Note that panels **a** and **b** are mutually symmetric with respect to time reversal $t \rightarrow -t$, interchanging states A and B. The apparent discrepancy with the respective flow patterns in Fig. 3.1a,b is resolved taking into account that the arrows in Fig. 3.1 symbolize actual information (negentropy), while here, the width of the distribution represents potential information (entropy).

The concept of dynamical systems that progressively lose the memory of their initial state is of course by far more general and applies way beyond dissipative physical systems. An appropriate framework, sufficiently wide to discuss this phenomenon, is control theory, including in particular the notion of feedback loops. They constitute a special class of information flows where information is cycled through a closed sequence of subsystems separated in space but not involving closed time-like curves (see above). Control systems designed to maintain the state of the total system as close as possible to a fixed set point or a predefined protocol are patent examples of finality (see Sect. 3.1.5). As central element, they involve negative feedback loops, which in fact are perfectly consistent with a causal information flow. A paradigm of negative feedback outside human technology is homeostasis in biological systems.

In negative feedback loops and more generally in control systems, the loss of a certain component of the input information is not just an unavoidable effect, as in dissipative systems, it is their purpose. The objective of their construction is to keep the state of the controlled system, represented by the *process variable* (possibly comprising two or more dimensions), at or near a desired optimal value, the *set value* $y(t)$ (it may consist of a time-dependent process protocol). As the system may be initiated in a different state and is exposed to external perturbations, which constitute an unwanted input, its actual state, the *measured value* $x(t)$, in general does not coincide with the set value. The function of the control loop is to attenuate the perturbation and thus suppress this unknown and unwanted input information as far as possible. Therefore, it couples the output signal, the deviation $w(t) = x(t) - y(t)$, back into the system by actuating the controller, an intended external impact that counteracts the perturbation. Closing a causal chain, the action of the controller should be opposed to that of the perturbation, but inevitably comes with a delay. Together, this constitutes a negative feedback loop.

The behaviour of the control system depends above all on the functional relation between the input, the measured disturbance $w(t) = y(t) - x(t)$, and the correcting variable $u(t)$, the output of the controller. In many simple control systems, e.g., an electric iron or a refrigerator, it consists in an on–off switch that is triggered if the disturbance exceeds a threshold value and deactivated if the disturbance falls below another, lower threshold. In linear control systems, such as the centrifugal governor of a steam engine, a transfer function determines how the controller reacts on the disturbance. Taking possible time delays into account, it takes the form of an integral kernel in an integro-differential equation,

$$\frac{d}{dt}u(t) = \int_0^\infty dt' G(t') w(t - t') \qquad (3.1.16)$$

where t' is the delay time. The transfer function is usually given by its Laplace transform $\tilde{G}(s) = \int_0^\infty dt\, G(t)e^{-st}$. For example, in the extremely simple case of an immediate negative response, $G(t) = -2\lambda\delta(t)$ and $\tilde{G}(s) = \text{const} = -\lambda$ a linear damping of the external perturbation would lead to its exponential decay, as in Eq. (3.1.15).

In general, this response is not only delayed but even nonlinear, which gives rise to complex dynamical phenomena characterizing control systems, including hysteresis, overshoot, self-sustained oscillations, and even chaos. They occur in technical as well as in natural feedback loops. Patent examples are predator–prey equations in population biology [Sch84, Str15] and the pork cycle in economy.

Divergent trajectories—chaotic systems

Deterministic chaos is characterized by an exponential divergence of initially close trajectories, manifest in a sensitive dependence on initial conditions, commonly known as "butterfly effect" [Lor63, Sha80],

$$|x_2(t) - x_1(t)| = |x_2(0) - x_1(0)| \exp(\lambda_+ t). \qquad (3.1.17)$$

The parameter λ_+, the *Lyapunov exponent*, has the meaning of an expansion rate and in general is different for each dimension of state space (for details see Sect. 6.1). A system prepared within a single cell of a given partition (state A in Fig. 3.9b), after an initial delay, will cover a number of cells that increases exponentially with an average rate λ_+ and contribute with the same rate to the KS entropy: Asymptotically, an additional information λ_+ per unit time is necessary to propagate the state by Δt into the future, or equivalently, the same rate is lost as to the value of the present

Fig. 3.10 Negative feedback is a basic feature of control systems. Any deviation $w(t) = x(t) - y(t)$ of the controlled system $x(t)$, exposed to external perturbations, from the set value $y(t)$ triggers a response $u(t)$ of the controller that counteracts the perturbation. Coupling this output back into the system closes a negative feedback loop.

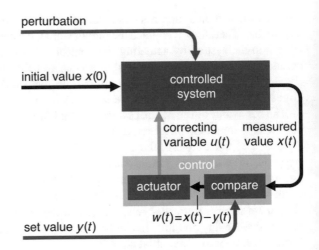

state as a predictor. Referring to the initial state, the actual information on the initial condition contained in the present state increases with this same rate [Sha80], while accordingly, the potential information approaches zero.

Figure 3.10 illustrates these opposing extremes with two standard models of dynamical systems, inspired by physics. The behaviour, for example of a damped pendulum, is described by the equation of motion of a harmonic oscillator with linear friction (also called "Ohmic friction", referring to the analogous case of resistances in electric circuits) [Kap03],

$$\ddot{x} + 2\lambda_- \dot{x} + \Omega^2 x = 0, \tag{3.1.18}$$

with a friction coefficient λ_- and natural frequency Ω. If the damping is not too strong so that oscillations still become manifest before the they are damped out, that is, for $\Gamma < \Omega$, the solution reads

$$x(t) = e^{-\lambda_- t}\left[\left(\cos(\omega t) + \frac{\lambda_-}{\omega}\sin(\omega t)\right)x(0) + \frac{1}{\omega}\sin(\omega t)\dot{x}(0)\right],$$
$$\dot{x}(t) = e^{-\lambda_- t}\left[\left(\cos(\omega t) - \frac{\lambda_-}{\omega}\sin(\omega t)\right)\dot{x}(0) - \frac{\Omega^2}{\omega}\sin(\omega t)x(0)\right]. \tag{3.1.19}$$

It corresponds to oscillations with an effective frequency $\omega = \sqrt{\Omega^2 - \lambda_-^2}$, spiralling with the rate λ_- from an initial state $x(0), \dot{x}(0)$ into the rest state $x = 0, \dot{x} = 0$, see Fig. 3.11a.

The opposite behaviour, trajectories spiralling outwards, can be achieved assuming negative friction, so that the system gains energy and is accelerated proportional to its present velocity. This situation is implemented in the van der Pol oscillator, defined by the equation of motion [GH83]

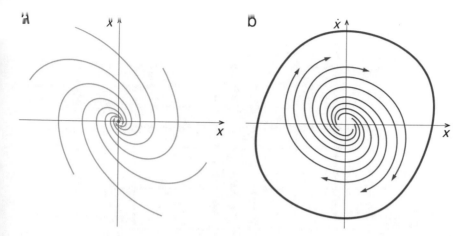

Fig. 3.11 Standard models of dynamical systems in a two-dimensional phase space with convergent (**a**) and divergent trajectories (**b**), as examples of the behaviour illustrated schematically in Fig. 3.9a and b, respectively. In the harmonic oscillator with linear (Ohmic) friction (**a**), Eq. (3.1.17), all trajectories spiral inwards, approaching the origin for $t \to \infty$ as a point attractor (green dot). The van der Pol oscillator (**b**), Eq. (3.1.20), a nonlinear oscillator with negative friction for small positions $|x| < 1$, possesses a limit cycle (closed bold red curve). Trajectories starting within the limit cycle spiral outwards to approach this attractor for $t \to \infty$.

$$\ddot{x} + 2\lambda_-(x)\dot{x} + \Omega^2 x = 0. \qquad (3.1.20)$$

The position dependent friction coefficient $\lambda_-(x)$ is chosen such that it is negative for $|x| < 1$, e.g., $\lambda_-(x) = x^2 - 1$. In this case, the system exhibits a one-dimensional attractor, a limit cycle, in the form of a closed, approximately circular curve centred in the origin, so that trajectories launched within this curve depart from the origin to asymptotically approach the limit cycle, see Fig. 3.11b. Close to the origin, they diverge exponentially with the rate $\lim_{x \to 0} \lambda_-(x) = -1$.

To be sure, exponentially divergent trajectories occur already in the simplest mechanical systems, even in systems conserving phase-space volume, such as a parabolic potential barrier. In order to give rise to deterministic chaos, another condition is required: Only if the exponential growth of structures in some directions of state space meets upper bounds, for example implied by the finiteness of the total available space, it must be compensated by a bending these structures back onto themselves: The combination of *stretching and folding* [Sha80] is a hallmark of deterministic chaos, see Sects. 5.5 and 6.3 and Fig. 6.4.

Systems with a state space of two or more dimensions generically combine contraction with expansion in different directions of this space (Sect. 6.3). One might expect that in this case, the overall entropy rate should be given by the sum of all the corresponding exponents, negative for contracting, positive for expanding directions. This is not so. As explained above, there is an asymmetry between the two cases: Converging trajectories generate a negative rate only for a transient time given

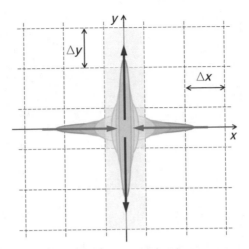

Fig. 3.12 Generic dynamical systems, with a two- or higher dimensional state space, combine the two cases illustrated in Fig. 3.9, showing contraction (green) in some directions of their state space (here x) and expansion (red) in others (here y). Nevertheless, the total entropy rate is not a sum of these tendencies but takes only diverging directions into account, cf. Eq. (3.1.18). The reason is that for a finite cell size $\Delta x \times \Delta y$ of the state-space partition, contraction can no longer be resolved after a transient time, while expansion continues indefinitely to cover an increasing number of cells (grey).

by the resolution of the partition, asymptotically their contribution to information loss or gain reduces to zero. By contrast, diverging trajectories show an initial delay before they start contributing a positive rate to the KS entropy, Eq. (3.1.14) (Fig. 3.12). Generally, therefore, the KS entropy is only sensitive to positive Lyapunov exponents. For deterministic systems combining chaos with dissipation, it is

$$
h_{\text{KS}} = \sum_{\substack{j = 1 \\ \lambda_j > 0}}^{D} \lambda_j, \tag{3.1.21}
$$

the sum of all *positive* Lyapunov exponents only (for a detailed derivation, see Sect. 6.1). Comparing Eqs. (3.1.15) and (3.1.17), it becomes evident, though, that they are related by a duality: Upon inverting time, $t \to -t$, contraction and expansion interchange their roles, Lyapunov exponents become dissipation coefficients, decay rates etc., and vice versa. The apparent violation of the time-reversal symmetry between expansion and contraction by the KS entropy has its root in another asymmetry: The state-space partitions underlying all information measures only know a minimum cell size as a lower bound of resolvable scales. Had it also an upper bound (see the discussion of smallest vs. largest scales in the context of Fourier transform, Sect. 2.3.1), a maximum "horizon width", the entropy generation in chaotic systems would then stall as soon as expanding structures reach this scale, and the symmetry between positive and negative Lyapunov exponents would be restored.

Processes leading to a contraction in state space have been ranged as a special case of negative feedback. Conversely, also state-space expansion generalizes to a broader category in cybernetics, positive feedback. By contrast to negative feedback loops, in positive feedback the signal is amplified, not damped, from the output back into the input. A patent example is audio feedback, the penetrating sound that arises if a microphone of a sound reinforcement system is placed too close to the loudspeaker, so that the output of the amplifier is coupled back to its entrance. Other paradigms of positive feedback are nuclear chain reactions, where the neutrons produced in a reaction induce further reactions with a higher neutron yield than is required to excite them, and the onset of an epidemic, when the rate of infections exceeds those of recovery and fatality. In all these cases, the initial exponential increase with a rate given by a Lyapunov exponent, as predicted by a simple linear differential equation, will eventually be curbed by global limiting factors, such as, respectively, the total mass of excitable nuclear material or the total susceptible population.

Situated on the borderline between stable and unstable, contracting and expanding dynamics, there exists a third category, *marginally stable systems*, with a vanishing entropy rate $h_{KS} = 0$. For deterministic systems, by contrast, which conserve the total information, such as in Hamiltonian mechanics (see Chap. 5), the KS entropy can well be positive, if the expansion in some directions of phase space is compensated exactly by contraction in others. Purely random processes, in turn, that lead to diffusive spreading, correspond to an entropy rate $\sim 1/t$, which vanishes for $t \to \infty$ but diverges for $t \to 0$, see Eq. (7.1.22). This suggests taking a positive entropy rate as a criterion for time running forward.

A time arrow can however not be attributed to the mere existence of processes increasing or reducing entropy. Rather, it bears on a difference in the relative magnitude of the two tendencies. Summarizing the foregoing arguments suggests a preliminary explication: Taking it for granted for the moment that a positive direction of time "in the background" can be fixed, define,

Processes are called causal if their entropy rate is negative, acausal if it is positive, going in the positive time direction.

The notion "acausal" here encompasses all forms of unpredictability from deterministic chaos to "true" randomness. Finer distinctions will be addressed in the following sections.

This is a local statement in that it refers to specific individual processes, time series, sequences of events,…, compared to the supposed global background direction. A time arrow can still be extracted uniquely, in a logical bootstrap, by widening the view towards a grand total, to averages over up to cosmic scales. This is actually the strategy used in thermodynamics to salvage the Second Law, in spite of the undeniable existence of processes in subsystems where order temporarily increases. They occur under conditions far from thermal equilibrium, from frying pans through living cells through the Earth's atmosphere. In all these cases, a local and/or temporal decrease of entropy in a subsystem is more than compensated by a simultaneous increase if this subsystem is embedded in a larger system assumed to be closed as a whole, ultimately the universe.

An independent definition of a positive direction in time, equivalent to the Second Law, can be based on this criterion,

The direction forward in time is defined such that on average over cosmic scales, processes predominate that increase entropy.

These definitions make it clear in which sense causality and the time arrow achieve a well-defined meaning only for a macroscopic observer: The local and global entropy losses and gains thus appear as emergent phenomena, which can only be comprehended from a sufficiently wide perspective. This leaves the question open how the time arrow can be reconciled with the bidirectional image physics gives at microscopic scales. It can be resolved only by taking the presence of a large number of microscopic degrees of freedom into account that interact with each other, to complement "vertical" information currents between scales with its "horizontal" interchange and redistribution among microscopic degrees of freedom, as will be elaborated in Chap. 7.

3.1.4 Records and Memory

Among causally related events, those commonly known as "memory" or "records" play a special role worth some special consideration. From Sumerian clay tablets to present-day USB sticks, they form an indispensable element of human culture. In the terminology of communication theory, records can be interpreted as messages directed to unknown future recipients, often but not always with an identifiable sender, hardly ever with a specific addressee. In a wider sense, memory is constitutive for life in general and also abounds in innate nature as processes that conserve initial conditions on extremely long time scales. Fossils tell us about the remote past of the biosphere on Earth, asteroids conserve the conditions of the emerging solar system.

Even if records appear as paradigms of fidelity and timelessness, they also break time-reversal symmetry in a particularly pronounced manner: "We can have records of the past, but not of the future" [Rei56]. From a physical point of view, records represent extreme cases of causality, without measurable loss of information or with a decay occurring only on timescales of the order of decades, centuries, or millenniums. The process of writing information into a memory should provide easy accessibility to intended change, on the one hand, and come as close a possible to irreversibility under normal operating conditions on the other. Safe storage appears to defy the tendency of macroscopic systems at, say, ambient temperatures on Earth, to approach equilibrium by degrading and eventually losing all their macroscopic structure (Sect. 7.1). The physics behind storage media therefore deserves a closer look.

Most of the materials used by human technology as memory media—stone, ceramics, metal, even toner on paper or chalk on a blackboard—are almost immune to the interchange of macroscopic information with thermal noise that characterizes the approach to equilibrium. A pertinent example is hysteresis, the phenomenon that the response of a material property, such as magnetization, to an external forcing, such as a magnetic field, cannot be removed by inverting that same forcing but requires a stronger opposite force to be undone. It is employed, e.g., to store memory, typically in binary code, in terms of the magnitude and direction of magnetization in ferromagnetic materials. Such materials require an external magnetic field to surpass a minimum strength in order to alter their magnetization. In a similar way, more primitive devices use static friction to store information: The relative position of two rigid bodies in mechanical contact remains unaltered as long as an external force does not reach a minimum strength, thus stores information on the most recent relative shift of the two bodies.

Unlike common dissipation, in these cases two or more macroscopic attractors coexist, points in the macroscopic state space that are approached from states within their basin of attraction and separated by energy barriers, significantly higher than thermal energies, along the basin boundaries. A standard case is an ordinary binary switch, to be studied in more detail in Sect. 9.1: Moving from one attractor to a neighbouring one requires investing a minimum amount of potential energy to climb the barrier. Upon falling into the adjacent potential well, it is not recovered as kinetic energy that could be reinvested to surmount the barrier in the opposite direction, but is dissipated into heat. It is the presence of these potential walls that protects the states encoding the memory against degradation by thermal processes. The condition that the barrier separating two states should surpass the mean thermal energy in height implies a lower bound per processed bit for the energy and entropy invested in irreversible recording and computing, as has been analyzed in particular by Landauer [Lan61, BL85, Lan91] and contemporaries [Ben03]. The basic physics of a simple digital memory element, a bistable system, will be elucidated in Sect. 9.2.1. This is a prototypical instance of *shielding*. It refers to the fact that in certain systems, including computing devices, the free exchange of information between different tiers of a hierarchy of structures (see Sect. 1.5) is blocked in the direction from smaller to larger scales, so that higher levels become independent of what is going on at lower levels. A similar phenomenon exists in quantum mechanics, in the form of subspaces of the total Hilbert space of a quantum system that are immune to or significantly less affected by decoherence than other subspaces [LW03, Lid13]. See Sects. 8.3.3 and 10.5.1.

Both genetic heritage and human long-term memory employ a different strategy: They do not defy thermalization but, on the contrary, make positive use of their condition far from thermal equilibrium to generate frequent recopying or refreshing, resp., with an accuracy that could hardly be achieved, except in living systems.

In the context of causal explanation, records assume a key role as witnesses of the past that allow us to identify correlations and order in past events. On the face of it, causal explanation relates later events to earlier ones. However, at least the cause in the past and mostly also its subsequent effect can only be accessed from the present (the time when the explanation takes place) through a record, that is, through another independent effect of the cause that gives more direct, lasting, and reliable evidence of it than the effect that is actually to be explained (Fig. 3.18). At the same time, memory can extend the causal power of past events almost indefinitely into the future, a phenomenon that becomes relevant in particular in the context of prediction, see Sect. 3.2 below.

3.1.5   Causality and Special Relativity Theory

The previous subsections proceeded from causal relations between distinct events, described in terms of probabilities for discrete states, to information flows, given by a continuous time evolution in a space endowed with a metric. For the last step towards considering causality in signal propagation in fields, the central reference is the Special Theory of Relativity. It is more commonly known as a reformulation of the concepts of space and time inherent in classical physics, starting from the Lorentz invariance of Maxwell's electrodynamics with all its consequences, instead of Galilei invariance [JS98, GPS02]. We do not know whether Einstein had any implications, specifically for the transmission of information, in mind or even thought of this concept when he conceived SRT. Still, it is exactly this what it has in common with quantum mechanics, the other major revolution of physics in the twentieth century: Both theories postulate fundamental restrictions referring to information—quantum mechanics with respect to its quantity, the storage density in phase space, Special Relativity with respect to the velocity of its transmission—and make them explicit till their most bizarre implications.

Einstein was concerned, however, about causality, both in quantum mechanics and relativity. The axiom that signals cannot travel faster than the velocity of light is a statement about spatio-temporal relations between cause and effect. In fact, in its most general and most appropriate form, it refers to the transmission of information. To be sure, cases of superluminal signal propagation are reported from time to time; till now, all of them referred to phenomena that eventually turned out to be unsuitable for the transmission of information. This applies notably to two situations: Superluminal wave propagation can be achieved only for certain features of a wave packet, such as the wave front, that cannot trigger an effect, as has been analyzed, e.g., by conceiving a proper signal velocity distinct from group and phase velocities [Bri60, Chi93]. A more fundamental class of apparent counterexamples are non-classical correlations, consistent with quantum mechanics, which are observed for example in Einstein–Podolsky–Rosen-like experiments [EPR35, Bel64, ADR82]. They contradict locality but cannot be used for communication, either, as will be sketched in Sect. 8.4.3.

As an elementary instance of signal propagation in a field, consider a stone thrown into a pond: From its point of impact on the water surface, it sends out a packet of circular waves spreading over the pond till they are absorbed at the shore. Electro-magnetic waves, as travelling solutions of Maxwell's equations, provide a suitable framework to analyse this phenomenon in quantitative detail. They are described by the Helmholtz equation [Jac73]

$$\nabla^2 \Psi(\mathbf{x}, t) - \frac{1}{c^2}\frac{\partial^2}{\partial t^2}\Psi(\mathbf{x}, t) = 0, \tag{3.1.22}$$

where $\nabla^2 = \partial^2/\partial\mathbf{x}^2$ denotes the Laplacian operator and c is the (vacuum) velocity of light. If there is a sender, its signal can be included as an external perturbation on the right-hand-side of Eq. (3.1.22) [Jac75], rendering it inhomogeneous,

$$\nabla^2 \Psi(\mathbf{x}, t) - \frac{1}{c^2}\frac{\partial^2}{\partial t^2}\Psi(\mathbf{x}, t) = -4\pi F(\mathbf{x}, t), \tag{3.1.23}$$

The field $\Psi(\mathbf{x}, t)$ can stand for the static potential or components of the vector potential. The source $F(\mathbf{x}, t)$ is assumed to be local in space and time, for example a Gaussian pulse

$$F(\mathbf{x}, t) = A \exp\left(-\frac{\mathbf{x}^2}{2\sigma^2}\right)\exp\left(-\frac{t^2}{2\tau^2}\right) \tag{3.1.24}$$

of width σ in space and duration τ in time, starting from $(\mathbf{x}, t) = (0, 0)$. The essence of its propagation with the field $\Psi(\mathbf{x}, t)$ is contained in the Green function that solves the inhomogeneous wave Eq. (3.1.23) for propagation in vacuum or any other medium where the wavenumber is given by $k = \omega/c$ [Jac75],

$$G^{(\pm)}(\mathbf{x}'', t''; \mathbf{x}', t') = \frac{1}{|\mathbf{x}'' - \mathbf{x}'|}\delta\left(t' - \left(t'' \mp \frac{|\mathbf{x}'' - \mathbf{x}'|}{c}\right)\right). \tag{3.1.25}$$

It serves as an integral kernel to calculate the time evolution of the pulse,

$$\Psi(\mathbf{x}'', t'') = \int_{-\infty}^{\infty} dt \int d^3x\, G^{(\pm)}(\mathbf{x}'', t''; \mathbf{x}', t') F(\mathbf{x}', t'). \tag{3.1.26}$$

The Green function comes in two versions, $G^{(+)}(\mathbf{x}'', t''; \mathbf{x}', t')$ and $G^{(-)}(\mathbf{x}'', t''; \mathbf{x}'t')$, the *retarded* or *causal* and the *advanced* or *anticausal* solutions of Eq. (3.1.22), indicated by the superscripts + and –, resp. They refer to the direction of time and are related by time reversal. For the moment, we choose the retarded Green function, as it applies to the boundary conditions of an outgoing wave from a local source, but return to the issue below.

The wave triggered by the pulse (3.1.25), after the decay of an initial sharp peak, will propagate as a spherical wave front of Gaussian radial profile, with its mean radius growing at the velocity of light. The so-called far-field solution, evaluating Eq. (3.1.23) with source (3.1.24) and propagator (3.1.25) for radii r far larger than width and duration of the pulse, i.e., than σ and $c\tau$, or equivalently for times far beyond τ and σ/c, is

$$\Psi(\mathbf{x}, t) = \sqrt{\frac{(4\pi)^3}{c^2\tau^2 + \sigma^2} \frac{A}{r}} \exp\left(-\frac{(r - ct)^2}{4(c^2\tau^2 + \sigma^2)}\right), \qquad (3.1.27)$$

denoting the radial coordinate as $r^2 := |\mathbf{x}|^2$.

From the amplitude (3.1.27), we obtain a probability density distribution as a normalized field intensity,

$$\rho(\mathbf{x}, t) = \frac{1}{N}|\Psi(\mathbf{x}, t)|^2, \ N := \int d^3x |\Psi(\mathbf{x}, t)|^2, \qquad (3.1.28)$$

giving

$$\rho(\mathbf{x}, t) = \frac{1}{\sqrt{2\pi(c^2\tau^2 + \sigma^2)}} \frac{\exp\left(-(r - ct)^2/2(c^2\tau^2 + \sigma^2)\right)}{4\pi r^2}. \qquad (3.1.29)$$

It translates to an information content, applying Shannon's definition adapted to continuous probability densities, Eq. (1.8.4),

$$\begin{aligned}
I(t) &= -\int d^3x \rho(\mathbf{x}, t) \ln\left(4\pi d_r^3 \rho(\mathbf{x}, t)\right) \\
&= 2\ln\left(\frac{ct}{d_r}\right) + \ln\left(\sqrt{2\pi(c^2\tau^2 + \sigma^2)}/d_r\right) + \frac{1}{2}.
\end{aligned} \qquad (3.1.30)$$

The volume integration has been evaluated assuming again $ct \gg \sigma, c\tau$, so that the radial coordinate r can be considered as approximately constant across the narrow spherical wave front, and specifying the spatial resolution (see Sect. 1.8), here in three dimensions, as $4\pi d_r^3$, d_r denoting the resolution in one dimension. The factor 2 multiplying the first term in Eq. (3.1.30) reflects the dimension $D - 1$ of the spreading wave front, in $D = 3$ dimensions.

As the total information increases logarithmically with time, Eq. (3.1.30) implies a positive entropy rate

$$\frac{d}{dt}I(t) = \frac{2}{t}. \qquad (3.1.31)$$

The positive growth rate of the potential information in the spreading wave is an exemplary case of the entropy increase proposed above as a general criterion for causal processes. At the same time, in agreement with the discussion in Sect. 3.1.3, tracing the wave back to the location of its source becomes increasingly difficult, observing the wave front only within a finite spatial range. As the curvature of the wave front decreases, measurements of the radius lose accuracy, and in the limit $r \to \infty$ of a planar wave, a localized source is no longer defined.

Indeed, using the advanced Green function $G^{(-)}(\mathbf{x}'', t''; \mathbf{x}', t')$ instead of the retarded one would merely invert the direction of radial propagation and thus the sign of the entropy rate. Returning to the example of circular waves spreading over a pond, it is actually the absorption (by a sandy beach, vegetation, etc.) on the shore that introduces irreversibility, not the radial spreading in itself [Zeh10]. In general terms, the outer boundary condition for the spherical wave (3.1.27) could be chosen as "absorption at $r \to \infty$". Absorption along the boundary can become irreversible only by coupling the wave to a large number of microscopic freedoms, which convert the macroscopic information in the field to entropy: Inverting their motion with sufficient precision to create an ingoing spherical wave, contracting towards the source, is practically impossible. If however the wave had been emitted in the centre of a likewise spherical cavity with perfectly reflecting inner surface and a finite radius $R \gg \sigma, c\tau$, the echo would return to the origin, as described by the *advanced* Green function, and continue indefinitely bouncing back and forth between the centre and the boundary.

In conclusion, classical electrodynamics is invariant under time reversal, as long as phenomena such as attenuation and absorption, occurring when light interacts with condensed matter, are excluded. In this sense, the terms "causal" and "anti-causal" for the two types of solutions of the time-dependent Maxwell equations, cf. Eq. (3.1.25), are misleading, as they refer implicitly to an underlying causal time arrow. Again, irreversible many-body phenomena prove essential for breaking the symmetry between past and future.

Time-reversal invariance is manifest in particular in the group of Lorentz transformations, valid for Maxwell's equations but of fundamental importance far beyond electrodynamics. They combine time t and three-dimensional space $\mathbf{x} = (x, y, z)$ into a four-vector $(x_0, x_1, x_2, x_3) = (t, \mathbf{x})$. Between two frames of reference moving with relative velocity v, four-vectors transform as [Jac75]

$$x_0' = \gamma (x_0 - \boldsymbol{\beta} \cdot \mathbf{x}), \mathbf{x}' = \mathbf{x} + \frac{\gamma - 1}{\beta^2} (\boldsymbol{\beta} \cdot \mathbf{x})\boldsymbol{\beta} - \gamma \beta x_0, \qquad (3.1.32)$$

abbreviating $\boldsymbol{\beta} = \mathbf{v}/c$, $\beta = |\boldsymbol{\beta}|$, $\gamma = (1 - \beta^2)^{-1/2}$. If, e.g., $\mathbf{u}_1$, the unit vector in the direction of the x_1-axis of the reference frame, is aligned with the relative velocity, $\mathbf{v} = v_1 \mathbf{u}_1$, Eq. (3.1.31) reduces to

$$x_0' = \gamma (x_0 - \beta x_1), x_1' = \gamma (x_1 - \beta x_0), x_2' = x_2, \quad x_3' = x_3. \qquad (3.1.33)$$

Time reversal $t \to -t, \mathbf{v} \to -\mathbf{v}, \boldsymbol{\beta} \to -\boldsymbol{\beta}$ then amounts to replacing $(x_0, x_1, x_2, x_3) \to (-x_0, x_1, x_2, x_3)$, $(x_0', x_1', x_2', x_3') \to (-x_0', x_1', x_2', x')$ (x_0 and x_0' represent time), $\boldsymbol{\beta} \to -\boldsymbol{\beta}, \gamma \to \gamma$, thus leaves Eqs. (3.1.32) and (3.1.33) invariant.

Special Relativity Theory goes a decisive step further in that it does not regard Lorentz-invariant space–time as the stage and condition where electrodynamics develops, but conversely organizes space and time according to the possibility or impossibility of events to interact, to exchange information in the first place. That is, it transfers the basic structure of an encompassing network of signal transmission to the geometry of space–time. The chains of nodes connected by links, forming Bayesian networks, thus become *world lines* in continuous space–time, one-dimensional manifolds forming ordered sequences of elementary events.

The crucial element in this construction is the horizon in space and time within which an event at $(t, \mathbf{x})$ can connect by world lines, i.e., be causally related, to other events at $(t', \mathbf{x}')$, disregarding the time direction insinuated by "causal". The maximum reach of information exchange with other events in the past or future is determined by a spherical wave front, as in Eq. (3.1.25), expanding in three-dimensional space towards the future or towards the past with the velocity of light. The two wave fronts divide four-dimensional space–time into three regions (Fig. 3.13a), a partition that is invariant under Lorentz transformation:

– **The backward cone**, comprising the interior of a sphere with radius $r(t) = -ct$ for $t < 0$. Events within the backward cone can be *causes* of the reference event at $(t, \mathbf{x}) = (0, 0)$. They definitely belong to its past, independently of their position and proper velocity, implying a Lorentz-invariant *time-like* relationship.

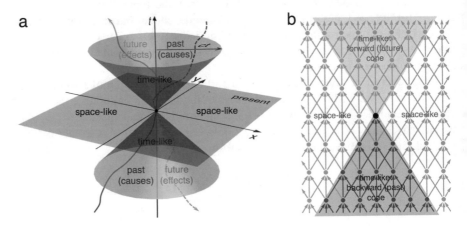

Fig. 3.13 Einstein's Special Theory of Relativity imposes a rigorous structure to space–time. as concerns the conditions for information exchange and causal relations. Reducing four-dimensional space–time to time t plus two spatial dimensions $\mathbf{x} = (x, y)$, panel a shows the three basic regions, the forward (future) cone $|\mathbf{x}(t)| < ct, t > 0$, the backward (past) cone $|\mathbf{x}(t)| < -ct, t < 0$, both implying *time-like* relations, and the rest of space–time that is *space-like* with respect to the origin and has no Lorentz invariant relation (earlier, later, or simultaneous) to it. All world lines emanating from the origin (curved lines) must remain within the time-like double cone and are thus uniquely divided into a past section comprising possible causes and a future section comprising possible effects. These attributes are symmetric with respect to time reversal, thus imply a linear order in time but not a time arrow. This qualitative structure can be mapped onto a directed acyclic graph (panel b, compare to Fig. 3.7d) with the same three zones, representing the basic topology of all physically allowed causal nets.

- **The forward cone**, comprising the interior of a sphere with radius $r(t) = ct$ for $t > 0$. Events within the forward cone can be *effects* of the reference event at $(t, \mathbf{x}) = (0, 0)$. Independently of their position and proper velocity, they always follow the reference event. Their relationship is then time-like as well.
- **The remainder**, comprising all of space–time outside the two cones. Events within this zone cannot have any causal relationship with the reference event. They are simultaneous to it and form the present in a comprehensive sense, that is, according to the Lorentz group (3.1.32), irrespective of their coordinates and proper velocity, they cannot precede or follow it, hence are in a *space-like* relationship to the point $(t, \mathbf{x}) = (0, 0)$.

Summarizing this Lorentz-invariant partition, all world lines passing through the origin of space–time must remain with their past tail within the backward cone and inside the forward cone with their future tail. In this way, Special Relativity induces a rigorous ordering of all events that implies necessary conditions for all causal relationships. Yet it is perfectly symmetric with respect to time reversal and cannot serve as a basis for defining the direction of time. Except for this symmetry, these qualitative features coincide with the general scheme of cause-to-effect information flow (Fig. 3.2). They can be translated back into the structure of directed acyclic graphs, as sketched in Fig. 3.13b, thus delineate a skeleton of all physically admissible DAGs.

3.1.6 Finality

"Why"-questions can be understood and answered in two almost antithetic manners: asking for a causal explanation, as discussed above, or, in the sense of "what for", expecting a justification. The latter category, also referred to as *teleological explanation* [Wri71], asks for the purpose of a tool, the function of an organ, the intention of an action, ..., and thus always points into the future. Teleological explanations pertain to the realms of technology, of biology, of psychology, law, and politics. In the natural sciences dedicated to the innate world, they are not only hardly ever accepted, they are even suspected of subjective bias and dismissed as paradigms of anthropomorphic, unscholarly reasoning (Fig. 3.14).

Another major difference, setting off final from causal explanations, is that they are often not free of normative connotations, in particular towards the far ends of justification chains: Ultimate objectives can be "good" or "evil". This is not the issue in the present context. To be sure, valuing intentions is natural in the context of human action. Where finality refers to biological processes, however, invoking values is unnecessary, can be misleading or even misused to lend objectivity to ideologies, as in social Darwinism [Cla00].

In the broad context of life, finality plays a special role, as it bears on a capacity shared by living organisms through all levels of their complexity: Being adapted to their physical, chemical, and biological environment and even prepared to react

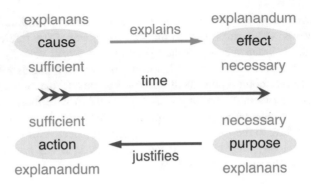

Fig. 3.14 Causal and final (teleological) explanations follow inverse time orders. While in the former, the cause is a sufficient condition for and precedes the effect it explains (top), an action or tool is justified by its purpose, representing a necessary consequence to be achieved in the relative future (bottom).

promptly and adequately to changing conditions, they appear to anticipate future events, from amoebae anticipating periodic deteriorations of their environmental condition [ST&08] to human technology predicting climate change. This capacity has even been proposed as a constituent feature of life, coining the notion of *anticipatory systems* [Ros85, Ros91]. Anticipation, however, seems to challenge the time order implied by causality. The present subsection is intended to demonstrate that and how the two principles can be reconciled [Wri71], and that final explanations even share important aspects of the time structure of causal explanations—albeit in inverse time. In the subsequent subsection, prediction and anticipation will be examined further as manifestations of information flow.

   A guiding idea will be that finality is not incompatible with causality but rather a special manifestation of it: They are two sides of the same coin. This becomes particularly evident in a rudimentary form of finality, not yet involving any conscious intention. The stabilization of a target state of a complex system by negative feedback, as sketched above, is realized in technological control as well as in homeostasis in living organisms. Here, the target state, defined, e.g., by the set value of a certain parameter, plays the role of an objective to be accomplished. If the control works effectively, the controlled system approaches this state asymptotically, in an iterative process. The control responds to external perturbations and to the effects of previous steps by new control pulses, designed such as to reduce deviations from the nominal condition. The set value therefore appears to act backward in time, attracting the state of the controlled system towards a future destination, as in a final explanation. In fact, it must be present already, at least implicitly and before the control sets in, in all corrective actions taken by the control. The alternation between checks and amendments characterizing control processes will reappear, albeit in much more elaborate form, in a causal analysis of evolutionary and personal learning that underlies biological function and purposeful action, cf. Fig. 3.15. The information flow behind this

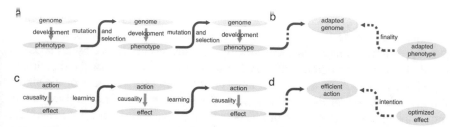

Fig. 3.15 Adaptation can be broken down into the accumulated effect of translation of genetic information into phenotypes, their selection by the environment and reproduction from a varied genome with mutations (**a**). The optimization resulting from the repeated action of these steps invites for being interpreted as finality, achieving a genome that encodes for a phenotype optimally adapted for survival in its environment (**b**). By an analogous feedback between human action (or design), its real, as compared to the intended, effect, and improved action, varied by trial and error (**c**), efficient action arises that can be explained as intentional for the effect it achieves (**d**). In both cases, a teleological explanation apparently defies causality but in fact summarizes a learning process that combines causality with creativity, to be understood as guided randomness.

process forms a loop within the components of the system. Yet it exhibits a unique directedness in time, absolutely consistent with causality.

In fact, finality and causality are even related by a close analogy. This can be seen by returning to the above intuitive explication of causes as sufficient conditions for their effects. Just as "it rains, but I have my umbrella closed" must be a sufficient condition for "getting wet" in order to explain it as a cause, "keeping dry while it rains" can be considered as a necessary consequence, hence an effect, of "opening the umbrella", which in turn should be a sufficient condition for that effect to occur. At the same time, in this case, it qualifies "keeping dry" as the purpose motivating "opening the umbrella" as a means. That is, action and purpose are related as cause to effect, but in a teleological explanation, the purpose is the *explanans*, the means is the *explanandum*, interchanging their roles. ("It rains" remains present in the background as a boundary condition also for the teleological explanation, which complements "keeping dry" to form a complete justification: "opening the umbrella to keep dry while it rains").

There is, therefore, a duality between the two types of explanation: In the former, the cause must be a sufficient condition preceding the effect to be explained, in the latter, the purpose must be a necessary condition, expected to follow the action it justifies in time [Wri71]. In short, causal and teleological explanations are transformed into one another by replacing sufficient by necessary conditions and inverting time (Fig. 3.14).

Concluding from the example, it might even appear as if, for a teleological explanation to apply, there should also be an objective causal relation between an action, a tool, etc., and the purpose motivating it: Opening the umbrella is an efficient way to keep the rain off, or else the argument would not go through. This is not correct, though, as will become clear in the sequel.

A decisive element in teleological reasoning is a well-founded expectation, based on evidence, that the action to be explained will indeed result in its objective. Learning in the widest sense is therefore a key to understanding finality. As a classic paradigm of learning, reduced to conditioning, consider Ivan Pavlov's dog experiment [Maz17]: The repeated presentation of meat to a dog, always accompanied by an acoustic signal, conditions the secretion of saliva, as a reaction of the dog with the function of digesting the food. After a sufficient number of learning cycles, the acoustic signal alone is able to trigger the secretion, even if the meat does not actually appear. A correct teleological explanation of this behaviour is that salivation upon perceiving the sound serves the purpose of facilitating the digestion of the expected food, and it is by no means invalidated in the case that this expectation proves unfounded. The same argument applies to explaining a crime by its motive. Whether a shot in an ambush has actually killed the victim, a bank raid has actually achieved its loot, is irrelevant for qualifying the former as murder, the latter as robbery.

The crucial condition for this kind of explanation to apply is therefore not the actual success of the action in the future, but the specific selection of this particular means in the present, based on experience gained in the past. In this way, the time order of causal explanation reappears at least partially in teleological reasoning. Still, the very aimed choice of an efficient means appears to anticipate its result, hence remains to be understood as a causal process.

Postponing a systematic analysis of how information is accumulated in learning and adaptation to Sect. 3.2, some basic principles can be extracted from genetic heritage and evolution, maybe the most profoundly studied instances of adaptation. Compared to man-made tools and technology, to human action, and the behaviour of other higher organisms with central nervous system, evolution lacks the element of a conscious intention. Yet the design of organs in the widest sense, from enzymes to organs of multicellular organisms, is amenable to teleological explanations pointing out their function for the next larger unit of organization, from cells to individuals and entire societies.

The sequence from DNA-based genome through transcription into RNA, translation into amino-acid chains, the folding of proteins and ultimately the development of fully fledged organisms has been highlighted in Sect. 2.2 as a case of directed information flow, it is perfectly consistent with causality. However, the other essential element in explaining function, the feedback from the device in action to the genome, is not as readily understood as a causal process: To the contrary, the Central Dogma [Cri58, WH&87], a generally accepted principle of genetics, excludes an inverse information flow directly from phenotype to genotype.

Rather, it is the interplay of random genetic variation at the origin of this sequence with selection at its opposite end that provides the missing link. More precisely, these two factors can be characterized as follows:

- **Undirected creativity** is fuelled by random variation at the genetic level due to diverse physico-chemical mutagens, including thermal fluctuation, chemical and radiation damage, quantum processes, ..., [Low63] as well as recombination in sexual reproduction. Often taken for granted, it is at this point that information is injected into the genome, in a direct quantitative sense as potential information or entropy. It drives a random walk (see Sect. 7.1) in the space of genotypes, i.e., in the extremely high-dimensional discrete space of DNA sequences [EW83, KL87, AK87].

- **Preferential heritage** results from exposing the phenotypes encoded by the varied genomes to an independent set of environmental conditions that increase or decrease the frequency of their reproduction. In this way, the undirected random walk in genomic space maps to strongly anisotropic diffusion exploring the space of phenotypes (Fig. 3.16, typically of lower dimension than that of genotypes, but at the same time of finer discretization or even continuous, and furnished with a strongly undulating, "rugged" [KL87, AK87] *adaptive landscape* that guides diffusion.

This process, roughly comparable with filling liquid metal or dough into a mould, allows transferring information on the shape of the mould to the casting or cake, even if the direct information flow goes in the opposite direction. The adaptive landscape playing the role of the mould for genetic diffusion is often called "fitness landscape" [KL87, AK87], a somewhat unfortunate term since "fitness" is not value-free. Valuing connotations are evidently inappropriate at least in the case of enzymes, where "fitness" would merely refer to the enzyme's efficiency at increasing the rate of the chemical reaction it catalyzes.

Repeating this process over many generations leads to an optimization, guiding diffusion into a local maximum of the adaptive landscape (or equivalently, a local minimum of a corresponding potential), see Fig. 3.16. This is clearly a directed process, combining causal with random steps (Fig. 3.15a), which approaches a final or at least metastable state. It can be summarized, condensing repeated adaptation forward in time into a single relation backward in time, as an explanation of the adapted genome by the survival of the phenotype it encodes in its environment (Fig. 3.15b).

This scheme, allowing for actual information being acquired in the genome "against the entropy current", is also effective, on a much shorter time scale, in individual and collective learning, not only in humans. An effect resulting initially from

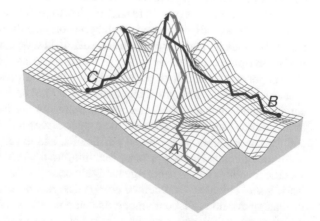

Fig. 3.16 Rugged fitness landscape on a two-dimensional parameter space with three fictitious random walks. Species *A* and *B* arrive at the same relative fitness maximum from different initial conditions, species *C* reaches a neighbouring secondary maximum where it is stuck in a metastable state.

a trial action, maybe without intending or even expecting it, may prove beneficial for some pre-existing necessity. It is even possible that the very failure of an intentional action—an irremovable stain indicates an efficient dye—inspires its reuse for a purpose different from the originally intended effect. If the action is then repeated and at the same time improved with respect to its beneficial effect, combining trial and error with memory of past failures and with creative modification (Fig. 3.15c), an optimization arises. Under the keyword of *evolutionary strategy*, this scheme is adopted in engineering in a broad range of optimization tasks [Nac02, Rec73]. In computer science, in particular, *evolutionary algorithms* form the basis of machine learning and artificial intelligence [Rec73, Bac96].

Resulting in very efficient actions, devices, recipes …, evolutionary learning by trial and error allows for a teleological explanation referring to an intended purpose to be achieved in the future (Fig. 3.15d). In this way, both on the level of biological evolution and of individual or collective behaviour, final explanation can be reconciled with causality, as a metaphorical shortcut that condenses a complex feedback process into a single relation but is consistent with, and can be broken down into, sequences of causal steps.

The shortcut of invoking a future target to interpret many situations in human life is not only legitimate but provides the only insightful explanation. In particular, the most sophisticated products of target-oriented collective human action hardly allow for any other approach. The most appropriate answers to questions such as "Whence the pyramids of Gizeh?", "Why that towering steel structure in the centre of Paris?", "How come that the Caribbean and the Pacific are connected across Central America?" is that these constructions have first sprouted as ideas and taken specific shape in the heads of inventive architects, engineers, and rulers, before any construction work even set in. This type of reasoning, which explains real facts by pre-existing mental images, plans, and concepts, raises the basic question in how far mental states can have causal power over stone blocks, rivets, and haul [MOE09, Ell16]. In a similar form, the problem of downward causation arises in the context of computing, in view of high-level algorithms controlling the movement of charge carriers in electronic chips [ED19], see Sect. 9.4.3.

While this kind of argument readily applies to societies, organisms, and technologies that have already developed an elaborate organization that assigns well-defined purposes to all its parts, the question remains how specific functions can emerge in the first place. Indeed, in biology, functions cannot generally be attributed to all components of a cell or an organism. Selection pressure appears to suppress "useless" features. Still, structures exist that do not serve any obvious purpose for survival at all, such as antlers and combs. Some organs, in turn, can be associated to two or more functions (in humans, the tongue serves for ingestion of food, for taste, and for speech, kidneys for blood purification and draining, several organs secrete hormones unrelated to their primary function, etc.). The specialization of subunits for specific functions seems to emerge in highly organized systems (multicellular organisms, insect colonies, commercial companies, human societies) as the result of an increasing division of labour among its components [US&18, KC&19]. It increases the efficiency of the whole on the expense of the survivability of the parts, but loses its specificity and significance towards smaller, simpler organized systems (cells, families).

An instructive example of specialization for different functions is the separation of two sexes in species with sexual reproduction. Gametes, the cells directly involved in reproduction, comply with two functions: dissemination, which serves the random mixing of genotypes (male), on the one hand, and the endowment of the developing embryo with energy and construction material (female) on the other. The one-dimensional fitness landscape related to preferential investment, under the condition of limited resources, in either of these functions proves to be bistable, that is, it exhibits two separate stable equilibria, each of them corresponding to a specialized strategy: invest either in the quantity and motility of gametes, on the expense of their size and endowment (male), or in their quality in terms of size and robustness on the expense of their number (female). The resulting anisogamy is then reflected in sexual dimorphism and propensity different social functions related to gender, which evolved in response to the differentiation of gametes.

Even where a function of a subsystem can be uniquely identified, asking for further, "higher", purposes eventually proves pointless. Concatenated final explanations tend to invoke increasingly more general objectives and end up sooner or later in an unspecific global goal such as "survival of the whole" or "prosperity for everyone", referring to the comprehensive entity that includes all subsystems involved (the cell, the individual organism, species, nation, humanity, …), or to close in circles, as is the case for the hypercycles which supposedly marked the origin of life [ES79, EW83]. This tendency represents another feature final explanations have in common with causal relations (Fig. 3.17). As pointed out above, causality is characterized by an increasing loss of actual information (gain of potential information) on the *explanans*, the initial condition, that is left in the *explanandum*, cf. Eq. (3.1.5). In an analogous fashion, final causes quoted to answer "what for" questions are at most as specific as the means they justify. Continuing to ask "what for" leads to some all-encompassing ultimate objective that contains a minimum of actual information, hence explains close to nothing. In the case of anisogamy addressed above, both sexes fulfil an obvious reciprocal function for the other: Neither sex can reproduce without the other. Their common purpose, maximizing reproduction, is far less self-evident as a positive value.

In the particular context of the evolution of life, this deficiency points to a deeper problem: Can we identify any long-term tendency, any ultimate goal in a phenomenon which so intricately entangles necessity with chance—as addressed explicitly in Jacques Monod's classic essay [Mon71]? The very nature of life as an open-end process, driven by thermal non-equilibrium and incessantly exploring new solutions, excludes the idea of a final aim in the sense of a stable state to be approached in the far future. It does not rule out, however, the existence of a prevailing tendency, a permanent driving force that keeps the process running in the same direction. If indeed there could be any positive answer in this sense, one of the most promising candidates is a steady increase in complexity of organisms and ecosystems [Kau93, Kau95, SM95]. The very notion of complexity, though, is not easy to grasp in sufficiently precise, if possible quantitative terms (for a comprehensive overview, see [BP97]).

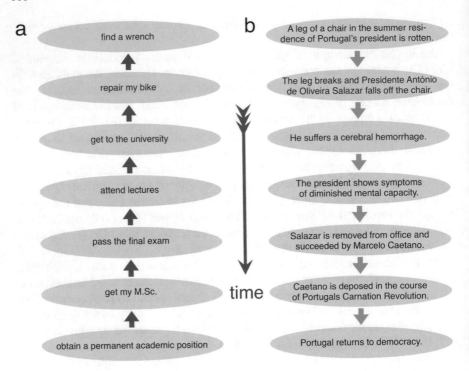

Fig. 3.17 A chain of final explications (justifications) (**a**) compared to a causal chain (**b**). While the final chain on the left is fictitious, the causal chain on the right recalls a real episode in the recent history of Portugal that developed between Salazar's accident in 1968 and the return to a democratic constitution in 1976. Even if it is rather more a caricature of a causal chain and would require considerable refinement to be substantiated, the comparison with the final chain (**a**) illustrates their common feature that forward in time, from first to last line (from *explanandum* to *explanans*, for final, from *explanans* to *explanandum* for causal explanation), the facts stated become progressively less specific, thus lose information on the initial state. Compare with Fig. 3.14.

The processes of adaptation and learning, introduced above as crucial factors to reconcile final explanation with causality, offer a plausible approach: They stand for an increasing amount of actual information, of knowledge on the environment, that keeps accumulating in different channels of heritage, as will be argued in Sect. 3.3. Life in general acts as an efficient catalyst of the down-conversion of low-entropy solar radiation into higher-entropy thermal radiation [Mic10, Kle10, Kle12, Kle16]. If learning and adaption stabilize life and thus contribute to its efficiency in this respect, the steep entropy gradient in the solar system could be considered as a motor behind the rising complexity of life as manifestation of increasing order on the planet.

3.2 Prediction

Till here, explanations of whatever kind, right or wrong, causal or final, have been considered as meta-statements, made by external observers from an independent standpoint outside the reality they refer to. This leaves an important point out of attention: Explanations, once conceived and established, constitute causal agents of their own right, which in turn can have consequences manifest in feedback on their subject. All this applies all the more to prediction, where the expected effect lies in the future.

Where on first sight, explanation and prediction appear to be related to one another by a mere time shift, they exhibit in fact a marked asymmetry, which in turn is a manifestation of the time arrow (Fig. 3.18): Explanations (with the exception of retrodiction) refer to confirmed facts, they relate a known event in the present or recent past to another occurrence in the remoter past, known at least through some reliable record. Predictions, by contrast, penetrate into *terra incognita*. Starting again from a known present or recent event, they construct a hypothetical fact in the future that has not yet occurred. Applied to continuous processes or developments, explications interpolate between two points, predictions extrapolate forward. This distinction is not exclusive. If explications and predictions refer to an ongoing slow development,

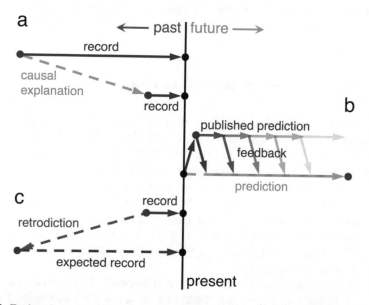

Fig. 3.18 Explanations, predictions, and retrodictions relate to events separated in time, but as propositions made in the present, always involve two simultaneous information flows or world lines. Causal explanations (**a**) invoke the record of a past event to explain a more recent event. Predictions (**b**), by becoming publicly known, generate a causal chain of their own that can have repercussions on the predicted event. Retrodictions (**c**), based on established records of the more recent past, hypothesize an event in the more recent past, of which less or nothing is known as yet.

such as climate change on Earth, every new climate data turns short-term forecasts into explanations, if they prove correct, and updates more long-term predictions.

The time arrow, manifest in the asymmetry between prediction and explanation, is even reflected in static spatial structures, notably in the body plan of higher organisms: They typically show a marked anterior–posterior asymmetry, the head, with the mouth to take up food and the main sensory organs that receive fresh data, pointing in the "forward" direction, the preferred direction of movement. This orientation clearly is an adaptation to the necessity to perceive and predict obstacles and events to be taken into account in the control of movement—in the same way as in a vehicle, the windshield and the drivers position with the steering wheel are oriented towards its default direction of motion.

Irrespective of their truth, predictions in themselves constitute facts that initiate new causal chains pointing into the future, they interfere with the very processes they anticipate, in particular if they are made by intelligent beings within a society of other intelligent beings. In many contexts of science, predictions are inevitably made "from within". This feedback complicates the information flow from cause to effect considerably, as will be explored below. Retroaction from a prediction to the process it refers to generally makes prediction more difficult or even prevents it, except for special situations where it is intended as a strategy, such as in self-fulfilling prophecy. Otherwise, we tacitly assume that predictions form independent processes, that they come true not by directly causing the anticipated event but by simulating the course of events in a parallel process that arrives at the same result, but significantly earlier.

3.2.1   Prediction, Anticipation, Simulation

Events predicted to occur in the future form the point of culmination of continuous processes, extending into the past and into the future. Towards the past, they are preceded by a widely ramified rootage of side effects and precursor events, which are strongly correlated to the approaching event without being its causes: Falling atmospheric pressure announces a tempest, shaking or releases of gas precede a volcanic eruption, fever heralds a flu. They form the objective basis of weather proverbs and country sayings, even of presage and oracles. Specific predictions are only the most conspicuous category within an entire class of phenomena that accompany causal chains and interfere with them, ranging from mere side effects through anticipation, simulation, to prediction proper.

If a prediction points out an event expected to happen in the future, it should better be available in due time before that occurrence, or else it would be futile. That is, it must be *faster* than the course of events it anticipates. Requiring this condition puts the two processes on the same footing: It compares two sequences of events, starting from the same present and leading to closely correlated final conditions, but one of them substantially earlier than the other. Otherwise, none of them is privileged. This suggests to analyze prediction as simulation, a concept emphasizing the equivalence of the two processes involved. Simulation amounts to a symmetric relation: If A

simulates B, then it is simulated in turn by B as well. There is a range of quite diverse cognitive capabilities that can be subsumed under this scheme:

1. Mere **anticipation** of one dynamical system by another [HP11] is observed already among relatively simple systems, such as in the case of synchronization with lag [RPK97]. If two identical self-sustained chaotic oscillators are coupled sufficiently strongly, their time evolution becomes nearly identical but one delayed with respect to the other, even if the coupling is symmetrical. In the case of one-way coupling in turn, it can even occur that the driven ("slave") system leads the driving ("master") system in time, hence anticipates its behaviour [Vos00], while the information flow actually goes from the driving to the driven system [HP11], as would be expected on physical grounds. In anticipation, the two processes form interacting parallel branches of the same causal chain.

2. **Simulation** refers to the more general case that the evolution of one dynamical system can be mapped to that of another, but is causally disconnected from it and proceeds independently on a different timescale. An elementary instance is a mechanical system (including fluid mechanics and aerodynamics) construed as a scale model of another. The timescale of its motion is then a function, given by the laws of similitude [Kli86], of the spatial scale, mass ratio, ratios of parameters like elasticity and viscosity etc.

    A more abstract class of similarity is found in **analogue simulation**. Here, the behaviour of any kind of dynamical system, described by a mathematical model in terms of differential equations, is reproduced, for example, by an electronic circuit containing mostly linear, sometimes also nonlinear, passive or active, elements connected such that the circuit follows the same set of model equations. In fact, most instances of analogy in science [HS13], serving to understand and predict phenomena, such as light waves compared to surface waves on water, could give rise to analogue simulations in the widest sense.

    The most versatile and sophisticated simulation strategy is **digital** or **numerical simulation**. It includes all the options offered by digital computers, for example to integrate sets of differential equations numerically. It is here where the inherent symmetry of the concept of two dynamical systems simulating one another proves most powerful: The term *emulation* refers to the case of one digital equipment capable of simulating another. A laptop can emulate a pocket calculator, but usually not a mainframe with multiple processors in parallel architecture, indicating that the ability to emulate induces a hierarchy of categories with respect to the capacity of information processing systems. The question in which sense a formal system can be regarded as superior or inferior to another formal system touches fundamental problems of mathematics, as will be elucidated in Chap. 4. The comparability of computing machines in particular will be addressed in Sects. 9.3 and 9.4. Even in numerical simulation, the computer serving as platform is to be considered as just another natural system with a behaviour, albeit extremely versatile, that resembles that of the system it simulates.

An intermediate stage, observed in intelligent natural systems, between simulation and the most sophisticated forms of anticipation is the ethological phenomenon of **imitation**. Apart from the imitation of static features (mimicry), imitating behavioural patterns is a strategy that offers a considerable survival value but requires advanced cognitive capacities, to be addressed in Sect. 3.3.2.

3. **Theoretical prediction**: The oldest form of quantitative prediction is probably the analytical calculation of the orbits of celestial objects, solving, for example, the orbit equations for masses interacting with a Kepler gravitational potential, to forecast special celestial constellations such as eclipses. The strategy reaches its climax with classical theoretical physics, where a small set of differential equations describes and summarizes entire classes of phenomena, such as Hamilton's equations for mechanics or Maxwell's equations for electrodynamics.

In theoretical prediction, the asymmetry between predictor and predicted system is most pronounced. A valid theoretical description provides a shortcut connecting initial to final state much more efficiently than the process to be described, and moreover applies with little or no modification to a nearly unlimited set of initial conditions. Theories are extreme cases of data compression, where incalculable quantities of past and future observational data are condensed into a few lines of mathematics. Here, the simulating system is not even a computer but an algorithm or more generally, a formal system (see Sect. 4.3.1), a mathematical structure that can be mapped to certain aspects of the natural process to be described. Theories achieve much more than simulations, in that they not only reproduce another phenomenon faithfully but even condense a large body of direct empirical observations into a much more compact essence. Providing efficient data compression, they form paradigms of sophisticated pattern recognition. A *theory of everything*, as the ultimate limit of this process, has been speculated of, but is not compatible with the emergence of phenomena, ascending to higher levels in the hierarchy of sciences [LP00].

A recent tendency in experimental and theoretical physics indicates, though, that the differentiation between reality, theory, and simulation are becoming more and more intricate. According to the traditional view, theory has the task of reproducing experimental observation as faithfully as possible. The vast increase of experimental possibilities, provided by latest high technologies entering the labs, opens a paradoxical option: It is versatile enough to simulate theoretical predictions in systems that have little to do with those that originally gave rise to the theory. For example, solid-state systems can be simulated in optical lattices, periodic potentials generated by counterpropagating laser beams, and quantum many-body systems can be simulated in the lab as networks of coupled qubits, using quantum computation technology.

Listing these categories one by one, a picture emerges of an almost continuous spectrum of instances of prediction, sharing the same constituent features. However, in order to qualify as a prediction in a more rigorous sense, a simulation must be causally separated from the process it simulates, otherwise it would just form a part of it. This condition is fulfilled at most by the second and third of these categories, but not by the first one. Notwithstanding, also cases of anticipatory synchronization deserve a closer look at the information flows involved, as they appear to violate causality:

Detailed numerical studies of the sign of the transfer information (3.1.6), as an indicator of the direction of information flow, for asymmetrically coupled chaotic maps [HP11] have revealed that the resolution Δx (see Eqs. (1.8.1), (3.1.10), (5.3.1)) used in extracting information from observational data on a state variable x is a crucial parameter: Only for sufficiently fine resolution, the sign of the measured information transfer agrees with the direction of the coupling, i.e., goes from the driving to the driven system. If the synchronization is anticipatory and the resolution too low, however, the measured information transfer may indicate a flow in the opposite direction, thus an anticausal process. A plausible interpretation of this result is that the information flow from drive to response, which enables synchronization with negative time lag in the first place, occurs on a smaller scale in the state variable x than the coarse-grained data that exhibit anticipation, so that with low resolution, only the latter becomes visible while the former remains concealed.

Issues of information transfer from the anticipating system, the predictor, to the predicted one, though, also arise in the case of "intelligent" prediction, based on observation and learning, without the need for direct physical coupling between the two. This will be addressed in the sequel. Conversely, anticipation in living systems does not constitute any information flow backwards in time, towards the predictor, either. It rests upon the same perfectly causal optimization process that underlies adaptation, invoked above (Fig. 3.16) to exorcise the demon of anticausality in final explanation.

3.2.2 Prediction from Within: Self-Fulfilling and Self-Destroying Prophecy

Prophecy and its possible back-action on the presaged occurrences was a phenomenon already pondered in ancient Greece and taken up in the most imaginative manner in Greek mythology, several centuries B.C. The most prominent example is the Oedipus saga. Reducing it to the key facts, stripping off poetic adornments and Freudian connotations, the story goes as follows [Smi67, Gra90]:

The life of Laios, King of Thebes, is under the curse, consequence of his former kidnapping of a young man, that, should he ever have a son, this son will kill his father and later marry his own mother. When Laios actually does have a son with his wife Iocaste, he consults the oracle of Delphi about the curse, which confirms the presage. In order to prevent being killed by this son, Laios exposes the baby to the wilderness, not before damaging his feet to impede his returning to Thebes. The baby is found by shepherds and then raised by the regional sovereign, who baptises him "Oedipus" ("swollen foot"). Reaching adulthood, Oedipus himself learns about the curse. Again in an attempt to prevent it from coming true, believing his stepfather to be his bodily father, he decides to leave his parents and ventures to the distance. In a skirmish for the right of way with the driver of a chariot and his noble passenger, who in fact is Laios, his actual father, he hurts both of them fatally. Not aware of his unwanted parricide, Oedipus arrives to Thebes, heroically liberates the city from the Sphinx, a monster threatening the kingdom, is appointed king of Thebes, and marries Iocaste, the widow of his late predecessor, thus fulfilling the last verdict of the curse. When the two find out later about the concatenation of sins behind their marriage, Iocaste commits suicide and Oedipus blinds himself.

Reducing the myth further to its skeleton of causal chains, it turns out that it contains two decisive turning points: The attempts, first of Laios and then later of Oedipus, to impede the curse precisely put the course of events on its track towards the inescapable catastrophe. From the point of view of these human agents, it would be interpreted as convincing evidence of their absolute powerlessness against divine will. From the viewpoint of the oracle that issued the curse, it appears rather as a plan, orchestrated with deadly precision: The presage not only proves true, it is precisely the effect it has on the human actors involved that is essential to let it become true in the first place. The Oedipus saga is a paradigm of *self-fulfilling prophecy*. On a closer look, it shows that as soon as there is a causal relation between a prediction and the system it refers to, if the prediction is issued from within the same system, it is no longer an independent meta-statement but the starting point of a proper causal chain that includes all kinds of feedback, negative as well as positive, to the predicted events. In conclusion, predictions from within, even if phrased like predictions, rather have the nature of statements of intention, which may coincide with the predicted (self-fulfilling) or the opposite state (*self-destroying prophecy*) or not be visibly correlated to it. Moreover, these retroactions are independent of whether the prediction later comes true or not.

Obviously, a feedback between anticipating and anticipated system exists already in the elementary case, mentioned above, of synchronization between coupled nonlinear oscillators. In the present context, however, their interaction is of a completely different nature: It includes as indispensable element that the prediction is received, understood, and interpreted as a message by the predicted system. Therefore, it constitutes semantic information, requiring some minimum insight also on the side of the predicted system—typically an individual or a group of humans—and the ability to take action in response to the prediction, thus placing self-fulfilling prophecy even in realm of pragmatic information (see Sect. 1.2). Under these conditions, practically all physical limitations are lifted that restrict the response to the prediction. For example, there is no longer any reason for anticipating and antici-pated actions to remain within a comparable energy scale, as in the case of coupled oscillators.

Were self-fulfilling prophecy merely a historical curiosity, it would hardly concern present-day information science. Far from it: It affects most areas of public life where prognosis and forecasts of any category are attempted. To name but a few prominent cases, including self-fulfilling prophecy as well as its antithetic counterpart, self-destroying prophecy:

- **Opinion polls** have become an important part of political life and political action. Substantial sums are invested by all agents in politics and economy to finance the most sophisticated methodology available in opinion polling. As their results are typically published in mass media with the broadest coverage, their effect on public opinion is undeniable and even intended. This feedback is well known and is reflected, for example, in legal restrictions, valid in most democratic countries, that ban the publication of opinion polls too close before elections [Bal02].
All evidence indicates that the sign of this feedback is mostly positive, that is, opinion polls induce self-fulfilling prophecy [Hit09]. A plausible explanation is a general inclination towards political opportunism: Most voters prefer to belong to the majority, to see the elections won by the party they voted for. This hypothesis, right or wrong, is at least reflected in the widespread tendency of actors in the political realm to consult preferentially those polling institutes they expect to publish polls biased in their own favour. However, strategic voting as well as a saturation effect ("My candidate will win anyway, why vote in the first place?") may lead to negative feedback as well.
- **Stock market prices** are known to be highly volatile and to respond sensitively to all kinds of news on the market. Investors orient their strategy in particular in the short- and mean-term market trends, that is, on price predictions and related economic information. Being aware themselves of this widespread tendency, they even take the expected effect of economic forecasts into account, irrespective of their truth. The result is strong positive feedback, which allures to be exploited positively: Announcing publicly an imminent increase (decrease) of the stock market value of a company will motivate investors to buy (sell) shares of that company, thus triggering the announced development. The efficiency of this mechanism is the reason why insider trading, the use of privileged access to internal information for stock transactions, is prohibited. Legal restrictions of insider trading, however, do not affect the publication of biased economic news in mass media, which may have a comparable effect.
- **Global warming**. For thousands of years of its existence, mankind could affect its environment and profit from its resources without feeling any significant reper-cussion, because the impact of human action remained negligible. As a result of major transformations of the economic model (industrialization) and of demo-graphic development, this state of innocence is lost. We are becoming increasingly aware of the global effects of our collective action. Almost from the first pioneering papers [Bro75] on, publications on climate change share the double face of sober neutral scientific forecast combined with serious warning: The daunting prog-nosis they contain is formulated with the clear intention *to prevent it* from coming true, thus leading to a trade-off between severity and credibility of the warning. In this sense, they are examples of self-destroying prophecy. This has remained so ever since and has been systematically included in the publication of systematic international studies [IPC21]. They contain conditional predictions of the type "If global carbon dioxide emission can be reduced by $X\%$ till the year A, compared to preindustrial times, global temperature increase will remain below $T°C$." This can be interpreted as a deliberate strategy to take the expected response in the

form of, e.g., national and international climate policies, already into account in the prediction.

Feedback in the context of global warming is usually understood as referring to geophysical and geochemical processes which, in response to increasing temperature, contribute positively (such as ice-albedo and the release of carbon dioxide and methane by melting permafrost soil) to the greenhouse effect. However, the feedback through climate studies and the reaction they entail in the public may prove by far stronger and more difficult to predict.

An important necessary condition that all predictions, causing feedback to the predicted system, must fulfil to be valid, is self-consistency. That is, the predicted state must coincide with the result of the response of the predicted system to the prediction: The consequences Laios and Oedipus took from the oracle's curse led precisely to the presaged outcome. How could the oracle achieve such perfect consistency? Daring a glimpse into its fictitious prophecy workshop, a surprising hint comes from present-day science: The knowledge of the social laws the presage is based on, in addition to familiarity with the individual behaviour of the actors involved, has an analogue in many-body physics.

A central task in solid-state theory is understanding the behaviour of electrons (more generally, quasiparticles) moving as almost free, negatively charged particles in the lattice formed by the positive ion cores of a metal or semiconductor. Also here, the challenge lies in including many-body interaction, given by the fact that by their negative charge, electrons feel the collective repulsive force exerted by the cloud of other electrons surrounding them. Any prediction of the state of an individual electron (the single-electron wavefunction) must therefore be consistent with the calculation of the collective state of the electrons in the sample (the N-electron wavefunction), based on this same single-electron state.

This condition of "social" symmetry (Immanuel Kant's Categorical Imperative, applied to electrons) can be turned into a recursive method for the calculation of the collective state of the ensemble of electrons, the *self-consistent field* or *Hartree–Fock method* [AM76, SO82] (Fig. 3.19a, suppressing all technical details):

1. Start with an electronic state for a single electron, subject only to the electric field of the ion lattice.
2. Assuming all the electrons to be in this same state, calculate the collective charge density and the additional electric field it generates.
3. With the total electric field updated accordingly, repeat the process until charge density and electric field are consistent with one another.

To be sure, self-consistency alone does not yet warrant a valid solution. In particular, it cannot exclude the existence of more than a single self-consistent solution. To which of them, if any, the algorithm will converge, depends on the trial density used as input and on other parameters. However, the method at least reduces the continuous family of possible densities to a few alternatives. Evidently, members of a social system are incomparably more complex individual agents than electrons. Notwithstanding, the Hartree–Fock method could inspire a self-consistent approach

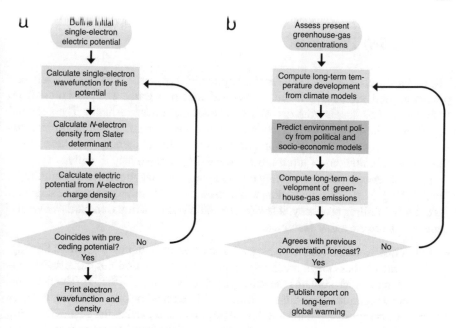

Fig. 3.19 Predictions on multi-agent systems have to comply with self-consistency between boundary conditions and collective behaviour. In the context of many-particle physics (**a**), the feedback between the electromagnetic field of the bulk background charge in a metal or semiconductor sample and the electron density can be used to construct a recursive scheme (Hartree–Fock method) for the computation of the electron density. Transferred to the global human society, it could inspire a recursive approach to climate modelling (**b**) that would take the feedback into account from climate forecasts to environmental decision-making affecting greenhouse-gas emissions. The close-to-impossible task this involves is not climate modelling but the modelling of the political and socio-economic processes that determine the response to predictions on global warming (pink block).

to predictions within a social medium, for example forecasts of global warming that anticipate their own impact on environmental decision-making, see Fig. 3.19b.

The three cases mentioned above have in common that the amplification factor from prediction—usually a scientific paper, a newspaper article, a TV program—to the response by the predicted systems—involving regional or international communities and institutions—is huge. In the case of global warming, the response may even be observable from outside the solar system, as variations in the total albedo of planet Earth or the spectral composition of the thermal radiation it emits. This is just one drastic example of how the particular processing of information in the biosphere, notably through adaptation and learning, affects the entropy balance of the planet, as will be argued in Sect. 3.3.

3.2.3 Self-Reference and Information-Theoretical Limits of Self-Prediction

The previous subsection considered the implications of causal relationships between predicting and predicted system, subsumed metaphorically as "prediction from within", but still keeping the two sides separate and distinguishable. Their mutual dependence, however, only reaches its extreme and its most drastic consequences when the two components are *identical*, reading "within" literally. It is here where fundamental limitations of information processing come to bear directly.

Self-reference is accompanied by countless abysses, paradoxes, and conundrums [Hof07] (Fig. 3.20). Self-prediction, as an instance of self-reference, is no exception. As did self-fulfilling prophecy, it has fascinated thinkers, writers, and artists over the centuries. Here are a few examples:

- **Laurence Sterne**, in his classical epic "The Life and Opinions of Tristram Shandy, Gentleman" does not immediately address self-prediction but exhaustive self-description, which contains the same paradox in a nutshell [Ste83]. The fictitious first-person narrator pretends recalling his life with all details, including the very phase of composition of the autobiography. When, reporting the past in temporal order, he arrives at the moment of initiation of the opus, he finds that the time span elapsed since then, with all the personal occurrences experienced during that period, still remains to be told. When terminating this subsequent task, he realizes that … etc. (Fig. 3.20b). The narrator concludes that completing the intended autobiography is an impossible task, since it cannot be accomplished within his lifetime.
- **Max Planck**, in an essay contributing to the debate around free will vs. determinism [Pla36], analyzes the fundamental requirements for self-prediction with the thoroughness and depth of a theoretical physicist. He starts from the assertion that causal determination of an action is equivalent to its predictability. Following a similar reasoning as in the preceding subsection, he argues that in predicting human action, even by an external observer-predictor not identical with the agent to be predicted, requires a sufficiently complete record of the initial condition, including motives, doubts, expectations of the agent, hence an intimate knowledge of his state of mind, so that it comes close to self-prediction. Once the prediction is consolidated and enters the consciousness of the acting person, it has the potential to alter intentions even into their opposite, thus rendering the original initial conditions of the prediction obsolete. An infinite regression results, not unlike the aporia contrived by Sterne. Planck's conclusion is that comprehensive self-prediction is impossible. Objective causal determination of human action therefore is not at variance with a subjective perception of an open future, unpredictable from within but susceptible to our decisions.
- **Alfred Gierer**, former director at the Max Planck Institute for Developmental Biology, reflects on self-prediction in a chapter, in [Gie85], on the physical description of mental processes. He applies the fundamental limitations of the analysis and prediction of a system by itself to brain research and arrives at the

A dog came in the kitchen
And stole a crust of bread.
Then cook up with a ladle
And beat him till he was dead.

Then all the dogs came running
And dug the dog a tomb
And wrote upon the tombstone
For the eyes of dogs to come:

"A dog came in the kitchen

. . .

For the eyes of dogs to come:

"A dog came . . .

" " "

Fig. 3.20 *Mise en abyme* (thrown into the abyss) is the technical term for images, texts, ...,
containing a replica of themselves. They are examples of the challenge to store the initial state
of a given system within this same system as input for a prediction of its own future behaviour.
Panel a shows the design of a packet of cocoa, including a reduced image of this same packet.
The popular song reproduced in b culminates in a verbatim selfcitation, resulting in repetition ad
infinitum.

verdict that even if brain processes follow causal physical law, based on this same
system with its inherent limitations, we shall not be able to understand and predict
our mental states in their entirety.

These three reflections on self-description and self-prediction, as well as many
similar analyses, even if they refer to quite different contexts, follow the same pattern.
They point out two types of basic limits, related to the finite capacity of the system
in question, for example regarding memory space and the computation time required
for the intended operation.

The exhaustive description of the present state of a system requires a minimum
of information, say m. If an image of the system is stored separately (Fig. 3.21a), for
example as a back-up, it will occupy additional memory space, say γm, assuming
a reduction by a factor $0 < \gamma \leq 1$, even if data compression and similar strategies
are employed. Repeating this process ad infinitum, the total memory size required is
given by a geometric series

$$m + \gamma m + \gamma^2 m + \ldots = m \sum_{n=0}^{\infty} \gamma^n = \frac{m}{1 - \gamma}. \qquad (3.2.1)$$

Now including the original system with all its images in a total memory M
(Fig. 3.21b) implies that its size must be at least by a factor $(1 - \gamma)^{-1}$ larger than

a   b

Fig. 3.21 Schematic rendering of the picture-within-the-picture phenomenon. Successive reductions of a TV or computer screen (**a**), here by a linear factor of 0.6 at each step, increasingly blur the image till it is reduced to a single pixel, whereupon the sequence ends. Notwithstanding, the total area of the original plus all reductions does not grow indefinitely but occupies a finite size given by a geometric series, in this case $\left[1 - (0.6)^2\right]^{-1} = 1.5625$ times the original screen size. If the reductions are shown on the same screen (**b**), as if recording the image with a camera and feeding it back to the same screen [Hof07], then a central part of the image has to be sacrificed, here comprising 36% of the total size, in addition to the loss of resolution. For a detailed study of video feedback, see Ref. [Cru84].

that of the nucleus m. This factor diverges for $\gamma \to 1$. That is, a finite total memory size can only be achieved if there is a substantial reduction at every step, resulting inevitably in a loss of resolution (see also Fig. 3.20a) till the point that only a single bit or a single pixel is left to store the entire image. This affects self-prediction fatally, since in intelligent systems, the finest detail of the initial condition could matter.

Even leaving the memory problem aside, given a comprehensive and sufficiently detailed account of the initial state s, the time evolution from there to the intended target time in the future will take a finite computation time t. Simultaneously, within this same time, the state of the system, following its own natural course, will have changed, say to $s' \neq s$. This requires updating the simulation underlying the prediction, taking the modified initial condition into account. The update may be faster than the full prediction by a factor $0 < \lambda \leq 1$, $t' = \lambda t$. Even so, the total computation time to achieve a definitive result, including all updates, reaches $T = (1 - \lambda)^{-1}t$, cf. Eq. (3.2.1). To be sure, for λ strictly less than unity, the hare can actually catch up with the tortoise within a finite time T. However, an update will require at least an overhead of a few mandatory steps requiring a constant time that cannot be reduced indefinitely, preventing t' from actually approaching zero. Therefore again, for $\lambda \to 1$, T diverges.

With these two quantitative arguments, it is clear that prediction from within a system is at least strongly restricted in its scope and precision and becomes downright pointless—turns into a declaration of intent—if an intelligent system pretends anticipating its own behaviour. A similar reasoning applies to groups and societies: Social sciences inevitably result in description and prediction from within. Sociological and economic theory always contains a normative component, related to political orientations in the former and taking fundamental economic objectives for granted ın the latter, such as full employment, substantial growth, and low inflation. Theories in these fields therefore inevitably are of a hybrid nature, combining objective description with subjective intention.

This verdict becomes particularly evident when prediction is applied to cognitive processes. Anticipating a substantial insight that may drastically alter the behaviour cannot be separated from having this intuition instantly. Insight therefore is unpredictable and at the same time irreversible: Once in the world, a bright idea cannot be undone but starts its own unforeseeable career.

3.3 Learning and Adaption

With the advent of life, a progressively detailed image of nature starts forming within nature itself [Lor68, Lor78]. The first catalytically active proteins, reproduced faithfully from primordial RNA [ES79, EW83, Kup90], conserved information on the existence of reaction paths, as efficient as improbable to be encountered by mere chance, in a storage medium that could last millions of years. Cells enclosed in a membrane already comprise a host of molecular machinery, optimized for specific functions within the cell. They bear an enormous thesaurus of knowledge on the physics and chemistry on subcellular scales, thus establishing the genome as the first and principal heritage channel for information on environmental conditions. Unicellular organisms, still lacking a nervous system, are already capable of recognizing correlations in their environment and to control their individual behaviour accordingly, so as to optimize their fitness [ST&08]. For higher organisms, the evolution of a nervous system, as a fast additional storage medium complementing genetic heritage, offers the advantage of adapting individual behaviour to short-term changes in the environment. Anticipation based on conserved experience is a constituent feature of life since its origin [Ros85].

3.3.1 Detectors of Correlation and Causality

At the heart of learning and adaption is the capability of living systems to detect and register correlations between external stimuli, perceptual data in the widest sense. A prototypical example is synaptic plasticity, the feedback from the frequency of activation of a synapse (the contact transmitting chemical signals from one neuron to another) to its strength [Hug58]. If the feedback is positive, the synapse will keep a lasting record, called long-term potentiation, in terms of its strength, of similar stimuli that tend to repeat [CB06]. Combined with a previous filtering of afferent (incoming) nerve signals, this mechanism allows detecting complex correlations among sensory data.

The advantage in fitness (survival probability) gained by correlation detectors is evident. A complementary sensitivity to time order among stimuli renders it even more powerful: Telling cause from effect and identifying links in causal chains as early as possible allow an organism to respond to threats in due advance. Time lags can be implemented in multiple fashions in neuronal signal processing. The detection of causal relations, more generally the development of a rudimentary concept of causality, evidently constitutes a positive selection factor. As argued in Sect. 3.1, a time arrow only exists for and can only be perceived by macroscopic systems. Only on the macroscopic level, records can be written and stabilized by dumping unwanted information, "noise", into invisible microscopic freedoms. Here, "macroscopic" includes the scale of neurons, even down to the size of macromolecules such as DNA and RNA. The same relation between lasting memory and information flow into microscopic freedoms applies to technical memory devices, see Sect. 3.1.4.

In the course of evolution, the complexity of correlations that can be identified by the perception apparatus has increased enormously. A patent example is image segmentation, a high-level feature of visual perception. Again, it is evident that the ability to detect the sight of relevant animals—prey or predators—from the surroundings is crucial for survival (as is, on the other hand, escaping detection by strategies like camouflage). On this level of sophistication, it is no longer plausible that complex patterns would be broken down to ever higher-order correlations between more elementary features and in this way be represented as excitation of specific neurons (dubbed "grandmother cells" [Gro02]). By sheer combinatorics, the number of possibilities grows just too large [Sin02]. Rather, it appears that nature has followed a constructivist, "top-down" strategy [GK&98, Sin02], resembling the trial-and-error scheme of learning (see Sect. 3.1.5): Patterns are generated, hypothesized, at random on higher levels of the visual system, to be matched with percepts on a lower level. If there is a hit, a coincidence, the neuron assembly that generated the matching pattern is selected and reinforced, otherwise it is inhibited or disassembled.

Another illustrative case is redundancy in language production and perception. Natural languages are characterized by a high level of redundancy, with the obvious function of optimizing fidelity and accuracy of communication, on all scales. They reach from few-letter correlations (syllables) through word patterns (syntax) through large-scale narratives that occur repeatedly in literature [Gle11]. Patterns in language not only support understanding, they also allow the listener or reader to anticipate subsequent parts of a text over a limited time or number of symbols. A prominent example from antiquity is the phrase "*Ceterum censeo…*", frequently used by the Roman politician Cato the Elder, which contemporary listeners would invariably have complemented by his "… *Carthaginem esse delendam*" (similar automatisms exist in present-day political debate).

As a consequence, active listening to speech is accompanied by a permanent foresight, a running extrapolation into the immediate future, based on correlations and patterns stored in memory. Where the actually perceived utterance coincides with the anticipated, it reinforces the expectation, where not, it serves to refine the prediction. Based on redundancy in a broad sense, this anticipation mostly improves communication. Often enough, though, the expectation is so strong that it supersedes the original message, leading to serious misunderstanding. The strategy of a running forecast probably applies not only to speech perception, but also to many

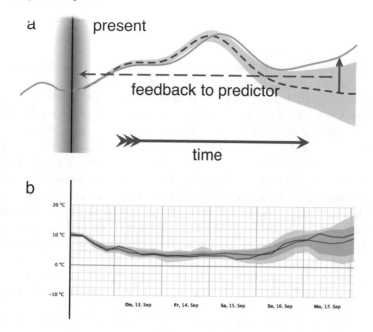

Fig. 3.22 Anticipatory systems often generate a running forecast that extrapolates a known past and an extended present into the immediate future (**a**). The prediction (crimson dashed), based on experience and learning, comes with an uncertainty that increases with the temporal distance (grey fringe). If it is continuously matched with the actual perception (green), deviations are fed back to improve the prediction. The situation is very similar to trend prognosis, e.g., in meteorology (**b**), here a temperature forecast for the city of Hamburg, as of 11–09–2018 (red: most probable forecast, blue: ensemble mean).

situations where planned action permanently interacts with experience, such as, e.g., manoeuvring a car through dense traffic. It means that for the anticipatory subject, an extended present of nearly certain occurrences merges continuously into an open future that can be foreseen only with decreasing accuracy (Fig. 3.22). Similarly, the reliability of forecast decreases also with the "causal distance" from self, from near infallibility in the motion of our own limbs through the manipulation of tools to more distant events where other agents interfere with the own intentions and actions.

This limitation of subjective predictability to a tight proximity in space and time is particularly evident and restrictive if the environment includes other subjects, pursuing their own intentions and performing their own predictions.

3.3.2 Predictors in Society

In parallel to the increasing capacity of living organisms to anticipate their environment, also the complexity of this environment has grown and with it the challenge it poses for anticipation. This is true in particular for the tropical and subtropical ecosystems where the typical environment of an organism consists largely of other organisms and their debris, while climatic and other physical conditions are of secondary relevance. In social species, successful interaction with conspecifics even becomes the decisive factor for survival and reproduction. The analysis of self-prediction in Sect. 3.2 concluded with the verdict that predictions from within a system are hampered by complex feedback phenomena and become impossible in rigorous terms if "within" means by intelligent systems on themselves. The situation becomes even more complicated when there are several anticipatory systems participating and interacting, each of them pretending to foresee its own behaviour as it interferes with that of others.

This situation adds an important element of complexity to those already mentioned in the context of self-prediction. As soon as two or more predictors of comparable capacity interact with each other, competition arises for the faster and more precise prediction. This is obvious in the case of predator and prey: The fox needs to catch up with the hare anticipating its path as closely as possible, the hare pretends to outmanoeuvre the fox by unpredictable doubling. A crucial step towards comprehensive prediction of behaviour is imitation. The ability to produce faithful imitations of sounds, movements, and other behavioural traits of conspecifics or members of other species provides a powerful tool for domination. Not surprisingly, it is a reliable indicator of intelligence.

When animals form social groups—colonies, packs, flocks, band societies—anticipating the actions of other members of the group becomes even more important. Successful participation in such a social network, in a primordial sense, then requires in particular a correct image of the relations inside the group: Which animal is the alpha? Which are its closest allies? Where are relations sufficiently fragile so that they could be broken to my own advantage? The more accurately a member of a group comprehends the structure of the network it belongs to, the more skilful it knows to manipulate and exploit it, the better chances it has to climb in the hierarchy. This creates a strong selection pressure in favour of social intelligence. Outstanding examples in mammals are wolves [Zim81] and primates [CSS86, Goo90] with their already elaborate and dynamical internal social organization.

Arguably, nature's present high-end product in this respect is *homo sapiens*. Optimizing the own capacity to anticipate and manipulate the behaviour of others and at the same time protecting ourselves against being imitated, penetrated, anticipated by others is a central issue in social human life. Espionage, counterespionage, eavesdropping, lie, deceit, fake news, hypocrisy, poker faces, … are standard tools in our rich repertoire of behavioural patterns dedicated to this purpose. Attempts to prevent eavesdropping of sensitive communication have stimulated a science of its own, cryptography, culminating in high-tech devices such as quantum cryptography.

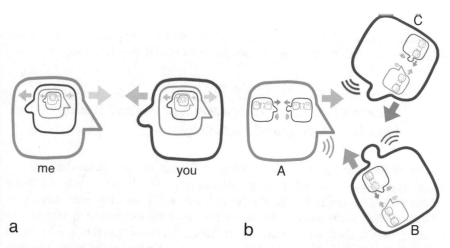

a   b

me   you   A

Fig. 3.23 Anticipatory systems in society face a challenge posed not only by the picture-in-the-picture problem (Figs. 3.20 and 3.21), but also in addition by the complex network of social relations to be comprehended. Already in the case of a pair of subjects (**a**), the interplay of mutual images, counter-images etc. is complicated. For three partners (**b**), each of them has to grasp and antici-pate the pair relation between the other two and its impacts on herself, resulting in an interaction structure of formidable complexity. Survival in larger social groups requires a huge degree of social intelligence, thus elicits a run-away competition among its members that probably has stimulated the rapid growth of brain volume in the transition from primates to hominids.

The challenge our intelligence faces when performing prediction from within a complex society is indeed formidable: It not only includes recognizing the structure of social relations among our contemporaries, it also requires to anticipate the image they have of ourselves and our social environment, their ability to predict and interfere with our behaviour, etc. [Lea94]. Pair relations with their endless complications of correct or erroneous mutual appraisal (Fig. 3.23a) fill large parts of our literature. Another illustrative example is chess: For a chess player, foreseeing the opponent's next turn is indispensable. The number of turns over which a player (or a chess program) is able to anticipate the most probable sequence of subsequent turns and counter turns is a decisive parameter for success. *Mélanges a trois* allow for even more complex patterns of prediction (Fig. 3.23b): "As I know how much my boss appreciates the psychological intuition of his secretary, gaining her sympathy might boost my career even more effectively than impressing the guy himself."

In a society whose members, if not kin, are at least genetically similar, the compe-tition for the fastest, most complete and accurate prediction of the behaviour of other members leads to strong positive feedback, to a runaway intra-species and inter-species arms race. Only limited by physiological restrictions, such as anatomic limits of skull size and energy resources available for brain function, it provokes a rapid growth of brain size and functionality. It is a plausible hypothesis that the over-whelming selection pressure towards social intelligence, capable of comprehending and exploiting the intricacies of social networks that formed in monkey groups and hunter-gatherer bands [CSS86], is responsible for the triplication of relative brain size (brain weight to body weight quotient) from non-human primates to hominids, maybe even for the development of consciousness [CS90, Jer91, Lea94].

3.3.3 Darwin's Demons: Anticipatory Systems and Entropy Flow in Ontogeny and Phylogeny

The preceding subsections offered ample evidence for the particular role the biosphere, composed of anticipatory systems, plays for the dynamics of entropy on our planet. It contributes massively, and with still increasing efficiency, to its catalyzing effect in the approach of the solar system towards thermal equilibrium, by down-converting short-wavelength solar radiation to thermal radiation in the infrared regime [Mic10, Kle10, Kle12, Kle16]. Beginning on the molecular level, the enzymatic activity of proteins renders reactions orders of magnitude more probable than they would be without these catalysts. One of the massive effects of their presence is the complete overturn of the composition of the atmosphere from a prebiotic phase, consisting largely of nitrogen and carbon dioxide, to the present biogenic composition with some 21% of oxygen [Lov00]. The increase of oxygen is essentially owed to photosynthesis, setting in with primordial cyanobacteria.

The far-from-equilibrium chemistry of the atmosphere, permanently kept on the verge of collapse by direct oxidation (fire), enabled the development of higher organized forms of life, which participate in their way in information processing. As do enzymes on the molecular level, they generate processes and structures that are highly improbable from the point of view of equilibrium statistical mechanics, yet obviously remain consistent with fundamental physical laws. Trees stand upright with their centre of mass meters above ground and stick even on steep slopes, animals move against strong force fields (gravity, air and water flow) and build organized stable structures (termite mounds, beaver dams) resisting these forces.

The advent of a single species accelerates this tendency even further: Endowed with outstanding cognition capabilities which themselves emerged from evolution, *homo sapiens* bridges valleys and penetrates mountains. Lakes appear where they have never existed, nights become as bright as daylight. The planet changes its face, visibly even from the distance. Objects made on Earth even reach remote celestial bodies (possibly causing major impact [Nas05]), rays of coherent light cross the solar system. Even global geophysical parameters such as surface temperature and composition of the atmosphere start changing under the influence of our species, so that it will presumably be detectable from outside the solar system. In our own terminology, these drastic anthropogenic effects gave rise to postulating the *anthropocene* as the most recent geological stratum [WZ&16].

Accompanying and underlying this process, a faithful replica of the planet, from a detailed inventory of its surface to a rough outline of its cosmic vicinity, is forming as the collective result of partial images, reduced, condensed, encoded in multiple ways

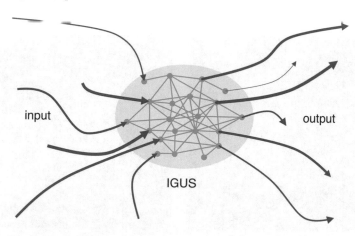

Fig. 3.24 Information Gathering and Utilizing Systems (IGUS) represent a new category of information processing in natural systems. While their closest kin, chaotic systems, just homogeneously expand input from a single source in its state space (Figs. 3.9b and 3.11b), lifting information from small to large scales, IGUS receive input from close-by as well as from the remotest sources, strong as much as arbitrarily weak, they store, link, and combine it internally in multiple ways, to generate output which again can reach arbitrarily far in space and time.

in countless genomes, brains, and external memory media (Fig. 3.25). Even more elementary than anticipation, adaption requires a lasting source of reliable information on the environment, independently of its direct manifestation. The fundamental role of this feature as a criterion for life, at least as universal as reproduction and metabolism, has motivated the notion of IGUS (Information Gathering and Utilizing System, Fig. 3.24) coined by Gell-Mann [Gel94] and Hartle [Har05]. Indeed, originally a biological concept, it is relevant also from a physical point of view. It indicates a novel way of information processing by natural systems, qualitatively different from any other model. Of the two categories defined above (Fig. 3.9), systems absorbing vs. systems generating entropy, IGUS are certainly closer to the latter.

With chaotic systems, IGUS have in common that minute variations in the initial conditions can entail drastic changes in the subsequent time evolution. The crucial difference, however, is in the way the input is being processed. While chaotic systems do not exhibit much more complexity than an approximately homogeneous amplification of initial conditions, see Sects. 5.3.3 and 6.3, IGUS permanently mix and recombine the incoming information in an extremely intricate manner (Fig. 3.24). They receive input from a multitude of extremely diverse origins, ranging from their immediate ambience (by their sensory organs) to sources almost arbitrarily remote in space and time (e.g., through telescopes receiving light from the edges of the universe). In order to construct an integral image of their environment, sufficiently comprehensive and reliable to support their survival, and to store it in short- and long-term memory, they have to accomplish a formidable task in data selection and compression. Features of this image are then selected and linked by a complex processing in unlimited combinations to generate output through actuators (limbs,

Fig. 3.25 Since the advent of humankind, the planet and all its inventory are being mirrored in countless visual representations and written documents emerging on Earth. Early evidence of this phenomenon is a pictograph (**a**) showing a fishery scene, a fisherman hunting fish with a harpoon. Approximately 10,000 BC, Serranía de la Lindosa, Río Guaviare, Colombia. Photo by Fernando Urbina Rangel, reproduced by courtesy of the photographer. Scale models of the entire planet—globes—can accumulate quite literally (**b**), as in the show window of the globe department of a geography bookshop. Photo by the author.

tools, machinery …) which again can reach out arbitrarily far in space and time (such as satellites leaving the solar system). In short, IGUS form nodes of natural information processing where the laminar flow of world lines is completely disrupted and entangled, where any input can interact with any other. From this perspective, computers pertain to the same category. Artefacts made by IGUS as tools to complement their information processing capabilities, as will be discussed in Chap. 9, they contribute further to their impact on the entropy balance of the planet.

As a bottom line, life has become a relevant factor in the way entropy is being processed on planet Earth, not just modifying marginal aspects but significantly altering the global entropy balance. Genetic and cultural evolution, accumulating increasingly comprehensive and reliable information on the planet and its spatial proximity, constitute an essential aspect of this role. With them, it is *semantic* information (Sect. 1.2) that comes into focus. In this way, it does not appear far-fetched to consider the highly complex way semantic information is acquired, processed, interchanged, communicated, and stored by human beings in particular, the subject of anthropology, arts and literature, psychology and sociology, as integral parts of a global information dynamics, inextricably linked to its physical aspects. Using the same term, information, on both levels is not just metaphorical or imprecise language, it refers to the same quantity. At the same time, going to the opposite extreme, to the inner workings of information processing at molecular and atomic scales, it is randomness of physical origin, thermal or quantum, which ultimately drives the overwhelming creativity of macroscopic phenomena on Earth, from clouds to life.

Chapter 4
Information and Randomness

The notion of randomness can be traced back to the same desire that has already been alluded to in the context of prediction within a social group: to prevent one's own intentions from being foreseen by others. In many games, concealing the own decisions to rivals is a decisive factor. In competitions, conflicts, games of luck, …, impeding predictability even becomes the central issue. Tools and techniques to generate events that cannot be foreseen by any participant of the game form a part of human culture from very early on. Finds of dice go back to the third millennium BC, but they were certainly not the first instruments to generate chance. They developed continuously from coin toss [DHM08] and dice up to more sophisticated mechanisms such as wheels of fortune, roulette wheels [Poi96], and lottery machines, digital random-number generators, through high-tech devices based on quantum mechanics (see Chap. 8), used for example in cryptography. In all these cases, the common objective is to impede as far as possible that human observers could anticipate the outcome produced by the device. Indeed, games of luck inspired pioneering mathematicians, such as Cardano, Fermat, Pascal, and Huygens, and served as models for the conception of probability theory. They were used as a metaphor in the description of natural phenomena, considered as impossible or too difficult to be characterized and predicted by precise deterministic laws, such as in statistical physics. Table 4.1 lists the principal subject areas where randomness plays a central role.

Besides science and technology, also fine arts and music were inspired by randomness, particularly in the twentieth century. In an attempt to reduce conscious decisions in the processes of design and composition as far as possible, the use of random features has been pioneered in music by Charles Ives and established as a category of its own, *aleatoric music*, by composers such as John Cage, Pierre Boulez, Karlheinz Stockhausen, and Witold Lutosławski. In the fine arts, the same objective to suppress intentional determination in the making has led artists such as Jackson Pollock, Mark Tobey, and contemporaries to distribute paint on the canvas in a partially uncontrolled pouring and dripping technique, *action painting* (Fig. 4.1, panels c, d). While they still adhered to traditional oil painting, artists such as François Morellet and Gerhard

T. Dittrich, *Information Dynamics*, The Frontiers Collection, https://doi.org/10.1007/978-3-030-96745-1_4

Table 4.1 Randomness plays a central role in many sciences as well as in daily life, such as in games of luck. The table lists some prototypical random phenomena in these areas, together with the principal mechanisms responsible for their unpredictability. It is only intended to outline the panorama, without pretending completeness.

Area	Random phenomenon	Origin of randomness
Games of luck	Coin toss, dice, dreidls, wheels of luck, roulette wheels, card shuffling	Initial condition too difficult to observe
IT security	Keys for cryptography	Pseudo-random generators
Physics	Macroscopic: turbulence	Deterministic chaos
	Microscopic: thermal noise, Brownian motion, shot noise	Complex many-body dynamics
	Frozen disorder: amorphous solid (glass)	Microscopic disorder in a supercooled liquid
	Polymers	Complex many-body dynamics
	Radioactive decay, quantum measurement	Quantum randomness
	Non-equilibrium structures: clouds, snow flakes	Instabilities in nonlinear dynamics
Geophysics	Weather, climate	Deterministic chaos
	Ocean waves	Instabilities and turbulence at the wind-water interface
	Seismic activity, volcanic eruptions	?
Chemistry	Chirality	?
	Chemical oscillators	Deterministic chaos
Biology	Mutation	?
	Recombination	?

Richter implemented the idea even more systematically, using random numbers to determine the colours in pixelated images (panels a, b).

It therefore does not surprise that randomness is predominantly defined negatively, in counter-distinction to determinateness, explicability and predictability. Where we cannot find a causal explanation, we tend to fall back to chance. In the previous section, the analysis of the concepts of explanation and prediction referred exclusively to the origin of the available information: Explaining one fact as an effect of another fact amounts to tracing the information it represents back to a source prior in time, to be identified as the cause. Quantitative formal definitions of causality, as in Sect. 3.1.2, refer to the *amount* of information that is transferred. Which *kind* of information flows from cause to effect remains out of sight. Upon a closer look, this proves unsatisfactory. In order to be content with an explanation, we expect some insight from it, an intellectual added value: "Reckoning does not imply understanding" [Lot88]. Specifically, we expect a tool that enables more efficient thinking and planning, "next time we know better", a gain in terms of Ernst Mach's *economy of*

Fig. 4.1 Randomness has inspired numerous works in the fine arts. François Morellet's "Random Distribution of 40,000 Squares Using the Odd and Even Numbers of a Telephone Directory, 50% Blue, 50% Red" (1960, panel **a**) visualizes a sequence of binary random numbers, the parity of the last digits of telephone numbers in a telephone directory, in a two colour code. Gerhard Richter's "4096 Farben" (1974, **b**), selects the colour of each pixel in a matrix of 64 × 64 at random from a set of 4096 hues. "Escape from Static" by Mark Tobey (1968, **c**), a continuous distribution of colours in continuous space, has been generated by a largely uncontrolled process of pouring paint onto the canvas. The technique of "action painting" has been pioneered by Jackson Pollock's drip paintings ("No. 41", 1950s, **d**).

thought [Mac82]. Explanations and predictions should provide a privileged access, a shortcut from *explanans* to *explanandum*, and should suggest how to generalize them to a wide class of similar situations.

This expectation is directly reflected also in the concept of randomness. Reducing explanation to extending causal chains backwards in time, tracing the information flow back to specific sources, identifies random events accordingly as initial points of world lines. They inject information into a flow that otherwise is causal, hence does not generate actual information (Fig. 3.4): Observing an unpredictable event, such as a radioactive decay, we gain information that was absolutely inaccessible beforehand. This accounts also for the element of surprise that is often associated with chance, and it suggests a useful first approach to randomness. In the context of randomness in quantum mechanics,, it will even prove necessary to fall back to this elementary criterion. Reductionist science therefore searches above all for the sources of randomness that escape direct observation, in particular on the microscopic level.

Confronted with a phenomenon that eludes a patent explanation, an immediate reaction is the question "How come?", asking for a hidden origin of the apparent contingency. This has been a fruitful attitude in many cases in science. It leads to a satisfactory analysis, which at least demystifies the phenomenon and, even if it does not permit predictions, indicates how to improve insecure prognoses. An illustrative instance of the endeavour to reduce chance on the macroscopic level to a microscopic source is biology. The interpretation of randomness in some traits of organisms (for example, fingerprints) was revolutionized by the advent of genetics. It postulated a source at the molecular level that determines such traits and even opens the perspective of a targeted intervention to modify and control them. On that molecular level, randomness entered again when genetics coalesced with Darwinism, now in the guise of mutations as a fundamental factor.

A comparable change of paradigm, regarding the interpretation of macroscopic randomness, was induced in physics by the theory of deterministic chaos. A theoretical account of phenomena such as turbulence and weather, avoiding any recourse to random forces, is possible. Epitomized in the butterfly effect [Lor63] (see Sect. 3.1.3, Eq. 3.1.17), it reveals initial conditions on the smallest, even microscopic, scales as the source of information that later becomes amplified to macroscopic effects. Deterministic chaos was then invoked to explain stochastic processes even outside physics, such as heartbeat in medicine, population dynamics in ecology [May74, May76], and cyclic development in economy.

It might appear that typical mechanical random generators, coin toss, wheels of luck, roulette and kin, unsuspicious of any hidden microscopic source of randomness, are emblematic instances of deterministic chaos. Ironically, this is not the case. Despite their relatively simple motion—basically, damped rotation in the plane or in three dimensions, fine differences in the initial conditions, too small to be observed by the players or the croupier, are still sufficient to be registered in the final state as a discrete flip in the result of the game (e.g., from head to tail in coin flipping, from one number to another in roulette) [DHM08].

Reductionist approaches to randomness in biology and in physics of course do not reveal any primordial origin of the produced entropy. Rather, they direct the view further down towards randomness on microscopic scales, in molecular and atomic dimensions. Relegating it further to even smaller scales does not solve the problem. Instead, other factors have to be considered in order to understand random processes on this level: While models of deterministic chaos comprise only a few degrees of freedom, the interaction of many subsystems now proves decisive. Towards such small dimensions, quantum mechanics becomes the obligatory framework, in particular where information processing is involved.

In molecular biology, the complexity of macromolecules, embedded in highly diverse subcellular structures, poses an almost insurmountable challenge for an analysis in terms of entropy dynamics. Research in this direction is only beginning. In physics, though, concerned with comparatively simpler structures such as gases and homogeneous solid-state systems, the understanding of these elementary cases in statistical mechanics and many-body physics is already quite advanced and has produced crucial insights. A central question is how, and under which conditions, physical systems approach a state of thermal equilibrium. It is inextricably related to a complementary phenomenon, thermal noise, as a prototypical manifestation of stochasticity in microscopic freedoms. In the phenomenon of Brownian motion (see Sect. 7.2), it becomes directly observable. One of the pioneering achievements, substantiating the intimate relationship between the loss of energy in the approach to equilibrium and the strength of thermal noise, is Einstein's Fluctuation–Dissipation Theorem, to be discussed in Sect. 7.2.

Randomness in many-body systems of classical mechanics is of a stronger quality even than deterministic chaos, because the underlying information flows ramify rapidly and get inextricably dispersed among the multitude of degrees of freedom. Even so, it remains an apparent indeterminacy, in the sense that all underlying microscopic motion follows strictly deterministic evolution laws. This condition is questioned only by quantum mechanics, the most rigorous and universal account of microscopic physics we have at our disposal. To be sure, unitary time evolution, the mathematical framework of quantum dynamics in closed systems (Sect. 8.3.1), is just as strictly deterministic as its classical counterpart, Hamiltonian mechanics. However, an important class of phenomena, such as radioactive decay and quantum measurement, remains that appear to be outside reach of unitary quantum mechanics (Sect. 8.4.5). *Quantum randomness* is considered as an instance of fundamental indeterminism [Zei99], i.e., of information created out of nothing, in stark contradiction to the conservation of entropy in unitary time evolution. Integrating it in the body of quantum theory is an open frontier of research. Its analysis in terms of information dynamics presently advances in quantum many-body physics [LB17] and in quantum optics and measurement.

However, the quest for the hidden sources of randomness does not exhaust the depth of the notion. What may well be appropriate for isolated point-like events remains unsatisfactory when dealing with continuous processes, time series and data sets. Here, it is the structure, not the source, that comes into view. Can we measure different degrees of randomness? What distinguishes deterministic time series from "truly random" noise [KS04]? Is there a pattern in the time series, does it even contain an encrypted message or will it remain an enigmatic sequence of signs? Complementarity between explanation and randomness will be the guiding line in Sect. 4.2, intended to refine the concepts of randomness and chance along with a deeper penetration of the concept of explanation.

Conceptualizing randomness more precisely even bears on the fundamental question how to adequately quantify information content. Basic definitions such as Shannon's ignore redundancy, they are blind to the degree of novelty of message has for its receiver. Rigorous criteria for randomness also help objectifying such a tighter condition for novelty and surprise, they allow us to carve out the irreducible core of a message. Fundamental limits of an unambiguous identification of random processes, implied by their very unpredictability, have been pointed out, inspired by Kurt Gödel's incompleteness theorem, as will be elucidated in Sect. 4.3.

4.1   Quantifying Randomness

An inspiring starting point for an analysis of randomness is the theory of deterministic chaos. It marked a major advance towards understanding stochastic processes in classical physics (for details, see Sects. 5.3.3 and 6.3), even promised to eliminate randomness altogether as a crucial element in several areas of physics. In essence, chaos replaces chance by an ordered flow of information from invisibly small to large scales, accessible to perception and measurement. The information surfacing in chaotic processes has been hidden in fine details of their initial condition [Sha80]. Reasoning like this, however, it remains out of sight how regular or irregular the time series looks that is generated by a chaotic system: The analysis applies equally well to periodic as to "truly irregular" orbits.

For example, the chaotic Bernoulli map, Eq. (5.3.25), can be considered as inspired by a popular card shuffling technique (Fig. 5.6). It provides an elementary mathematical model of a random generator. For example, applied to a typical initial condition, an irrational number such as $x_0 = \left(\sqrt{5} - 1\right)/2 = 0.618034\dots$ (the Golden Mean): it then produces a time series

$$1001111000110111011111001101110 01 \dots, \qquad (4.1.1)$$

obtained by reducing each output x_n at the nth step to its leading binary digit $a_n = \text{int}(2x_n)$. For a rational number such as $x_0 = 1/3$, however, it generates the time series

$$101010101010101010101010101010 \dots. \qquad (4.1.2)$$

Both sequences result in exactly the same way as records in discrete time of the discretized state of a deterministic chaotic dynamical system. Intuitively, we would accept only the first one as a random sequence, but qualify the second as periodic and perfectly predictable. This immediate perception is a striking example of the power of pattern recognition alluded to in previous sections.

A fundamental touchstone for randomness could be hard evidence for the absence of causal relations. They have been defined quantitatively in Sect. 3.1.1 in terms of cross correlations and information flow. Standard mechanical random generators have the objective in common to inhibit too obvious causal relations between subsequent trials: A coin toss, for instance, generates a binary sequence (e.g., tail $= 0$, head $= 1$) that is considered random [DHM08], owing to two decisive features: (i) Heads and tails occur to very good approximation with the same probability of 0.5, and (ii) subsequent tosses are, for all practical purposes, causally unrelated. Precisely for this reason, however, all possible binary sequences occur with the same probability. A sequence such as (4.1.2), even if it occurs only with a frequency of 2^{-32} in one, in itself should not cause concern for bias or manipulation. That is, blocking causal relations does not prevent sequences we would not accept as random.

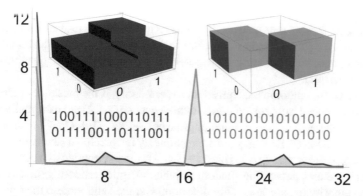

Fig. 4.2 Elementary methods to detect patterns, applied to the 32-symbol binary sequences (4.1.1) (crimson) and (4.1.2) (green). The power spectrum (2-dim. plot) is almost featureless for (4.1.1) but shows a marked peak at period $(16/32)^{-1} = 2$ for (4.1.2). Two-point correlations (3-dim. histograms) are nearly absent for (4.1.1), for (4.1.2) they are negative, peaked at pairs (0, 1) and (1, 0).

Another diagnostic for randomness that is readily quantified is the degree of periodicity of a time series. Periodicity can be measured by spectral analysis, that is, by Fourier transforming the time series. Periodic structures show up in the power spectrum as marked peaks, while their absence is reflected in a featureless flat spectrum for all frequencies > 0, as is evident for the spectra of the series (4.1.1) and (4.1.2) (Fig. 4.2).

Remarkably, for the distinction of periodic versus disordered spatial structures, there even exists a surprising physical, as opposed to mathematical, benchmark, independent of any measurement or intelligent observer: a quantum effect known as *Anderson localization* [And58, LR85]. Quantum mechanical states of spatially extended systems, represented as wave functions, are highly sensitive to periodic features in their boundary conditions, specifically in the dependence of the potential energy on position and time. For instance, if the potential $V(x)$ is periodic, $V(x + a) = V(x)$ for some fixed spatial period a, as is the case in particular for the crystal lattices of monocrystalline metals, the corresponding energy eigenfunctions (the temporally stable states the system can assume) are periodic with the same period, that is, they extend without attenuation throughout the entire sample ("from $-\infty$ to ∞"). This gives rise to the conductivity of metals. It can be understood as a consequence of multiple scattering processes at the ion cores of the metal. By their periodic superposition, they interfere constructively. If however the periodicity is imperfect, as in alloys, or altogether absent, as in amorphous materials such as glass, this interference becomes predominantly destructive, the conductivity is reduced or lost, and the material turns isolating. Its eigenstates are then no longer extended but localized: They only extend over a finite interval of length L, the *localization length*, that depends above all on the degree of disorder characterizing the aperiodic potential. Considering spatial disorder as frozen randomness suggests using this effect, directly observable as conductivity, to assess the quality of pseudorandom numbers, for example, by constructing model potentials with this same randomness and computing the eigenfunctions with their localization properties [BF92]. This physical test, free of any observer's subjectivity, however proves to be not very sensitive. Even a pseudorandom sequence that fails in simple statistical randomness tests is able, once converted into static potential disorder, to localize quantum energy eigenstates [BF92].

A criterion for randomness, equivalent to periodicity, is the presence or absence of correlations. Indeed, subsequent digits in the sequence (4.1.2) are maximally anticorrelated (the one-step autocorrelation approaches -1 as $-1 + (1/N)$ with sequence length N), while in (4.1.1), they are approximately uncorrelated (the autocorrelation approaching 0 faster than $1/N$). This criterion is routinely applied as a benchmark for numerical routines generating random numbers. Digital random generators, being based on algorithms, are inevitably deterministic, the randomness they produce is therefore basically fake, and are often referred to as *pseudorandom* number generators. They require a *seed* as input, analogous to the initial value x_0 for the Bernoulli map: Once the seed is known, they are perfectly reproducible.

A crucial parameter controlling the rigour of a correlation test is the order of the correlations taken into account. Patterns in a data set are often hidden in higher-order correlations, but remain invisible in low-order moments and correlations. An efficient technique to reveal patterns is therefore representing data in spaces of higher dimension, for example by constructing delay vectors as introduced in Sect. 3.1 (Fig. 3.3). In this way, a time series originating in deterministic chaos can be reliably distinguished from "true" randomness [KS04]. More generally, going to higher dimensions allows to visualize the degree of correlation that remains concealed in projections to a lower dimension: Non-zero mean values (zeroth-order correlations) can be removed by a shift in one dimension, two-point (first-order) correlations can be concealed by rotation in two dimensions, etc.

The absence of correlations and of marked structures in the spectrum, as indicators of randomness, are in fact closely related. In a continuous signal $\eta(t)$, a vanishing correlation time, that is, a delta-like auto-correlation function,

$$C_{\eta\eta}(s) := \langle \eta(t+s), \eta(t) \rangle \sim q\delta(s), \qquad (4.1.3)$$

is associated by the Wiener-Khintchin theorem [BP97, KS04, CX19] to a flat (constant) spectral density,

$$S_{\eta\eta}(\omega) := \int\limits_{-\infty}^{\infty} \mathrm{d}s\, C_{\eta\eta}(s) \exp(-\mathrm{i}\omega s) = \text{const} = q. \qquad (4.1.4)$$

Alluding to the spectrum of visible white light, delta-correlated random processes are called *white noise*. Deviations of the power spectrum from homogeneity, on the other hand, result in *coloured noise* and indicate some degree of regularity, hence of redundancy, in the signal. In particular, so-called $1/f$ noise [Sha80, BP97], refers to processes with a spectral density decaying as an inverse-power law

$$S_{\eta\eta}(\omega) \approx \omega^{-\alpha}, \quad 0.8 < \alpha < 1.4. \tag{4.1.5}$$

It is observed in the time dependence as well as in the spatial structures of many non-equilibrium phenomena, such as hydrodynamic, meteorological, biological, and economic data, even in time series measured in music and language.

This observation already touches the question of sophisticated pattern recognition by intelligent subjects. Imagine a "mechanical" composition process that employs a random generator to produce stochastic symbol strings (texts or musical notes) [Gle11]. Even if it takes high-order correlations into account, adopted from "genuine" music or literature, it will generate correlated noise, but hardly symbol sequences we would perceive as meaningful in a cultural context. Automatic composition including correlations up to the scale of word lengths can generate texts with a strikingly authentic appeal [Gle11], but does not produce literature. Sampling the statistics of frequencies, intervals, durations, rhythms will not culminate in the composition of a new Bach Toccata or Beethoven symphony. Human observers perceive complex patterns that cannot be comprehended in terms of correlations alone—a phenomenon that is used systematically for example in online access controls based on CAPTCHAs (*Completely Automated Public Turing test to tell Computers and Humans Apart*) [AB&03]. Just as the existence of such patterns proves indispensable for explanation and prediction, excluding them is essential for a comprehensive definition of randomness. It requires advancing to a yet deeper analysis of patterns in data sets.

4.2 Randomness According to Structure: Redundancy, Data Compression, and Scientific Induction

Data analysis is at the heart of science. Already at pre-scientific levels, finding simple patterns in apparently disordered observations is a central attainment, as is manifest, for example, in popular weather lore. Systematic observation and generalization set in with Greek science and philosophy, Islamic alchemy, astronomy, and mathematics, and the Aztec calendar, among other disciplines. At the same time, researchers started reflecting on the fundamental rules they routinely apply to distinguish valid laws from spurious generalizations and superstition. Section 3.2 focussed on the causal consequences a prediction can entail, leaving the question aside how to come to a valid prediction in the first place. Also here, the same rules concerning information flow in causal chains prove relevant. Where explanations pretend to reveal the origin of the information manifest in a given event, prediction evaluates the information presently available, including records of the past, to anticipate future developments. The way relevant information is extracted and processed to achieve predictions is essential.

4.2.1 Induction

A procedure characterizing most of empirical science is induction: How can the step from a finite set of observations to a general rule, applying to a potentially unlimited number of cases, be justified, when it certainly does not qualify as logical inference? Besides a host of metaphysics, also pragmatic rules were proposed very early to support induction. All sophisticated theory in the empirical sciences is based on induction, from theoretical physics through theoretical biology through economy. In terms of the efficient representation of observed data, quantitative theory, where it can be applied, is superior by a huge margin to more elementary methods such as finding correlations.

As an advocate of induction as a rationale of empirical science, Francis Bacon, the "father of empiricism", was even preceded by William of Ockham. With *Occam's razor*, he introduced the principle of conceptual parsimony into scientific reasoning: Whatever concept, assumption, or precondition that is not indispensable is to be avoided (*entia non sunt multiplicanda praeter necessitate*, entities must not be multiplied beyond necessity). It is the same idea that re-emerges, for example, in Mach's economy of thought [Mac82] as well as in technical objectives that appear obvious for our times, such as saving memory and reducing CPU time. Occam's razor may appear appealing, plausible, useful, and in accord with similar tenets outside science, in economy and even in the arts. Still, it enjoys as much or as little fundamental justification as any other heuristic principle.

To be sure, austerity in constructing hypotheses stands out by one particular virtue: Its application leads to unique results while the opposite strategy opens Pandora's box for a proliferation of alternatives. There are always unlimited possibilities how to augment a given hypothesis by additional concepts and assumptions: The Loch Ness monster is always waiting to be invoked as the true culprit. Adding items in this way to the list of premises does not affect the logical validity of an argument, see Sect. 2.1.4. This is the easy direction. Inversely, purifying the hypotheses, i.e., finding elements that can be relinquished without losing the essence, is by far more difficult; wherever it can be achieved, it increases the universality of a hypothesis. As has been demonstrated above for logical inference as well as for causal explanation, actual information decreases (potential information increases) with every step in a logical inference or causal chain. Reducing the information content of the *antecedent* or *explanans* therefore amounts to a purification that cannot go arbitrarily far without losing the final conclusion. There is, therefore a lower bound for the list of premises. An upper bound, by contrast, does not exist.

In this way, Occam's razor anticipates the rule for causal chains highlighted in the previous section: The actual information on the initial state decreases, later links in a causal chain must not introduce more contingent information in the form of additional facts that would not already be contained, at least implicitly, in the initial condition. Applied to prediction, this implies that elements added without necessity to the hypothesis, the initial state, are arbitrary.

4.1.1 Pattern Recognition and Algorithmic Complexity

This argument, as reasonable as it appears, leaves a central question open: Is there any systematic way to distil the information that is actually relevant for a given phenomenon, how can the boundless repertoire of potentially significant information be reduced to its essence? This defines the basic task in data compression. A quantitative answer came into reach only with the advent of computation as a rigorous science. The key to systematize and quantify data compression by a more abstract description is the concept of algorithm. Algorithms condense processes for the manipulation of numbers, more generally of symbol series, into finite sequences of elementary operations. The decisive advantage of measuring compressibility of data in terms of the efficiency of algorithms lies in the fact that they themselves can be encoded as symbol sequences, as occurs for example in programming in an established programming language. This opens the possibility to measure also the size of an algorithm in bits, by counting the number of symbols the program code comprises, and to compare it with the lengths of the symbol sequences it generates or operates on.

The string (4.1.1) for example, could be generated by a simple FORTRAN code as follows:

```
      PROGRAM  SEQUENCE 411
      CHARACTER*32  SEQ411
   10 SEQ411 = '10011110001101110111100110111001'
      WRITE(6,A32) SEQ411
      STOP
      END
```
$$(4.2.1)$$

It is evidently not an example of clever programming. By contrast, sequence (4.1.2) allows for a smarter solution involving a loop,

```
      PROGRAM  SEQUENCE  412
      CHARACTER*2   PER412
      CHARACTER*32  SEQ412
      INTEGER  N
   10 PER412 = '10'
      N = 16
      SEQ412 = PER412
      DO 20  I = 1,N - 1
   20 SEQ412 = SEQ412//PER412
      WRITE(6,A32) SEQ412
```

STOP

END   (4.2.2)

On the face of it, the code (4.2.2) is longer than (4.2.1). The decisive difference lies in the dependence of the respective program sizes on the length of the symbol sequence that has to be entered as input in line 10: As (4.2.1) contains a verbatim specification of the sequence it generates, it grows linearly, up to a finite overhead, with the number of symbols in the sequence, while (4.2.2) only requires stating, once and forever, the period "10" to be repeated, together with the desired number N of repetitions, which can be specified with $\log(N)$ symbols (their precise number depends on the system used—binary, decimal, etc.). Therefore, defining the program size as $I(C)$, to be measured as the code length in bits in a given programming language, we can conclude that

$$\lim_{N\to\infty} \frac{I((4.2.2), N)}{I((4.2.1), N)} \sim \lim_{N\to\infty} \frac{\ln(N)}{N} = 0. \qquad (4.2.3)$$

By this criterion, sequence (4.1.1) would qualify as random (but wait!), while (4.1.2) definitely proves to be non-random (but see below!).

An analogous reasoning can indeed be invoked as well to justify induction on a technically sound basis [Sol64]. Considering sequence (4.1.2) as a time series of binary data, the code (4.2.2) allows to continue the sequence periodically, "by induction", just by following the loop 20 identically further in the discrete "time variable" I, the loop counter. Any deviation from this minimal scheme requires a significant modification of the code, inevitably increasing its length.

Data compression, as quantified in Eq. (4.2.3), wherever it applies, leads to a clear separation of lawfulness and contingency. It identifies that part of the algorithm that does not depend on any specific input, such as the DO-loop 20 in (4.2.2), with the general law, the pattern to be detected, and defers the individual input, such as line 10 in (4.2.2), to the individual initial condition, that is, to contingency.

Generalizing this idea suggests to define the *algorithmic complexity* of a symbol sequence as the shortest algorithm capable of reproducing this sequence [Cha75, Cha82, Kol98]:

> **The algorithmic complexity of a symbol sequence is given by the size of the shortest possible program that can generate this sequence.**

Randomness thus reduces to incompressibility in terms of algorithmic complexity:

> **A series of symbol sequences $S(N)$, comprising N symbols each, is random if and only if its algorithmic complexity grows linearly with N.**

This definition comes with two caveats: The length of a code, hence algorithmic complexity, depends on the programming language used. That is not a severe problem. The translation from one language to another can be encoded in a translation program, a compiler, to be included as an extra part of the code, a subroutine, of a size independent of N.

A much more serious objection concerns the fact that the random nature of a sequence S, depending on its algorithmic complexity, implicitly hinges on finding the shortest possible algorithm that determines the entire sequence, that is, a *lower* bound of its length. In order to define an infimum rigorously, the range has to be specified over which it is to be sought. In the present case of program codes C, it becomes well defined only if the code size $I(C)$ remains bounded *from above* by a fixed maximum length ($\forall C$: $I(C) \leq I_{max}...$). Without such an upper bound, the minimum condition inherent in algorithmic complexity ($...\neg \exists C'$: $I(C') \leq I(C) \wedge [C'$ generates $S]$) would lose its restrictive effect. That is, if all algorithms shorter than a threshold I_{max} have failed to reproduce S, an algorithm longer than I_{max} might still exist that does the job. The consequences of this argument reach far beyond the question of rigorous criteria for randomness and will be elucidated in the following section.

By the same token, any algorithm that *is* able to reproduce a given sequence imposes an *upper* bound to its algorithmic complexity. If its size grows less progressively than proportional to N, the existence of this algorithm proves that the sequence can be compressed, hence is not random. As is evident from example (4.2.1), verbatim reproduction requires a program comprising the sequence of length N itself plus a constant overhead N_O. The number of possible programs C of size $I(C) \leq N - N_O$, even with positive N_O, can be estimated by combinatorics to be of the order of $\exp(N - N_O)$. Since each code defines at most a single sequence, the number of sequences of algorithmic complexity $< I(C)$ is of the same order of magnitude, so that according to this criterion, only a fraction of about $\exp(-N_O)$ of all sequences of length N is definitely non-random, the rest is potentially random—that is, the huge majority of all other sequences *could* qualify as random. Inversely, the comparability of codes and sequences implies that a string encoding a minimal program is itself necessarily a random string, irrespective of the symbol sequences this code may generate. Otherwise, a different, even shorter algorithm existed that is capable of generating the code in question and with it all the sequences produced by this code.

While accordingly, the non-random nature of a symbol string can verified, proving minimality, hence randomness, in this framework encounters insurmountable difficulties [Cha75, Cha82]. This gives the concept of randomness a distinctly subjective component: A sequence is qualified as random as long as we do not find an underlying algorithm that allows to compress it.

4.3 Gödel's Theorem and Incompleteness

The consequences of these considerations around randomness reach in fact much further, they bear on fundamental issues of mathematics. The link is given by an analogy between algorithms and formal systems: In mathematics, a theory is ideally organized as an axiomatic system. Prominent instances are Euclidean and Riemannian geometry, set theory, and number theory. Essential elements of this structure have already been addressed with the example of formal logics (Sect. 2.1). Starting from a set of propositions, the premises, a list of rules of inference allows deducing other propositions by recursive application of these rules. In mathematical terminology, the premises are *axioms*, the derived propositions are *theorems*. This time-honoured concept, going back to the paradigm of Greek geometry, is highly relevant also for randomness as it allows to cast compressibility in the rigorous terms of mathematical logics.

4.3.1 Formal Systems

Not every collection of propositions is acceptable as a set of axioms. In order to qualify as such, they have to comply with two conditions:

- They must be **consistent**, free of contradictions between them. A single contradiction suffices to reduce the potential information content of a set of axioms A to zero, $I_{pot}(A) = 0$, so that any arbitrary proposition could be derived from it (see Sect. 2.1.4): *ex falso quodlibet*.
- They should be **non-redundant**, i.e., none of the axioms should be deductible from the others. This condition guarantees that a set of axioms is minimal in a well-defined sense, as explained above in the context of Occam's razor.

Besides the list of axioms, a complete formal system comprises three more elements, defining its syntax:

- An **alphabet**, that is, a list of all elementary symbols to be used to compose strings within the system.
- A **grammar**, encompassing all rules that restrict the concatenation of elements of the alphabet.
- A set of **rules of inference** which specify how, from a given valid symbol string, other valid strings can be composed.

The analogy with algorithms is established by associating axioms with input, the inference rules with the algorithm proper, and theorems with output.

Another instructive analogy arises if we compare formal structures, defined by algorithms and axiomatic systems, with dynamical systems. Now axioms become initial conditions, inference rules become evolution laws, and theorems compare to final states. This analogy suggests a point of view converse to that of compressibility: How far can we reach out in the forward direction into the space of valid theorems, starting from a set of axioms as initial conditions and equipped with rigorous rules how to navigate from there. In this sense, define the range of a formal system as follows

> The range $R(A)$ of a formal system A is the set of all symbol strings that can be deduced within A, including its axioms.

For example, considering chess as a formal system, the range of chess would comprise the set of all configurations that can be reached from the initial configuration through sequences of allowed moves. For chess, e.g., this is a finite set, comprising all possible configurations of pieces on the checkerboard. For most formal systems it will be infinite, but remains countable. In a similar fashion, we can define a qualified meaning of "truth", applicable within a given formal system, as derivability from the axioms and through the inference rules of that system,

A symbol string S is called "true within the formal system A" if and only if it can be deduced within A,

$$t(S, A) \leftrightarrow S \in R(A). \tag{4.3.1}$$

Specified like this, truth within a formal system amounts to a narrowed meaning of mathematical truth. Mathematical truth without qualification would then be the most comprehensive version of this concept.

These concepts insinuate the possibility of defining a hierarchy among formal systems: System A is more powerful than system B if the range of A includes and exceeds that of B, $R(B) \subset R(A)$, that is, if every symbol sequence derivable within B can also be derived within A, $t(S, B) \rightarrow t(S, A)$. Returning to algorithms, more generally to computing systems, it is evident that this notion of computational power is a suitable model for that of emulation, alluded to in the context of prediction (Sect. 3.2.1): Platform A is sufficient to emulate platform B thus amounts to $R(B) \subset R(A)$.

4.3.2 Gödel's Incompleteness Theorem and Provability of Randomness

The development that culminated in the conception and publication of Kurt Gödel's article "*Über formal unentscheidbare Sätze der Principia Mathematica und verwandter Systeme I*" ("On Formally Undecidable Propositions of Principia Mathematica and Related Systems I") originates in a historical controversy on the foundations of mathematics. The emergence of new axiomatic systems in the nineteenth century, for example replacing Euclidean by Riemannian geometry, raised serious questions how to demonstrate their consistency, in a situation where intuition and other "soft" criteria could not be trusted any longer. Interminable discussions on the validity of proofs among his contemporaries in the 1920s provoked the German mathematician David Hilbert to proposing a radical solution, following Whitehead's and Russell's trailblazing attempt to found mathematics in logics and set theory [WR10]: Reduce mathematics to formal systems, proofs to formal derivations, so that their validity as well as the consistency of the entire system can be decided "mechanically". Referring to axiomatic geometry, Hilbert is cited with the sentence that "Instead of talking about 'points, lines, and planes', we could just as well use the words 'tables, chairs, and beer mugs'." There would be a high price to be paid, though, depriving mathematics of all tangible meaning. Still, he considered it acceptable in trade for the huge gain in certainty and clarity.

In terms of the concepts defined above, Hilbert's strategy could be summarized in the assertion that for every branch of mathematics that can be formalized as an axiomatic system A, mathematical truth is unconditionally equivalent to truth within A, or that the range of A comprises the full set of true theorems in this field. He found an opponent of equal rank in Kurt Gödel. A convinced advocate of the Platonist view of mathematics, Gödel had no doubt that mathematics enjoyed a reality of its own right, was exploring *terra incognita*. In contrast to Hilbert, who saw mathematics reduced to a purely formal game of manipulating symbol sequences, void of any relation to reality, Gödel was convinced that new and surprising facts could be discovered by doing mathematics [Gol05].

Faced with Hilbert's program, Gödel's objective was to show that there existed mathematical truth beyond the reach of formal systems. He succeeded in converting this conviction into a rigorous mathematical proof that was to overthrow all previous views of the groundwork of mathematics, and that moreover added a number of revolutionary techniques to the toolbox of mathematical logic [NN56, NN85]. His proof is not only one of the greatest intellectual achievements of the twentieth century, together with relativity theory and quantum mechanics, it is also a unique piece of absurd mathematical theatre, comparable in its desperate beauty to Beckett's "En attendant Godot".

Gödel knew very well that with *Principia Mathematica*, Hilbert's program had already been implemented successfully in the realm of logics and set theory—he himself had contributed to this enterprise in the case of predicate logic. He therefore had to proceed to a richer field of mathematics to demonstrate the limitations of formal systems, and found a promising instance in arithmetic or number theory. It is sufficiently powerful to give rise to theorems that are mathematically true but out of reach of any formal system.

The objective of the proof was therefore to drive a wedge between the range of provability within a formal system and that of mathematical truth in number theory. A patent strategy would have been trying to find a particular theorem of undeniable truth, which at the same time could be shown not to be derivable from the axioms of arithmetic. However, Gödel was quite aware of the fact that such a particular counterexample could be readily invalidated, simply by adding it to the list of axioms, thus returning to the same situation as before with respect to the augmented formal system. Instead he opted for a proof based on a generic argument, sufficiently comprehensive to include all possible ad-hoc extensions of the formal system from the outset. This required working simultaneously on the level of symbol strings *within* the formal system, axiomatic arithmetic, as the object language (cf. Sect. 1.2) and on the level of statements *about* number theory, representing the metalanguage. The central claim of the theorem thus takes the form of a negative, almost self-referential statement: The theorem asserts of itself not to be deducible within the framework of axiomatic arithmetic.

Like this, it is inspired by a pattern of negative self-referential statements that has puzzled thinkers from the antiquity (Fig. 4.3), yet modifies them in a subtle but decisive point. The model is Epimenides' paradox, "All Cretans are liars", uttered by a Cretan, or in its most concise version,

Epimenides' paradox
Epimenides the Cretan says, "All the Cretans are liars."

Barber paradox
A barber announces that he shaves all men in town who do not shave themselves, and only them. Does this barber shave himself?

Russell's paradox
Call sets "normal" that do not contain themselves as an element, those who do "non-normal". Is the set of all normal sets itself normal?

Berry's paradox
What is the smallest natural number that can only be specified with more characters than this sentence comprises?

Richard's paradox
Assign natural numbers to arithmetic properties of positive integers (e.g., being even, odd, square, etc.). Call numbers that do not themselves share the property assigned to them "Richardian". Is the number to be associated to the property of being Richardian itself a Richardian number?

Bohr's paradox
"Professor Bohr, you had a horseshoe fixed over the entrance of the Copenhagen Institute of Theoretical Physics. Are you supersticious?" – "Not at all. But I have been told that horseshoes do help, even if you do not believe in them."

Fig. 4.3 Paradoxical statements involving a negative self-reference have perplexed thinkers since antiquity. Examples reach from Epimenides' paradox to creations of the twentieth century that accompanied Bertrand Russell's attempt to found mathematics in set theory or served as models for Kurt Gödel's proof of his Incompleteness Theorem. In many cases, the paradox resides in the implicit attempt to assume a higher viewpoint, manifest in the use of metalanguage, which however is forced down to object level by self-reference.

This sentence is false.

It has been succeeded by a long list of paradoxes, all following the same basic scheme and considered as paradigms for Gödel's theorem. Given the intended structure of negative feedback between object and metalanguage, Gödel faced a formidable constructive challenge: Instead of the improvised self-reference inherent in those paradoxes, which requires an intelligent and benevolent listener to be comprehended, he had to phrase the statement in a logically rigorous, unequivocal form. He needed to introduce a code that could be interpreted simultaneously as symbol strings, on the level of the formal system, and as assertions referring to natural numbers, on the level of number theory, creating a short circuit between object and metalanguage. The attempt to talk about oneself from a higher point of view, but within the same system, a "mental bootstrap", is common to the above paradoxes and is also at the heart of Gödel's proof (Fig. 4.4).

a   b

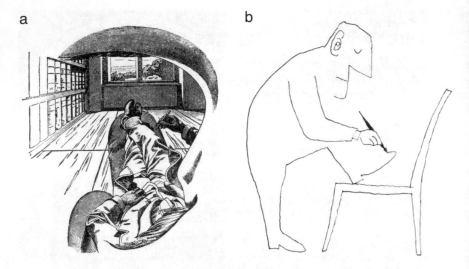

Fig. 4.4 Two artists' views of self-rendering. The physicist Ernst Mach, in a pseudo-realistic illustration to his "Antimetaphysical Introductory Remarks" to *Contributions to the Analysis of Sensations* [Mac97], draws the limits of his self-portrayal so far inside that they border on self-inspection (**a**). Saul Steinberg, cartoonist with substantial philosophical background, achieves a faithful visualization of self-referential statements such as the liar's paradox, models for Gödel's proof (Fig. 4.3). In his drawing, the graphical "metalanguage level" (the hand with the drawing pen) couples directly to the object level (the shoe that is being drawn), thus literally closing a loop (compare to Fig. 3.16). His playful entanglement of object and image recalls Magritte's "Ceci n'est pas une pipe" (Fig. 1.3). Reproduced with permission from [Ste45].

For this purpose, he invented a technical device that has become known as *Gödel numbering* (Table 4.2). The relation he sought to establish between symbols and numbers had to be invertible, that is, every symbol sequence had to be identified with a unique code number, and no number should encode more than a single symbol string. Arithmetic provides a singular option for it: prime factorization. The same unique fingerprint that makes prime numbers a crucial asset in cryptography [Jev79, Sta90] also served Gödel to obtain a one-to-one reconstruction of symbol strings from their numerical code. Moreover, a partition of a symbol string into various substrings can be incorporated by applying the same technique to the substrings, using their individual Gödel numbers as exponents for the numbering of the total string.

As an additional profit Gödel achieved with this mapping, he also obtained a unique relationship even between assertions in the metalanguage, such as claims on the derivability of a symbol string, and arithmetic properties of the corresponding Gödel numbers. For example, the property that a string C is composed by concatenating the two partial strings A and B translates into the arithmetical fact that the Gödel number for C factorizes into those for A and B, $N(C) = N(A) \times N(B)$.

Combining these ingredients, Gödel reached codifying the statement (call it $G(n)$)

$$\texttt{The string with the Gödel number } n$$
$$\texttt{is not derivable.} \qquad\qquad (4.3.2)$$

Table 4.2 Simplified version of the symbol numbering Gödel introduced as a mapping from a formal system comprising axiomatic arithmetic back into the natural numbers. Besides a number code for the elementary symbols of the formal system, it comprises a code for symbol strings that uses prime factorization to achieve uniqueness and reversibility of the mapping.

Symbol	Gödel number	Meaning
¬	1	NOT
∨	2	OR
⊃	3	IMPLIES
∃	4	THERE IS
=	5	equals
0	6	zero
s	7	successor of
(	8	opening bracket
)	9	closing bracket
,	10	comma
+	11	plus
×	12	times

Logical variable	Gödel number
a	13
b	16
...	$13 + 3(n-1)$

Numerical variable	Gödel number
x	14
y	17
...	$14 + 3(n-1)$

Predicate variable	Gödel number
p	15
q	18
...	$15 + 3(n-1)$

A symbol string S, composed of symbols $s_1, s_2, s_3, \ldots, s_m$ with individual Gödel numbers $N_1, N_2, N_3, \ldots, N_m$ (see above), is encoded as a whole by the Gödel number

$$N(S) = \prod_{n=1}^{m} p_n^{N_n},$$

where p_n stands for the nth prime number ≥ 2 ($p_1 = 2, p_2 = 3, p_3 = 5, p_4 = 7, \ldots$).

Unlike the liar's paradox, $G(n)$ does not claim downright falsehood, a semantic proposition, it only denies derivability, a syntactic feature, within axiomatic number theory. Since derivability (and its negation) can be mapped to arithmetic properties by Gödel numbering, the Gödel number of $G(n)$, call it $N(G(n))$, defines an arithmetic function $N_G(n)$ of the integer argument n,

$$N_G : \mathbb{N} \to \mathbb{N}, \quad n \mapsto N_G(n) := N(G(n)), \tag{4.3.3}$$

(in the framework of axiomatic arithmetic, the integer n, as part of the string G, has to be expressed recursively by applying the relation "successor of" n times to the seed 0, $n = s(s(s(\ldots s(0)\ldots)))$, before applying the procedure as in Table 4.1).

The final step consists in closing the self-referential loop by choosing a value for n, say $n = n_G$, such that it equals the Gödel number of $G(n_G)$. That this is indeed possible is ascertained by the *diagonal lemma* [Raa15]. Its validity, a cornerstone of Gödel's theorem, is implicit in Gödel's proof itself: Within axiomatic arithmetic, no more can be said than that the theorem

$$\exists n_G : n_G = N(G(n_G)), \tag{4.3.4}$$

or in plain text,

> There exists an argument n_G that
> encodes the formula "The formula with
> the Gödel code n_G is not derivable.".   (4.3.5)

can be constructed. It can be true or false only if referring to a specific model, in particular by defining the range of n_G in formula (4.3.4) to be the natural numbers,

$$\exists n_G \in \mathbb{N} : n_G = N(G(n_G)), \tag{4.3.6}$$

or in plain text,

> There exists a natural number n_G which, as a
> Gödel number, encodes the string "The string
> with the Gödel number n_G is not derivable.".   (4.3.7)

Now should (4.3.6) be *false*, that would imply that the string $G(n_G)$ can be derived within axiomatic arithmetic. If a false statement can be formally proven in arithmetic, that would condemn arithmetic to inconsistency—a mathematical apocalypse. If, on the contrary, (4.3.6) is *true*, it constitutes an instance of mathematical truth that cannot be corroborated by a formal proof within an axiomatic system. In this way, it turns out that the statement (4.3.6), while evidently involving self-reference, does not belong to the class of self-referential statements with negative feedback, such as Epimenides' paradox (Fig. 4.3). It is precisely by nesting itself in the fringe between derivability and the limits of mathematical truth that it escapes the perpetual oscillation between truth and falsehood the negative feedback would otherwise induce (Fig. 4.5).

4.3.3   Interpretations and Consequences of Gödel's Incompleteness Theorem

With his proof, Gödel not only contributed a strong argument against Hilbert's program, with an immediate corollary he overthrew it directly: The truth of (4.3.6) follows by logical inference (*reductio ad absurdum*) from the consistency of arithmetic, but not by formal deduction! Therefore, if consistency could be derived formally within the axiomatic system, so could statement (4.3.6): It would be deducible, hence would be false. The consequence, namely that the consistency of arithmetic cannot be formally proven, is known as Gödel's second Incompleteness Theorem.

A similar corollary concerns an obvious immediate response to Incompleteness: "Just add the 'missing mathematical truth' as an additional item to the original list of axioms!" as alluded to above. This strategy, however, is precisely prevented by the generic nature of Gödel's construction. It applies as is to *any* particular list of axioms for arithmetic whatsoever, hence cannot be invalidated by an ad hoc amendment of the list.

A striking feature of Gödel's proof and evidence of its ingenuity is that it implies the existence of mathematical truth outside the axiomatic system *without explicitly stating any such truth.* With (4.3.4) he merely claims the *existence* of an $n_G \in \mathbb{N}$ with the alleged property, but he does not state this number and in fact does not need to. Gödel's theorem demonstrates that there is truth beyond formal provability, that the range of axiomatic arithmetic has an open frontier, without actually trespassing this borderline. Are there any specific candidates for formally unprovable yet true mathematical theorems? Among the highly suspects is the Goldbach conjecture (every even number larger than two can be expressed in at least one way as a sum of two primes), which up to now could neither be refuted by a counterexample nor proven by a formal derivation.

Returning to randomness, from a wider point of view the Incompleteness Theorem bears on the issue of compressibility of information, calling attention to the crucial question of its dependence on the system that achieves the compression. It demonstrates that the system of natural numbers possesses patterns that cannot be detected from within axiomatic arithmetic. Interpreted in this way, its relevance for randomness becomes evident. Gödel's strategy offers a scheme that can be transferred directly to the provability of randomness, more precisely of algorithmic complexity [Cha75, Cha82]: Instead of Gödel's statement G, consider a string S of size N_S or larger, and instead of its truth, its randomness. In order to prove that its algorithmic complexity is of the order of N_S, it must be shown that no algorithm A of size N_A exists that reproduces S but is significantly shorter, $N_A < N_S$. Such a proof would have to check systematically all possible algorithms of size up to N_S in the programming language used (all possible derivations in the formal system at hand) in order of their length, whether they are able to reproduce S. Such an algorithm, see the code (4.2.2), would consist of a constant overhead, say of size N_O, and a part increasing at least logarithmically with the length N_S of the expected output string, see again (4.2.2), thus would have a total size of at least

$$N_A = N_O + a \log_2(N_S). \tag{4.3.8}$$

with some positive constant a. If it finds an algorithm of length $N_A < N_S$, the string S is not random and the proof failed. In order to proceed till the target size $N_A \approx N_S$ (an algorithm of this size that outputs S can always be constructed, based on verbatim reproduction, see (4.2.1)), the proof-checking program itself would have to be at least of this size.

As a consequence, the complexity of a string can only be proven using algorithms at least as large as the string itself. A proof of the randomness of S, shorter than

S itself, must itself be able to compress S to a shorter size, if only to handle it. Therefore, could a shorter proof checker be found, it would amount to proving the randomness of S with an algorithm shorter than S, hence compressing it, in contradiction to the premise. However, since the definition of algorithmic complexity (cf. Equation (4.2.3)) requires letting $N_S \to \infty$, no proof of finite size can accomplish this goal. For strings generated as an infinite series, a conclusive proof of their randomness is therefore out of reach. In plain words, the undecidability of randomness means that, if no pattern compressing S has been found to date, this may simply mean that the genius still has to be born who will discern that pattern.

To state a simple example: The sequence (4.1.1) has been given as an instance of obvious randomness. In fact, it is not random at all, since it has been constructed systematically as the output of a deterministic chaotic map, starting from an initial value, $x_0 = \left(\sqrt{5} - 1\right)/2$, which is not a rational number. Even so, however, it can be computed by an algorithm. Also irrational numbers allow for systematic approximations by rationals, for example by means of recursive computation. (In fact, the notation $\sqrt{a}$ can only be understood adequately as referring implicitly to an algorithm calculating the square root). For example, the Babylonian method, a corollary of Newton's method [SB10], is a root-finding algorithm that consists in iterating the recursion formula

$$x_{n+1} = \frac{1}{2}\left(x_n + \frac{a}{x_n}\right),\tag{4.3.9}$$

which converges towards $\sqrt{a}$. Combined with a simple code to extract a binary expansion, it allows writing an algorithm that generates the sequence (4.1.1). It only depends on the desired length N of the sequence through a single input, the intended accuracy of the approximation in terms of valid binary digits, and therefore grows only as $\log(N)$.

Algorithmic complexity can be interpreted as a refined measure of information content, a measure that takes genuine novelty into account, revealing any kind of redundancy with the most sophisticated available methods. Paradoxically, the bedrock of information content, the information that by any objective measure is novel and surprising, the opposite of Shannon's indiscriminate symbol counting, proves to be characterized by randomness. Only random strings according to algorithmic complexity are incompressible. Therefore, axioms of formal systems must be random symbol strings in this sense.

Gödel's Incompleteness Theorem and algorithmic complexity provide rigorous models for the notion of a hierarchy of information processing systems, alluded to in Sects. 3.2.1 and 4.3.1, and indicate how to phrase it more precisely: Metalanguage statements referring to features of a formal system, an algorithm, even a symbol string, can only be made from the higher viewpoint of a system that is more powerful than the one it is talking about. They require an Archimedean point outside the system "to lift it". In the case of algorithmic complexity, that simply means it must be larger in terms of information content. Number theory, as demonstrated by Gödel, constitutes an even higher category, superior to formal systems of any size. It represents truth that is not accessible from any kind of formal system. Indeed, a key ingredient of Gödel's proof, Gödel numbering based on prime factorization, could not be achieved in a formal system since the very notion of prime numbers is beyond its reach. The paradoxes presented in Figs. 4.2 and 4.4 all have in common that they pretend ascending to the metalanguage level from within, that is, without actually providing the superiority of a larger system. They refer to themselves in a way that in fact would require being superior in capacity to themselves. Such attempts prove futile for the same reasons that restrict self-prediction, as pointed out in Sect. 3.2.3.

The theory of computing systems founded by Alan Turing (see Chap. 9) suggests that formal systems can be regarded as models for digital computers. Interpreted on basis of this analogy, the Incompleteness Theorem has far-reaching consequences, for instance for artificial intelligence: It would imply that theoretical bodies, including number theory or even more powerful, can generate truths that are out of reach of digital computing. Phrased more pathetically even, human minds conceiving and understanding proves such as Gödel's are uncatchably superior to digital computers and can never be simulated by them. The fact that the Incompleteness Theorem demonstrates the hopelessness of attempts such as Hilbert's to ground mathematics on a rigorous base, is sometimes considered as an irrevocable failure of the discipline as a whole. At the same time, however, it can be interpreted in a more optimistic sense [NN56, NN85]: It shows that mathematics will not run out of surprising truths to be discovered, and that mathematicians cannot be replaced completely by digital machines.

This view raises a challenging question: What capability might actually distinguish the human mind as information processor so favourably from its own creations, digital computers? There is at least one promising candidate: creativity. As pointed out in Sect. 3.1.5, learning and even high-performance perception require mental creativity as a fundamental resource, in the same way as genetic evolution is fueled by random mutations. High-level pattern recognition is not possible without the creation of random patterns from scratch, generated top-down, to be matched with structures in perceived data [GK&98, EFS01, Sin02]. Considerations around Gödel's theorem seem to indicate that here, creativity has to be understood more precisely as selective randomness: optimized to search preferably in the most promising directions for a specific purpose and qualified in the sense of irreducibility in terms of algorithmic complexity. Since unpredictable behaviour is a vital factor of fitness in various species, including notably primates (see Sect. 3.3.2), it is evident that indeed, a strong selection pressure towards creativity in this sense must have acted through the entire evolution of hominids.

Another unique feature distinguishing human minds from artificial intelligence might be the capacity of self-perception. Self-reference, such a vital resource in Gödel's construction, is already implemented in the human brain, in that the brain is able to watch its own present state as if observing an external stimulus. It is even speculated that this feature, a recent achievement of brain evolution located in the neocortex, is the essential ingredient of consciousness [Flo91, Flo95, Rot97]. In any case, it is plausible that only systems capable of self-perception could understand arguments such as Gödel's and even conceive them in the first place.

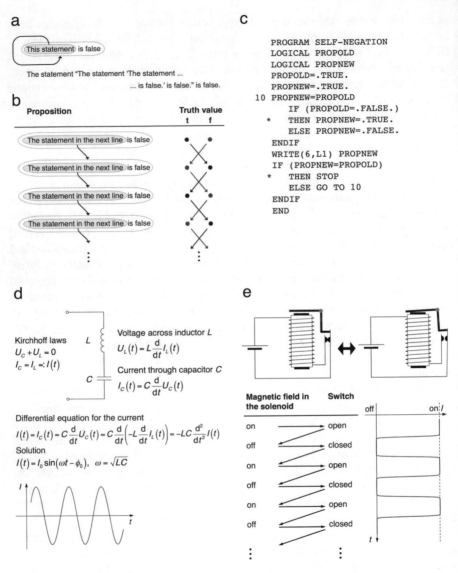

a

This statement is false

The statement "The statement 'The statement ...
... is false.' is false." is false.

b

Proposition	Truth value
	t f

The statement in the next line is false

The statement in the next line is false

The statement in the next line is false

The statement in the next line is false

c

```
PROGRAM SELF-NEGATION
LOGICAL PROPOLD
LOGICAL PROPNEW
PROPOLD=.TRUE.
PROPNEW=.TRUE.
10 PROPNEW=PROPOLD
   IF (PROPOLD=.FALSE.)
*      THEN PROPNEW=.TRUE.
       ELSE PROPNEW=.FALSE.
   ENDIF
   WRITE(6,L1) PROPNEW
   IF (PROPNEW=PROPOLD)
*      THEN STOP
       ELSE GO TO 10
   ENDIF
   END
```

d

Kirchhoff laws
$U_C + U_L = 0$
$I_C = I_L =: I(t)$

Voltage across inductor L
$$U_L(t) = L\frac{d}{dt}I_L(t)$$

Current through capacitor C
$$I_C(t) = C\frac{d}{dt}U_C(t)$$

Differential equation for the current
$$I(t) = I_C(t) = C\frac{d}{dt}U_C(t) = C\frac{d}{dt}\left(-L\frac{d}{dt}I_L(t)\right) = -LC\frac{d^2}{dt^2}I(t)$$
Solution
$$I(t) = I_0\sin(\omega t - \phi_0), \quad \omega = \sqrt{LC}$$

e

Magnetic field in the solenoid	Switch	
on	open	
off	closed	
on	open	
off	closed	
on	open	
off	closed	

off   on

Fig. 4.5 Self-referential statements such as the liar's paradox (Fig. 4.3) merge meta- and object language in a logically inconsistent manner. Written explicitly, they in fact amount to infinitely nested propositions where each outer segment is a metastatement with respect to the next inner segment (**a**). Like this, a negative self-referential statement can be "unrolled" (**b**) to form a recursive sequence that generates alternating truth values. It is equivalent to a computer code (**c**), here in FORTRAN, that contains an infinite loop. The alternation of truth values it generates is similar to oscillations in analogue physical systems involving negative feedback, such as an LC circuit that simulates a harmonic oscillator (**d**) or a relays circuit as it is used in electromechanical doorbells to generate vibrations (**e**).

Chapter 5
Information in Classical Hamiltonian Dynamics

After a long excursion through applications of the concept of information in various disciplines, venturing as far afield as genetics, epistemology, and formal logics, the present chapter will lead us in the opposite direction, complementing the wide scope with a solid formal basis in physics. The purpose is not only to provide the necessary quantitative concepts, anchoring the dynamics of information in basic physical dynamics, but also to present a central feature of entropy that is decisive for all discussions of its role in physics: A conservation law, assuring that it forms a constant of motion analogous to energy but valid even in time-dependent Hamiltonian systems, has far-reaching consequences in particular if confronted with thermodynamics. Starting from a brief review of Hamiltonian mechanics and symplectic geometry, we shall arrive at clear symptoms of a fundamental incompleteness of classical mechanics, related to the concept of information, which ultimately induced and motivated its replacement by quantum mechanics.

5.1 Review of Hamiltonian Dynamics and Symplectic Geometry

Classical mechanics comes in a variety of guises, formats, and presentations, as elementary as Newton's equations, as sophisticated as fibre bundle theory. For the present purpose, Hamilton's formulation is ideally suited as an interface between mechanics and measures of information and their time evolution [JS98]. Symplectic geometry, the mathematical framework for the treatment of phase space without any reference as to the interpretation of its coordinates, matches well with the basic concept of distinguishable states underlying Boltzmann's entropy. Shannon's generalization towards probabilities associated to states then leads naturally to the dynamics of continuous distributions in phase space.

The state space appropriate for Hamiltonian mechanics is *phase space* [GPS01, JS98], i.e., a real vector space of even dimension $2f$, $\mathbf{r} \in \mathbb{R}^{2f}$. Usually, half of

© Springer Nature Switzerland AG 2022
T. Dittrich, *Information Dynamics*, The Frontiers Collection,
https://doi.org/10.1007/978-3-030-96745-1_5

these dimensions are allocated to position coordinates, the other half to the momenta canonically conjugate to these positions, that is

$$\mathbf{r} = (\mathbf{p}, \mathbf{q}) = \left(p_1, \ldots, p_f, q_1, \ldots, q_f\right) \in \mathbb{R}^f \oplus \mathbb{R}^f, \qquad (5.1.1)$$

but this particular interpretation already breaks the isotropy of phase space, so essential for symplectic geometry. Exceptions are only those cases where, for example, an angle as coordinate is associated (canonically conjugate) to an angular momentum, $(p, q) = (\phi, I) \in S^1 \otimes \mathbb{R}$, so that the topology of phase space is that of a cylinder. Wherever it is necessary keeping positions and momenta apart, we shall adopt the convention to assign the first f components of the phase space vector to the momenta, the remaining f to positions.

Introducing time as a continuous parameter counting subsequent states, the transition rules take the form of Hamilton's equations of motion [GPS01, JS98],

$$\frac{d\mathbf{p}}{dt} = -\frac{\partial H(\mathbf{p}, \mathbf{q})}{\partial \mathbf{q}}, \frac{d\mathbf{q}}{dt} = \frac{\partial H(\mathbf{p}, \mathbf{q})}{\partial \mathbf{p}}. \qquad (5.1.2)$$

The Hamiltonian $H(\mathbf{r})$, a real-valued scalar function on phase space, contains all the information required to determine its time evolution uniquely. In this sense, initial conditions and Hamiltonian are analogous to input and program, resp., of a computer. In the unified symplectic notation, Eq. (5.1.1), the equations of motion reduce to

$$\frac{d\mathbf{r}}{dt} = \mathbf{J}^t \frac{\partial H(\mathbf{r})}{\partial \mathbf{r}}. \qquad (5.1.3)$$

While Eq. (5.1.2) appears to break the isotropy of phase space, the symplectic version (5.1.3) shows that this is not really the case: The symplectic unit matrix $\mathbf{J}$ effects a $\pi/2$ rotation within each of the f freedoms in phase space,

$$\mathbf{J} = \begin{pmatrix} \mathbf{0}^{(f)} & \mathbf{I}^{(f)} \\ -\mathbf{I}^{(f)} & \mathbf{0}^{(f)} \end{pmatrix}, \mathbf{0}^{(f)} = \begin{pmatrix} 0 & \cdots & 0 \\ \vdots & \ddots & \vdots \\ 0 & \cdots & 0 \end{pmatrix}, \mathbf{I}^{(f)} = \begin{pmatrix} 1 & \cdots & 0 \\ \vdots & \ddots & \vdots \\ 0 & \cdots & 1 \end{pmatrix}, \qquad (5.1.4)$$

where $\mathbf{0}^{(f)}$ and $\mathbf{I}^{(f)}$ denote the $(f \times f)$ zero and unit matrices, resp., that is, it replaces p_n by q_n and q_n by $-p_n$, $n = 1, \ldots, f$. The properties

$$\mathbf{J}^t = \mathbf{J}^{-1} = -\mathbf{J}, \mathbf{J}^2 = -\mathbf{I}^{(2f)}, \mathbf{J}^4 = \mathbf{I}^{(2f)}, \det(\mathbf{J}) = 1, \qquad (5.1.5)$$

substantiate the interpretation of $\mathbf{J}$ as a $\pi/2$ rotation.

As innocuous as it may appear, Eq. (5.1.3) implies a wealth of consequences that are crucial for the physical interpretation of Hamiltonian mechanics. To state but the most striking ones [JS98],

1.   Hamilton's equations constitute a coupled set of $2f$ first-order partial differential equations.
2.   For a given initial condition $\mathbf{r}(t')$, the equations of motion determine uniquely the states $\mathbf{r}(t'')$ of the system at all other times t'', before or after t': Preparing the system at identical initial conditions, also the final state will be reproduced identically. In geometric terms, this means that exactly one trajectory passes through every point in phase space.
3.   Equivalently, (i) the $2f$-dimensional phase space is a unique and complete representation of the state of a mechanical system, and (ii) Hamiltonian time evolution is *Markovian*, i.e., the present state $\mathbf{r}(t)$ alone determines all past and future states of the system completely. With this property, Hamiltonian mechanics satisfies all the criteria for a deterministic time evolution, illustrated in Fig. 3.1f, and indeed it is a paradigm of determinism.
4.   Unlike a general dynamical system with states $\mathbf{x}(t) \in \mathbb{R}^d$, where a time evolution might be determined, e.g., by a differential equation such as

$$\frac{d\mathbf{x}}{dt} = \left(\frac{\partial}{\partial t} f_1(\mathbf{x}), \ldots, \frac{\partial}{\partial t} f_d(\mathbf{x}) \right), \tag{5.1.6}$$

with a set of d separate functions $((f_1(\mathbf{x}), \ldots, f_d(\mathbf{x}))$ determining the fate of the system, it is the single scalar function $H(\mathbf{r})$ that takes this role in Hamiltonian dynamics (Fig. 5.1).
5.   Even if trajectories do not cross, split, or merge (see 2), they might still converge to or diverge from a common asymptote. However, not even that is possible, owing to a property of the Hamiltonian flow, i.e., the field of phase-space velocities,

$$\dot{\mathbf{r}} = \frac{d\mathbf{r}}{dt} =: \mathbf{F}(\mathbf{r}), \mathbf{F}(\mathbf{r}) = \mathbf{J}^t \frac{\partial H(\mathbf{r})}{\partial \mathbf{r}}. \tag{5.1.7}$$

Its divergence vanishes, that is, the flow is free of sources and sinks,

$$\operatorname{div}\mathbf{F}(\mathbf{r}, t) = 0. \tag{5.1.8}$$

6.   In integral form, the same feature assumes the form of a conservation law for phase-space volume, the quantity of phase space occupied by a set of initial conditions,

$$\frac{dV}{dt} = 0. \tag{5.1.9}$$

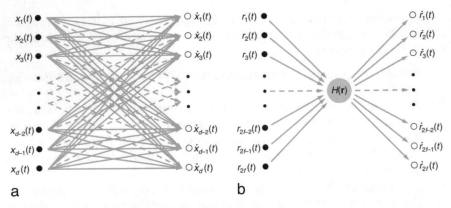

Fig. 5.1 In a typical d-dimensional dynamical system, if the time evolution is determined by a set of first-order differential equations (**a**), all the d time derivatives depend separately on all the d momentary values of the variables. By contrast, in Hamiltonian mechanics (**b**), the information contained in the momentary $2f$-dimensional phase-space vector $\mathbf{r}(t)$ is processed through the "bottleneck" of a single scalar function, the Hamiltonian $H(\mathbf{r})$, to generate the $2f$ time derivatives $\dot{\mathbf{r}}(t)$.

In hydrodynamic terms, this conservation of volume can be interpreted as incompressibility of the fictitious fluid formed by a set of initial conditions in phase space. It will prove pivotal for the conservation of information in Hamiltonian dynamics.

Both properties, Eqs. (5.1.8) and (5.1.9), are readily verified: Invoking the definitions of the Hamiltonian flow, Eq. (5.1.7), and of the symplectic unit matrix, Eq. (5.1.4), implies

$$\operatorname{div}\mathbf{F}(\mathbf{r}, t) = \frac{\partial}{\partial \mathbf{r}} \cdot \mathbf{F}(\mathbf{r}, t) = \begin{pmatrix} \partial/\partial \mathbf{p} \\ \partial/\partial \mathbf{q} \end{pmatrix} \cdot \begin{pmatrix} \mathbf{0} & -\mathbf{I} \\ \mathbf{I} & \mathbf{0} \end{pmatrix} \begin{pmatrix} \partial H/\partial \mathbf{p} \\ \partial H/\partial \mathbf{q} \end{pmatrix} = 0. \qquad (5.1.10)$$

Fig. 5.2 The conservation of phase-space volume with the Hamiltonian flow, Eq. (5.1.9), amounts to the incompressibility of a fictitious fluid. The volume initially occupied in phase space by a set of initial conditions may vary arbitrarily in shape, (without splitting or merging, cf. Fig. 3.2e, f), but not change its total measure as it moves with the Hamiltonian flow (arrows).

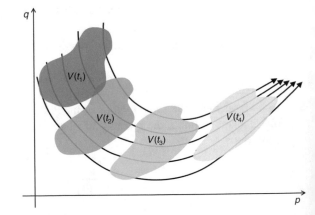

The final identity follows from the structure of the triple product in the penultimate member of the chain. In order to prove Eq. (5.1.9), define the volume V as the phase-space measure $dV = d^{2f} r = d^f p\, d^f q$, integrated over the interior of a closed surface $S = \delta V$. In terms of the characteristic function

$$\chi_V(\mathbf{r}) = \begin{cases} 1 & \mathbf{r} \in V, \\ 0 & \text{else}, \end{cases} \tag{5.1.11}$$

the volume change is

$$\frac{d}{dt} V = \frac{d}{dt} \int d^{2f} r\, \chi_V(\mathbf{r}) = \int d^{2f} r\, \frac{\partial \chi_V}{\partial \mathbf{r}} \frac{d\mathbf{r}}{dt}. \tag{5.1.12}$$

The characteristic function is constant everywhere, with the exception of its surface δV where it jumps from 0 (outside) to 1 (inside). The gradient $\partial \chi_V / \partial \mathbf{r}$ therefore is a vector pointing into the interior of δV, weighted by a delta function on the surface ($d\mathbf{S}$ denoting a directed infinitesimal surface element),

$$\frac{d}{dt} V = \oint_{\delta V} d\mathbf{S} \cdot \dot{\mathbf{r}} = \oint_{\delta V} d\mathbf{S} \cdot \mathbf{F}(\mathbf{r}). \tag{5.1.13}$$

Applying Gauss' Theorem to the surface integral and taking Eq. (5.1.10) into account,

$$\frac{d}{dt} V = \int_V d^{2f} r\, \mathrm{div}\mathbf{F}(\mathbf{r}) = 0. \tag{5.1.14}$$

For the discussion of entropy and entropy flow in Hamiltonian mechanics, the evolution over a finite time, obtained by integrating the equations of motion, is as relevant as the infinitesimal time evolution. It can be considered in terms of trajectories $\mathbf{r}(t)$, parameterized by time running from $-\infty$ to ∞ and determined by an anchor point $\mathbf{r}(t_0)$. As a complementary point of view, one considers the entire set of "final" points $\mathbf{r}(t)$ for a fixed propagation time t (which can be positive or negative), but as a function of the initial point running over the full phase space. Mathematically, this amounts to a transformation of phase space, parameterized by the initial and final times t', t'',

$$T_{t',t''} : \mathbb{R}^{2f} \to \mathbb{R}^{2f}, \mathbf{r}' \mapsto \mathbf{r}'' = T_{t',t''}(\mathbf{r}'). \tag{5.1.15}$$

If a transformation is induced by a Hamiltonian flow (if it is *canonical* [JS98]), its properties will reflect the characteristics of Hamiltonian dynamics pointed out above. In particular, both the initial and the final vector should obey Hamilton's Equation (5.1.3), with the same Hamiltonian. This means, on the one hand,

$$\dot{\mathbf{r}}' = \mathbf{J}^t \frac{\partial H(\mathbf{r}')}{\partial \mathbf{r}'}, \dot{\mathbf{r}}'' = \mathbf{J}^t \frac{\partial H(\mathbf{r}'')}{\partial \mathbf{r}''}. \tag{5.1.16}$$

On the other hand, $\mathbf{r}'$ and $\mathbf{r}''$ are related by a transformation, $\mathbf{r}'' = \mathbf{r}''(\mathbf{r}')$, so that

$$\dot{\mathbf{r}}'' = \frac{\partial \mathbf{r}''}{\partial \mathbf{r}'} \dot{\mathbf{r}}' = \frac{\partial \mathbf{r}''}{\partial \mathbf{r}'} \mathbf{J}^t \frac{\partial H(\mathbf{r}')}{\partial \mathbf{r}'} =: \mathbf{M}_{t',t''} \mathbf{J}^t \frac{\partial H(\mathbf{r}')}{\partial \mathbf{r}'}. \tag{5.1.17}$$

This introduces the stability or monodromy matrix [JS98],

$$\mathbf{M}_{t',t''} = \frac{\partial \mathbf{r}''}{\partial \mathbf{r}'} = \begin{pmatrix} \partial r_1'' / \partial r_1' & \cdots & \partial r_{2f}'' / \partial r_1' \\ \vdots & \ddots & \vdots \\ \partial r_1'' / \partial r_{2f}' & \cdots & \partial r_{2f}'' / \partial r_{2f}' \end{pmatrix}, \tag{5.1.18}$$

a $(2f \times 2f)$ matrix of derivatives of the image with respect to the original vector, or more explicitly, in terms of momentum and position components,

$$\mathbf{M}_{t',t''} = \begin{pmatrix} \partial p_1'' / \partial p_1' & \cdots & \partial p_f'' / \partial p_1' & \partial q_1'' / \partial p_1' & \cdots & \partial q_f'' / \partial p_1' \\ \vdots & \ddots & \vdots & \vdots & \ddots & \vdots \\ \partial p_1'' / \partial p_f' & \cdots & \partial p_f'' / \partial p_f' & \partial q_1'' / \partial p_f' & \cdots & \partial q_f'' / \partial p_f' \\ \partial p_1'' / \partial q_1' & \cdots & \partial p_f'' / \partial q_1' & \partial q_1'' / \partial q_1' & \cdots & \partial q_f'' / \partial q_1' \\ \vdots & \ddots & \vdots & \vdots & \ddots & \vdots \\ \partial p_1'' / \partial q_f' & \cdots & \partial p_f'' / \partial q_f' & \partial q_1'' / \partial q_f' & \cdots & \partial q_f'' / \partial q_f' \end{pmatrix}. \tag{5.1.19}$$

Using the same matrix to transform back from $\mathbf{r}'$ to $\mathbf{r}''$ in the last member of Eq. (5.1.17),

$$\dot{\mathbf{r}}'' = \mathbf{M}_{t',t''} \mathbf{J}^t \left(\frac{\partial \mathbf{r}''}{\partial \mathbf{r}'} \right)^t \frac{\partial H(\mathbf{r}'')}{\partial \mathbf{r}''} = \mathbf{M}_{t',t''} \mathbf{J}^t \mathbf{M}_{t',t''}^t \frac{\partial H(\mathbf{r}'')}{\partial \mathbf{r}''}, \tag{5.1.20}$$

and comparing with Eq. (5.1.13) leads to an explicit condition for the stability matrix,

$$\mathbf{M}_{t',t''} \mathbf{J}^t \mathbf{M}_{t',t''}^t = \mathbf{J}^t \tag{5.1.21}$$

(equivalently, $\mathbf{M}_{t',t''}^t \mathbf{J}^t \mathbf{M}_{t',t''} = \mathbf{J}^t$, $\mathbf{M}_{t',t''} \mathbf{J} \mathbf{M}_{t',t''}^t = \mathbf{J}$, or $\mathbf{M}_{t',t''}^J \mathbf{J} \mathbf{M}_{t',t''} = \mathbf{J}$). Combined with Eq. (5.1.5), this implies for the determinant that

$$\left| \det\left(M_{t',t''} \right) \right| = 1. \tag{5.1.22}$$

(If $\det(\mathbf{M}_{t',t''}) = -1$, the transformation includes spatial reflections in an odd number of dimensions.) The most important consequence of Eq. (5.1.22) concerns the Jacobian of phase-space integrals of arbitrary functions $f(\mathbf{r})$, upon changing variables by canonical transformation,

$$\int d^{2f} r'' f(\mathbf{r}''(\mathbf{r}')) = \int d^{2f} r' \left| \det\left(\frac{\partial \mathbf{r}''}{\partial \mathbf{r}'}\right) \right| f(\mathbf{r}') = \int d^{2f} r' |\det(\mathbf{M}_{t't''})| f(\mathbf{r}')$$

$$= \int d^{2f} r' f(\mathbf{r}'). \tag{5.1.23}$$

This is another manifestation of volume conservation in phase-space,

$$dV'' = d^{2f} r'' = \left| \det(\partial \mathbf{r}''/\partial \mathbf{r}') \right| d^{2f} r' = d^{2f} r' = dV'. \tag{5.1.24}$$

5.2 Hamiltonian Dynamics of Continuous Density Distributions

Trajectories tracing the motion of a point in phase space over time are a useful mathematical concept, but physically unrealizable. "Point particles" can be localized in phase space, at best, with a finite error margin for their position and momentum. Already in the present classical context, delta functions in phase space are excluded by the fact that they correspond to an infinite information, as is clear from the general discussion in Sect. 1.8 and will be worked out in detail below: If the minimum resolvable scale appearing in Eq. (1.8.4) is allowed to take the value $d_x = 0$, the information diverges. Similarly, in statistical physics where we are dealing with ensembles of systems so large that they only permit a probabilistic treatment, probability density distributions in phase space are the appropriate mathematical tool. Indeed, with Shannon's definition (1.3.9) in mind, the information content of physical systems should be calculated in terms of likelihoods.

While the volumes defined by characteristic functions, as in Eq. (5.1.11), still amount to the discrete alternative "inside" versus "outside", we here need smooth distributions that assign a probability density $\rho(\mathbf{r})$ to each phase-space point $\mathbf{r}$,

$$\rho : \mathbb{R}^{2f} \to \mathbb{R}^+, \mathbf{r} \mapsto \rho(\mathbf{r}). \tag{5.2.1}$$

They can be considered as differential quotients measuring probability per unit volume,

$$\rho(\mathbf{r}) = \frac{dP(\mathbf{r})}{dV} = \frac{dP(\mathbf{r})}{d^{2f} r}. \tag{5.2.2}$$

Conversely, probabilities proper are related to the measure by

$$dP(\mathbf{r}) = \rho(\mathbf{r})d^{2f}r. \tag{5.2.3}$$

In particular, the total probability is normalized,

$$\int d^{2f}r\rho(\mathbf{r}) = 1, \tag{5.2.4}$$

where the integral extends over all of phase space. If the probability $dP(\mathbf{r})$ is conserved even locally, Eq. (5.2.3) implies a general rule for the transformation of densities under transformations of the underlying space: The conservation law $\rho(\mathbf{r}'', t'')dV'' = dP(\mathbf{r}'') = dP(\mathbf{r}') = \rho(\mathbf{r}', t')dV'$, taking Eq. (5.1.22) into account, is equivalent to [Gas98]

$$\rho(\mathbf{r}'', t'') = \rho(\mathbf{r}', t')\frac{dV'}{dV''} = \rho(\mathbf{r}', t')\left|\det\left(\frac{\partial \mathbf{r}''}{\partial \mathbf{r}'}\right)\right|^{-1} = \rho(\mathbf{r}'(\mathbf{r}'', t', t'')), \tag{5.2.5}$$

where $\mathbf{r}'(\mathbf{r}'', t', t'')$ denotes the phase-space point reached by tracing the trajectory from $\mathbf{r}''$ at t'' backwards till t'. Probability densities in position and in momentum space, for example, are obtained as "marginal distributions", that is, as projections of $\rho(\mathbf{r})$ over all momenta or positions, resp.,

$$\rho_p(\mathbf{p}) = \int d^f q \; \rho(\mathbf{p}, \mathbf{q}), \; \rho_q(\mathbf{q}) = \int d^f p \; \rho(\mathbf{p}, \mathbf{q}). \tag{5.2.6}$$

Likewise, the probability density in the partial phase space $\mathbf{r}_n = (p_n, q_n)$ of the nth degree of freedom is found by integrating over all other freedoms,

$$\rho_n(\mathbf{r}_n) = \int_{-\infty}^{\infty} d^2r_1 \cdots \int_{-\infty}^{\infty} d^2r_{n-1} \cdots \int_{-\infty}^{\infty} d^2r_{n+1} \cdots \int_{-\infty}^{\infty} d^2r_{2f}\rho(\mathbf{r}) = \prod_{\substack{n'=1 \\ n'\neq n}}^{2f} \int_{-\infty}^{\infty} d^2r_{n'}\rho(\mathbf{r}). \tag{5.2.7}$$

Mean values of observables $A(\mathbf{r})$ that can be expressed as functions on phase space are calculated as weighted averages,

$$\langle A \rangle = \int d^{2f}r\,A(\mathbf{r})\rho(\mathbf{r}). \tag{5.2.8}$$

The evolution in time of phase-space densities, induced by Hamilton's equations of motion, will reflect the general features of the Hamiltonian flow pointed out above.

Contact with Eq. (5.1.3) is made by differentiating with respect to time,

$$\frac{d}{dt}\rho(\mathbf{r}, t) = \frac{\partial}{\partial \mathbf{r}}\rho(\mathbf{r}, t)\frac{d\mathbf{r}}{dt} + \frac{\partial}{\partial t}\rho(\mathbf{r}, t) = \left(\frac{\partial}{\partial \mathbf{r}}\rho(\mathbf{r}, t)\right)^{t}\mathbf{J}^{t}\frac{\partial H(\mathbf{r})}{\partial \mathbf{r}} + \frac{\partial}{\partial t}\rho(\mathbf{r}, t).$$

(5.2.9)

The partial time derivative $\partial\rho/\partial t$ refers to the time dependence of the phase-space density in the laboratory (fixed) reference frame (not in the comoving frame as the total time derivative), that is, to the *Eulerian specification* [Bat73] of the probability flow. The combination of two gradients of phase-space functions as in the preceding term is an instance of a *Poisson bracket* [LS98, GPS01], defined generally as

$$\{f, g\} := \left(\frac{\partial}{\partial \mathbf{r}}g(\mathbf{r})\right)^{t}\mathbf{J}^{t}\frac{\partial}{\partial \mathbf{r}}f(\mathbf{r}).$$

(5.2.10)

Poisson brackets constitute a Lie product between the two phase-space functions forming their arguments; as a particular consequence, $\{f, g\} = -\{g, f\}$. This allows writing Eq. (5.2.8) as

$$\frac{d}{dt}\rho(\mathbf{r}, t) = \{\rho, H\} + \frac{\partial}{\partial t}\rho(\mathbf{r}, t).$$

(5.2.11)

If the total time derivative vanishes, as in the absence of death and birth processes, Eq. (5.2.11) reduces to

$$\frac{\partial}{\partial t}\rho(\mathbf{r}, t) = \{H, \rho\}.$$

(5.2.12)

This evolution equation for phase-space densities, a corollary of Hamilton's equations of motion, is known as the *Liouville equation*.

Advancing from infinitesimal to finite time, the time-evolved density $\rho(\mathbf{r}'', t'')$ is formally related to the initial $\rho(\mathbf{r}', t')$ by a relation of the form

$$\rho(\mathbf{r}'', t'') = \int d^{2f}r' G(\mathbf{r}'', t''; \mathbf{r}', t')\rho(\mathbf{r}', t'),$$

(5.2.13)

involving the *propagator* $G(\mathbf{r}'', t''; \mathbf{r}', t')$ as integral kernel. In deterministic Hamiltonian mechanics, see Eq. (5.2.5), it is readily found to be $G(\mathbf{r}'', t''; \mathbf{r}', t') = \delta(\mathbf{r}'' - \mathbf{r}''(\mathbf{r}', t'', t'))$, where $\mathbf{r}''(\mathbf{r}', t'', t')$ is the result of integrating Hamilton's equations of motion from the initial condition $\mathbf{r}'$ forward in time from t' to t'' (the propagation inverse to that figuring in Eq. (5.2.5)), so that

$$\rho(\mathbf{r}'', t'') = \int d^{2f} r' \delta(\mathbf{r}'' - \mathbf{r}''(\mathbf{r}', t'', t')) \rho(\mathbf{r}', t'). \tag{5.2.14}$$

Alternatively, consider Eq. (5.2.12) as an operator equation [Gas98],

$$\frac{\partial}{\partial t} \rho(\mathbf{r}, t) = \hat{L} \rho(\mathbf{r}, t), \tag{5.2.15}$$

with the linear differential operator $\hat{L} := \{H, \cdot\}$. The formal solution of this first-order differential equation is

$$\rho(\mathbf{r}'', t'') = \exp\left(\hat{L}(t'' - t')\right) \rho(\mathbf{r}', t'), \tag{5.2.16}$$

where the exponential of the operator $\hat{L}$ can be constructed iteratively as a nested Poisson-bracket expansion, see Eq. (5.2.13),

$$\exp\left(\hat{L}(t'' - t')\right) \rho(\mathbf{r}', t') = \sum_{n=0}^{\infty} \frac{\hat{L}^n}{n!} \rho(\mathbf{r}', t')$$

$$= \rho(\mathbf{r}', t') + \{H, \rho\}(t'' - t') + \frac{(t'' - t')^2}{2} \{H, \{H, \rho\}\} + \frac{(t'' - t')^3}{6} \{H, \{H, \{H, \rho\}\}\} + \cdots. \tag{5.2.17}$$

In terms of the probability density, the invariance of the phase-space volume, Eq. (5.1.24), takes the form of a similar conservation law. In a closed system, the probability to be in a volume element dV cannot change as the volume element moves with the flow. Therefore,

$$\rho(\mathbf{r}'') = \frac{dP(\mathbf{r}'')}{dV''} = \frac{dP(\mathbf{r}')}{dV'} = \rho(\mathbf{r}'), \tag{5.2.18}$$

that is, the density remains constant along each trajectory (Fig. 5.3).

Probabilities moving along trajectories in phase space suggest to introduce the concept of a probability density flow as the phase-space velocity $\dot{\mathbf{r}}$, weighted with the density $\rho(\mathbf{r})$,

$$\mathbf{j}(\mathbf{r}) := \rho(\mathbf{r}) \dot{\mathbf{r}}. \tag{5.2.19}$$

By difference with the phase-space velocity, we cannot expect the probability flow to have a vanishing divergence. Rather, invoking Eqs. (5.1.8 and 5.1.9),

Fig. 5.3 Continuous
probability density
distributions in phase space
are carried with the
Hamiltonian flow, as are the
volumes depicted in Fig. 5.2.
Here, the conservation of
volume becomes manifest as
the invariance of the
probability contained in each
infinitesimal volume element
dV as it moves with the flow:
Along each trajectory, the
density $\rho(\mathbf{r}(t))$ is constant.

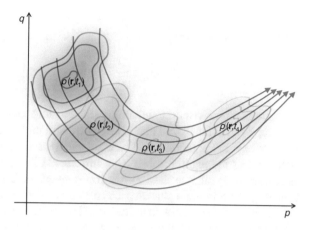

$$\mathrm{div}(\mathbf{j}(\mathbf{r})) = (\dot{\mathbf{r}} \cdot \nabla)\rho(\mathbf{r}, t) + \rho(\mathbf{r}, t)\nabla \cdot \dot{\mathbf{r}} = \left(\mathbf{J}^{\mathrm{t}} \frac{\partial H}{\partial \mathbf{r}} \cdot \frac{\partial}{\partial \mathbf{r}} \right)\rho(\mathbf{r}, t) = \{\rho, H\}.$$

$$(5.2.20)$$

A comparison with Eq. (5.2.12) reveals that

$$\frac{\partial}{\partial t}\rho(\mathbf{r}, t) + \mathrm{div}(\mathbf{j}(\mathbf{r})) = \{H, \rho\} + \{\rho, H\} = 0. \qquad (5.2.21)$$

This identity, known as the *continuity equation*, is another manifestation of probability conservation in a closed Hamiltonian system: Where there is a net flux of probability into or out of a volume, $\mathrm{div}(j(\mathbf{r})) \neq 0$, it must be reflected in a corresponding
increase or decrease of the probability inside that volume, $\partial\rho(\mathbf{r}, t)/\partial t = -\mathrm{div}(j(\mathbf{r}))$.

5.3 Information Density, Information Flow, and Conservation of Information in Hamiltonian Systems

The concept of a probability density distribution in phase space provides us with
a comprehensive tool to formalize Boltzmann's idea of complexions of a mechanical system, and to assign probabilities to them, as required to employ Shannon's
definition of information. A suitable version of his definition, applicable to probability densities depending on a continuous argument, is Eq. (1.8.4). Substituting the
phase-space density for the input probability distribution, it reads

$$I := -k_{\mathrm{B}} \int \mathrm{d}^{2f}r \; \rho(\mathbf{r}) \ln\left(d_{\mathbf{r}}^f \rho(\mathbf{r})\right). \qquad (5.3.1)$$

Like this, it is equivalent to a phase-space average, as in Eq. (5.2.8), of the observable $A(\mathbf{r}) = -k_B \ln\left(d_{\mathbf{r}}^f \rho(\mathbf{r})\right)$,

$$I = -k_B \langle \ln(d_{\mathbf{r}}^f \rho(\mathbf{r})) \rangle. \tag{5.3.2}$$

Adopting units as in thermodynamics, the undetermined constant c in Eq. (1.8.4) has been replaced in Eqs. (5.3.1 and 5.3.2) by Boltzmann's constant k_B (in all that follows, the more neutral notation c will be used again). The prefactor $d_{\mathbf{r}}$ that appears in the argument of the logarithm has been introduced in Sect. 1.8 as a finite resolution, necessary to regularize the information content. In the present context, it has the meaning of a minimum phase-space cell that can be resolved to tell two states of the system apart. It provides a quantitative criterion for Boltzmann's otherwise imprecise explication of "complexions" as "distinguishable states of the system", and already at this basic level anticipates the difficulties encountered when pursuing this classical definition of entropy to all consequences. Formally, at the present stage, $d_{\mathbf{r}}$ can be given any positive value determined, e.g., by practical limitations of measurements of positions and their canonically conjugate momenta. One of the implications of quantum mechanics, pioneered by Planck, will be replacing $d_{\mathbf{r}}$ with an absolute minimum value, a fundamental constant of nature, his quantum of action $h = 2\pi\hbar$ for each degree of freedom, i.e., $d_{\mathbf{r}} = h$ [Weh78].

The expression (5.3.1) bears the same structure as, for example, Shannon's Equation (1.3.9), hence it will share the same principal properties, listed in Sect. 1.6. Specifically, the entropy, also as it is defined here, will reach a maximum for a flat probability distribution (usually interpreted as "maximum disorder") and decreases as the distribution assumes more structure. The lower bound is reached if the system is known with certainty to be found within a volume of size $d_{\mathbf{r}}$. Indeed, for a distribution

$$\rho(\mathbf{r}) = \begin{cases} 1/d_{\mathbf{r}} & \text{if } \mathbf{r} \in V, \\ 0 & \text{otherwise,} \end{cases} \tag{5.3.3}$$

with a compact subset V of volume $d_{\mathbf{r}}$ in phase space, Eq. (5.3.2) gives

$$I := -c \int_V d^{2f} r \frac{1}{d_r} \ln(1) = 0. \tag{5.3.4}$$

For distributions localized in a volume even smaller than $d_{\mathbf{r}}$, Eq. (5.3.2) would imply a negative information value, i.e., it is not properly defined.

As defined in Eq. (5.3.1), the information has the form of a total average of a quantity distributed over phase space. This suggests defining an information density as [Gas98]

$$\iota(\mathbf{r}, t) := -c\rho(\mathbf{r}, t) \ln\left(d_{\mathbf{r}}^f \rho(\mathbf{r}, t)\right) \tag{5.3.5}$$

and an information flow, in analogy to the probability flow $\mathbf{j}(\mathbf{r},t)$, Eq. (5.2.19), as [Gas98]

$$\mathbf{i}(\mathbf{r}, t) := \iota(\mathbf{r}, t)\,\dot{\mathbf{r}} = -c\mathbf{j}(\mathbf{r}, t)\ln\left(d_{\mathbf{r}}^f \rho(\mathbf{r}, t)\right). \tag{5.3.6}$$

Obviously $I = \int d^{2f}r\,\iota(\mathbf{r}, t)$. These two quantities may appear as immediate and obvious extensions of the underlying information concept, Eq. (5.3.1). That they are far from trivial will again become apparent in the context of quantum mechanics: It turns out that quantum information cannot, in fact, be considered as a sum of (more precisely, an integral over) independent local contributions, associated to phase-space cells of size $d_{\mathbf{r}}$, but is irreducibly delocalized over larger regions in phase space.

Contrary to the Hamiltonian probability flow, the information flow in general does not even comply with a continuity equation, cf. Eq. (5.2.21). There can be local sources and sinks of entropy, also in Hamiltonian mechanics. Only globally, the total information content of a Hamiltonian system is conserved, as we shall see below.

To be sure, an information content per subsystem can be devised, where "subsystem" may refer to as small a subunit as a single degree of freedom. Recalling the definition of a probability density $\rho_n(\mathbf{r}_n)$ per freedom n, Eq. (5.2.7), we can define the corresponding information density

$$\iota_n(\mathbf{r}_n, t) := -c\rho_n(\mathbf{r}_n, t)\ln(d_{\mathbf{r}}\rho_n(\mathbf{r}_n, t)) \tag{5.3.7}$$

and the partial information pertaining to the nth freedom as

$$I_n = \int d^2r_n \iota_n(\mathbf{r}_n, t) = -c\int d^2r_n \rho_n(\mathbf{r}_n, t)\ln(d_{\mathbf{r}}\rho_n(\mathbf{r}_n, t)). \tag{5.3.8}$$

It has been discussed in Sect. 1.7 that the total information contained in a system may be less than the sum of the partial information contents, owing to correlations among the subsystems, cf. Eq. (1.7.8). This applies notably here. Instead of an exact sum rule, we can only state an inequality,

$$I \leq \sum_{n=1}^{f} I_n, \tag{5.3.9}$$

where equality applies if the subsystems are statistically independent. As a consequence, it does not make sense defining information currents between subsystems, as if in a system of communicating vessels: Subsystems or degrees of freedom are not associated to regions but to dimensions of phase space. While we could readily define a complete partition of phase space into disjoint regions V_j, $j = 1, \ldots, J$, so that

$$I_j = \int_{V_j} d^{2f} r \iota(\mathbf{r}) \tag{5.3.10}$$

and evidently, $I = \sum_{j=1}^{J} I_j$, such a subdivision into disjunct contributions is not possible for individual or groups of degrees of freedom or subsystems.

As the Hamiltonian dynamics induces an evolution in time of the phase-space density, Eq. (5.2.11), it also induces the time evolution of the entropy. According to the definition (5.3.1), its total time derivative comprises two terms,

$$\frac{dI}{dt} = -c \int d^{2f} r \frac{d}{dt} \left(\rho(\mathbf{r}, t) \ln(d_{\mathbf{r}}^f \rho(\mathbf{r}, t)) \right)$$

$$= -c \int d^{2f} r \left(\left(\frac{d}{dt} \rho(\mathbf{r}, t) \right) \ln(d_{\mathbf{r}}^f \rho(\mathbf{r}, t)) + \rho(\mathbf{r}, t) \left(\frac{d}{dt} \ln(d_{\mathbf{r}}^f \rho(\mathbf{r}, t)) \right) \right). \tag{5.3.11}$$

The second term reduces to

$$\int d^{2f} r \rho(\mathbf{r}, t) \left(\frac{d}{dt} \ln(d_{\mathbf{r}}^f \rho(\mathbf{r}, t)) \right) = \int d^{2f} r \frac{d}{dt} \rho(\mathbf{r}, t) = \frac{d}{dt} \int d^{2f} r \rho(\mathbf{r}, t) = 0, \tag{5.3.12}$$

owing to the normalization of phase-space probability, Eq. (5.2.4). The first term,

$$\frac{dI}{dt} = -c \int d^{2f} r \left(\frac{d}{dt} \rho(\mathbf{r}, t) \right) \ln(d_{\mathbf{r}}^f \rho(\mathbf{r}, t)), \tag{5.3.13}$$

generally has a non-zero value but can vanish as well, namely if the total time derivative of the phase-space density so does. As argued in the previous subsection, this is the case in systems that are closed with respect to loss or gain of particles.

This indicates that under Hamiltonian dynamics, the entropy is a conserved quantity. A more transparent analysis is obtained by considering finite time steps, from a state $\rho(\mathbf{r}', t')$ at t' to a state $\rho(\mathbf{r}'', t'')$ at t''. The time-evolved entropy,

$$I(t'') = -c \int d^{2f} r'' \rho(\mathbf{r}'', t'') \ln(d_{\mathbf{r}}^f \rho(\mathbf{r}'', t'')), \tag{5.3.14}$$

can be expressed in terms of the original density

$$I(t'') = -c \int d^{2f} r'' \rho(\mathbf{r}'(\mathbf{r}'', t', t''), t') \ln(d_{\mathbf{r}}^f \rho(\mathbf{r}'(\mathbf{r}'', t', t''), t')). \tag{5.3.15}$$

A concomitant change of the integration variable from $\mathbf{r}''$ to $\mathbf{r}'$ not only requires including the Jacobian of this transformation but also a rescaling of the phase-space density, as in Eq. (5.2.5),

$$I(t'') = -c \int d^{2f}r' \left|\det\left(\frac{\partial \mathbf{r}''}{\partial \mathbf{r}'}\right)\right| \det\left(\frac{\partial \mathbf{r}''}{\partial \mathbf{r}'}\right)^{-1} \rho(\mathbf{r}',t') \ln\left(d_{\mathbf{r}}^f \left|\det\left(\frac{\partial \mathbf{r}''}{\partial \mathbf{r}'}\right)\right|^{-1} \rho(\mathbf{r}',t')\right),$$

(5.3.16)

Both steps involve the determinant of the stability matrix, which is unity, see Eq. (5.1.22),

$$I(t'') = -c \int d^{2f}r' \rho(\mathbf{r}',t') \ln\left(d_{\mathbf{r}}^f \left|\det \mathbf{M}(t'',t')\right|^{-1} \rho(\mathbf{r}',t')\right)$$

$$= c \int d^{2f}r' \rho(\mathbf{r}',t') \ln\left(\left|\det \mathbf{M}(t'',t')\right|\right) - c \int d^{2f}r' \rho(\mathbf{r}',t') \ln\left(d_{\mathbf{r}}^f \rho(\mathbf{r}',t')\right)$$

$$= -c \int d^{2f}r' \rho(\mathbf{r}',t') \ln\left(d_{\mathbf{r}}^f \rho(\mathbf{r}',t')\right) = I(t').$$

(5.3.17)

The invariance of phase-space volume under Hamiltonian dynamics, Eqs. (5.1.14, 5.24), therefore induces another universal conservation law,

In a closed system that evolves according to Hamilton's equations of motion, information is a conserved quantity.

Referring to evolution over finite instead of infinitesimal time, it is equivalent to say,

Information is invariant under canonical transformations.

A few comments are in order to appraise the far-reaching significance of this conservation law:

1. The way this conservation law has been derived, from Eq. (5.3.14) through Eq. (5.3.17), it is clear that the decisive argument is the conservation of phase-space volume, manifest in the Jacobian $\left|\det(\partial \mathbf{r}''/\partial \mathbf{r}')\right| = 1$. By reverse conclusion, entropy is not conserved where phase-space volume varies in time, as in dissipative systems, see Sect. 6.2.

2. The conservation of entropy is analogous to that of energy, and in closed systems is equivalent to it. By way of contrast, in open systems, the two conservation laws need not coincide. For example, phase-space volume, hence also the entropy, are constant even in time-dependent Hamiltonian systems, in particular driven systems, where energy is *not* conserved (see the example in Sect. 5.4). This is not the case, though, if processes like dissipation are involved that lead to a contraction of phase-space volume. Indeed, the derivation of Eq. (5.3.17) goes through identically if an explicit time dependence of the Hamiltonian is allowed for (a time argument, $H(\mathbf{r},t)$, has been suppressed in most cases to avoid technicalities). This shows that invariance of energy and of

entropy are two independent conservation laws, and that entropy conservation is the more fundamental one.

3. Entropy conservation is an important, maybe the decisive, aspect of the time-reversal invariance of Hamilton's equations of motion for a closed system. In plain terms, it means that "What is known about a system at a given time (i.e., the negentropy) is sufficient to predict its state at any other moment before or after the reference time." The argument embodied by *Laplace's demon* [Lap51] actually refers exactly to this point: "An intellect which at a certain moment would know all forces that set nature in motion ... for such an intellect nothing would be uncertain and the future just like the past would be present before its eyes." Eq. (5.3.17) can therefore be considered as an updated version of the demon, in terms of an exactly defined quantity that had not been available to Laplace.

4. An important corollary of volume and entropy conservation, pointing out a particularly drastic manifestation, is *Poincaré's recurrence theorem* [Bar06]: If in addition, the phase-space volume accessible to a system is bounded, i.e., if the system has only bounded orbits, it will return to states arbitrarily close to its initial state for an infinite number of times. The proof hinges on the argument that otherwise, the system would have to explore phase-space regions covering an accumulated volume that increases indefinitely.

5. Equation (5.3.17) seems to be at variance with the Second Law of Thermodynamics. The apparent contradiction owes to the fact that Hamiltonian dynamics amounts to a complete description in terms of microscopic degrees of freedom, while the Second Law refers to coarse-grained macroscopic observables. Sections 6 and 7 will reconsider this contradiction in more detail.

6. Given the numerous conceptual differences between classical physics, including Hamiltonian dynamics, and quantum mechanics, one might suspect that by advancing to a quantum mechanical description, also the incompatibility between the time evolution of entropy and the Second Law might be resolved or at least mitigated. The opposite is true: It turns out that an equivalent conservation law can be derived in the framework of unitary quantum dynamics, with arguments very similar to those used on the classical level (Sect. (8.3.2)). The reasoning is even more stringent in quantum mechanics, as it treats informational aspects from the beginning in a rigorous and self-consistent manner, see Sect. 8.1.1.

7. Only thanks to the conservation of entropy in total phase space, we can meaningfully talk of information exchanged among different phase-space regions and information currents flowing between them. A partition into subsystems, however, owing to possible cross-correlations, is not reflected in a corresponding sum rule for the entropy, so that information currents between subsystems are generally not well defined, cf. Eq. (5.3.9). An elementary example is illustrated in Fig. 5.4.

8. Analogous to the information exchange between phase-space regions, it can also be interchanged between scales. Applying Fourier analysis to structures in phase space, it becomes plausible that information can be shifted between

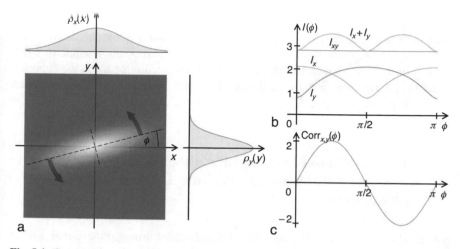

Fig. 5.4 Consider a system with two state variables x, y, prepared as a Gaussian distribution centred in the origin, with different variances along its two principal axes (a). If the distribution rotates rigidly around the origin (as, for example, the phase-space density of a harmonic oscillator), the total information I_{tot} remains constant (blue line, panel b). However, the projected distributions for x and for y (upper and right margins of panel a, resp.) vary in time, as do the corresponding partial information contents I_x, I_y (green and red curves in b, resp.). The sum of the two partial information contents $I_x + I_y$ (yellow curve in b) may well exceed the total information I_{xy}. This is readily explained by the existence of correlations $\text{Corr}_{x,y}(\phi)$ between x and y (c), which oscillate as a function of the rotation angle ϕ and take maximal positive or negative values when the distribution is oriented along the diagonal ($\phi = \pi/4$) or the second diagonal ($\phi = 3\pi/4$) of the state space.

low-frequency and high-frequency sectors of the spectrum, without loss or gain of the total information content (see the example below).

9. Deterministic chaos, characterized by positive Lyapunov exponents λ_+ indicating an expansion of phase space in some of its directions, see Eq. (3.1.17), is well compatible with Eq. (5.3.7). Each expanding phase-space direction, however, must be compensated by a complementary direction that contracts phase space at the same rate, with a negative Lyapunov exponent $\lambda_- = -\lambda_+$. A patent example is the baker map, see Eq. (5.5.17) and Figs. 5.10 and 5.11.

10. In numerical evaluations of Eq. (5.3.7), based on simulations of ensembles of trajectories of a Hamiltonian system, the unavoidable discretization of phase space, for example by subdividing it into bins, can lead to artefacts that violate the invariance of entropy. In particular, as discussed in Sect. 3.1.3, asymptotically in the long-time limit, only expanding phase-space directions contribute to the information balance, while contraction along other directions, which compensates for the expansion, will eventually be obscured by the finite resolution.

In the following subsections, the concepts developed above for the analysis of information dynamics in phase space will be applied to a number of instructive sample cases, representing emblematic phenomena of information processing in Hamiltonian systems. The independence of the invariance of entropy from energy conservation will be demonstrated for a pertinent example, a driven harmonic oscillator, in Sect. 5.4. Section 5.5 is dedicated to chaotic systems where currents of information, up and down the spectra of time and space scales within a single or a few degrees of freedom, dominate. Interactions between freedoms that, to the contrary, exchange information between freedoms but keep it exclusively on the same scale will be illustrated in Sect. 5.6.

5.4 Conservation of Information Without Energy Conservation: Harmonic Oscillator Driven at Resonance

A particularly striking example of energy being injected through an external driving force into a mechanical system is resonance. As soon as the amount of energy absorbed surpasses a certain threshold, dissipation and other irreversible processes will become dominant, but in order to contrast energy non-conservation with conservation of entropy, a simple Hamiltonian model with a single degree of freedom is sufficient: The Hamiltonian

$$H(p, q, t) = \frac{1}{2m}p^2 + \frac{m\omega^2}{2}q^2 + Aq\cos(\omega t) \tag{5.4.1}$$

describes a harmonic oscillator with mass m and natural frequency ω, driven harmonically with exactly the same frequency and with amplitude A. Hamilton's equations of motion (5.1.2) read

$$\dot{p} = -m\omega^2 q - A\cos(\omega t),$$
$$\dot{q} = \frac{p}{m}. \tag{5.4.2}$$

They are equivalent to a second-order Newton equation

$$\ddot{q} = -\omega^2 q - \frac{A}{m}\cos(\omega t) \tag{5.4.3}$$

and are solved, for initial conditions $p(t') = p', q(t') = q'$, by

$$p'' = p(t'') = p'\cos(\omega t) - m\omega q'\sin(\omega t) - \frac{A}{2\omega}(\sin(\omega t) + \omega t\cos(\omega t)),$$
$$q'' = q(t'') = \frac{p'}{m\omega}\sin(\omega t) + q'\cos(\omega t) - \frac{A}{2m\omega}t\sin(\omega t), \tag{5.4.4}$$

abbreviating $t := t'' - t'$. The stability matrix corresponding to this solution is the same as for a harmonic oscillator without driving force,

$$\mathbf{M}(t'', t') = \begin{pmatrix} \partial p''/\partial p' & \partial q''/\partial p' \\ \partial p''/\partial q' & \partial q''/\partial q' \end{pmatrix} = \begin{pmatrix} \cos(\omega t) & \frac{1}{m\omega}\sin(\omega t) \\ -m\omega\sin(\omega t) & \cos(\omega t) \end{pmatrix}, \qquad (5.4.5)$$

and indeed has the determinant

$$\det(\mathbf{M}(t'', t')) = 1. \qquad (5.4.6)$$

As an immediate consequence, making use of the second line of Eq. (5.3.17) with $\ln(|\det \mathbf{M}(t'', t')|) = \text{const} = 0$, the total entropy is constant (Fig. 5.5a),

$$I(t'') = c\ln(|\det M(t'', t')|) + I(t') = I(t'). \qquad (5.4.7)$$

The energy balance at time t'' is readily calculated by substituting the solution (5.4.4) in the Hamiltonian (5.4.1),

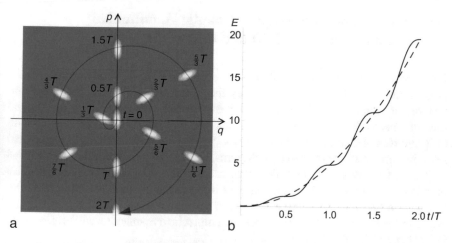

Fig. 5.5 If a harmonic oscillator is driven at resonance, that is, with its own natural frequency, it absorbs energy without bound. Starting at rest, $(p, q) = (0, 0)$, its trajectory spirals out in phase space, the radius increasing on average linearly with time (a, continuous curve). The initial distribution (colour code, from blue – zero to yellow – positive) moves along the trajectory and rotates with it (snapshots at integer multiples of $T / 6$, with $T = 2\pi/\omega$, the period of the oscillator) but without changing shape or size: The entropy is constant. Unlike information, with amplitude and momentum growing linearly with time, the energy (b, thick line) increases as a quadratic function of time if periodic oscillations are smoothed out (dashed). This open system violates energy conservation, but the invariance of entropy continues to hold.

$$H(p'', q'', t'') = H(p', q', t') + \frac{A^2}{8m\omega^2}\left(\omega^2 t^2 - 2\omega t \sin(\omega t)\cos(\omega t)\right)$$
$$+ A\left(q'\left(1 + \frac{1}{2}(\cos(\omega t))^2\right) + \frac{p'}{m\omega}\left(\frac{1}{2}\sin(\omega t)\cos(\omega t) - \omega t\right)\right).$$

(5.4.8)

It contains monotonically increasing terms as well as contributions oscillating with the driving. The latter can be eliminated by averaging over a period $T = 2\pi/\omega$ of the driving. If the system is prepared at $(p, q) = (0, 0)$, the remaining smooth time dependence of the energy, for $t \gg T$, is

$$E(t) = \frac{A^2}{8m^2}\frac{t^2}{T^2}.$$

(5.4.9)

That is, it increases as a quadratic function of time, apart from terms of lower order (Fig. 5.5b).

5.5 Information Processing in Chaotic Hamiltonian Systems: Bernoulli Shift and Baker Map

Hamiltonian systems with a chaotic time evolution play a prominent role in the present context. With the particular way they process information, they highlight features of Hamiltonian dynamics that prove crucial when analyzing the Second Law of Thermodynamics, see Chap. 7, and confronting classical mechanics with the principles of quantum mechanics, see Chap. 8. Hamiltonian chaos is of practical relevance above all for microscopic systems, i.e., for a detailed physical description on the level of atoms and molecules. Macroscopic chaos, by contrast, epitomized by turbulence and weather, does not conserve information as Hamiltonian systems do. Dissipation becomes essential, for example for the formation of strange attractors, and requires to abandon the framework of symplectic dynamics, as will be discussed in Chap. 6.

In the study of chaotic dynamics, systems in discrete time [CE80, Sch84, Ott93] and sometimes even with a discrete state space [HH85, BR87, Dom05] have proven enormously fruitful, despite their unphysical, often highly artificial, nature. Iterated maps on the unit interval enabled the first rigorous results in the theory of dynamical systems [CE80]. They are just as useful if the focus is more specifically on the way these systems process information.

A particularly striking case is the Bernoulli shift, maybe the simplest example of a map of the unit interval onto itself, with a markedly nonlinear behaviour. It can be motivated by comparing it with an example of a random generator taken from daily life, a popular technique for shuffling playing cards. In terms anticipating its conversion into a mathematical model, it can be described as follows (Fig. 5.6a):

1. Form a deck of J playing cards, J even, of total thickness d.
2. Divide the deck in the middle, forming two decks of equal size $J/2$ and thickness $d/2$ each (Fig. 5.6a, step 2).
3. Open them up and fan them out evenly, leaving slits of the thickness of a card between all the cards, so that each of the two half decks reaches the original thickness d of the full deck (Fig. 5.6a, step 3).
4. Shift the two card decks into one another, intercalating the cards one by one, as in a zipper, so as to form a single deck of size J and thickness d again (Fig. 5.6a, step 4).
5. Return to step 1 and repeat.

By normalizing the total thickness of the card deck to unity and representing the position of each card by a continuous variable $x \in [0, 1[$, this procedure is cast in the form of a map (Fig. 5.6b),

$$\text{ber} : [0, 1[\rightarrow [0, 1[, \, x \mapsto x' = \text{ber}(x) = 2x \, (\text{mod } 1)$$

$$= \begin{cases} 2x & 0 \leq x < 1/2, \\ 2x - 1 & 1/2 \leq x < 1. \end{cases} \quad (5.5.1)$$

In Eq. (5.5.1), the multiplication by 2 corresponds to the stretching performed in step 2 of the shuffling procedure. The *modulo* operation, in turn, combines the actions of division and merger of the card decks, performed in steps 2 and 4, resp. It is

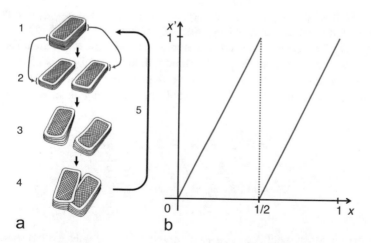

Fig. 5.6 The Bernoulli map is one of the simplest mathematical models of a chaotic system. It can be motivated by a popular card shuffling technique (a): Converted into a map on the unit interval, it takes the form of Eq. (5.5.1). The discrete nature of the cards even allows to reproduce quantum suppression of chaos in a rudimentary form.

this operation that renders the mapping highly nonlinear. Moreover, it prevents the map from being reversible (bijective): For continuous coordinate x, each image has exactly two preimages.

At every application of the Bernoulli map, the information is lost whether the preimage of x' is in the right or in the left half of the unit interval, $x = x'/2$ or $x = (x' + 1)/2$, suggesting that in this way, one bit of information on the position of the preimage is discarded. This analysis can be substantiated by encoding the position as a binary sequence,

$$x = \sum_{n=1}^{N} a_n 2^{-n} = 0.a_1 a_2 a_3 a_4 a_5 \ldots a_N, \, a_n \in \{0, 1\}, \qquad (5.5.2)$$

and using an obvious notation to visualize its information content in bits explicitly. The upper limit N of the summation defines the resolution to which the position can be determined. It must not go to infinity, in order to maintain a finite information content, but otherwise needs not to be specified here.

In terms of this binary code, the action of the map is

$$x' = \left(2 \sum_{n=1}^{N} a_n 2^{-n} \right) \bmod 1 = \left(\sum_{n=0}^{N} a_{n+1} 2^{-n} \right) \bmod 1$$

$$= \sum_{n=1}^{N} a_{n+1} 2^{-n} = 0.a_2 a_3 a_4 a_5 a_6 \ldots = \sum_{n=1}^{N} a'_n 2^{-n}, \, a'_n := a_{n+1}. \qquad (5.5.3)$$

At each iteration, the sequence of binary digits is rigidly shifted by one position towards the left (towards the lower exponents), an operation referred to as *Bernoulli shift* (Fig. 5.7). This clearly shows that not only one bit is discarded, the information contained in a_1. At the same time, if the total number N of digits is kept constant, another bit is gained at the lower end of the sequence, the digit $a'_N = a_{N+1}$ that was not previously contained in the specification of x because it could not yet be resolved.

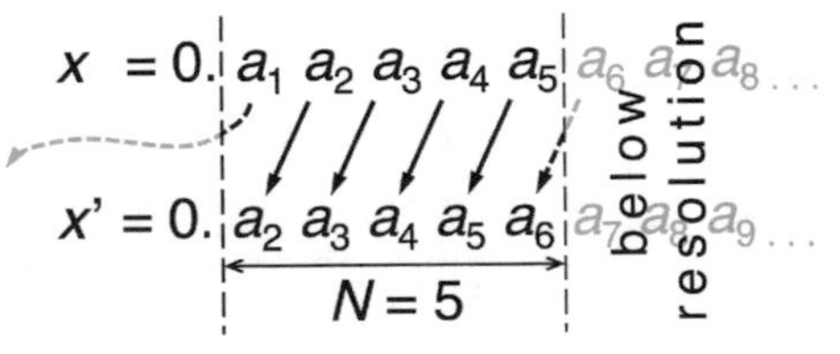

Fig. 5.7 The "Bernoulli shift" characterizing the map (5.5.1) refers to the fact that encoding the variable x in binary digits, the action of the map amounts to a rigid shift of the sequence of binary digits by one position to the left, Eq. (5.5.3). As a consequence, the most significant digit a_1 is lost. Another digit is added, here a_6, entering across the border of resolution at the opposite, least significant side of the sequence.

In other words, while information is lost "globally", at the "macroscopic" extreme of the spectrum of scales, the same amount is gained "locally", at the "microscopic" extreme. Summarizing the entropy flow, it amounts to a Lyapunov exponent of $\lambda = \ln(2)$, which is also the Kolmogorov-Sinai entropy, Eq. (3.1.17), of the Bernoulli map.

This continuous entropy production, lifting information initially hidden in small scales to macroscopic visibility, is in fact a hallmark of chaotic dynamics [Sha80]. It is equivalent to the sensitive dependence on initial conditions, usually stated as constituent property of chaos and embodied in the metaphorical butterfly effect. Dynamical systems that can be represented as a shift by a finite number of positions of a symbol sequence are generally classified as *Bernoulli systems* and form the top level of a hierarchy of chaotic behaviour [Sch80].

In the present context, "entropy production" does not signify more than the emergence of a single binary digit of the initial condition x_0 per time step in the window of resolution of the current coordinate $x_n \in [0, 1[$. It does not imply anything concerning the random character of the resulting sequence. One of the principal conclusions of the discussion of randomness in Chap. 4 is that among all possible symbol strings, say of binary digits, the vast majority are indeed "random numbers" according to quantitative criteria. But there are notable exceptions. Sequences certainly not considered random are in particular periodic repetitions, such as $x_0 = 0.011011011011011\cdots$, Fig. 5.8. In the context of dynamical systems, such an initial condition would indicate a periodic point, in this case of period 3, that is, a fixed point of the map $\mathrm{ber}^3(x) = \mathrm{ber}(\mathrm{ber}(\mathrm{ber}(x)))$. Indeed, the fixed-point condition for this map,

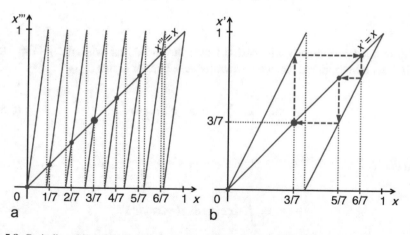

Fig. 5.8 Periodic orbits of period 3 of the Bernoulli map are fixed points of the triply iterated map, $x''' := \mathrm{ber}^3(x) = x$, Eq. (5.5.4) (a). Solutions are $x_j^* = j/7$, $j = 0, 1, \ldots, 7$ (blue dots). (b) Tracking the iteration step by step graphically in the single map $x' = \mathrm{ber}(x)$, Eq. (5.5.1), for the periodic point $x_3^* = 3/7$ (bold dot in panel a) reveals that the full trajectory is $3/7 \to 6/7 \to 5/7 \to 3/7 \to \ldots$, corresponding to a periodic binary symbol sequence 011011011....

$$\mathrm{ber}^3(x^*) = 2^3 x^* \bmod (1) = \begin{cases} 8x^* & x^* \in [0, 1/8[\\ \vdots & \vdots \\ 8x^* - j & x^* \in [(j-1)/8, j/8[\\ \vdots & \vdots \\ 8x^* - 8 & x^* \in [7/8, 1[\end{cases} = x^* \quad (5.5.4)$$

has eight solutions $x_j^* = j/7$, $j = 0, \ldots, 7$, one for each of the options in Eq. (5.5.4), including $x_3^* = 3/7 = 0.\overline{011}$ (bold blue dot in Fig. 5.8a), in binary notation, the overbar indicating repetition ad infinitum. It is readily shown that every rational number $p/q \in [0, 1[\cap \mathbb{Q}, p, q \in \mathbb{N}$, has a periodic representation in any n-adic number system [HW08], in particular a periodic binary code, and thus corresponds to a periodic point of the Bernoulli map. However, they are of measure zero among the real numbers in the unit interval, that is, choosing an initial condition $x_0 \in [0, 1[$ at random, it will not be a rational with probability 1. Moreover, periodic orbits of the Bernoulli map are unstable, with the same Lyapunov exponent $\lambda = \ln(2)$ as for the map in general.

Changes in scale become most clearly manifest in Fourier space. The Bernoulli map becomes amenable to Fourier analysis if the discrete variable x is replaced by a continuous probability density, normalized, say, on the unit interval,

$$\rho : [0, 1[\to \mathbb{R}^+, x \mapsto \rho(x), \int_0^1 \mathrm{d}x \rho(x) = 1, \quad (5.5.5)$$

and continued periodically with period 1 over the entire real axis, $\rho(x+1) = \rho(x)$. With these assumptions, it can be expanded in a Fourier series

$$\rho(x) = \sum_{n=-\infty}^{\infty} c_n \exp(2\pi i n x) \quad (5.5.6)$$

with coefficients

$$c_n = \int_0^1 \mathrm{d}x \rho(x) \exp(-2\pi i n x) \quad (5.5.7)$$

Since ρ is a real probability density, these coefficients form a Hermitian sequence, that is $c_{-n} = c_n^*$ (see Eq. (2.3.3)).

Applied to a density distribution, the operation of the Bernoulli map consists in (i), stretching the position by 2, $\rho(x) \to \rho(x/2)$, and (ii), superposing the images of the intervals $[0, 1/2[$ and $[1/2, 1[$. Together,

$$\rho(x) \rightarrow \rho'(x') = \frac{1}{2}\left(\rho\left(\frac{x'}{2}\right) + \rho\left(\frac{x'+1}{2}\right)\right), \tag{5.5.8}$$

The prefactor 1/2 serves to maintain the distribution normalized (see Eq. (5.2.5)). The Bernoulli map for continuous distributions can thus be expressed as an integral operator with a kernel $G_{Ber}(x'', x')$

$$\rho''(x'') = \int_0^1 dx' G_{Ber}(x'', x') \rho'(x'),$$

$$G_{Ber}(x'', x') = \frac{1}{2}\left(\delta\left(x' - \frac{x''}{2}\right) + \delta\left(x' - \frac{x''+1}{2}\right)\right). \tag{5.5.9}$$

Applying the same Fourier expansion as above leads to new coefficients

$$c'_{n'} = \int_0^1 dx' \rho'(x') \exp(-2\pi i n' x)$$

$$= \frac{1}{2}\int_0^1 dx' \left(\rho\left(\frac{x'}{2}\right) + \rho\left(\frac{x'+1}{2}\right)\right) \exp(-2\pi i n' x). \tag{5.5.10}$$

To perform the integration, subdivide the integration range into the two intervals mentioned above and change the integration variable to $x = x'/2$ and $x = (x'+1)/2$, resp.,

$$c'_{n'} = \int_0^{1/2} dx \rho(x) \exp(-4\pi i n' x) + \int_{1/2}^1 dx \rho(x) \exp(-4\pi i n'(x - 1/2))$$

$$= \int_0^1 dx \rho(x) \exp(-2\pi i (2n') x) = c_{2n'}, \tag{5.5.11}$$

that is, only coefficients $c_{2n'}$ with even index become new coefficients $c'_{n'}$.

In conclusion, in Fourier space, the Bernoulli map contracts the spatial frequency scale by a factor 1/2 and deletes all coefficients with odd indices. What appears as a much more drastic reduction of the information content than the loss of 1 bit, found above for the original mapping, owes itself to the fact that here, the map is operating on a much larger state space: A density distribution on an interval is equivalent to an infinite-dimensional vector of discrete coefficients c_n, as compared to a single real number x. Contracting the spectrum towards low spatial frequencies amounts to a deletion of fine structures, thus to a low-pass filtering.

For long times, that is, in the limit $m \rightarrow \infty$ of m iterations of the Bernoulli map, its action corresponds to an eventual leveling out of all structures in the distribution.

The stationary state approached in this limit from a smooth initial distribution is

$$\lim_{m \to \infty} \rho^{(m)}(x) = \text{const} = 1, \tag{5.5.12}$$

so that all information on the initial state, except for the normalization, is lost: If the number of non-zero Fourier coefficients in the initial distribution has been finite (which notably excludes Dirac delta functions), all of them vanish, except for c_0.

This is an example of how the permanent entropy production in the Bernoulli map exhausts the information content of the initial state, if it is limited. What if the original state space, the position x, is discretized from the beginning? Similarly to a smooth initial distribution, this amounts to restricting the number of binary digits in Eq. (5.5.3) or the number of Fourier coefficients in Eq. (5.5.6) to finite values. Indeed, it is also suggested by taking the analogy with card shuffling, referred above to motivate the Bernoulli map, literally and revising the model, so as to adapt it to a finite number J of playing cards.

In this setting, with both time and space discretized, the map takes the form of a matrix. If the available positions are

$$x_j = \frac{j-1}{J} \in [0, 1[, \, j = 1, 2, \ldots, J, \, J \in \mathbb{N}, \tag{5.5.13}$$

the probability distribution becomes a vector

$$\boldsymbol{\rho} = (\rho_1, \rho_2, \ldots, \rho_J), \sum_{j=1}^{J} \rho_j = 1, \tag{5.5.14}$$

and the map operates as multiplication with a $(J \times J)$ permutation matrix, $\boldsymbol{\rho}' = \mathbf{B} \boldsymbol{\rho}$ [BS93]. For example, if $J = 2^M$, for $M = 3$, the matrix $\mathbf{B}$ takes the form

$$\mathbf{B}_3 = \begin{pmatrix} 1\,0\,0\,0\,0\,0\,0\,0 \\ 0\,0\,0\,0\,1\,0\,0\,0 \\ 0\,1\,0\,0\,0\,0\,0\,0 \\ 0\,0\,0\,0\,0\,1\,0\,0 \\ 0\,0\,1\,0\,0\,0\,0\,0 \\ 0\,0\,0\,0\,0\,0\,1\,0 \\ 0\,0\,0\,1\,0\,0\,0\,0 \\ 0\,0\,0\,0\,0\,0\,0\,1 \end{pmatrix}, \tag{5.5.15}$$

with a structure of non-zero elements along skew diagonals that evidently resembles the graph of the Bernoulli map, Fig. 5.6b. The matrices $\mathbf{B}_M$ share a remarkable property: They are M^{th} roots of the $(2^M \times 2^M)$-unit matrix,

$$\mathbf{B}_M^M = I^{(2^M)}. \tag{5.5.16}$$

For the Bernoulli map, this means that after M iterations, the initial state is recovered (Fig. 5.9a). The cards are unshuffled! This surprising result is readily explained by looking at the binary code for the index j (Fig. 5.9b): The discrete Bernoulli map merely cycles the M relevant digits through the "window of resolution" (cf. Fig. 5.6), also comprising M positions. As an example, card no. 4 in a deck of eight (position $x_j = 3/8$) goes through positions $4 \rightarrow 7 \rightarrow 6 \rightarrow 4$ (Fig. 5.9a). In binary code, the sequence is $011 \rightarrow 110 \rightarrow 101 \rightarrow 011$ (Fig. 5.9b). Due to the discrete character of the state space, the information contained in the most significant digit a_1 is not lost as in the continuous case, but becomes encoded in the parity (even vs. odd) of the position in the subsequent step, that is, it returns through the back door of the smallest scale, encoded in the least significant digit a_M. In this sense, the discrete map *is* reversible.

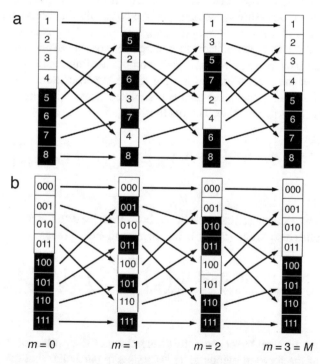

Fig. 5.9 If the discrete nature of the playing cards is included in the model of a card shuffling procedure as illustrated in Fig. 5.6, a discrete Bernoulli map results. Mathematically, it is equivalent to multiplication with a permutation matrix. As a striking consequence, the mixing process eventually undoes itself, after a finite time, the initial order of the card deck is recovered (a). If the total number J of cards is a power of 2, $J = 2^M$, the period of the process is $\mathrm{lb}(J) = M$. This is easily explained in terms of a binary code for the position of each card in the deck (b): The action of the map is thus revealed as a mere cycling of the M binary digits through the M available positions.

As a general conclusion, the example of the discrete Bernoulli map suggests that a discrete state space comprising a finite number J of distinguishable states is incompatible with chaotic behaviour. With an entropy production of $\lambda = \ln(2)$ per time unit, the system explores its entire state space in a time of the order of $\ln(J)/\ln(2)$ (in the above example, $J = 8$ results in a recurrence time of 3 steps), giving way to a repetitive behaviour that no longer produces novelty. This result has been demonstrated in early studies of chaos in discrete spaces such as digital computers, anticipating later results in quantum chaos.

As useful as the Bernoulli map is as a mathematical model, it not only lacks any physical motivation, it also fails satisfying a basic condition for Hamiltonian systems: Its state space comprises only a single dimension, a momentum canonically conjugate to the position is missing. This deficiency, however, can be repaired by a straightforward extension of the model, giving rise to what is known as the *baker map*.

If the convenient restriction to the unit interval is applied also to the additional variable, the state space becomes a unit square, $[0,1[\times[0,1[$. The construction of the mapping for the momentum is restricted by the condition that phase-space volume should be conserved. As the Bernoulli map, applied to the position x, implies an amplification by 2, it suggests itself to compensate for this expansion, contracting the momentum p by the same factor. Again, complementing the Bernoulli map with an operation on the p-coordinate, converse to the superposition of the images of $[0,1/2[$ and $[1/2,1[$ in x, the baker map is completed by stacking the corresponding images in p on top of each other. The result is (Fig. 5.10)

$$\text{bak} : [0, 1[\times[0, 1[\to [0, 1[\times[0, 1[, \; \begin{pmatrix} x \\ p \end{pmatrix} \mapsto \begin{pmatrix} x' \\ p' \end{pmatrix} = \begin{pmatrix} 2x(\text{mod } 1) \\ (p + \text{int}(2x))/2 \end{pmatrix}.$$

$$(5.5.17)$$

In Hamiltonian mechanics, time evolution is invertible. By difference to the Bernoulli map, this applies also to Eq. (5.5.15): The inverse mapping reads

$$\text{bak}^{-1} : [0, 1[\times[0, 1[\to [0, 1[\times[0, 1[, \; \begin{pmatrix} x \\ p \end{pmatrix} \mapsto \begin{pmatrix} x' \\ p' \end{pmatrix} = \begin{pmatrix} (x' + \text{int}(2p'))/2 \\ 2p'(\text{mod } 1) \end{pmatrix}.$$

$$(5.5.18)$$

Evidently, the inverse baker map just interchanges the roles of position and momentum of the forward mapping. This implies in particular that the momentum component of the baker map coincides with the inverse of its position component. The alternating operations of stretching and of cutting-and-gluing of state space, generically referred to as "stretch and fold" (Fig. 6.4), form constituent elements of chaotic systems. While the stretching is indispensable for the entropy production,

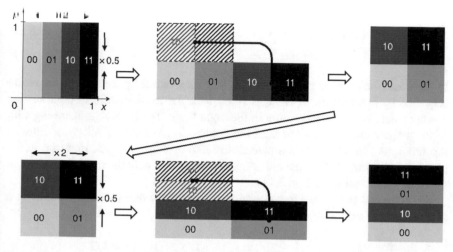

Fig. 5.10 The baker map, Eq. (5.5.17), is a function of the unit square onto itself, which combines the expansion by 2 of the Bernoulli map for the spatial coordinate x with a contraction by 0.5 for the momentum p (upper left), so that in total, the baker map preserves volume. The initial shape of a square is recovered by stacking the two halves of the 2×0.5 rectangle resulting from the previous rescaling on top of one another (upper right). The alternation of stretching and folding (here, cutting) operations is a constituent feature of chaotic dynamics. As artificial as it appears, the baker map nevertheless has a physical interpretation, specifically in ray optics [HKO94].

it cannot be performed in an altogether compact state space without concomitant folding.

In the context of the Bernoulli shift, it has been argued that the irreversibility of this map is a consequence of the loss of one bit of position information per time step. Similarly, an analysis in terms of binary digits explains why the baker map *is* invertible: If both the position and the momentum are written in binary code,

$$x = \sum_{n=1}^{N} a_n 2^{-n} = 0.a_1 a_2 a_3 a_4 a_5 \ldots a_N, a_n \in \{0, 1\},$$

$$p = \sum_{n=1}^{N} b_n 2^{-n} = 0.b_1 b_2 b_3 b_4 b_5 \ldots b_N, b_n \in \{0, 1\}, \tag{5.5.19}$$

the momentum part of the baker map also undergoes a shift, but in the opposite direction,

$$p' = \frac{1}{2} \left(\sum_{n=1}^{N} b_n 2^{-n} + \mathrm{int}\left(2 \sum_{n=1}^{N} a_n 2^{-n} \right) \right) = \sum_{n=2}^{N+1} b_{n-1} 2^{-n} + \frac{1}{2} a_1$$

$$= 0.a_1 b_1 b_2 b_3 b_4 \dots b_{N-1} = \sum_{n=1}^{N} b_n' 2^{-n}, \ b_n' = \begin{cases} a_1 & n = 1 \\ b_{n-1} & 2 \le n \le N - 1. \end{cases} \quad (5.5.20)$$

In this case, the leftmost position, which becomes vacant by the shift of the entire sequence by one unit to the right, is filled with the most significant digit a_1 that has been deleted at the same time in the code for x. The least significant digit b_N of the momentum is lost, since after the contraction, it remains below the limit of resolution. All in all, we find a paternoster-like cycling of digits, upwards in the position and downwards in the momentum component, with the revolving wheel at the most significant level (Fig. 5.11).

Applying the baker map to a continuous probability density $\rho(p,x)$ results in a straightforward extension of Eq. (5.5.8),

$$\rho'(p', x') = \begin{cases} \rho(2p', x'/2) & 0 \le p' < 1/2 \\ \rho(2p' - 1, (x' + 1)/2) & 1/2 \le p' < 1 \end{cases}$$
$$= \rho(2p'(\bmod 1), (x' + \mathrm{int}(2p'))/2). \quad (5.5.21)$$

Since it is based on a one-to-one relationship between (x,p) and (x',p'), this map is invertible,

$$\rho(p, x) = \begin{cases} \rho'(p/2, 2x) & 0 \le x < 1/2 \\ \rho'((p + 1)/2, 2x - 1) & 1/2 \le x < 1 \end{cases}$$
$$= \rho'((p + \mathrm{int}(2x))/2, 2x(\bmod 1)). \quad (5.5.22)$$

The same equivalence between inversion and exchange of arguments, $p \leftrightarrow x$, also indicates how to construct a discrete baker map, analogous to the discrete Bernoulli

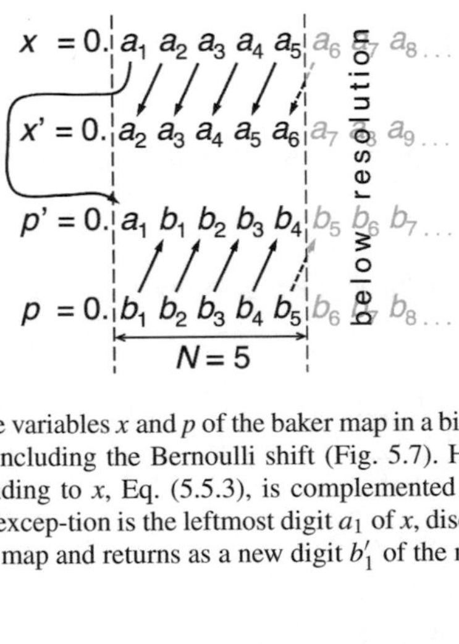

Fig. 5.11 Writing both state variables x and p of the baker map in a binary code, the map translates into a rigid shift of digits, including the Bernoulli shift (Fig. 5.7). Here, the upward shift of the symbol sequence corresponding to x, Eq. (5.5.3), is complemented by a downward shift of the digits encoding p. The only excep-tion is the leftmost digit a_1 of x, discarded by the Bernoulli shift. It is "recycled" in the baker map and returns as a new digit b_1' of the momentum code.

slllfl, Eq. (5.5.14). If the momentum is discretized in the same way as the position, the density distribution becomes a matrix,

$$\rho : \{1, \ldots, N\} \times \{1, \ldots, N\} \to \mathbb{R}^+, (n, m) \mapsto \rho_{n,m}, \quad \sum_{n,m=1}^{N} \rho_{n,m} = 1. \quad (5.5.23)$$

While the rows (position index m) are reordered by the permutation matrix $\mathbf{B}$, Eq. (5.5.12), the columns (momentum index n) are permuted by the inverse matrix,

$$\mathbf{B}_3^{-1} = \mathbf{B}_3^t = \begin{pmatrix} 1\,0\,0\,0\,0\,0\,0\,0 \\ 0\,0\,1\,0\,0\,0\,0\,0 \\ 0\,0\,0\,0\,1\,0\,0\,0 \\ 0\,0\,0\,0\,0\,0\,1\,0 \\ 0\,1\,0\,0\,0\,0\,0\,0 \\ 0\,0\,0\,1\,0\,0\,0\,0 \\ 0\,0\,0\,0\,0\,1\,0\,0 \\ 0\,0\,0\,0\,0\,0\,0\,1 \end{pmatrix}. \quad (5.5.24)$$

With these permutations of rows and columns, the full discrete baker map takes the form of a similarity transformation,

$$\rho' = \mathbf{B}^{-1}\rho\,\mathbf{B}^{\mathrm{T}} = \mathbf{B}^{\mathrm{T}}\rho\,\mathbf{B}^{\mathrm{T}}. \quad (5.5.25)$$

The remarkable observation that the permutation effected by $\mathbf{B}$ undoes itself after a finite number of iterations, see Fig. 5.9, now applies to rows and columns alike. Specifically, for $N = 2^M$, $\mathbf{B}^M = \mathbf{I}$, so that any pixelated image encoded in the matrix ρ will be recovered identically after applying the discrete baker map M times (Fig. 5.12).

The interpretation of the Bernoulli and baker maps as rigid shifts of binary sequences extends readily to the discrete baker map: In this case, the revolving mechanism that returns the most significant digit to the back end of the sequence is complemented by another such mechanism at the least significant (N^{th}) digit, so that the entire cyclic chain of 2 N digits returns to its initial position after N mappings.

The foregoing analyses demonstrate that a permanent production of entropy, hallmark of deterministic chaos, depends on the availability of an unlimited supply of information. If that source is finite, be it due to a smooth distribution of initial conditions or to a discretized state space, chaotic entropy production is doomed to stall eventually, giving way to a stationary or periodic asymptotic state. In Sects. 8.5.2 and 8.5.3, this classical phenomenon will be compared with the closely related suppression of chaos in quantum systems.

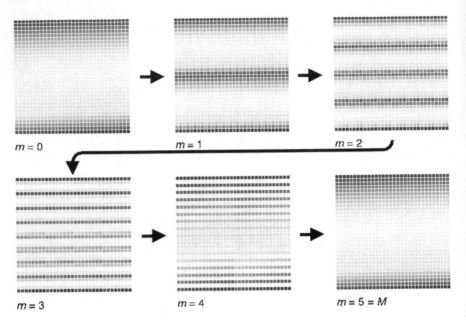

Fig. 5.12 A discrete baker map, which extends the discrete Bernoulli map to two dimensions, can be construed by combining the discrete Bernoulli map applied to the position (column index) with its inverse, applied to the momentum (row index), cf. Eq. (5.5.25). For a size $N \times N$, with $N = 2^M$ (here $M = 5$), of the array, both operations undo themselves after M iterations. As a visually striking consequence, an image pixelated as an $N \times N$ array will reappear identically after M iterations of the map.

5.6   Information Exchange Between Degrees of Freedom: Normal Modes in Pairs and Chains of Harmonic Oscillators

The previous subsection has shown how information can be transferred "vertically" between different scales in phase space. Another important mode of information flow in a Hamiltonian system, now in a "horizontal" direction, is its exchange between degrees of freedom. However, while the conservation of entropy in Hamiltonian dynamics allows us to track the flow of entropy between scales, such a detailed balance does not apply to the entropy interchanged among degrees of freedom. Information cannot be attributed uniquely to parts of a composed system. If they are correlated, information is shared among them, hence belongs to two or more freedoms simultaneously.

5.6.1 Two Coupled Harmonic Oscillators in Resonance

A classic example of horizontal information flow is the reversible interchange of energy and indeed of entropy between two coupled harmonic oscillators. It is manifest in particular in the phenomenon of beats, occurring if the frequencies are not too different and the coupling is not too strong. If initially only one of the two oscillators is excited, energy as well as information will gradually be transferred to the other oscillator till the first one is at rest, whence the reverse process sets in till the initial state is recovered, etc. (Fig. 5.13).

A rudimentary case of information exchange between different dimensions of a state space is shown in Fig. 5.4. Applying the idea to a bipartite Hamiltonian system allows us to analyze it in quantitative detail. A suitable starting point is the model of a pair of linearly coupled harmonic oscillators (frequencies ω_1, ω_2, identical masses m, coupling constant c) with the Hamiltonian

$$H(\mathbf{r}_1, \mathbf{r}_2) = H_1(\mathbf{r}_1) + H_2(\mathbf{r}_2) + V_{12}(q_1, q_2),$$

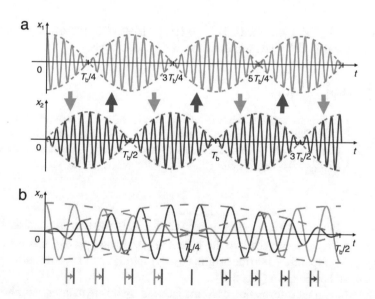

Fig. 5.13 Beats characterize the motion of two weakly coupled harmonic oscillators in resonance. If only one of them is initially excited (x_1 in panel (a)), its amplitude will gradually diminish (dashed envelope) while that of the other oscillator (x_2 in the figure) increases till, at one quarter of the full period T_b of the beats, the first one comes to a complete rest and the second one oscillates with the full initial amplitude of the first one, whence the same process evolves in inverse order. It amounts to an alternation of energy and information transfer from oscillator 1 to oscillator 2 (green arrows) and back (pink arrows), twice per period T_b. Panel (b) superposes the two oscillations (b) to reveal their phase relation (zeros of x_1 and x_2 with $\dot{x}_n > 0$ are marked with green and pink vertical strokes, resp.). During the first quarter period of the beats, with information transfer from oscillator 1 to 2, x_1 precedes x_2 by $\pi/2$, during the second quarter, with information transfer from oscillator 2 to 1, the phase relation is reversed.

$$H_n(\mathbf{r}_n) = \frac{p_n^2}{2m} + \frac{m}{2}\omega_n^2 q_n^2, n = 1, 2, \quad V_{12}(q_1, q_2) = mkq_1q_2, \quad (5.6.1)$$

considered at resonance, $\omega_1 = \omega_2 =: \omega$, and in the limit of weak coupling, $k \ll \omega^2$.

Its two normal modes $q_+(t)$, $q_-(t)$, correspond to oscillation in phase and in antiphase, resp., of the two oscillators,

$$q_+(t) = \frac{1}{\sqrt{2}}(q_1(t) + q_2(t)), \quad q_-(t) = \frac{1}{\sqrt{2}}(q_2(t) - q_1(t)), \quad (5.6.2)$$

with frequencies $\omega_\pm = \sqrt{\omega^2 \pm k}$. Combining momenta in the same way results in a canonical transformation between Cartesian coordinates $(\mathbf{r}_1, \mathbf{r}_2)$ and normal modes $(\mathbf{r}_+, \mathbf{r}_-)$ in phase space,

$$\begin{pmatrix} \mathbf{r}_+ \\ \mathbf{r}_- \end{pmatrix} = \mathbf{H}\begin{pmatrix} \mathbf{r}_1 \\ \mathbf{r}_2 \end{pmatrix}, \begin{pmatrix} \mathbf{r}_1 \\ \mathbf{r}_2 \end{pmatrix} = \mathbf{H}^{\mathrm{T}}\begin{pmatrix} \mathbf{r}_+ \\ \mathbf{r}_- \end{pmatrix}, \mathbf{H} = \frac{1}{\sqrt{2}}\begin{pmatrix} \mathbf{I}^{(2)} & \mathbf{I}^{(2)} \\ -\mathbf{I}^{(2)} & \mathbf{I}^{(2)} \end{pmatrix}. \quad (5.6.3)$$

The orthogonal (4×4)-matrix $\mathbf{H}$, $\mathbf{H}^{-1} = \mathbf{H}^{\mathrm{T}}$, consists of four blocks, given by the (2×2)-unit matrix $\mathbf{I}^{(2)}$. The normal modes give rise to beats,

$$\begin{pmatrix} \mathbf{r}_1(t) \\ \mathbf{r}_2(t) \end{pmatrix} = \mathbf{M}(t)\begin{pmatrix} \mathbf{r}_1(0) \\ \mathbf{r}_2(0) \end{pmatrix},$$

$$\mathbf{M}(t) = \begin{pmatrix} \cos(\omega t)\cos(\omega_b t) & -m\omega\sin(\omega t)\cos(\omega_b t) & \sin(\omega t)\sin(\omega_b t) & -m\omega\cos(\omega t)\sin(\omega_b t) \\ \frac{\sin(\omega t)\cos(\omega_b t)}{m\omega} & \cos(\omega t)\cos(\omega_b t) & \frac{\cos(\omega t)\sin(\omega_b t)}{m\omega} & \sin(\omega t)\sin(\omega_b t) \\ \sin(\omega t)\sin(\omega_b t) & -m\omega\cos(\omega t)\sin(\omega_b t) & \cos(\omega t)\cos(\omega_b t) & -m\omega\sin(\omega t)\cos(\omega_b t) \\ \frac{\cos(\omega t)\sin(\omega_b t)}{m\omega} & \sin(\omega t)\sin(\omega_b t) & \frac{\sin(\omega t)\cos(\omega_b t)}{m\omega} & \cos(\omega t)\cos(\omega_b t) \end{pmatrix}.$$

$$(5.6.4)$$

with the low frequency $\omega_b = \sqrt{(\omega_+^2 - \omega_-^2)/2} = \sqrt{k}$. In the four-dimensional phase space (p_1, q_1, p_2, q_2), projected to the planes (p_1, q_1) and (p_2, q_2), this means that while (p_1, q_1) is spiralling into the origin, (p_2, q_2) spirals out till maximum amplitude is reached, and back again (Fig. 5.14).

In order to extract information currents induced by this dynamics, an inhomogeneous probability distribution that serves as a tracer has to be propagated with the phase space flow. As a standard choice, prepare the bipartite system in a Gaussian distribution in all the four phase-space variables, initially uncorrelated between the two degrees of freedom. Rescaling momenta and positions to a common dimension $\sqrt{\text{action}}$,

$$\mathbf{r} = (\mathbf{r}_1, \mathbf{r}_2) \rightarrow \boldsymbol{\zeta} = (\boldsymbol{\zeta}_1, \boldsymbol{\zeta}_2),$$
$$\boldsymbol{\zeta}_n := (\eta_n, \xi_n), \eta_n := (m\omega_n)^{-1/2}p_n, \xi_n := (m\omega_n)^{-1/2}q_n, n = 1, 2, \quad (5.6.5)$$

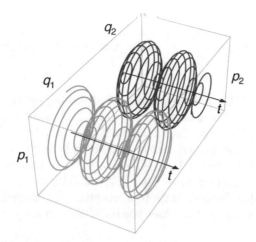

Fig. 5.14 Beats in phase space. If the four-dimensional phase space (p_1,q_1,p_2,q_2) of a pair of coupled harmonic oscillators 1, 2 is projected onto the planes (p_1,q_1), (p_2,q_2), beats become manifest as an alternating contraction and expansion of the otherwise circular trajectories: While oscillator 1 (green) spirals into the origin, oscillator 2 (pink) gains amplitude from zero till it reaches the maximum, and back, cf. Fig. 5.13.

the four-dimensional Gaussian takes the form

$$\rho\big(\zeta_1,\zeta_2,0\big) = \rho_1\big(\zeta_1,0\big)\rho_2\big(\zeta_2,0\big), \ \rho_n\big(\zeta_n,0\big) := \frac{1}{2\pi S_n}\exp\left(-\frac{|\zeta_n|^2}{2S_n}\right), n = 1,2,$$

(5.6.6)

where the two actions S_1, S_2, measure the widths in the two respective sectors of phase space. It is appropriate to choose the widths symmetrically within each freedom, as in Eq. (5.6.6), but markedly different between them, say $S_1 > S_2$, in order to trace the corresponding variation of partial entropies with the time evolution.

The propagated density $\rho'(\zeta',t)$ is obtained from the initial distribution by following the trajectory $\mathbf{r}(t)$ back to its initial point $\mathbf{r}(0)$, cf. Eq. (5.2.13),

$$\rho'\big(\zeta'(t),t\big) = \rho\big(\zeta(\zeta',t,0),0\big),$$

(5.6.7)

denoting with $\zeta(\zeta',t,0)$ the result of tracing $\zeta'(t)$ back till the initial time 0. Taking the "breathing" alternation of phase-space contraction and expansion in the two freedoms into account, cf. Eq. (5.6.4) and Fig. 5.14, we expect the distribution to concentrate towards the origin in the first freedom and to expand in the other before the process runs backwards. Invoking Eq. (5.6.5) to evaluate Eq. (5.6.7) gives

$$\rho\big(\zeta,t\big) = \rho_1\big(\zeta_1,t\big)\rho_2\big(\zeta_2,t\big)\exp\left(\frac{-\zeta_1\wedge\zeta_2}{4S_{12}(t)}\right), \ S_{12}(t) := \frac{S_1 S_2}{(S_2 - S_1)\sin(\omega_b t)},$$

$$\rho_n(\zeta_n, t) = \frac{e^{-|\zeta_n|^2/2S_n(t)}}{2\pi S_n}, \ S_n(t) := \frac{S_1 S_2}{S_n(\cos(\omega_b t))^2 + S_{3-n}(\sin(\omega_b t))^2}, n = 1, 2.$$

$$(5.6.8)$$

The first two factors in the time-evolved density, depending only on $\mathbf{r}_1$ and $\mathbf{r}_2$, respectively, corroborate the above surmise. The third factor, containing the symplectic product $\mathbf{r}_1 \wedge \mathbf{r}_2$ [JS98], comes less expected. It modifies the distribution only if the system has non-zero components simultaneously in both sectors. It vanishes periodically at integer multiples of $T_b/4 = \pi/2\omega_b$, that is, at each node of the beats of either one of the two oscillators, see Figs. 5.14 and 5.15.

Partial distributions for each of the two subsystems are extracted by tracing over the respective other degree of freedom, see Eq. (5.2.6). In the present case,

$$\overline{\rho}_n(\zeta_n, t) = \int d^2 \zeta_{3-n} \rho(\zeta_1, \zeta_2, t) = \frac{1}{2\pi \overline{S}_n(t)} \exp\left(-\frac{|\zeta_n|^2}{2\overline{S}_n(t)}\right),$$

$$\overline{S}_n(t) := S_n(\cos(\omega_b t))^2 + S_{3-n}(\sin(\omega_b t))^2, n = 1, 2,$$

$$(5.6.9)$$

with the initial widths S_1, S_2, as in Eq. (5.6.6) (note that the widths $\overline{S}_1(t)$, $\overline{S}_2(t)$ do not coincide with $S_1(t)$, $S_2(t)$ defined in Eq. (5.6.7)).

The second crucial ingredient to interpret the beats in terms of information flows, besides the total and the partial densities, is the correlation between the two freedoms. Their cross-correlations are defined as

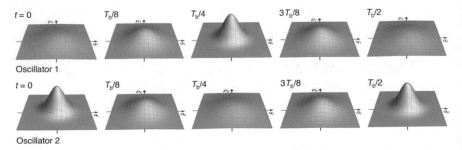

Fig. 5.15 Beats reflected in the redistribution of phase-space densities. Prepare the pair of coupled harmonic oscillators, as in Fig. 5.13, in Gaussian probability density distributions in their respective phase space sectors, see Eq. (5.6.6), here with $S_1 = 4S_2$. The phase-space flow then leads to an alternation of complementary contraction and expansion processes of the partial densities describing the two oscillators (Eq. (5.6.8)), consistent with the trajectories spiralling into and out of the origin (Fig. 5.14). The figure shows snapshots of the phase-space density distributions for oscillators 1 (upper row) and 2 (lower row) at integer multiples of $T_b/8 = \pi\omega_b/4$, covering the first period of the swapping process.

$$C_{12}(t) = \frac{\langle \mathbf{r}_1(t) \wedge \mathbf{r}_2(t) \rangle}{\sqrt{\langle |\mathbf{r}_1(t)|^2 \rangle \langle |\mathbf{r}_2(t)|^2 \rangle}} = \frac{\mathrm{Covar}(\mathbf{r}_1(t), \mathbf{r}_2(t))}{\sqrt{\mathrm{Var}(\mathbf{r}_1(t))\,\mathrm{Var}(\mathbf{r}_2(t))}}. \tag{5.6.10}$$

With the time-dependent density (5.6.55), the correlation evolves in time as

$$C_{12}(t) = \frac{1}{2}\sin(2\omega_b t)\left(\frac{1}{S_1} - \frac{1}{S_2}\right)\sqrt{S_1(t)S_2(t)} \tag{5.6.11}$$

($S_1(t)$ and $S_2(t)$ as in Eq. (5.6.8)). It vanishes for $S_1 = S_2$, when the distribution has spherical symmetry in the four-dimensional phase space, and at integer multiples of $T_b/4$, that is, at the nodes of the beats, where the distribution factorizes between the two subsystems. In between these periodic zeros, the correlation oscillates between positive and negative values. This behaviour is easily explained by inspecting sections of the density that include one phase-space coordinate of each of the two freedoms, e.g., p_1 and q_2 (Figs. 5.16 and 5.17b).

With the total and partial phase-space densities we have all ingredients required to calculate the corresponding information contents as well as their respective distributions in phase space. As a global measure, the total information, cf. Eq. (5.3.1), associated to the phase-space density (5.6.8),

$$I_{12} = -c\int d^2 r_1 \int d^2 r_2\, \rho(\mathbf{r}_1, \mathbf{r}_2)\ln(h_1 h_2 \rho(\mathbf{r}_1, \mathbf{r}_2))$$

$$= -c\left(\ln\left(\frac{h_1}{2\pi S_1}\right) + \ln\left(\frac{h_2}{2\pi S_2}\right) - 2\right), \tag{5.6.12}$$

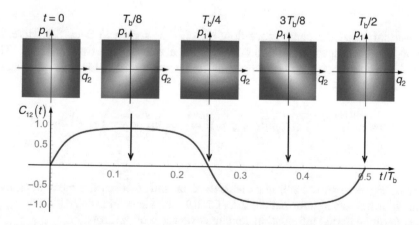

Fig. 5.16 Cross-correlation (5.6.11) between the two coupled harmonic oscillators of Figs. 5.13, 5.14 and 5.15, over half a period T_b of the beats (red curve). Its periodic oscillation between positive and negative correlations with period $T_b/2$ is a consequence of the rotation of the density in phase-space planes across the two sectors, e.g., (p_1,q_2) (upper row), in analogy to the rotation of a two-dimensional Gaussian distribution, featured in Fig. 5.4.

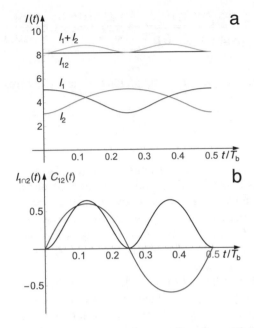

Fig. 5.17 Total, partial, and shared information vs. trajectory cross-correlation function in the pair of coupled harmonic oscillators of Figs. 5.12, 5.13, 5.14 and 5.15. (**a**) The time evolution of the total information I_{tot} (black), the partial information contents I_1 and I_2 for oscillators 1 (blue) and 2 (orange), and their sum $I_1 + I_2$ (green) over half a period T_b of the beats shows that $I_1 + I_2$ is not conserved and generally exceeds I_{tot}. (**b**) Comparing the surplus $I_{1 \cap 2}(t) = I_1 + I_2 - I_{12}$, Eq. (5.6.14) (black), with the cross-correlation, Eq. (5.6.11) (red) reveals the intimate relationship between the two quantities, a direct consequence of the rotation of a Gaussian distribution in phase space that is not circular symmetric (cf. panels a and b with panels b and c, resp., of Fig. 5.4).

is invariant under the phase-space flow (5.2.19), as shown in Sect. 5.2. Here, h_1 and h_2 are the respective minimum resolvable actions in the two freedoms. The partial entropies, however, corresponding to the reduced densities per subsystem, Eq. (5.3.8), do depend on time,

$$I_n(t) = -c \int d^2 r_n \, \overline{\rho}_n(\mathbf{r}_n) \ln\big(h_n \overline{\rho}_n(\mathbf{r}_n)\big) = -c\left(\ln\left(\frac{h_n}{2\pi \, S_n}\right) - 1\right), \, n = 1, 2.$$

$$(5.6.13)$$

Comparing these partial entropies with the total information, the most significant quantity is the shared information, Eq. (1.7.10), i.e., the surplus $I_1(t) + I_2(t) - I_{12}$ of the sum of the partial information contents over the total information,

$$I_{1 \cap 2}(t) = I_1(t) + I_2(t) - I_{12} = c \ln\left(\frac{S_1(t) S_2(t)}{S_1 S_2}\right)$$

$$= -c\ln\left(\left(\tfrac{1}{2}\sin(2\omega_b t)\right)^2 \left(\frac{S_1}{S_2} + \frac{S_2}{S_1}\right) + (\cos(\omega_b t))^4 + (\sin(\omega_b t))^4 \right) > 0,$$

(5.6.14)

which is evidently positive definite. Total, partial, and shared entropies for this system are shown in Fig. 5.17a and compared to the cross correlation in Fig. 5.17b.

Beats in a pair of coupled harmonic oscillators provide a striking example of the phenomenon discussed in Sect. 5.3 and illustrated in Fig. 5.4: Distinct degrees of freedom do not correspond to disjoint parts of phase space; in the presence of correlations, information cannot be uniquely assigned to different subsystems. The surplus $I_{1 \cap 2}(t)$, in particular, can be interpreted as that part of the total information that is shared among the two freedoms.

Where the sum of partial contents is not conserved, talking of currents between the parts of a system is not meaningful. However, it still makes sense to consider information flows, as defined in Eq. (5.3.6), within each subsystem.

$$\mathbf{i}_n(\mathbf{r}_n, t) := \iota_n(\mathbf{r}_n, t)\, \dot{\mathbf{r}}_n(\mathbf{r}_n, t), n = 1, 2.$$

(5.6.15)

with the information density $\iota_n(\mathbf{r}_n, t)$ per freedom n as given in Eq. (5.3.5). As a consequence of the non-conservation of the partial information contents, this current density does not even fulfil a continuity equation but exhibits sources and sinks, which in turn depend on time. The deviation from continuity,

$$\mathrm{div}(\mathbf{i}_n(\mathbf{r}_n, t)) + \frac{\partial}{\partial t}\iota_n(\mathbf{r}_n, t) = c\rho_n(\mathbf{r}_n, t)\mathrm{div}\big(\dot{\mathbf{r}}_n(\mathbf{r}_n, t)\big),$$

(5.6.16)

measures the local entropy loss or gain at $\mathbf{r}_n$ in the phase space of subsystem n. Since the divergence of the *partial* phase-space flow $\dot{\mathbf{r}}_n$ does not necessarily vanish, the l.h.s. of Eq. (5.6.16) can be non-zero as well. Figure 5.18 shows the partial information densities (first and third row) and the corresponding current density fields (second and fourth row) for both subsystems of Eq. (5.6.1), superposed on the deviation from continuity, Eq. (5.6.15), at the first five integer multiples of $T_b/8$. It is evident that as the distribution of oscillator 1 contracts at $t = T_b/8$, its partial entropy diminishes and the information flow forms a vortex, spiralling into a sink, while the partial entropy in oscillator 2 increases and its information flow exhibits a source. At $t = 3T_b/8$, the roles are interchanged.

Additional insight into the information dynamics in a pair of coupled harmonic oscillators is gained by returning to the normal modes defined in Eq. (5.6.3). The Hamiltonian (5.6.1), transformed to normal phase-space coordinates,

$$H(\mathbf{r}_+, \mathbf{r}_-) = H_+(\mathbf{r}_+) + H_-(\mathbf{r}_-), H_\pm(\mathbf{r}_\pm) = \frac{p_\pm^2}{2m} + \frac{m}{2}\omega_\pm^2 q_\pm^2$$

(5.6.17)

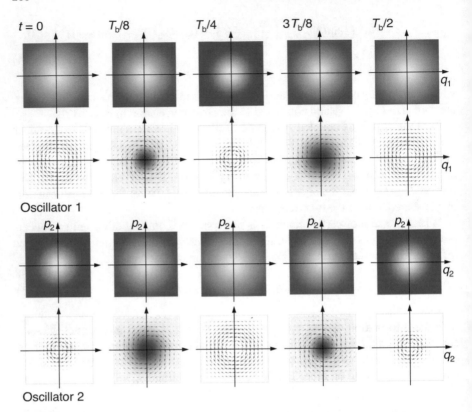

Fig. 5.18 Partial information densities (first and third rows, colour code ranging from blue to yellow for increasing density) and density currents (arrows, second and fourth rows), at $t = nT_b/8$ for n = 0,...,5, as they accompany the beats featured in Figs. 5.12, 5.13 and 5.14. At $t = T_b/8$, during the initial phase of the first beat, the Gaussian distribution for oscillator 1 contracts, its partial entropy reduces, and the information flow spirals into a sink (negative deviation from continuity, Eq. (5.6.16), second row, colour coded from white (zero) to red for increasing negativity). At the same time, oscillator 2 gains entropy, and the information flow spirals out of a source (positive deviation from continuity, Eq. (5.6.16), fourth row, colour coded from white (zero) to blue for increasingly positive values). At $t = 3 Tb/8$, the same process repeats itself in the opposite sense.

gives rise to uncoupled equations of motion for the two modes. They are solved by

$$\mathbf{r}_\pm(t) = \mathbf{M}_\pm(t)\mathbf{r}_\pm(0), \mathbf{M}_\pm(t) = \begin{pmatrix} \cos(\omega_\pm t) & -m\omega_\pm \sin(\omega_\pm t) \\ \sin(\omega_\pm t)/m\omega_\pm & \cos(\omega_\pm t) \end{pmatrix}, \quad (5.6.18)$$

that is, by independent rotations with the normal frequencies $\omega_\pm$ in the two respective sectors of phase space.

The probability density distribution, transformed to normal modes, cf. Eq. (5.6.34),

$$\rho_{+-}(\mathbf{r}_+, \mathbf{r}_-, t) = \rho\left(\mathbf{H}^T(\mathbf{r}_1, \mathbf{r}_2)^T, t\right), \tag{5.6.19}$$

rotates anticlockwise with the angular frequencies ω_+, ω_-, resp., along elliptic trajectories $\mathbf{r}_\pm(t)$, given in Eq. (5.6.18), independently in each of the two phase-space sectors,

$$\rho_{+-}(\mathbf{r}_+, \mathbf{r}_-, t) = \rho_{+-}\left(\mathbf{M}_+^T(t)\mathbf{r}_+, \mathbf{M}_-^T(t)\mathbf{r}_-, 0\right), \tag{5.6.20}$$

with $|\det(\mathbf{M}_\pm(t))| = 1$. In particular, a factorizing initial distribution $\rho_{+-}(\mathbf{r}_+, \mathbf{r}_-, 0) = \rho_{0+}(\mathbf{r}_+)\rho_{0-}(\mathbf{r}_-)$, for example into two Gaussians with different widths S_+, S_-, resp., analogous to that given in Eq. (5.6.6) for the original degrees of freedom,

$$\rho_{0\pm}(\mathbf{r}_\pm) = \frac{1}{2\pi S_\pm} \exp\left(-\frac{(m\omega_\pm)^{-1}p_\pm^2 + m\omega_\pm q_\pm^2}{2S_\pm}\right), \tag{5.6.21}$$

will then continue to factorize, with the two Gaussians propagated accordingly,

$$\rho_{+-}(\mathbf{r}_+, \mathbf{r}_-, t) = \rho_{0+}\left(\mathbf{M}_+^T(t)\mathbf{r}_+\right)\rho_{0-}\left(\mathbf{M}_-^T(t)\mathbf{r}_-\right), \tag{5.6.22}$$

Partial probability density distributions per normal mode,

$$\overline{\rho}_\pm(\mathbf{r}_\pm) := \int d^2 r_\mp \rho(\mathbf{r}_+, \mathbf{r}_-), \tag{5.6.23}$$

then do not mix, either, and evolve in time independently,

$$\overline{\rho}_\pm(\mathbf{r}_\pm, t) = \rho_{0\pm}\left(\mathbf{M}_\pm^T(t)\mathbf{r}_\pm\right). \tag{5.6.24}$$

Based on the partial densities $\overline{\rho}_\pm(\mathbf{r}_\pm, t)$, we can define the partial information in each mode as

$$I_\pm(t) := -c \int d^2 r_\pm \overline{\rho}_\pm(\mathbf{r}_\pm, t) \ln\left(h_\pm \overline{\rho}_\pm(\mathbf{r}_\pm, t)\right). \tag{5.6.25}$$

with minimum resolvable action cell sizes h_+ and h_-. Their time dependence, transforming the phase-space integral from propagated to initial normal coordinates and using $|\det(\mathbf{M}_\pm(t))| = 1$, is

$$I_\pm(t) = -c \int d^2 r_\pm \rho_{0\pm}\left(\mathbf{M}_\pm^T(t)\mathbf{r}_\pm\right) \ln\left(h_\pm \rho_{0\pm}\left(\mathbf{M}_\pm^T(t)\mathbf{r}_\pm\right)\right)$$

$$= -c \int d^2r_\pm |\det(\mathbf{M}(t))| \rho_{0_\pm}(\mathbf{r}_\pm) \ln\left(h_\pm |\det(\mathbf{M}(t))| \rho_{0_\pm}(\mathbf{r}_\pm)\right)$$

$$= -c \int d^2r_\pm \rho_{0_\pm}(\mathbf{r}_\pm) \ln\left(h_\pm \rho_{0_\pm}(\mathbf{r}_\pm)\right) = I_\pm(0). \tag{5.6.26}$$

Partial information is conserved in each of the two normal modes! Moreover, if the two distributions factorize initially, hence are statistically independent, the total information reduces to their sum,

$$I_{12} = -c \int d^2r_+ \int d^2r_- \rho(\mathbf{r}_+, \mathbf{r}_-, t) \ln(h_+ h_- \rho(\mathbf{r}_+, \mathbf{r}_-, t)) = I_+(t) + I_-(t), \tag{5.6.27}$$

(assuming $h_+ h_- = h_1 h_2$) and the shared information vanishes,

$$I_{+\cap-}(t) = I_+(t) + I_-(t) - I_{+-} = 0. \tag{5.6.28}$$

This substantiates that the normal modes are in fact the most appropriate definition of degrees of freedom for this system of coupled harmonic oscillators.

The exchange of information among the two oscillators even allows for conclusions concerning a causal relationship. According to the basic criterion developed in Sect. 3.1.2, the direction of information flow indicates that during the first quarter period of the beats, when oscillator 1 loses energy and information, see Figs. 5.13a and 5.17a, it plays the role of the cause, the motion of oscillator 2 that of the effect, while during the second quarter period, roles are interchanged.

Another test for causal relations among oscillating systems is phase lags: Fig. 5.13b shows that the oscillator initially losing information also precedes the other by a phase shift of $\pi/2$. This phase relation reverses itself at the first node of the beats at $t = T_b/4$ and then alternates in synchrony with the information transfer. It coincides with the phase lag of a harmonic oscillator, driven periodically at resonance, with respect to the driving force, which allows to identify the driving osillator with the cause and the driven one with the effect. This shows that even in Hamiltonian systems, under conditions of time-reversal invariance and entropy conservation, we can temporarily assign the roles of cause and effect to parts of a system, if a global time arrow is predefined. Evidently, time reversal leaves the solutions of the equations of motion invariant but interchanges cause and effect.

5.6.2 Chains of N Coupled Harmonic Oscillators

The analysis carries over readily to larger systems, specifically sets of N coupled harmonic oscillators. A configuration of couplings that allows for a particularly transparent treatment is a chain of oscillators, interacting only through identical nearest-neighbour couplings with periodic boundaries (i.e., closed to a ring), Fig. 5.19a. The Hamiltonian for such a system is

$$H(\mathbf{p}, \mathbf{q}) = T(\mathbf{p}) + V(\mathbf{q}), \mathbf{p} = (p_0, \ldots, p_{N-1}), \mathbf{q} = (q_0, \ldots, q_{N-1}),$$

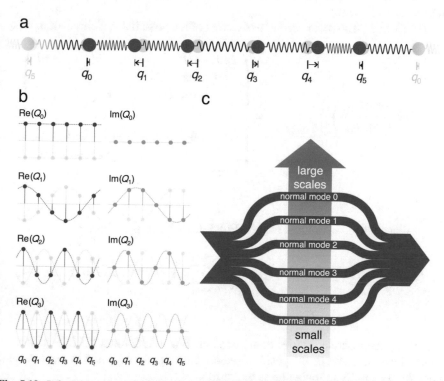

Fig. 5.19 Information dynamics in a system of coupled harmonic oscillators. Arranging six masses in a linear chain with periodic boundary conditions, interacting between nearest neighbours by identical elastic springs (a) results in coupled equations of motion for the positions q_n, $n = 0,\ldots,5$. They are uncoupled by transforming to normal modes Q_ν, $\nu = 0,\ldots,5$, (Eq. (5.6.31)). The schematic presentation (b) shows the six positions for real (pink) and imaginary part (green) at phase 0 (bold) and phase π (pale), indicating the underlying continuous waves as dashed lines, for six linearly independent normal modes. Their wavelengths are 1 (mode 3), 2 (modes 2, 4), 6 (modes 1, 5) units of discrete positions along the chain, and undefined (mode 0). They filter incoming information according to frequency and wavelength (c) and transmit it, conserving the component in each normal mode without any mixing among them.

$$T(\mathbf{p}) = \sum_{n=0}^{N-1} \frac{p_n^2}{2m}, \ V(\mathbf{q}) = \frac{mk}{2} \sum_{n=0}^{N-1} (q_{n+1} - q_n)^2 = mk \sum_{n=0}^{N-1} \left(q_n^2 - q_n q_{n+1}\right).$$

$$(5.6.29)$$

Here we count the index n modulo N, that is, identifying $n = N$ with $n = 0$. The potential $V(\mathbf{q})$ comprises only linear position-position coupling terms $mk q_n q_{n+1}$, no on-site potentials $m\omega_n^2 q_n^2 / 2$, so that the Hamiltonian is invariant under rigid spatial shifts $q_n \rightarrow q_n + \Delta q$ of all oscillators. (Note that in this way, the Hamiltonian (5.6.29) does not reduce for $N = 2$ to the form (5.6.1) for a pair of weakly coupled harmonic oscillators.)

The $(N \times N)$-matrix of second derivatives of the potential is a circulant matrix,

$$\mathbf{V}'' = \left(\frac{\partial^2 V(\mathbf{q})}{\partial q_n \partial q_{n'}}\right)_{n,n'=0,\ldots,N-1} = \begin{pmatrix} 2mk & -mk & & & & & -mk \\ -mk & 2mk & -mk & & & & \\ & -mk & 2mk & & \mathbf{0} & & \\ & & & \ddots & & & \\ & \mathbf{0} & & & 2mk & -mk & \\ & & & & -mk & 2mk & -mk \\ -mk & & & & & -mk & 2mk \end{pmatrix}$$

$$(5.6.30)$$

and is diagonalized by normal modes [FL85, LSV08] (Fig. 5.19b),

$$P_\nu = \frac{1}{\sqrt{N}} \sum_{n=0}^{N-1} p_n e^{-2\pi i \nu n / N}, \ Q_\nu = \frac{1}{\sqrt{N}} \sum_{n=0}^{N-1} q_n e^{-2\pi i \nu n / N}, \ \nu = 0, \ldots, N-1,$$

$$(5.6.31)$$

that is, by the discrete Fourier transforms, cf. Eqs. (2.3.24 and 2.3.25), of the N-dimensional momentum and position vectors $\mathbf{p}$ and $\mathbf{q}$. Transformed to these normal coordinates, the Hamiltonian takes the form

$$H(\mathbf{P}, \mathbf{Q}) = T(\mathbf{P}) + V(\mathbf{Q}), \ \mathbf{P} = (P_0, \ \ldots, \ P_{N-1}), \ \mathbf{Q} = (Q_0, \ \ldots, \ Q_{N-1}),$$

$$T(\mathbf{P}) = \frac{1}{2m} \sum_{\nu=0}^{N-1} P_\nu^2, \ V(\mathbf{Q}) = \frac{m}{2} \sum_{\nu=0}^{N-1} \Omega_\nu^2 Q_\nu^2, \qquad (5.6.32)$$

with normal-mode frequencies

$$\Omega_\nu = 2\sqrt{k} \sin\left(\frac{\pi \nu}{N}\right). \qquad (5.6.33)$$

In this form, it separates into a sum of N uncoupled Hamiltonians,

$$H(\mathbf{P}, \mathbf{Q}) = \sum_{\nu=0}^{N-1} H_\nu(P_\nu, Q_\nu), \quad H_\nu(P_\nu, Q_\nu) = \frac{P_\nu^2}{2m} + \frac{m}{2}\Omega_\nu^2 Q_\nu^2. \quad (5.6.34)$$

The normal modes include in particular the zero mode, that is, rigid motion (center-of-mass motion) of the total set of oscillators with constant velocity. The other modes form discrete waves, of wavelength

$$\lambda_\nu = \frac{Na}{\nu}, \quad (5.6.35)$$

if the spatial distance (e.g., the rest length of connecting springs) between neighbouring oscillators is a. That is, we can associate a spatial scale to each normal mode that decreases with the mode number ν as $1/\nu$.

In the presence of an external perturbation, for example an oscillating force applied to one of the degrees of freedom, they give rise to waves propagating with a velocity

$$v_\nu = \frac{\lambda_\nu \Omega_\nu}{2\pi} = a\sqrt{k}\,\frac{\sin(\pi\nu/N)}{\pi\nu/N}, \quad \nu = 1, \ldots, N-1. \quad (5.6.36)$$

As in the case of a pair of coupled oscillators, we can define a probability density distribution $\rho_\mathbf{r}(\mathbf{r}_0, \ldots, \mathbf{r}_{N-1})$ depending on the phase-space coordinates $\mathbf{r}_n = (p_n, q_n)$ of each oscillator, and transform it to normal modes $\rho_\mathbf{R}(\mathbf{R}_0, \ldots, \mathbf{R}_{N-1})$, $\mathbf{R}_\nu = (P_\nu, Q_\nu)$. Partial distributions $\overline{\rho}_\nu(\mathbf{R}_\nu, t)$ for individual normal modes ν, given as averages over all the other modes, evolve in time according to simple rotation in the phase space of this normal mode. The total information in the oscillator chain is then

$$I_{\text{tot}} = -c \int d^2 R_{N-1} \cdots \int d^2 R_0 \rho_\mathbf{R}(\mathbf{R}_0, \ldots, \mathbf{R}_{N-1}) \ln\left(h_\mathbf{R}^N \rho_\mathbf{R}(\mathbf{R}_0, \ldots, \mathbf{R}_{N-1})\right).$$

$$(5.6.37)$$

It reduces to information contents per mode as

$$I_\nu(t) = -c \int d^2 R_\nu \rho_\nu(\mathbf{R}_\nu, t) \ln(h_\mathbf{R}\rho_\nu(\mathbf{R}_\nu, t)). \quad (5.6.38)$$

The analysis in terms of normal modes now allows for reasoning along the same lines as for a pair of coupled oscillators. Separating into N uncoupled normal modes, which can be treated as independent degrees of freedom, the harmonic oscillator chain is identified as an integrable system [JS98, GL&06], possessing the same number of constants of the motion, the single-mode energies $H_\nu(P_\nu, Q_\nu)$, cf. Eq. (5.6.34). For the same reason, also the individual information in each of the N normal modes is conserved,

$$I_\nu(t) = \text{const} = I_\nu(0). \tag{5.6.39}$$

Moreover, it is transmitted within this normal mode with the propagation velocity (5.6.36).

A picture emerges where incoming information is filtered and injected into normal modes, according to its scales in space and time, and transmitted independently in each mode, without any crosstalk to other modes (Fig. 5.19c). This is a striking example of *shielding*, the separation of different levels of a structural hierarchy, see Sect. 1.5, by an information processing within each tier that prevents communication with smaller or larger scales. The normal modes of a set of coupled harmonic oscillators are in fact a strongly reduced instance of a broader concept, *collective modes*, dynamical phenomena in many-body systems that comprise a macroscopic number or all of the degrees of freedom but, for their particular robustness and stability, supersede single particles. For the information dynamics, they represent therefore the most appropriate way of partitioning the many-body system into nearly independent subsystems. However, complete isolation of the information component in each mode, as in harmonic potentials, is untypical. As soon as interactions between particles include anharmonicities, information starts leaking into neighbouring modes, till the concept loses its meaning and the exchange of information between scales becomes the dominating process, as in the chaotic systems discussed above.

In this section, a close-up view of Hamiltonian dynamics revealed its central characteristic in terms of information: Entropy is a constant of the motion, more fundamental even than energy, reflecting the conservation of phase-space volume with the Hamiltonian flow, which implies for example that information flows from small to large scales must be compensated by simultaneous flows in the opposite direction.

Another decisive feature is the continuous character of space and momentum, which allows for an unlimited information density in classical phase space. Its consequences, in particular for chaotic dynamics, are brought out drastically if the state space is discretized. It restricts the information density to finite values, enforces a periodic time evolution, and thus anticipates the effects of quantum mechanics.

The following two sections will rather take a long shot of physical information flows, extending the view to systems that comprise a major number of degrees of freedom, and the interplay of the "vertical" flow of information between scales within a few degrees of freedom and its "horizontal" exchange among neighbouring freedoms becomes the dominant process. Where this environment becomes too large to allow for an explicit inclusion of its degrees of freedom in the equations of motion as was possible above, it is appropriate to take it into account instead in terms of an emerging collective phenomenon: dissipation. It no longer conserves phase-space volume and information, violates time-reversal invariance, and requires a more versatile approach to information flows.

Chapter 6
Information in Classical Dissipative Dynamics

The Hamiltonian approach to classical mechanics reviewed in the previous section represents a decidedly microscopic view of physical reality: It assumes implicitly that we can take all degrees of freedom fully into account that are involved in a dynamical process, that in particular we have access, through measurement and control, to the quantities determining their dynamical state. In the vast majority of physical situations, this is not nearly the case. Most of the degrees of freedom are not visible, let alone measurable, yet indirectly they do affect the dynamics: Their existence becomes manifest in phenomena like friction, heat, and fluctuations. These ambiguous conditions require a different treatment also in an information theoretic approach. In particular, the conservation of information, as a fundamental tenet of Hamiltonian dynamics, ceases to be valid, with drastic consequences for the information flow in dissipative systems. The present section is dedicated to global macroscopic aspects, contrasting dissipative dynamics with chaos as antagonistic phenomena in this system class. Evidently, the information exchange between scales will play a dominant role. The subsequent Chap. 7 will discuss noise and fluctuations, manifestations of randomness on microscopic scales that form a central concept in information theory, of general importance also beyond the physical context.

6.1 Lyapunov Exponents Measure Vertical Information Flows

In order to obtain a first impression of what to expect if information is not conserved as in Hamiltonian systems, it is instructive to consider the rate of change of the entropy in general terms. Starting from Boltzmann's original definition, Eq. (1.3.2), $S = c \log(\Omega) - C$, where Ω is the phase-space volume occupied by the system, we already conclude that

© Springer Nature Switzerland AG 2022
T. Dittrich, *Information Dynamics*, The Frontiers Collection,
https://doi.org/10.1007/978-3-030-96745-1_6

$$\frac{dS}{dt} = c\frac{\dot{\Omega}}{\Omega}, \tag{6.1.1}$$

that is, the entropy changes with the *relative* variation of phase-space volume: Extended to a more general context than mechanics, the metric of the state space proves crucial. In order to look into more detail, return to Shannon's information for continuous probability density distributions, cf. Eqs. (1.8.4) and (5.3.1). Taking the time derivative yields

$$\frac{d}{dt}I(t) = -c\int d^D x\left(1 + \ln\left(d_x^D \rho(\mathbf{x},t)\right)\right)\frac{\partial}{\partial t}\rho(\mathbf{x},t)$$

$$= -c\int d^D x\,\frac{\partial \rho(\mathbf{x},t)}{\partial t}\ln\left(d_x^D \rho(\mathbf{x},t)\right). \tag{6.1.2}$$

(The first term inside the bracket in the first is the time derivative of the total probability. It vanishes since $\rho(\mathbf{x},t)$ is normalized.)

From Eq. (6.1.2), it is not yet obvious how the metric of the state space $\{x\}$ would be involved. It becomes clearer, considering the entropy increment over a finite time interval Δt, instead of the derivative,

$$\Delta I = I(t') - I(t) = I(t + \Delta t) - I(t)$$

$$= -c\left(\int d^D x'\rho(\mathbf{x}',t')\ln\left(d_x^D \rho(\mathbf{x}',t')\right) - \int d^D x\,\rho(\mathbf{x},t)\ln\left(d_x^D \rho(\mathbf{x},t)\right)\right). \tag{6.1.3}$$

If the *total* probability is conserved, that is, in the absence of birth or death processes (if no particles enter or leave), the probability dP contained in a volume element $dV = d^D x$ remains constant. It is here that the state-space metric enters,

$$\rho(\mathbf{x}',t')d^D x' = dP' = dP = \rho(\mathbf{x},t)d^D x \Leftrightarrow \rho(\mathbf{x}',t') = \left|\det\left(\frac{\partial \mathbf{x}'}{\partial \mathbf{x}}\right)\right|^{-1}\rho(\mathbf{x},t), \tag{6.1.4}$$

Equation (6.1.4) increases the probability density, compensating for the reduction of the bin size $d^D x$. It generalizes Eq. (5.2.5), obtained for Hamiltonian dynamics evolving in phase space. In that context, the matrix $\mathbf{M}(t',t) = \partial \mathbf{x}'(t')/\partial \mathbf{x}(t)$ is identified as the stability matrix, introduced in Eq. (5.1.18). Its determinant proves decisive to quantify the entropy increment. Interpreted as a Jacobian, it suggests performing a transformation of the integration variable from x to x' in Eq. (6.1.4),

$$\Delta I = -c\int d^D x\rho(\mathbf{x},\mathbf{t})\left(\ln\left(d_x^D|\det(\mathbf{M}(t',t))|^{-1}\rho(\mathbf{x},t)\right) - \ln\left(d_x^D \rho(\mathbf{x},t)\right)\right)$$

$$= c\int d^D x\rho(x,t)\ln(|\det(\mathbf{M}(t',t))|). \tag{6.1.5}$$

As it stands, the stability matrix does not provide any clue how to calculate its eigenvalues. In the present context, however, it is possible to construct a Hermitian matrix from it,

$$\mathbf{H}(t', t) := \mathbf{M}^\dagger(t', t)\mathbf{M}(t', t), \tag{6.1.6}$$

such that $\mathbf{H}^\dagger(t', t) = \mathbf{H}(t', t)$. The eigenvalues $h_n(t', t)$ of $\mathbf{H}(t', t)$ are real and positive,

$$\mathbf{H}(t', t)\mathbf{x}_n(t', t) = h_n(t', t)\mathbf{x}_n(t', t), \, h_n(t', t) \in \mathbb{R}^+, \, n = 1, \dots, D. \tag{6.1.7}$$

Eigenvectors $\mathbf{x}_n(t', t), \mathbf{x}_{n'}(t', t), n \neq n'$, pertaining to two non-degenerate eigenvalues, are orthogonal. The determinant in Eq. (6.1.5) can now be written in terms of $h_n(t', t)$ as

$$\left|\det\left(\mathbf{M}(t', t)\right)\right| = \left|\det\left(\mathbf{H}(t', t)\right)\right|^{1/2} = \prod_{n=1}^{D} \sqrt{h_n(t', t)}. \tag{6.1.8}$$

Typically, these eigenvalues are exponential functions of the time period Δt with initial value h_n^0,

$$h_n(t', t) = h_n^0 \exp(\lambda_n(t' - t)) = h_n^0 e^{\lambda_n \Delta t}. \tag{6.1.9}$$

As a consequence of the initial condition $\det(\mathbf{M}(t, t)) = 1$, $\prod_{n=1}^{D} h_n^0 = 1$. In terms of the *Lyapunov exponents*,

$$\lambda_n = \lim_{\Delta t \to \infty} \frac{1}{2\Delta t} \ln\left(h_n(\Delta t, 0)/h_n^0\right), \tag{6.1.10}$$

the entropy increment takes the suggestive form

$$\Delta I = c \int d^D x \rho(\mathbf{x}, \Delta t) \sum_{n=1}^{D} \lambda_n \Delta t. \tag{6.1.11}$$

Considering the integral as a weighted average over the system's state space, the rate of change of information can now be expressed as

$$\frac{dI}{dt} = c \int d^D x \, \rho(\mathbf{x}, \Delta t) \sum_{n=1}^{D} \lambda_n = c \left\langle \sum_{n=1}^{D} \lambda_n \right\rangle. \tag{6.1.12}$$

Equation (6.1.12) is readily interpreted in the sense that Lyapunov exponents measure entropy fluxes between degrees of freedom or subsystems, gains at freedom n if λ_n

is positive, losses where it is negative. Conversely, in terms of actual information (negentropy, see Sect. 1.4), positive Lyapunov exponents indicate a decrease, negative ones an increase of specific knowledge on the state of the system. As illustrated in Sect. 3.1.3, they can be associated to chaotic and dissipative motion, respectively. However, there it is also argued that in the long-time limit, only positive exponents contribute to the entropy balance. This justifies relating the mean entropy gain to the Kolmogorov–Sinai entropy, see Eq. (3.1.17),

$$\lim_{t \to \infty} \frac{dI}{dt} = c \langle h_{KS} \rangle = c \left\langle \sum_{\substack{n=1 \\ \lambda_n > 0}}^{D} \lambda_n \right\rangle. \tag{6.1.13}$$

Equation (6.1.13) applies to Hamiltonian as well as to dissipative systems. The special case of Hamiltonian dynamics with a zero overall rate of volume contraction readily fits in the frame of Eq. (6.1.13). Symplectic geometry implies moreover that this balance applies not only to the sum, but individually, term by term, to each of the f dimensions of configuration as well as of momentum space: The D eigenvalues of the symmetrized stability matrix, $D = 2f$, come in pairs, say $h_n(t', t)$ and $h_{2f+1-n}(t', t)$, with $h_{2f+1-n}(t', t) = 1/h_n(t', t)$, so that accordingly, $\lambda_{2f+1-n} = -\lambda_n$ A pertinent example is the baker map discussed in Sect. 5.3.2. In systems where also energy is conserved, there is neither expansion nor contraction in the direction perpendicular to the energy shell (the hypersurface in phase space with constant energy), so that associating this direction to the fth degree of freedom, $\lambda_f = 0$. The direction canonically conjugate to it is path length along the trajectory measured in units of time. Indeed, the corresponding Lyapunov exponent vanishes as well, $\lambda_{f+1} = 0$, since the time separating phase-space points on the same trajectory is also invariant.

6.2 Entropy Loss into Microscales: The Dissipative Harmonic Oscillator

In Sect. 5.3.4, the interchange of entropy between parts of a system has been illustrated with the example of beats exhibited by a pair of coupled harmonic oscillators. What happens if we increase the number N of oscillators participating in the system, towards thermodynamic values $\gg 1$? Focussing on, say, oscillator 1 as the subsystem excited initially, the initial phase of loss of energy and entropy, before the process reverses its direction, increases in length till it becomes irreversible for $N \to \infty$. In this limit, the reduced dynamics of the first oscillator has turned dissipative.

An elementary example of a dissipative dynamical system, which nevertheless allows to demonstrate all relevant aspects in detail, is a single harmonic oscillator with a linear friction term. It combines rotation in phase space, the hallmark of harmonic oscillators [GL&06], with contraction towards a point attractor, the origin.

All parameters are introduced with the equation of motion for the canonical coordinate q,

$$\ddot{q} = -2\Gamma\dot{q} - \Omega^2 q. \tag{6.2.1}$$

The natural frequency of the oscillator is Ω, the coefficient Γ controls the velocity proportional (Ohmic) friction. Comparing the two parameters, we discard the two cases of overdamped motion, $\Gamma > \Omega$, and of critical damping, $\Gamma = \Omega$, since they do not preserve oscillations as characteristic dynamical feature [Kap03]. In the under-damped regime, $\Gamma < \Omega$, the effective frequency $\omega := \sqrt{\Omega^2 - \Gamma^2}$ and the dimensionless damping $\gamma := \Gamma/\omega$ are defined to write the solution of Eq. (6.2.1) in succinct form,

$$q(t) = e^{-\Gamma t}\left[(\cos(\omega t) + \gamma \sin(\omega t))q(0) + \frac{1}{\omega} \sin(\omega t)\dot{q}(0) \right]. \tag{6.2.2}$$

Adopting phase space from Hamiltonian dynamics as the canonical stage for mechanical systems, even for systems that do not conserve its volume, replacing velocity by momentum as second coordinate, the solution takes the form of a linear mapping

$$\begin{pmatrix} p(t) \\ q(t) \end{pmatrix} = M(t)\begin{pmatrix} p(0) \\ q(0) \end{pmatrix} \tag{6.2.3}$$

with

$$M(t) = e^{-\Gamma t}\begin{pmatrix} \cos(\omega t) - \gamma \sin(\omega t) & -m\omega(1 + \gamma^2) \sin(\omega t) \\ \sin(\omega t)/m\omega & \cos(\omega t) + \gamma \sin(\omega t) \end{pmatrix}. \tag{6.2.4}$$

(Terms of order γ^3 or higher have been neglected, assuming a very weak damping $\Gamma \ll \Omega$.) Eq. (6.2.4) is already sufficient to calculate the rate of entropy loss. The contraction of phase-space volume, given by

$$\frac{dV(t)}{dV(0)} = |\det(M(t))| = e^{-2\Gamma t}, \tag{6.2.5}$$

is constant over phase space. It directly yields the time-dependent information content for an initial probability distribution $\rho(\mathbf{r}, 0)$,

$$I(t) = -c \int d^{2f} r' \rho(\mathbf{r}'(t), t) \ln\left(d_\mathbf{r}^f \rho(\mathbf{r}'(t), t)\right)$$

$$= -c \int d^{2f} r \rho(\mathbf{r}(0), 0) \ln\left(d_\mathbf{r}^f |\det(M(t))|^{-1} \rho(\mathbf{r}(0), 0)\right), \tag{6.2.6}$$

where the contraction is taken into account by inserting $|\det(\mathbf{M}(t))|^{-1}\rho(\mathbf{r}(0), 0)$ for the time-evolved density $\rho(\mathbf{r}'(t), t)$. Substituting Eq. (6.2.5) for the determinant,

$$I(t) = -c \int d^{2f} r \rho(\mathbf{r}, t) \left(\ln\left(d_r^f \rho(\mathbf{r}, 0)\right) + 2\Gamma t \right) = I(0) - 2c\Gamma t, \qquad (6.2.7)$$

gives the rate of information loss, equivalent to a negative Lyapunov exponent,

$$\frac{dI}{dt} = -2c\Gamma t. \qquad (6.2.8)$$

Equation (6.2.8) only gives a global rate, it does not provide any indication where in phase space this loss occurs—it does not identify the "entropy sink(s)". A spatially resolved image of the contraction is obtained by looking at the phase-space velocity

$$\dot{\mathbf{r}}(t) = \begin{pmatrix} \dot{p}(t) \\ \dot{q}(t) \end{pmatrix} = \mathbf{M}(t) \begin{pmatrix} p(t) \\ q(t) \end{pmatrix}, \qquad (6.2.9)$$

with

$$\dot{\mathbf{M}}(t) = \omega e^{-\Gamma t} \times$$
$$\begin{pmatrix} -2\gamma \cos(\omega t) - (1 + \gamma^2) \sin(\omega t) & -m\omega(1 + \gamma^2)(\gamma \sin(\omega t) - \cos(\omega t)) \\ (\cos(\omega t) - \gamma \sin(\omega t))/m\omega & -(1 + \gamma^2) \sin(\omega t) \end{pmatrix}.$$
$$(6.2.10)$$

Equations (6.2.9) and (6.2.10) are only of limited value, though, since they express the velocity field in the frame of reference of the original coordinates $(p(0), q(0))$. Transforming them to the time-evolved frame $(p(t), q(t))$ requires an additional transformation with $\mathbf{M}(t)$,

$$\dot{\mathbf{r}}(t) = \dot{\mathbf{M}}(t)\mathbf{M}^{-1}(t) \begin{pmatrix} p(t) \\ q(t) \end{pmatrix} = \omega \begin{pmatrix} -2\gamma & -m\omega \\ 1/m\omega & 0 \end{pmatrix} \begin{pmatrix} p(t) \\ q(t) \end{pmatrix}. \qquad (6.2.11)$$

The deviation from a Hamiltonian system, where $\gamma = 0$ and $\dot{\mathbf{r}}$ is tangential to the energy contours, is manifest in the diagonal element -2γ which lets the flow lines spiral inwards, approaching the origin (Fig. 6.1).

As is to be expected from Eq. (6.2.5), the divergence of this velocity field is no longer zero, as for a symplectic dynamics, Eq. (5.1.10), but takes a negative value,

$$\text{div } \dot{\mathbf{r}}(t) = -2\Gamma. \qquad (6.2.12)$$

Invoking again Eq. (6.2.8), Lyapunov exponents can be calculated as eigenvalues of the Hermitian matrix $\mathbf{H} = \mathbf{M}^\dagger \mathbf{M}$ (cf. Eqs. (6.1.7–6.11)). To leading order in γ,

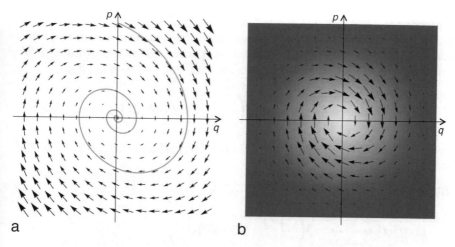

Fig 6.1 The phase-space velocity (panel **a**, black arrows) for an underdamped harmonic oscillator (Eqs. (6.2.9, 6.2.10) shown at $t = 0$) combines the rotation around the origin of the conservative harmonic oscillator with a radial component directed inwards (towards the origin) that reflects the loss of kinetic energy due to friction. Individual trajectories form exponential spirals (blue curve). For a Gaussian initial probability distribution (Eq. (6.2.15), panel **b**, colour code ranging from blue (zero) upwards through grey through yellow), the phase-space density flow (Eq. (6.2.18), black arrows) has the same direction, but decays away from the origin, as does the probability distribution.

$$\mathbf{H}(t) = \begin{pmatrix} e^{-2\Gamma t} & 0 \\ 0 & e^{-2\Gamma t} \end{pmatrix} \tag{6.2.13}$$

so that its eigenvalues and the Lyapunov exponents are, respectively,

$$h_1(t) = h_2(t) = e^{-2\Gamma t}, \; \lambda_1 = \lambda_2 = -\Gamma, \tag{6.2.14}$$

with a total sum $\lambda_1 + \lambda_2 = -2\Gamma$.

Calculating information densities and flows requires to define a specific initial probability distribution and to find its time evolution. A standard choice is a Gaussian centred in the origin,

$$\begin{aligned} \rho(\mathbf{r}(0), 0) &= \frac{1}{2\pi S_0} \exp\left(\frac{-1}{2S_0} \left(\frac{p^2}{m\omega} + m\omega q^2 \right) \right) \\ &= \frac{1}{2\pi S_0} \exp\left(\frac{-1}{2} \mathbf{r}^\dagger(0) \cdot \mathbf{S}(0)\mathbf{r}(0) \right), \\ \mathbf{S}(0) &= \begin{pmatrix} 1/m\omega S_0 & 0 \\ 0 & m\omega / S_0 \end{pmatrix}. \end{aligned} \tag{6.2.15}$$

The action S_0 controls the size of the Gaussian. In order to propagate this distribution in time, the current position $\mathbf{r}(t)$ has to be traced back to the corresponding $\mathbf{r}(0)$ to find the initial probability density at the starting point, and the normalization is to be corrected to compensate for phase-space contraction, $S_0 \rightarrow S_0|\det(\mathbf{M}(t))|$,

$$
\begin{aligned}
\rho(\mathbf{r}(t), t) &= \frac{1}{2\pi S_0 |\det(\mathbf{M}(t))|} \exp\left(-\frac{1}{2S_0}\mathbf{r}^\dagger(0)\mathbf{S}(0)\mathbf{r}(0)\right) \\
&= \frac{e^{2\Gamma t}}{2\pi S_0} \exp\left(-\frac{1}{2S_0}\mathbf{r}^\dagger(t)\cdot(\mathbf{M}^{-1}(t))^\dagger\mathbf{S}(0)\mathbf{M}^{-1}(t)\mathbf{r}(t)\right). \quad (6.2.16)
\end{aligned}
$$

At the same level of accuracy as Eq. (6.2.13),

$$
\rho(\mathbf{r}(t), t) = \frac{e^{2\Gamma t}}{2\pi S_0} \exp\left(-\frac{1}{2}e^{2\Gamma t}\mathbf{r}^\dagger(t)\cdot\mathbf{S}(t)\mathbf{r}(t)\right),
$$

$$
\mathbf{S}(t) = \begin{pmatrix} (1+\gamma\sin(2\omega t))/m\omega S_0 & \gamma(1-\cos(2\omega t)) \\ \gamma(1-\cos(2\omega t)) & m\omega(1-\gamma\sin(2\omega t))/S_0 \end{pmatrix}. \quad (6.2.17)
$$

From the time evolution of the probability density, various related quantities can be obtained. Combining it with the phase-space velocity (6.2.11), it directly gives the phase-space flow

$$
\begin{aligned}
\mathbf{j}(\mathbf{r}, t) &= \rho(\mathbf{r}, t)\dot{\mathbf{r}} \\
&= \frac{\omega e^{2\Gamma t}}{2\pi S_0} \exp\left(-\frac{1}{2}e^{2\Gamma t}\mathbf{r}^\dagger(t)\cdot\mathbf{S}(t)\mathbf{r}(t)\right)\begin{pmatrix} -2\gamma & -m\omega \\ 1/m\omega & 0 \end{pmatrix}\begin{pmatrix} p(t) \\ q(t) \end{pmatrix}. \quad (6.2.18)
\end{aligned}
$$

The continuity Eq. (5.2.20) for this flow is readily verified.

We cannot expect the information density and the associated information flow to obey the same conservation law. The information density (Fig. 6.2a)

$$
\begin{aligned}
\iota(\mathbf{r}, t) &= -c\ln\left(d_r\rho(\mathbf{r}, t)\right) \\
&= -c\left(\ln\left(\frac{d_r}{2\pi S_0}\right) + 2\Gamma t - \frac{1}{2}e^{2\Gamma t}\mathbf{r}^\dagger(t)\cdot\mathbf{S}(t)\mathbf{r}(t)\right)\rho(\mathbf{r}, t) \quad (6.2.19)
\end{aligned}
$$

comprises two contributions, besides the constant $-c\ln(d_r/2\pi S_0)$, a spatially homogeneous term $\sim \Gamma t$ that arises from the time-dependent renormalization of the density, and a term that depends strongly on the position in phase space.

Transforming momentum and position such that the contours of the Gaussian (6.2.15) form circles in phase space, $(p, q) \rightarrow (\eta, \xi) = (p/\sqrt{m\omega}, q\sqrt{m\omega})$, this term depends only on the radius $r = \sqrt{\eta^2 + \xi^2}$, as $\sim r^2\exp(-r^2/2S_0)$: It is sharply peaked around a ring with radius $r = \sqrt{2S_0}$, where the slope of the Gaussian is steepest and hence the distribution deviates strongest from a homogeneous density.

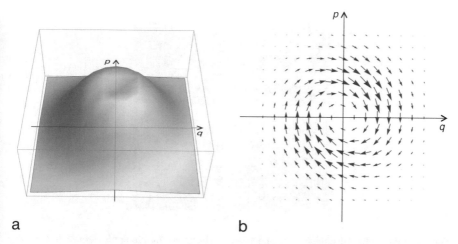

Fig. 6.2 The information density resolves the total entropy in a system into contributions associated to points in state space. For a Gaussian probability density distribution centred in the origin of phase space, Eq. (6.2.16) and Fig. 6.1b, the information density, Eq. (6.2.19), exhibits a shoulder, forming a maximum along a ring around the origin (panel **a**). Weighting the phase-space velocity with the information density defines the information flow. For an underdamped harmonic oscillator, Eq. (6.2.11) with the initial distribution as in part **a**, the information flow (red arrows, panel **b**) forms a vortex around the origin and is strongest along the ring-shaped maximum of the information density.

The information flow, Eq. (5.3.4), is the phase-space velocity weighted with the information density. For the initial Gaussian (6.2.15), it takes the form (Figs. 6.2b and 6.3)

$$
\begin{aligned}
\mathbf{i}(\mathbf{r}, t) = \iota(\mathbf{r}, t)\dot{\mathbf{r}} &= -c\ln(d_r\rho(\mathbf{r}, t))\mathbf{j}(\mathbf{r}, t) \\
&= -c\left(\ln\left(\frac{d_r}{2\pi S_0}\right) + 2\Gamma t - \frac{1}{2}e^{2\Gamma t}\mathbf{r}^{\dagger}(t) \cdot \mathbf{S}(t)\mathbf{r}(t)\right) \\
&\quad \times \rho(\mathbf{r}, t)\begin{pmatrix} -2\gamma & -m\omega^2 \\ 1/m & 0 \end{pmatrix}\begin{pmatrix} p(t) \\ q(t) \end{pmatrix}.
\end{aligned}
\tag{6.2.20}
$$

It allows us to check in how far the information flow deviates from the continuity equation. Using the continuity equation for the probability density, one finds

$$
\text{div}\,\mathbf{i}(\mathbf{r}, t) = c\rho(\mathbf{r}, t)\text{div}\,\dot{\mathbf{r}} - \frac{\partial}{\partial t}\iota(\mathbf{r}, t) = -2c\Gamma\rho(\mathbf{r}, t) - \frac{\partial}{\partial t}\iota(\mathbf{r}, t),
\tag{6.2.21}
$$

The factor $-2c\Gamma$ is identified as the entropy loss rate, cf. Eq. (6.2.8), so that

$$
\text{div}\,\mathbf{i}(\mathbf{r}, t) + \frac{\partial}{\partial t}\iota(\mathbf{r}, t) = -2c\Gamma\rho(\mathbf{r}, t) = \rho(\mathbf{r}, t)\frac{\partial}{\partial t}I(t).
\tag{6.2.22}
$$

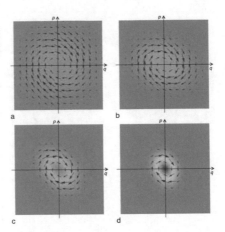

Fig 6.3 With the time evolution of an underdamped harmonic oscillator, the information density starting with the distribution as in Fig. 6.2a contracts towards the origin of phase space, as shown for $t = 0$ (panel **a**), $0.25\,T$ (**b**), $0.5\,T$ (**c**), and $0.75\,T$ (**d**), $T = 2\pi/\omega$. The colour code ranges from blue (negative) through grey (zero) through yellow (positive). At the same time, the initially shallow central minimum becomes deeper and narrower, accounting for the unbounded linear decrease of the total entropy (Eq. 6.2.8). Accordingly, the information flow (red arrows) approaches a narrow vortex at the origin.

In the present case, therefore, the region of peak probability coincides with the information sink. Integrating over phase space, we verify that

$$\int d^2r \left(\operatorname{div} \mathbf{i}(\mathbf{r}, t) + \frac{\partial}{\partial t} \iota(\mathbf{r}, t) \right) = -2c\Gamma = \frac{d}{dt} I(t). \qquad (6.2.23)$$

6.3   The Generic Case: Coexistence of Chaos and Dissipation

The preceding Chap. 5 on information dynamics in Hamiltonian systems focused particularly on chaotic systems, where expansion and contraction of phase space go along with exact conservation of phase-space volume. Hamiltonian mechanics with its symplectic phase-space geometry is particularly suited for a precise description on the microscopic level and is of basic importance for the foundation of statistical mechanics. Historically though, the first models, for which the concept of deterministic chaos has been coined, were dedicated to strongly dissipative macroscopic phenomena, notably weather. They expand their state space in some directions but contract it even faster in others, resulting in a process termed *stretching and folding* (Fig. 6.4), so that fundamental symmetries such as the conservation of energy and information do not nearly apply. Distributions that initially occupy a fraction of positive measure of the total state space collapse onto manifolds of measure zero.

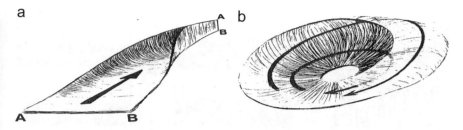

Fig. 6.4 Chaotic dynamics is characterized by two complementary processes occurring simultaneously: While the state space is stretched, injecting information locally, it is also folded as a whole, erasing information globally (**a**). Their repeated action (**b**) leads to a continuous rise of information from small to large scales. Reprinted from [Sha80] with permission.

Hamiltonian mechanics therefore is not an appropriate framework in this realm. To be sure, in some relatively simple cases in classical and quantum physics, notably in optics, molecular, and solid-state physics, it is possible to devise a complete description of dissipative processes in terms of microscopic models, see, e.g., Sect. 8.3.3. However, important instances of chaotic dynamics occur in highly complex systems outside physics, such as in turbulence and climate, in physiology, e.g. heartbeat, in population dynamics, and even in economy, where a microscopic account is out of the question. Instead, a sufficiently broad framework to cover this huge range of phenomena is provided by the theory of dynamical systems [CE80, GH83, Sch84, LL92, Ott93, Str15]. The present section introduces a few of the most important of its concepts and tools, some of them inspired in turn by deterministic chaos, specifically strange attractors and fractal dimensions.

The essentials of dissipative chaos can already be illustrated by returning to the baker map discussed in Sect. 5.3.2, complementing it by a global contraction in the momentum direction [Ott93, Sch84]. If instead of occurring simultaneously with the chaotic stretching and folding, the contraction is treated as a separate step preceding the baker map proper, the map decomposes into two subsequent operations, the first of which introduces a dissipative reduction of momentum by a contraction factor a, $0 \leq a \leq 1$,

bak:$[0, 1[\times [0, 1[\rightarrow [0, 1[\times [0, 1[$,

$$\begin{pmatrix} x \\ p \end{pmatrix} \mapsto \begin{pmatrix} x' \\ p' \end{pmatrix} = \begin{pmatrix} x \\ ap \end{pmatrix}, \begin{pmatrix} x' \\ p' \end{pmatrix} \mapsto \begin{pmatrix} x'' \\ p'' \end{pmatrix} = \begin{pmatrix} (2x') \bmod 1 \\ (p' + \mathrm{int}(2x'))/2 \end{pmatrix}, \quad (6.3.1)$$

Figure 6.5 shows the action of this map over the first five steps, starting with a homogeneous distribution over the unit square, for $a = 0.5$. While at each step m, the number of separate strips doubles, the volume of each strip shrinks by $a/2 = 0.25$, so that the overall volume reduces by 0.5. For $m \rightarrow \infty$, the distribution collapses to a sequence of infinitely many strips of infinitesimal thickness, of overall measure zero. The fractal dimension of the point sequence of their positions in p can be calculated by comparing the scale factor relating two levels of the self-similar pattern to the

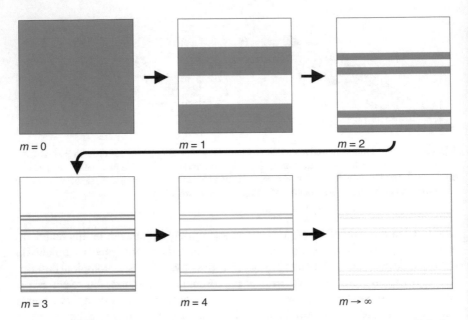

Fig. 6.5 If the volume-conserving baker map of Eq. (5.3.40) is complemented by step of dissipative momentum reduction, the dissipative baker map, Eq. (6.3.1), results. Starting from a homogeneous distribution ($m = 0$), the reduction of phase-space volume by $a = 0.5$ per step contracts the distribution onto a self-similar strange attractor ($m \to \infty$) of fractal dimension, here $D = 1.5$.

ratio of the respective lengths covered by them,

$$D_0 = \frac{\log(\text{volume ratio})}{\log(\text{scale factor})} = \frac{\ln(1/2)}{\ln(a/2)} = \frac{\ln(2)}{\ln(2) - \ln(a)}, \tag{6.3.2}$$

that is, $D_0 = 0.5$ for $a = 0.5$, halfway between points ($D = 0$) and lines ($D = 1$). The dimension of the line pattern in phase space, Fig. 6.5, panel $m \to \infty$, is then $D = D_0 + 1 = 1.5$.

Further insight into this process is gained by reconstructing the pattern of bidirectional information flows, illustrated for the volume conserving baker map in Fig. 5.10, Sect. 5.3.2, for the dissipative map. In the special case $a = 0.5$, the dissipative map fits smoothly in the scheme of rigid symbol shifts applied to the conservative map. Adopting the same binary coding with N significant digits as in Eq. (5.3.42),

$$x = \sum_{n=1}^{N} a_n 2^{-n} = 0.\, a_1\, a_2\, a_3\, a_4\, a_5 \ldots a_N, a_n \in \{0, 1\},$$
$$\tag{6.3.3}$$
$$p = \sum_{n=1}^{N} b_n 2^{-n} = 0.\, b_1\, b_2\, b_3\, b_4\, b_5 \ldots b_N, b_n \in \{0, 1\},$$

the dissipative component of the map takes the form

$$p' = \frac{1}{2}p = \frac{1}{2}\sum_{n=1}^{N} b_n 2^{-n} = \sum_{n=1}^{N-1} b_n 2^{-n-1}$$

$$= 0.0\, b_1\, b_2\, b_3\, b_4\, b_5 \ldots b_{N-1} = \sum_{n=1}^{N} b'_n 2^{-n},\ b'_n = \begin{cases} 0 & n = 1, \\ b_{n-1} & 2 \le N \le 1, \end{cases} \qquad (6.3.4)$$

of a rigid shift of the sequence by one position to the right, with the leftmost digit filled with a 0. Combined with a similar shift generated already by the conservative map, this amounts to a shift of the momentum code but now by two positions per application of the map, with the original digits b_n of the momentum replaced alternatingly by 0 and by the most significant digit a_1 of the position code, as it is erased by the Bernoulli map applied to x. This means that after a transient phase during which the distribution contracts onto its strange attractor, the map is equivalent to a combination of thinning out and shifting of symbols as in Fig. 6.6. We can draw two major conclusions:

Contrary to the conservative case, where the uplift of positional information, 1 bit per time step, by the Bernoulli shift is compensated exactly by the momentum contraction, the dissipative map returns only half of this information current to the small momentum scales of this same degree of freedom, while the other half is diverted to ambient degrees of freedom, as discussed in Sect. 6.2. The structure of the symbol sequence encoding the momentum component after n time steps,

$$p = 0.\, a_{n-1}\, 0\, a_{n-2}\, 0\, a_{n-3}\, 0\, a_{n-4}\, 0 \ldots$$

$$= 2\sum_{m=1}^{N/2} a_{n-m} 4^{-m} = \sum_{m=1}^{N} b_m 2^{-m},\ b_m = \begin{cases} 0 & m \text{ even}, \\ a_{n-(m+1)/2} & m \text{ odd}, \end{cases} \qquad (6.3.5)$$

even allows us to reconstruct the strange attractor: At infinite resolution, it is given by an enumerably infinite set of horizontal lines $p = \text{const}$, at momentum values p of the form

$$p = 2\sum_{n=1}^{\infty} a_n 4^{-n},\ a_n \in \{0, 1\} \qquad (6.3.6)$$

and labelled by the infinite sequence $\{a_n | n \in \mathbb{N}\}$.

This appears as an expansion in ternary numbers, i.e., with exponent 4. However, the fact that the digits may only take the values 0 or 1, not 0, 1, 2, 3, implies that these numbers do not represent the entire real interval $[0,1[$ but only a subset of measure zero, with fractal dimension $D = 0.5$. In fact, Eq. (6.3.6) has been used to generate the panel $m \to \infty$ of Fig. 6.5.

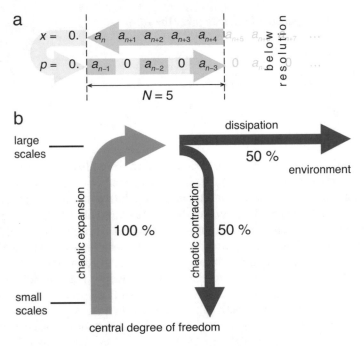

Fig. 6.6 Information currents in the dissipative baker map with $a = 0.5$. (**a**) In the presence of dissipative phase-space contraction, the symbol shift in the p-component is modified by inserting a 0 at every second digit, Eq. (6.3.5), reducing the information flow into small scales from 1 bit to 0.5 bit per time step. (**b**) Broken down to overall information currents, this amounts to an interplay of vertical currents between different scales within the central degree of freedom, and horizontal currents, transferring information to the environment. In the case of the baker map with $a = 0.5$, the ratio of downward and sideward currents is 50%: 50%.

6.4 Fractals, Dimension, and Information

Strange attractors in dissipative chaotic systems offer a suitable context to address the close kinship between the concepts of information and dimension. That they should be related is already clear from the fact that for a set of D uncorrelated systems of the same size and structure, the total information content is D times the individual one—like this, the concept has been conceived by Boltzmann, see Sect. 1.3. The number of system states increases exponentially with the number of subsystems, as it does with the spatial dimension. Both quantities depend logarithmically on measures of the sheer system size (e.g., volume in position space).

This idea is already manifest in Eq. (6.3.2). Generally, we may define dimension as [HP83, Ott93]

$$D = \frac{\log(\text{measure of system size})}{\log(\text{scale factor})}. \tag{6.4.1}$$

To be more specific, the imprecise notion "measure of system size" can be based operationally on a box-counting procedure: Subdivide the space in question into bins ("boxes") of size ε and count the number of boxes that overlap with the geometric object to be analyzed, resulting in a number $N(\varepsilon)$. Find how this number grows as the resolution is increased by scaling ε towards 0. The *box-counting* or *Hausdorff dimension* [HP83, Ott93, Str15] is then defined as

$$D_0 = \lim_{\varepsilon \to \infty} \frac{\ln(N(\varepsilon))}{\ln(1/\varepsilon)}. \tag{6.4.2}$$

The subscript 0 is attached to distinguish this definition from other versions, see below.

Like this, it applies directly to calculate the fractal dimensions of self-similar objects. A standard example is the *middle-third Cantor set* (Fig. 6.7). Each step in this recursive construction consists in contracting the distribution on the unit interval reached at step n by a factor 1/3 towards 0 and adding a replica of this same contracted distribution on the other side of the interval. Starting from a homogeneous distribution on the unit interval, repeating the process *ad infinitum*, leads to an object that comprises an infinite number of line segments of length zero and covers a fraction of measure zero of the unit interval.

As opposed to objects of integer dimension like points, lines, planes, which are invariant under continuous scaling, self-similarity in fractals is a discrete symmetry.

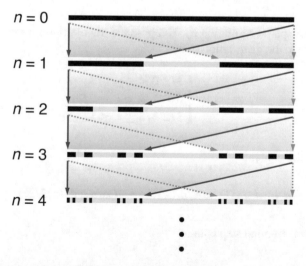

Fig. 6.7 Construction of the middle-third Cantor set. Starting from a homogeneous distribution, say on the unit interval, cut out the part between 1/3 and 2/3. Repeat this procedure for the two remaining intervals of length 1/3 each. Apply the same operation to the resulting distribution, and so on, *ad infinitum*. The final structure for $n \to \infty$ comprises an infinite number of line sections of zero length and covers a fraction of measure zero of the unit interval. It is a self-similar object of fractal dimension $\ln(2)/\ln(3)$, situated between points (dimension 0) and lines (dimension 1).

Specifically, by construction, the middle-third Cantor set is invariant under reductions by 1/3 and integer powers of it. Each such step retains exactly one half of the point set and erases the other, cf. Fig. 6.7, read backwards. The numerator in Eq. (6.4.1) is therefore $\ln(1/2)$, the denominator is $\ln(1/3)$, so that

$$D_0 = \frac{\ln(1/2)}{\ln(1/3)} = \frac{\ln(2)}{\ln(3)}, \qquad (6.4.3)$$

independently of ε. The corresponding values for the fractal formed by the vertical cross section of the strange attractor of the dissipative baker map, see Fig. 6.5, are $\ln(1/2)$ and $\ln(1/4)$, resp., resulting in the dimension $D_0 = 0.5$ for this set.

In order to calculate the information contained in a distribution with a fractal support, we have to associate probability densities to each component of the set. Following the same construction steps for the middle third Cantor set as in Fig. 6.7, we start with a normalized homogeneous probability distribution over the unit interval,

$$\rho_0(x) = 1, x \in [0, 1], \qquad (6.4.4)$$

corresponding to an information content

$$I_0 = -c \int_0^1 dx \rho_0(x) \ln(d_x \rho_0(x)) = -c \ln(d_x), \qquad (6.4.5)$$

introducing the spatial resolution d_x. Keeping the distribution normalized, the first step then leads to the distribution

$$\rho_1(x) = \begin{cases} 3/2 & 0 \leq x \leq 1/3, \\ 0 & 1/3 < x < 2/3, \\ 3/2 & 2/3 \leq x \leq 1. \end{cases} \qquad (6.4.6)$$

with information content

$$I_1 = -c \int_0^1 dx \rho_1(x) \ln(d_x \rho_1(x)) = -c(\ln(d_x) + \ln(3/2)). \qquad (6.4.7)$$

The general construction step is then

$$\rho_{n+1}(x) = \frac{3}{2}(\rho_n(3x) + \rho_n(1 - 3x)), \qquad (6.4.8)$$

resulting in a reduction of the information content,

$$l_{n+1} = -c \int_0^1 dx \frac{3}{2}(\rho_n(3x) + \rho_n(1-3x)) \ln\left(\frac{3}{2}d_x(\rho_n(3x) + \rho_n(1-3x))\right)$$

$$= -c \int_0^1 dx \rho_n(x)(\ln(d_x\rho_n x) + \ln(3/2)) = I_n - c\ln(3/2). \qquad (6.4.9)$$

That is, at each step the entropy decreases by an amount $-c\ln(3/2)$ and the actual information increases by $c\ln(3/2)$. For the final fractal set, obtained in the limit $n \to \infty$, it diverges without bound!

Of course, for any finite resolution d_x, the process ends as soon as the size of the segments is no longer resolved, i.e., when $d_x \geq 1/3^n$, or at

$$n \geq n_{max} = \text{int}(\ln(d_x)/\ln(3)). \qquad (6.4.10)$$

It inserts a bottom to self-similarity (much as the finite size of the object trivially imposes an upper bound) and in this way, keeps the information content finite (Fig. 6.8).

In fact, the very definition of dimension can be modified so as to relate it more directly to Shannon's information, by allowing for different probabilities associated to the bins underlying the box-counting dimension, Eq. (6.4.2). This leads to the Balatoni–Renyi dimension [BR56, HP83, Ott93],

$$D_q = \frac{1}{1-q} \lim_{\varepsilon \to \infty} \frac{1}{\ln(1/\varepsilon)} \ln\left(\sum_{n=1}^{N(\varepsilon)} \mu_n^q\right), \qquad (6.4.11)$$

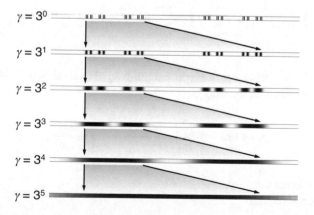

Fig. 6.8   A finite spatial resolution imposes a lower bound to the self-similarity of fractals. If the middle-third Cantor set, see Fig. 6.7 above, is smoothed on a scale d_x, subsequent magnifications by its self-similarity factor $\gamma = 3$ will eventually reveal the absence of more detail on smaller scales and terminate in a featureless distribution, corresponding to zero actual information.

where μ_n is a weight (measure) associated to box n, and q, with $q \geq 0$, parameterizes a continuous family of dimension definitions.

If the measures are identical, $\mu_1 = \ldots = \mu_{N(\varepsilon)} = 1/N(\varepsilon)$, the sum in Eq. (6.4.11) simplifies to

$$\sum_{n=1}^{N(\varepsilon)} \mu_n^q = N(\varepsilon) \frac{1}{(N(\varepsilon))^q} = (N(\varepsilon))^{1-q}, \tag{6.4.12}$$

and the Balatoni–Renyi dimension reduces to the box-counting dimension,

$$D_q = \frac{1}{1-q} \lim_{\varepsilon \to 0} \frac{1}{\ln(1/\varepsilon)} \ln\big((N(\varepsilon))^{1-q}\big) = \lim_{\varepsilon \to 0} \frac{\ln(N(\varepsilon))}{\ln(1/\varepsilon)} = D_0. \tag{6.4.13}$$

Likewise, in the limit $q \to 0$, the box-counting dimension is recovered,

$$\lim_{q \to 0} D_q = \lim_{\varepsilon \to 0} \frac{1}{\ln(1/\varepsilon)} \ln\left(\sum_{n=1}^{N(\varepsilon)} \mu_n^0\right) = \lim_{\varepsilon \to 0} \frac{\ln(N(\varepsilon))}{\ln(1/\varepsilon)} = D_0. \tag{6.4.14}$$

However, in the limit $q \to 1$,

$$D_1 = \lim_{q \to 1} D_q = \lim_{\varepsilon \to 0} \frac{1}{\ln(1/\varepsilon)} \left(\lim_{q \to 1} \ln\left(\sum_{n=1}^{N(\varepsilon)} \mu_n^q\right)\right)$$

$$= -\lim_{\varepsilon \to 0} \frac{1}{\ln(1/\varepsilon)} \left(\sum_{n=1}^{N(\varepsilon)} \mu_n\right)^{-1} \sum_{n=1}^{N(\varepsilon)} \mu_n \ln(\mu_n) = \lim_{\varepsilon \to 0} \frac{\sum_{n=1}^{N(\varepsilon)} \mu_n \ln(\mu_n)}{\ln(\varepsilon)}. \tag{6.4.15}$$

If we identify the weights μ_n with probabilities p_n, the relationship with Shannon's entropy $I(\varepsilon)$ becomes compelling,

$$D_1 = \lim_{\varepsilon \to 0} \frac{\sum_{n=1}^{N(\varepsilon)} p_n \ln(p_n)}{\ln(1/\varepsilon)} = \lim_{\varepsilon \to 0} \frac{I(\varepsilon)/c}{\ln(\varepsilon)}, \tag{6.4.16}$$

and gives rise to the term *information dimension* for D_1.

We can conclude that

The actual information contained in a fractal set diverges, if it is not regularized by a finite resolution.

In classical statistical mechanics, there is no fundamental reason to introduce an absolute limit of resolution, notably of structures in phase space. This statement therefore reveals a serious deficiency of classical mechanics, in particular in the context of chaotic systems with strange attractors. Lacking a fundamental limit of the resolution of phase-space structures, the sheer existence of strange attractors in classical mechanics implies a divergence problem, similar in nature to Gibbs' paradox [Gib02, Gib93, Rei65, Jay92] and even to the ultraviolet catastrophe in the classical (Rayleigh–Jeans) description of black-body radiation, which gave rise to Planck's postulation of a quantum of action [Pla49].

Fractal patterns in phase space providing an unlimited supply of information in their small scales are by no means restricted to dissipative chaos. They arise also in Hamiltonian systems, e.g., as self-similar structures in systems with a mixed phase space, which combine integrable with chaotic dynamics [Ber78, Ozo88, LL92, Ott93, JS98] (Fig. 6.9). They appear as a consequence of the KAM (Kolmogorov–Arnol'd–Moser) theorem [LL92, Ott93, JS98, GPS01], concerning weak perturbations that render an otherwise integrable system chaotic, and even constitute the generic case.

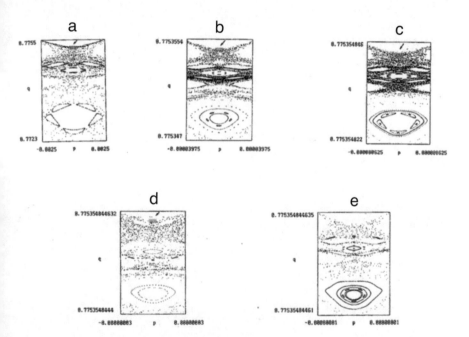

Fig. 6.9 Fractal structures in phase space do not only arise as strange attractors in dissipative chaos (see Fig. 6.5). Also in Hamiltonian systems with a mixed phase space where regular and chaotic dynamics coexist, generically exhibit nested self-similar structures. Panels (**a–e**) show subsequent amplifications, each by a factor of ca. 400, of the regions marked in pink (arrows) in the respective previous panel, for the conservative Hénon map. The similarity of the patterns is evident. The process could continue *ad infinitum*, revealing an unlimited supply of information upon closing in to ever finer resolution. After [HA90].

The diverging information density in fractal phase-space patterns sheds doubt on the consistency and completeness of microscopic Hamiltonian dynamics, in a way that again resembles Gibbs' paradox in statistical mechanics: In both cases, the fact is brought to bear that in Classical mechanics, arbitrarily small differences between their states suffice to render systems distinguishable. It is only in the framework of quantum mechanics that this problem finds a satisfactory solution, see Chap. 8.

Chapter 7
Fluctuations, Noise, and Microscopic Degrees of Freedom

The foregoing sections presented two almost antithetic views of the way physical systems organize the flow of information: In the myopic picture drafted by Hamiltonian mechanics, entropy is a strictly conserved quantity, processes expanding or contracting phase space must exactly compensate each other. The impressionist view inspired by macroscopic phenomena includes top-down flows as in dissipation, as well as information flows bottom-up as in particular in chaotic systems and other that create information on the macroscopic level. The present section serves to reconcile these contrasting views with one another as well as with the Second Law of Thermodynamics, culminating in an argument why the conservation of information in the microscopic equations of motion can well be compatible with the global tendency of increasing entropy predicted by thermodynamics.

It focuses on a phenomenon that links microscopic and macroscopic scales and has not yet been envisaged in the previous sections: thermal fluctuations. As they smooth and eventually erase structures on the macroscopic level, they constitute an upward current of entropy from the microscopic bottom, complementing the chaotic information flow: Diffusion, a paradigm of irreversibility, will be discussed in Sect. 7.1 as a manifestation of noise. Statistical mechanics allows us to relate thermal fluctuations to another irreversible process, dissipation. General laws such as the Einstein relation and Nyquist's theorem couple the opposing information flows associated to fluctuations and dissipation, as will be shown in Sect. 7.2.

Chaos, dissipation, diffusion, they all pertain to the realm of nonequilibrium thermodynamics. Section 7.3 will outline a synthesis of these phenomena, as an integrated account of the dynamics of entropy in systems far from thermal equilibrium. In particular, the Second Law will be interpreted as the manifestation of a global imbalance between ascending and descending information currents, exchanging entropy between small and large scales.

© Springer Nature Switzerland AG 2022
T. Dittrich, *Information Dynamics*, The Frontiers Collection,
https://doi.org/10.1007/978-3-030-96745-1_7

7.1   Noise, Diffusion, and Information Loss

Noisy channels are a nuisance of everyday life. Radio signals arrive distorted, copies of documents lose detail, oral tradition perverts facts. In all these cases, the original information content is not only erased but gradually replaced by other information not related to it and therefore considered as "random". In terms of the criteria for randomness introduced in Chap. 4, no more than the decay or complete absence of correlations is implied in the present context. A paradigm of such processes is diffusion. It manifests itself in spatial structures losing contours on increasingly larger scales and approaching asymptotically a featureless homogeneous distribution (Fig. 7.1).

An elementary model for diffusion processes is the random walk [Fel68]: a point moving in (one-, two-, ..., n-dimensional) space by random increments. An equation of motion describing this process is

$$\frac{\mathrm{d}}{\mathrm{d}t}x(t) = \xi(t), \qquad (7.1.1)$$

where $\xi(t)$ is a random function of time, with characteristics to be specified further below. Even like this, in continuous space and time, random walks require sophisticated mathematics, such as Wiener processes [Fel68], to be analyzed. More accessible in elementary terms are discrete random walks, sequences of unit steps on a discrete grid (Fig. 7.2), say with equal probability $p = 0.5$ for steps to the left and to the right (drifts induced by an imbalance $p_{left} \neq p_{right}$ are discarded). A suitable mechanical implementation would be a Galton board [HM&88]. The series of binary

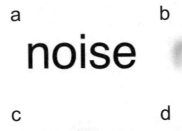

a   b

c   d

Fig. 7.1 Diffusion, a paradigm for irreversibility, blurs and eventually wipes out distinct patterns in a given distribution, replacing the initial information gradually by random noise. It occurs in thermodynamics as well as in technical processes like repeated photocopying or communication via noisy channels. Panels **a–d** show a density distribution, in the initial state (**a**) and after successive applications of a filter simulating diffusion (**b–d**).

Fig. 7.2 A discrete version of a random walk, consisting of unit steps per unit time, with equal probability 0.5 to the left and right, is exemplified by the Galton board. The resulting probabilities to reach the site l of the board after n steps, $l = -n, -n+2, \ldots, n-2, n$, follow the same recursion relations as the binomial coefficients and are given by the binomial distribution (grey blocks). In the limit $n \to \infty$, the binomial distribution approaches a Gaussian, as in a continuous diffusion process.

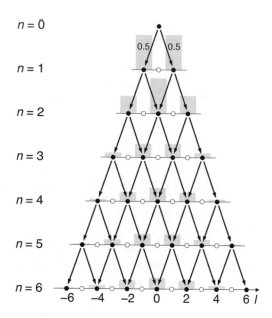

left/right decisions is assumed to be random, i.e. here, lacking two-point correlations even between two subsequent steps. It is this condition that renders the process irreversible, since at each step, the memory of the previous steps is immediately lost.

The random walk is characterized by probabilities $p(l, n)$ to reach the site $l, l = -n, -n + 2, \ldots, n - 2, n$, after n steps. They follow the recursion relation

$$p(l, n) = 0.5(p(l - 1, n - 1) + p(l + 1, n - 1)) \tag{7.1.2}$$

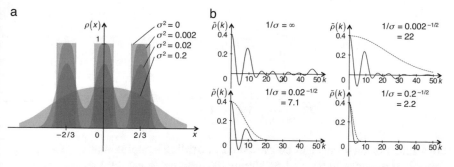

Fig. 7.3 Diffusion in position and in frequency space. A probability density $\rho(x)$, prepared as a sharply defined distribution (three boxes of width 1/3 centred at $x = -2/3$, 0, and 2/3, resp.) is increasingly blurred by Gaussian smoothing (panel **a**). With growing variance σ^2 of the Gaussian, cf. Eq. (7.1.18), characteristic features of the distribution—the sharp edges, the subdivision into three parts, and the overall rectangular outline—disappear one after the other. In the frequency domain (panel **b**), a Gaussian low-pass filter (dashed) suppresses corresponding structures in the characteristic function $\tilde{\rho}(k)$ (full line) outside a spectral window of decreasing width $1/\sigma$.

for internal sites, $|l| < n$, and $p(l, n) = 0.5p(l \mp 1, n - 1)$ for $l = \pm n$, hence are given by a binomial distribution. For an even number of steps, $n = 2k$, $k \in \mathbb{N}$, it reads

$$p(2j, 2k) = 2^{-n} \binom{n}{k + j} = 2^{-2k} \frac{(2k)!}{(k + j)!(k - j)!}, \quad j = -k, \ldots, k, \quad (7.1.3)$$

and for an odd number $n = 2k - 1$,

$$p(2j - 1, 2k - 1) = 2^{-n} \binom{n}{k + j - 1} = \frac{2^{-2k+1}(2k - 1)!}{(k + j - 1)!(k - j)!},$$
$$j = -k + 1, \ldots, k. \quad (7.1.4)$$

By construction, this distribution is symmetric, $p(l, n) = p(-l, n)$, and has zero mean, $\langle l \rangle = 0$. For large n, the binomial distribution approaches a normal one,

$$p(l, n) = \frac{1}{\sqrt{2\pi n}} \exp\left(\frac{-l^2}{2n}\right). \quad (7.1.5)$$

It is readily verified that the variance of l increases as the number of steps,

$$\langle l^2 \rangle = \sum_{j=-n/2}^{n/2} (2j)^2 p(j, n) = n. \quad (7.1.6)$$

For a discrete distribution such as $p(l, n)$, calculating Shannon's information is straightforward,

$$I(n) = -c \sum_{j=-n/2}^{n/2} p(2j, n) \ln(p(2j, n))$$
$$= -\frac{c}{2^n} \sum_{j=-n/2}^{n/2} \binom{n}{k + j} \ln\left(2^{-n} \binom{n}{k + j}\right). \quad (7.1.7)$$

Logarithms of factorials can be evaluated in Stirling's approximation if n is large enough, as assumed in Eq. (7.1.5), giving the asymptotic expression

$$I(n) = \frac{c}{2}\left(1 + \ln\left(n\frac{\pi}{2}\right)\right). \quad (7.1.8)$$

Unlike the full binary tree where the information content increases linearly with the level n, here it grows only as $\log(n)$.

If n and l are interpreted as discrete time, $t_n = n\Delta t$, and discrete space, $x_l = l\Delta x$, Eq. (7.1.6) already resembles Fick's law for diffusion, $\langle x^2 \rangle = 2Dt$. The agreement becomes complete in the continuous limit of the discrete random walk, letting spatial steps Δx and time steps Δt approach zero while

$$\frac{(\Delta x)^2}{\Delta t} = \text{const} = 2D. \tag{7.1.9}$$

Combined with Eq. (7.1.6), this confirms the expected Fick's law,

$$\frac{\langle x^2 \rangle}{t} = \frac{\langle I^2 \rangle (\Delta x)^2}{n \Delta t} = 2D. \tag{7.1.10}$$

This analogy now allows us to be more specific about the stochastic process $\xi(t)$. As in the discrete case, it should have zero mean, while correlations decay within the infinitesimally short time Δt and on a spatial scale Δx related to Δt by Eq. (7.1.9),

$$\langle \xi(t) \rangle = 0, \ \langle \xi(t')\xi(t) \rangle = 2D\delta(t' - t). \tag{7.1.11}$$

Correlations decaying as a delta function are to be understood as an approximation to smooth correlation functions, valid for times much larger than their correlation time. They qualify the stochastic process as *white noise*, alluding to its flat spectral density, see Eq. (4.1.4),

$$S(\omega) := 2 \int_{-\infty}^{\infty} dt \, \langle \xi(t)\xi(0) \rangle e^{-i\omega t} = 4D \int_{-\infty}^{\infty} dt \delta(t) e^{-i\omega t} = 4D. \tag{7.1.12}$$

The stochastic differential equation (7.1.1) for the position x is equivalent to a differential equation for the time-dependent probability density $\rho(x, t)$, assumed to be normalized. Since the stochastic term $\xi(t)$ tends to flatten the distribution without any preferred direction (Fig. 7.3a), it is clear that it leaves distributions invariant that depend only linearly on x, but affects concave or convex functions. This suggests to surmise

$$\frac{\partial}{\partial t}\rho(x, t) \sim \frac{\partial^2}{\partial x^2}\rho(x, t), \tag{7.1.13}$$

at least to leading order in $\partial/\partial x$. The missing proportionality constant can be found by substituting the asymptotic form (7.1.5) of the binomial distribution as a solution of Eq. (7.1.13). The corresponding probability density, using Eq. (7.1.10)

$$\rho(x, t) = \frac{1}{\sqrt{4\pi Dt}} \exp\left(\frac{-x^2}{4Dt}\right), \tag{7.1.14}$$

solves Eq. (7.1.13) in the form

$$\frac{\partial}{\partial t}\rho(x, t) = D\frac{\partial^2}{\partial x^2}\rho(x, t). \tag{7.1.15}$$

It represents the simplest case of a *Fokker–Planck equation* for a diffusion process. A more versatile version will be discussed in Sect. 7.2. Equation (7.1.15) is equivalent to the continuity Eq. (5.2.21) if the probability current $j(x, t) = \langle \dot{x} \rangle \rho(x, t)$ is assumed to be given by Fick's law in the form [Gas98],

$$j(x, t) = -D \frac{\partial}{\partial x} \rho(x, t). \tag{7.1.16}$$

The Fokker–Planck equation is linear in the density function $\rho(x, t)$ and therefore can be solved with the help of a propagator or Green function, cf. Eq. (5.2.13). It is defined implicitly as an integration kernel,

$$\rho(x'', t'') = \int_{-\infty}^{\infty} dx' G(x'', t''; x', t') \, \rho(x', t'). \tag{7.1.17}$$

The obvious initial condition

$$G(x'', t'; x', t') = \delta(x'' - x') \tag{7.1.18}$$

is required to make sure that at $t'' - t'$, Eq. (7.1.17) maps $\rho(x', t')$ onto itself. For Eq. (7.1.15), the propagator reads

$$G(x'', t; x', 0) = \frac{1}{\sqrt{4\pi Dt}} \exp\left(\frac{-(x'' - x')^2}{4Dt} \right), \quad \text{for } t > 0. \tag{7.1.19}$$

It is not surprising that convolving a given distribution with a Gaussian, as implied by Eq. (7.1.15) with (7.1.19), will increase its entropy. The amount of detail lost by Gaussian smoothing, however, depends on the initial distribution the smoothing is applied to. If initially it was already a Gaussian,

$$\rho(x, 0) = \frac{1}{\sqrt{2\pi}\sigma_0} \exp\left(\frac{-x^2}{2\sigma_0^2} \right), \tag{7.1.20}$$

diffusion increases its variance by $2Dt$, cf. Eq. (7.1.14),

$$\rho(x, t) = \frac{1}{\sqrt{2\pi}\sigma(t)} \exp\left(\frac{-(x'' - x')^2}{2(\sigma(t))^2} \right), \quad (\sigma(t))^2 = \sigma_0^2 + 2Dt. \tag{7.1.21}$$

The entropy thereby rises from an initial $I(0) = c\left(\frac{1}{2} \ln(\sigma_0/d_x) + 1 \right)$ to

$$I(t) = c\left(\ln\left(\sqrt{\sigma_0^2 + 2Dt/d_x}\right) + 1\right) \xrightarrow[t \gg \sigma_0^2/2D]{} \frac{c}{2} \ln\left(\frac{2Dt}{d_x^2}\right), \tag{7.1.22}$$

in accordance with Eq. (7.1.8) for the discrete random walk. The asymptotic entropy growth rate, for times far beyond σ_0^2/D,

$$\frac{d}{dt}I(t) = \frac{c}{2t}, \tag{7.1.23}$$

is positive but decreases arithmetically. Evidently, this global increase is not compatible with a continuity equation, such as Eq. (5.2.21) for the density, see Eqs. (5.3.5) and (5.3.6), for the entropy density. Yet it can be reconciled with it, if a source term $s_t(x, t)$ is included on the right-hand side, as in Eq. (5.6.16) [Gas98]

$$\frac{\partial}{\partial t}\iota(x, t) + \frac{\partial}{\partial x}i(x, t) = s_t(x, t) = D\frac{(\partial\rho(x, t)/\partial x)^2}{\rho(x, t)} \geq 0. \tag{7.1.24}$$

In order to find out where in the spectrum this loss of detail occurs, look at the characteristic function of the distribution, i.e., its spatial Fourier transform (cf. Eq. (2.3.15)),

$$\tilde{\rho}(k, t) = \frac{1}{2\pi}\int_{-\infty}^{\infty} dx\rho(x, t)\exp(-ikx). \tag{7.1.25}$$

For the characteristic function, the convolution with a Gaussian becomes a mere multiplication with the Fourier transform of that Gaussian. If Eq. (7.1.17) propagates $\rho(x, t)$ from 0 to t, then $\tilde{\rho}(k, t)$ is propagated by

$$\tilde{\rho}(k'', t) = \tilde{\rho}(k'', 0)\tilde{G}(k'', t, 0), \tag{7.1.26}$$

if the propagator depends on x' and x'' only through their difference, $G(x'', t; x', 0) = G(x'' - x', t, 0)$, as indeed is true for Eqs. (7.1.18 and 7.1.19). In that case,

$$\tilde{G}(k, t, 0) = \int_{-\infty}^{\infty} dx \, G(x, t, 0)\exp(-ikx), \tag{7.1.27}$$

which means for the diffusion propagator, Eq. (7.1.19), that

$$\tilde{G}(k, t, 0) = \exp\left(-\frac{Dt}{2}k^2\right). \tag{7.1.28}$$

Multiplication with this Gaussian envelope restricts the spatial frequencies contained in the characteristic function to a spectral window of width $1/\sqrt{Dt}$ around zero (Fig. 7.3b), that is, it acts like a low-pass filter. The entropy in the characteristic function thus *decreases* as

$$\tilde{I}(t) = \tilde{I}(0) - \frac{c}{2}\ln(Dt), \tag{7.1.29}$$

corresponding to a rate (compare to Eq. (7.1.23))

$$\frac{\mathrm{d}}{\mathrm{d}t}\tilde{I}(t) = -\frac{c}{2t}. \tag{7.1.30}$$

The simultaneous processes of smoothing in position space and contraction in frequency space are in accord with the uncertainty relation for Fourier transformation, Eq. (2.3.42).

7.2 Fluctuation–Dissipation Theorems: Einstein's Relation and Nyquist's Theorem

The previous subsection discussed diffusion and noise as general concepts that apply to a vast range of subjects, from telecommunication through image processing through epidemiology and economy. A much more detailed picture arises if we focus on a specific physical setting, a massive particle immersed in a dense medium at some finite temperature. It will feel the ambient medium in two distinct ways: On small scales in space and time, the random impact of surrounding particles becomes manifest as a noisy quivering of its trajectory, known as *Brownian motion* [Rei65, Ris89]. Averaging over space and time, the presence of the medium leads to damping. The two-way exchange of energy, kinetic energy dissipating into heat and Brownian motion preventing the particle from coming to rest, applies likewise to information: The loss of information into microscopic freedoms by dissipation, addressed in Sect. 6.2, and the entropy generated by diffusion as analyzed in the foregoing subsection, are two sides of the same medal.

Since both of them are consequences of the coupling of a central particle to a large number of ambient degrees of freedom, it is not surprising that the two phenomena should be related. In fact, the relationship is even immediate: The random character of the permanent impacts directly implies damping. Indeed, random collisions erase the memory of the preceding motion, thus break time-reversal invariance. Their accumulated action tends to bring the Brownian particle to rest in a specific reference frame, defined by the spatial position of the centre of mass of the medium, in agreement with friction.

This relationship can be made explicit by solving the stochastic equation of motion for Brownian particles. It is analogous to Eq. (7.1.1), but refers to velocities instead

ol positions. Including both fluctuations and friction, a similarly elementary version
reads [Rei65]

$$\frac{d}{dt}v(t) = -\gamma v(t) + \eta(t),$$ (7.2.1)

assuming Ohmic friction with damping constant γ. The last term is now a fluctuating *acceleration*, (i.e., $m\gamma$ is a fluctuating force), again with zero mean and
delta-correlated as in Eq. (7.1.11),

$$\langle \eta(t) \rangle = 0, \qquad \langle \eta(t)\eta(t') \rangle = q\delta(t' - t),$$ (7.2.2)

with noise strength q. Equation (7.2.1) is a standard example of a *Langevin equation*
[Rei65, Ris89].

By formally integrating from t' to t'' [Ris89],

$$v(t'') = v(t')\exp(-\gamma(t'' - t')) + \int_{t'}^{t''} dt\, \eta(t)\exp(-\gamma(t - t')),$$ (7.2.3)

the velocity-velocity correlation function is obtained as a double integral,

$$\langle v(t_1'')v(t_2'') \rangle = (v(t_1'))^2 \exp(-\gamma(t_1'' + t_2''))$$

$$+ \int_{t'}^{t_1''} dt_1 \int_{t'}^{t_2''} dt_2 \eta(t)\exp(-\gamma(t_1'' - t_1 + t_2'' - t_2))q\delta(t_2 - t_1)$$ (7.2.4)

It can be evaluated as one integration across and another along the diagonal $t_2 =
t_1$, up to $\min(t_1'', t_2'')$. As we are interested only in the asymptotic behaviour at large
times $t_1'', t_2'' \gg \gamma^{-1}$ when the initial momentum has been damped out, exponentials
depending on the sum $t_1'' + t_2''$ can be neglected against those depending on the
difference, resulting in [Rei65]

$$\langle v(t_1'')v(t_2'') \rangle = \frac{q}{2\gamma} \exp(-\gamma|t_2'' - t_1''|).$$ (7.2.5)

From the velocity-velocity correlation, another two integrations over time lead to the
position-position correlation function, to relate it with diffusion,

$$\langle (x(t'') - x(t'))^2 \rangle = \left\langle \left(\int_{t'}^{t''} dt\, v(t) \right)^2 \right\rangle = \int_{t'}^{t''} dt_1 \int_{t'}^{t''} dt_2 \langle v(t_1)v(t_2) \rangle.$$ (7.2.6)

In the same long-time limit as underlies Eq. (7.2.5), we can substitute the right-hand-side of Eq. (7.2.5) for the velocity correlation in the last member of Eq. (7.2.6), to obtain

$$\left\langle \left(x(t'') - x(t') \right)^2 \right\rangle = \frac{q}{\gamma^2} (t'' - t').$$

(7.2.7)

Equation (7.2.7) has the form of Fick's law, as in Eq. (7.1.11), with diffusion constant

$$D = \frac{q}{2\gamma^2}.$$

(7.2.8)

It constitutes a remarkably simple relationship between diffusion constant and friction coefficient. Equation (7.2.8) manifests the complementarity of the two phenomena involved: Dissipation represents the loss of entropy into ambient degrees of freedom, while fluctuations manifest the entropy absorbed from the environment.

To be more specific about the microscopic parameter q, the noise strength, we can understand the noise as thermal fluctuations. The velocity correlation function, at equal times, is then related to the mean kinetic energy, cf. Eq. (7.2.5), as

$$\langle E \rangle = \frac{m}{2} \left\langle (v(t))^2 \right\rangle = \frac{mq}{4\gamma}.$$

(7.2.9)

At the same time, for a thermal system in equilibrium at temperature T, the mean energy is $\langle E \rangle = k_B T / 2$, so that

$$q = 2\gamma k_B T / m.$$

(7.2.10)

This allows us to eliminate the noise strength q in the relations obtained above. Specifically, it expresses the two-point correlations of the fluctuating force, Eq. (7.2.2), by macroscopic quantities

$$\int_{-\infty}^{\infty} dt \, \langle \eta(t) \eta(0) \rangle = \frac{2\gamma k_B T}{m}.$$

(7.2.11)

Combined with Eq. (7.2.8), it implies

$$D = k_B T / m\gamma.$$

(7.2.12)

Equation (7.2.12) is the celebrated *Fluctuation–Dissipation Theorem* (FDT), due to Einstein [Ein05]. It marks a milestone in statistical mechanics, as it relates the diffusion constant, a measure of microscopic noise, to friction, an irreversible macroscopic phenomenon, and to a thermodynamic quantity, temperature.

Further evidence for the universal character of the FDT comes from the fact that it can be transferred from the mechanical context to electricity [Rei65]. In a circuit comprising an electromotive force V, an impedance L, and an Ohmic resistance R at temperature T, the current $I(t)$ will generate thermal noise in the resistance. The direct analogue of Eq. (7.2.1) applying to this case is a Langevin equation that includes a fluctuating electromotive force $v(t)$ [Rei65],

$$L\frac{d}{dt}I(t) = V - RI(t) + v(t). \qquad (7.2.13)$$

Identifying the damping coefficient γ in Eq. (7.2.1) with R/L in (7.2.13) and the fluctuating force $\eta(t)$ with $v(t)/L$, Eq. (7.2.11) translates to

$$\int_{-\infty}^{\infty} dt \langle v(t)v(0) \rangle = 2Rk_{\mathrm{B}}T. \qquad (7.2.14)$$

If instead of the two-point correlation we consider its spectral density,

$$J(\omega) = \int_{-\infty}^{\infty} dt \langle v(t)v(0) \rangle \exp(-i\omega t). \qquad (7.2.15)$$

Equation (7.2.15), combined with Eq. (7.2.14), proves equivalent to

$$J(0) = \frac{Rk_{\mathrm{B}}T}{2\pi}. \qquad (7.2.16)$$

However, for a δ-correlated noise as in Eq. (7.2.2), $J(0)$ in Eq. (7.2.16) can be extended to $J(\omega)$ for arbitrary frequencies ω. Assuming instead a positive but small correlation time Δt as a width of the correlation function, this approximation remains valid for frequencies below some cut-off $\omega_{\max} = 2\pi/\Delta t$,

$$J(\omega) = \frac{Rk_{\mathrm{B}}T}{2\pi}, \quad \omega < \omega_{\max}. \qquad (7.2.17)$$

In this form, the FDT is known as *Nyquist's theorem* [Rei65].

To identify information currents behind the interplay of dissipation and diffusion, it is not enough to know moments or correlation functions. We have to look at full probability density distributions and their time evolution. A suitable starting point is the Fokker–Planck equation, now for the velocity distribution instead of the spatial density as in Eq. (7.1.13). To include damping, the second-order derivative generating diffusion in Eq. (7.1.13) has to be complemented by a first-order term that leads to a directed drift of the distribution. For Ohmic damping, proportional to the velocity, the complete Fokker–Planck equation reads

$$\frac{\partial}{\partial t}\rho(v,t) = \gamma\frac{\partial}{\partial v}(v\rho(v,t)) + D\frac{\partial^2}{\partial v^2}\rho(v,t). \qquad (7.2.18)$$

More explicitly, writing out the first term on the right-hand side of Eq. (7.2.18),

$$\frac{\partial}{\partial t}\rho(v,t) = \gamma\rho(v,t) + \gamma v\frac{\partial}{\partial v}\rho(v,t) + D\frac{\partial^2}{\partial v^2}\rho(v,t), \qquad (7.2.19)$$

the second term can be interpreted as the drift proper towards smaller velocities, while the first term compensates for the contraction of the distribution by increasing its amplitude.

Both effects can be anticipated by transforming the velocity dependence of the distribution as

$$\rho(v,t) \to e^{-\gamma t}\rho_0(ue^{-\gamma t}, t). \qquad (7.2.20)$$

Indeed, for $\rho_0(u, t)$ only the diffusion term in Eq. (7.2.18) remains,

$$\frac{\partial}{\partial t}\rho_0(u,t) = \gamma D e^{-2\gamma t}\frac{\partial^2}{\partial u^2}\rho_0(u, t), \qquad (7.2.21)$$

however with a renormalized diffusion constant $\gamma D e^{-2\gamma t}$.

Scaling also the time,

$$t \to \tau := \frac{1}{2\gamma}(e^{2\gamma t} - 1), \qquad (7.2.22)$$

the equation reduces further to a pure diffusion process,

$$\frac{\partial}{\partial \tau}\rho_0(u, \tau) = \gamma D\frac{\partial^2}{\partial u^2}\rho_0(u, \tau), \qquad (7.2.23)$$

and is solved by $\rho_0(u, \tau) = (4\pi\gamma D\tau)^{-1/2}\exp(-(u-u_0)^2/4\gamma D\tau)$. For an initial condition $\rho_0(u, \tau) = \delta(\tau)$, the solution is equivalent to the Green function. In terms of the original variables it reads

$$G(v', t; v, 0) = \sqrt{\frac{m}{2\pi k_B T(1 - e^{-2\gamma t})}}\exp\left(-\frac{m(v' - ve^{-\gamma t})^2}{2k_B T(1 - e^{-2\gamma t})}\right). \qquad (7.2.24)$$

Propagating an initial Gaussian $\rho(v, 0) = (2\pi\sigma_0^2)^{1/2}\exp(-mv^2/2\sigma_0^2)$ results in

$$\rho(v',t) = \frac{1}{\sqrt{2\pi}\sigma(t)} \exp\left(-\frac{(v' - ve^{-\gamma t})^2}{2(\sigma(t))^2}\right), \tag{7.2.25}$$

with a time-dependent variance

$$(\sigma(t))^2 = \sigma_0^2 + \sigma_{\mathrm{th}}^2(1 - e^{-2\gamma t}). \tag{7.2.26}$$

It shows a crossover on the timescale $1/2\gamma$ between two Gaussians of different widths, from the initial width σ_0 to the final thermal velocity scale $\sigma_{\mathrm{th}} = k_B T/m$. The same crossover is reflected in the information content (Fig. 7.4a),

$$I(t) = c\left[\ln(\sqrt{2\pi}e) + \ln\left(d_v\sqrt{\sigma_0^2 e^{-2\gamma t} + \sigma_{\mathrm{th}}^2(1 - e^{-2\gamma t})}\right)\right] \tag{7.2.27}$$

and in the rate of loss or gain (Fig. 7.4b),

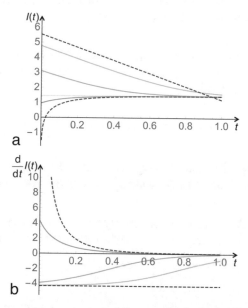

a

b

Fig. 7.4 In the simultaneous presence of dissipation and diffusion, as contemplated in Einstein's Fluctuation–Dissipation Theorem, the exchange of information between micro- and macroscales is dominated by the competition between diffusive spreading and contraction by dissipation. From very broad (blue dashed lines) or very narrow (red dashed lines) initial velocity distributions, compared to the thermal scale $k_B T/m$, the system loses or gains information till it thermalizes on a timescale $1/2\gamma$ (γ denoting the dissipation rate, panel **a**), cf. Eq. (7.2.27). In terms of information currents, Eq. (7.2.28), this amounts to initial losses $\sim -\gamma$ or gains $\sim 1/2t$, resp. (panel **b**). The curves correspond to parameter values $\sigma_0^2/\sigma_{\mathrm{th}}^2 = \infty$ (blue dashed), 100 (light blue), 10 (green), 1 (yellow), 0.5 (orange), 0 (red dashed). Information is measured in bits, i.e., $c = 1/\ln(2)$.

$$\frac{d}{dt}I(t) = \frac{c}{2(\sigma(t))^2}\frac{d}{dt}(\sigma(t))^2 = c\gamma\frac{(\sigma_{th}^2 - \sigma_0^2)e^{-2\gamma t}}{\sqrt{\sigma_0^2 e^{-2\gamma t} + \sigma_{th}^2(1 - e^{-2\gamma t})}}. \qquad (7.2.28)$$

It is instructive to consider two asymptotes, regarding the initial width σ_0:

1 $\sigma_0 \gg \sigma_{th}$

Prepared with an uncertainty much larger than the thermal scale (blue and green lines in Fig. 7.4), the system starts losing entropy due to dissipative contraction, at a rate

$$\frac{d}{dt}I(t) = -c\gamma \text{ for } \sigma_0 \gg \sigma_{th} \text{ and } t \ll 1/2\gamma, \qquad (7.2.29)$$

as in Eq. (6.2.8). Diffusive spreading then increasingly compensates contraction, till on the time-scale $1/2\gamma$, equilibrium between the two tendencies is reached, at a finite variance σ_{th}^2 [ZP94]. This shows that even without invoking quantum effects, notably uncertainty, classical statistical physics inserts a bottom to the otherwise unlimited entropy loss due to dissipation.

2 $\sigma_0 \ll \sigma_{th}$

If the initial uncertainty is much smaller than the thermal width (red and orange lines in Fig. 7.4), the short-time limit is dominated by diffusive spreading, with an increase of information

$$\frac{d}{dt}I(t) = \frac{c}{2t} \text{ for } \sigma_0 \ll \sigma_{th} \text{ and } t \ll 1/2\gamma. \qquad (7.2.30)$$

With increasing mean kinetic energy, dissipative loss into the medium reduces diffusion and confines it to the scale $\sigma_{th}^2 = k_B T/m$, corresponding to thermal equilibrium.

Obviously, if the velocity distribution already starts with the equilibrium width σ_{th}^2 (yellow lines in Fig. 7.4), losses compensate gains from the beginning and no transient information currents occur.

7.3 The Second Law of Thermodynamics in the Light of Information Flows

The information currents in physical systems, as elucidated in the foregoing subsections, can be attributed to two grand categories:

– **Vertical flows** transmit information between different spatial scales within a single degree of freedom or a few freedoms. Within this class, descending currents (top-down flows) by contracting phase-space structures, dump information, from large to small scales, as in dissipation. Ascending currents (bottom-up flows), by contrast, expand structures in phase space, thus lift microscopic information to macroscopic visibility. They occur, for example, in chaotic systems and more generally in all those nonequilibrium phenomena where thermal fluctuations become manifest macroscopically as random features.

– **Horizontal flows** distribute information among different degrees of freedom. They characterize interacting many-body systems and in particular form the second crucial ingredient of dissipation.

The interplay of vertical and horizontal information currents allows us to interpret the Second Law of Thermodynamics (SLT) as a global consequence of way information is processed in the universe. In particular, it provides decisive ingredients to reconcile macroscopic irreversibility with the time-reversal invariance of the microscopic equations of motion, central issue of countless controversies around the fundaments of statistical mechanics [Rei56, Zeh10].

7.3.1 Mixing and Thermalization

The concept of *mixing* [Sch84, LL92, Ott93, Gas98] constitutes a crucial link between the macroscopic phenomenon of systems approaching thermal equilibrium and the underlying dynamics of the microscopic degrees of freedom. In the context of complex dynamics, it is introduced as one of several levels of irregular motion in systems with a few degrees of freedom, forming together a hierarchy of stochasticity [Sch84, LL92, Ott93]. It starts with ergodic systems as the lowest level, and culminates in systems that can be described by a continuous shift of a rigid symbol sequence, such as in the Bernoulli map and the baker map, discussed in Sect. 5.3.2.

Within this hierarchy, mixing plays a pivotal role in that it relates deterministic irregular dynamics directly to concepts of statistical mechanics such as thermalization and irreversibility. In the guise of "molecular chaos", it formed a cornerstone already of Boltzmann's derivation of his H-theorem. (The H-theorem is Boltzmann's version of the Second Law, based on a statistical measure of disorder similar to Shannon information [Bec67].) Mixing contains the above inconsistency in a nutshell and therefore is a suitable laboratory to analyze it in depth.

In the context of statistical mechanics, mixing is defined in terms of the dynamics of phase-space density distributions. In qualitative terms, it requires that any inhomogeneous initial distribution will be stirred so intensely by the dynamics that for long times, it approaches an equilibrium in the form of a homogeneous distribution over the accessible phase space: If an ensemble, initially occupying a subset S_0 of some compact region of its the total state space U, is propagated by m iterations of a discrete area-preserving map $\mathbf{M}$, $S_m = \mathbf{M}^m(S_0)$, the map is called mixing if it meets the condition [Ott93, Sch84, Gas98]

$$\lim_{m \to \infty} \frac{\mu(S' \cap S_m)}{\mu(S')} = \frac{\mu(S_0)}{\mu(U)}, \tag{7.3.1}$$

or equivalently, $\lim_{m \to \infty} \mu(U)\mu(S' \cap S_m) = \mu(S_0)\mu(S')$, where μ is a measure on state space and S' is a fiducial reference set that may or may not coincide with S_0. It requires that asymptotically, the relative overlap of S_m with any arbitrary subset S' coincides with the relative measure S_0 occupies within the total accessible space U. Mixing is a stronger condition than (and implies) ergodicity in that it requires a uniform coverage of the accessible state space not only on average over long times, as in ergodic systems, but asymptotically *at each instant* of time.

In the context of dynamical systems, in terms of trajectories of discrete points, Eq. (7.3.1) translates to a condition on autocorrelation functions. Contracting S_0 and S' to points $\mathbf{r}_0$ and $\mathbf{r}'$ in the state space U, resp., and identifying $\mathbf{r}' = \mathbf{r}_0$, Eq. (7.3.1) implies that for $\mathbf{r}_m = \mathbf{M}^m(\mathbf{r}_0)$,

$$\lim_{m \to \infty} \langle \mathbf{r}_0 \mathbf{r}_m \rangle = \lim_{m \to \infty} \int_S d^{2f} r_0 \rho(\mathbf{r}_0) \mathbf{r}_0 \mathbf{r}_m = \int_S d^{2f} r_0 \rho(\mathbf{r}_0) \mathbf{r}_0^2 = \langle \mathbf{r}_0 \rangle^2. \tag{7.3.2}$$

As a particular consequence of the factorization required by Eq. (7.3.1), the autocorrelation decays to zero [Sch84, Gas98],

$$\lim_{m \to \infty} \langle (\mathbf{r}_m - \langle \mathbf{r}_0 \rangle)(\mathbf{r}_0 - \langle \mathbf{r}_0 \rangle) \rangle = \lim_{m \to \infty} \langle \mathbf{r}_0 \mathbf{r}_m \rangle - \langle \mathbf{r}_0 \rangle^2 = 0. \tag{7.3.3}$$

In both formulations, (7.3.1) as well as (7.3.2), a steady state is approached that is independent of the initial condition. More explicitly even, writing Eq. (7.3.3) as

$$\lim_{m \to \infty} \langle \mathbf{r}_0 \mathbf{r}_m \rangle = \langle \mathbf{r}_0 \rangle \lim_{m \to \infty} \langle \mathbf{r}_m \rangle, \tag{7.3.4}$$

it becomes evident that initial and final state become statistically uncorrelated: The mutual information (cf. Sect. 1.7) between $\mathbf{r}_0$ and $\mathbf{r}_m$, which initially coincides with the full positional information, is completely lost, a blatant violation of time-reversal invariance. Once this information is no longer available, the system will be unable to find the way back to its initial state upon inversion of time.

At the same time, the map $\mathbf{M}$ is assumed to be deterministic and measure preserving. It is invariant under time reversal and cannot be the origin of broken time-reversal invariance. The only remaining option to explain it is an asymmetry contained already in the initial conditions, represented by S_0 and S'. On first sight, the choice of the comparison set S' appears as innocuous, a mere matter of convention, but things are not that simple. In fact, the coarse-graining implicit in the measure $\mu(S' \cap S_m)$ defines a scale of resolution of fine details in S_m. Together with an inherent tendency of the chaotic dynamics, this bias in the choice of the initial conditions results in a violation of time-reversal invariance. As has been illustrated with

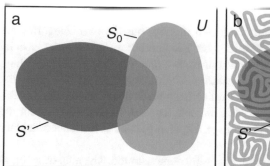
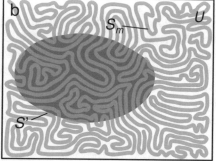

Fig. 7.5 Mixing in phase space. A dynamical process, generated by a map **M**, qualifies as mixing if it distributes an initial subset S_0 (panel a) so homogeneously over the accessible phase space U that with increasing number m of iterations, the overlap of $S_m = \mathbf{M}^m(S_0)$ (panel b, here, a caricature) with a fiducial set S' approaches an asymptotic value that coincides with the global fraction of U occupied by S_0, cf. Eq. (7.3.1). Mixing requires the combination of stretching and folding, cf. Fig. 6.4, that characterizes chaotic systems and leads to a striation of phase space by ever-finer filaments of S_m.

the example of the baker map, chaotic Hamiltonian motion contracts phase space in some of its directions and expands it in others. Conservation of phase-space area requires that the two tendencies exactly compensate one another. If the total accessible phase space U is compact, stretching has to be complemented by folding, while for contraction there is no lower limit. As a result, the set S_m striates phase space with increasingly finer layers, interspersed with correspondingly narrow voids (Fig. 7.5b). The initial state of the system is encoded in the fine structure of these filaments and can be recovered, tracing back the time evolution, if and only if this fine structure is completely conserved.

The intersection with the fiducial set S', implicit in the measure $\mu(S' \cap S_m)$, amounts to averaging over a part of S_m. By blurring the density variations across the filaments, it deletes the information they contain and thus leads to an irreversible approach to equilibrium (Fig. 7.8, panels a, b, to be compared with Fig. 7.5a, b). This means in turn that at least one dimension of S' must be large as compared to the width of the filaments. Indeed, if for instance, instead of an arbitrary compact blob as in Fig. 7.5, we choose S' specifically as $S' = S_{m'}$, i.e., identical to the initial set S_0 propagated till some reference time $m' \gg 1$, the overlap $S' \cap S_m$ will initially *increase* till at $m = m'$, it reaches a maximum and the ratio $\mu(S' \cap S_m)/\mu(S') = \mu(S_{m'} \cap S_{m'})/\mu(S_{m'})$ approaches unity, before it starts decaying again towards its asymptotic value $\mu(S_0)/\mu(U)$. Hence it is the macroscopic nature of both, the initial state S_0 and the fiducial state S', lacking microscopic fine structure, that is required for the time arrow to become manifest.

Mixing has been studied in various model systems, preferably in the same kind of dynamical systems that are also popular as models for chaotic motion. Standard examples in the realm of discrete maps of the unit square are the baker map, see Eq. (5.3.40), the Arnold cat map [LL92, Ott93, Sch84, Gas98], see Eq. (8.5.4), and other systems with a positive Lyapunov exponent, constant over phase space. Most measure-conserving chaotic systems are also mixing, but the two categories are neither necessary nor sufficient conditions for one another. In this context, special fame has been gained by the Sinai billiard or hard-sphere gas [LL92, Ott93, Sch84, Gas98], a simple mechanical system in continuous time that consists of a reflecting circular disk inside a square with cyclic or reflecting boundary conditions (Fig. 8.31a). For this model, Yakov G. Sinai could demonstrate features that are essential for the approach to equilibrium, based on the convex dispersing shape of the inner boundary [Sin70].

Returning to the baker map, mixing is readily demonstrated. The map transforms continuous probability density distributions $\rho(p, x)$ according to Eq. (5.3.44). The fact that it expands only along x and contracts only in the p-direction (Fig. 5.9) suggests considering a coarse-graining by integration over p,

$$\rho_x(x) = \int_0^1 \mathrm{d}p \rho(x, p). \qquad (7.3.5)$$

The mapping for $\rho_x(x)$ then takes the form

$$\rho_x(x) \mapsto \rho_x'(x') = \tfrac{1}{2}\left(\rho_x\left(\tfrac{x'}{2}\right) + \rho_x\left(\tfrac{x'+1}{2}\right)\right) \qquad (7.3.6)$$

and coincides with the Bernoulli map for continuous distributions, Eq. (5.3.31). This is not surprising, since the baker map has been constructed as a combination of the Bernoulli map in the x-direction with the inverse Bernoulli map in p, which is annihilated by the projection.

The combined actions of stretching, cutting, and superposition by the transformation (7.3.6) amount to a smoothing of $\rho_x(x)$ and will eventually blur all inhomogeneities in the distribution. This becomes even more evident if $\rho_x(x)$ is written as a Fourier series, see Eq. (5.3.29),

$$\rho_x(x) = \sum_{n=-\infty}^{\infty} c_n \exp(2\pi i n x). \qquad (7.3.7)$$

The map then replaces Fourier coefficients c_n according to (cf. Eq. (5.3.34))

$$c_{n'}' = c_{2n'}, \qquad (7.3.8)$$

that is, only coefficients c_{2n} with even index $2n$ are retained, those with odd index $2n-1$ are deleted, thinning out the spectrum and compressing it by $1/2$. In the limit $m \to \infty$ of the discrete time m, only c_0 will survive, so that

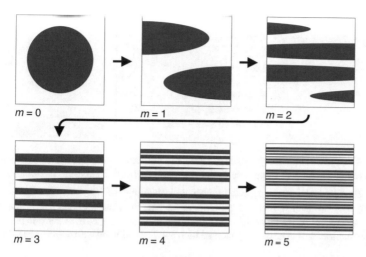

Fig. 7.6 Mixing in the baker map. If an initially compact set (red disk, panel $m = 0$) on the unit square is transformed by subsequent iterations of the baker map, Eq. (5.3.40), it shows the foliation of phase space characteristic of mixing dynamics. The combined action of stretching and folding (in this case cutting) leads to the formation of ever finer filaments, increasingly vulnerable to deletion by coarse-graining, be it by averaging over the p- (vertical) direction or by a physical mechanism such as diffusion.

$$\rho_x^\infty(x) = \lim_{m \to \infty} \mathrm{ber}^m(\rho_x(x)) = \mathrm{const} = 1, \qquad (7.3.9)$$

a homogeneous distribution over the unit interval. At the same time, the projection along x, $\rho_p(p)$, develops a dense sequence of singularities, as is evident from Fig. 7.6.

To conclude, the loss of autocorrelation implied by mixing can be explained by density variations penetrating into finer and finer scales, as it occurs also in time-reversal invariant systems, if at the same time, they are deleted below a certain limit of resolution. This argument, pointing out how mixing and coarse-graining conspire to break time-reversal invariance, is known as Gibbs' ink drop analogy [Gib02, Zeh10]. A related complementary reasoning says that the subset of microstates (states specified to microscopic detail) featuring pronounced variations of the probability density on macroscopic scales represents a minute minority within the total state space, that is, the probability to hit these states in a random sample is negligible. The huge majority of states, by contrast, show structure only on microscopic scales but are nearly homogeneous on larger ranges, hence appear close to thermal equilibrium. In other words, the number of microstates compatible with equilibrium is vastly larger than that of states showing macroscopic inhomogeneities. A similar argument is discussed in quantum mechanics as the *eigenstate thermalization hypothesis* [Deu91, Sre94, GL&06, PSW06, RDO08]: Among the eigenstates of a generic many-particle Hamiltonian, the overwhelming majority comes close to an equidistribution over the energy shell, a feature called typicality, see Sect. 8.6.2.

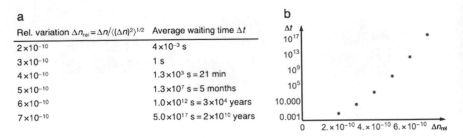

a

Rel. variation $\Delta n_{\mathrm{rel}} = \Delta n / \langle (\Delta n)^2 \rangle^{1/2}$	Average waiting time Δt
2×10^{-10}	4×10^{-3} s
3×10^{-10}	1 s
4×10^{-10}	1.3×10^{3} s = 21 min
5×10^{-10}	1.3×10^{7} s = 5 months
6×10^{-10}	1.0×10^{12} s = 3×10^{4} years
7×10^{-10}	5.0×10^{17} s = 2×10^{10} years

Fig. 7.7 Average waiting time for large fluctuations of the particle density in many-particle systems, as a function of the relative magnitude of the variation. (Table **a**) Increasing the relative variation $\Delta n_{\mathrm{rel}} = \Delta n / \sqrt{\langle (\Delta n)^2 \rangle}$ of the particle number n, in the range of 10^{-10}, by less than an order of magnitude in a subset of a macroscopic volume of gas, the time for such a variation to occur rises from microseconds to more than the age of the universe. (Panel **b**) A semi-logarithmic plot reveals that this amounts to an overexponential increase. Data adopted with permission from [Bec67, Bre94].

As a manifestation of this apparent preference in the time domain, the system will spend most of its time in configurations that are hardly distinguishable from thermal equilibrium. In particular, fluctuations far away from equilibrium, where macroscopic order emerges temporarily, will become exceedingly rare with increasing magnitude. This is illustrated in Fig. 7.7 for the example of thermal fluctuations of the local particle density in a macroscopic volume of gas. Even for minute relative variations, they take times of the order of the age of the universe to occur.

An reasoning frequently put forth to explain the time arrow [Rei56, Zeh10] concludes from here directly that a system prepared in one of the unlikely states of low entropy will evolve with probability close to unity into a "disordered" state of higher entropy, almost never vice versa. However, based on a macroscopic viewpoint, this argument inherits a subjective nature from the concept of macrostate, which masks out all microscopic information but incorporates all order on larger scales. Indeed, even if a system is prepared already in a microstate compatible with thermal equilibrium, distinguished from similar microstate only by small-scale variations, its autocorrelation will decay due to mixing as markedly as if it had started from a macroscopically ordered state. That is, the time arrow becomes manifest only if the macroscopic inhomogeneity has been prepared by a corresponding initial condition, be it the Big Bang or an intervention by a likewise macroscopic agent [Zeh10].

However, this asymmetry among scales does not imply a time arrow: The decay of the autocorrelation is in itself symmetric in time. From a major fluctuation, it takes place into the past in the same way as into the future. A preferred direction of the time evolution of many-particle towards disorder therefore cannot be derived from it. The argument can be refined, though, by taking an asymmetry between small and large scales into account that does have an objective physical basis: Interaction processes, which tend to smooth out structures by distributing energy and information among coexisting single-particle degrees of freedom, act predominantly on microscopic scales, although they become manifest also on the macroscopic level in phenomena such as noise and diffusion, as will be discussed in the sequel.

7.3.2 Diffusion and Coarse-Graining

The two decisive ingredients of the irreversible nature of mixing dynamics are therefore the continuous expansion of phase space in some of its directions and the averaging over a finite scale in the other, contracting directions, a process termed *coarse-graining* in the literature on the subject [Gru73, Bre94, Zeh10]. In theoretical accounts, imposing a finite resolution, be it, for instance, by convolution with a smoothing function or by partitioning the state space into cells [Gib02], is indispensable in order to avoid a divergence of information a probability density distributions, and it is in the spirit of Boltzmann's complexions, see Sect. 1.3. At the same time, it also occurs naturally in physical systems, owing to various mechanisms that tend to blur sharp variations in the probability density, notably all kinds of noise of whatever origin.

Which crucial role coarse-graining plays is highlighted also by a patent objection to resolve an argument, accredited to a contemporary of Ludwig Boltzmann: Loschmidt's paradox, also known as the *reversibility objection* [Rei56, Bre94, Zeh10], insists that the SLT is incompatible with the time-reversal invariance of the microscopic evolution laws, because upon inversion of time, any system approaching equilibrium will accurately return to its initial state, thereby reversing the previous increase of entropy. Attempts to implement an inversion of time in a real system, to bring about a "Loschmidt echo", be it in the laboratory or numerically by simulation backward in time, poses practically insurmountable difficulties: Registering and inverting the final state reached by the forward evolution as initial condition of the subsequent inversion has to be achieved with absolute precision. Any experimental or numerical inaccuracy invariably impedes this step. The invariably chaotic nature of the many-particle dynamics in particular implies that minute imprecisions in the inversion of all the microscopic momenta will amplify exponentially in the inverse time evolution.

Typical unstable Hamiltonian dynamical systems contract their phase space in part of its dimensions. This amounts to a boundless transport of information towards ever-smaller scales in position space or to higher Fourier components in reciprocal space. Coarse-graining, however, inserts a bottom to such a transport by definitely erasing information when it arrives at some lower bound. In Hamiltonian systems with a finite number of degrees of freedom, there is no inherent mechanism that could cause such a deletion. On the contrary, in systems with a discrete state space, such as the discrete baker map featured in Sect. 5.3.2 (see Fig. 5.11), the downward information current in the momentum component ends at the scale of pixels, but the information reenters "through the backdoor" at the largest scale. A similar recycling of small-scale information, resulting in recurrences, is observed in other discrete chaotic systems as well, see Sect. 8.5.2. While in classical mechanics with its continuous variables,

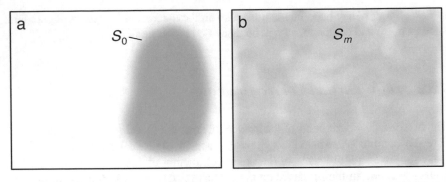

Fig. 7.8 Conspiracy of mixing and diffusion in phase space. The same amount of diffusive smoothing that hardly alters a compact initial distribution S_0 (**a**, compare with Fig. 7.5a), is sufficient to completely erase the information contained in the fine structure of the filaments the distribution develops after the repeated action of a mixing dynamics (**b**, compare with Fig. 7.5b).

such a revolving flow does not exist, analogous phenomena do occur in quantum mechanics.

In many-body systems comprising a large number of degrees of freedom, however, another effect comes into play that interferes with information transfer into small scales: diffusion. In Sect. 7.1, it was demonstrated that it entails a definite loss of information at scales comparable to or below the characteristic length $\sqrt{Dt}$ (D and t denoting the diffusion constant and time, resp.). More precisely, random noise, injected from the environment, gradually supersedes the information the system conserves from its initial state. In this sense, diffusion amounts to an interchange of information with neighbouring degrees of freedom, taking effect at the bottom of the length spectrum.

The way mixing and diffusion conspire to produce coarse-graining on a certain minimum scale can be made more comprehensible, looking at a specific example. For the present context, mixing can be reduced to the coexistence of an expanding (unstable) and a contracting (stable) direction in phase space, that is, of a hyperbolic flow pattern [LL92, Ott93, LS98]. It is generated already by a simple linear instability in the form of a parabolic potential barrier [ZP94],

$$V(x) = -\frac{1}{2}m\lambda^2 x^2, \qquad (7.3.10)$$

where λ gives the rates of spatial contraction and expansion and is analogous to the Lyapunov exponent in the context of chaotic dynamics. Combining this hyperbolic flow with diffusion requires including a term in the Fokker–Planck equation, given by Eq. (5.2.11), that represents the Hamiltonian dynamics induced by the potential $V(x)$. Assuming for the sake of simplicity that diffusion affects position and momentum with the same strength, results in an evolution equation for the probability density,

$$\frac{\partial}{\partial t}\rho(p, x, t) = \{H(p, x), \rho(p, x, t)\} + D\left(\frac{\partial^2}{\partial p^2} + \frac{1}{m^2\lambda^2}\frac{\partial^2}{\partial x^2}\right)\rho(p, x, t),$$

$$(7.3.11)$$

(for the Poisson bracket $\{f(p, x), g(p, x)\}$, see Eq. (5.2.10)). For a standard Hamiltonian $H(p, x) = p^2/2m + V(x)$, that means

$$\frac{\partial}{\partial t}\rho = -\left(m\lambda^2 x\frac{\partial}{\partial p} + \frac{p}{m}\frac{\partial}{\partial x}\right)\rho + D\left(\frac{\partial^2}{\partial p^2} + \frac{1}{m^2\lambda^2}\frac{\partial^2}{\partial x^2}\right)\rho. \qquad (7.3.12)$$

Anticipating the hyperbolic flow, it is appropriate to transform to coordinates $s :=$ $p/\sqrt{m\lambda}-\sqrt{m\lambda}x$ along the stable (contracting) and $u := p/\sqrt{m\lambda}+\sqrt{m\lambda}x$ along the unstable (expanding) direction of phase-space flow. In this way, the Fokker–Planck equation reduces to the simpler form,

$$\frac{\partial}{\partial t}\rho = \lambda\left(s\frac{\partial}{\partial s} - u\frac{\partial}{\partial u}\right)\rho + \frac{2D}{m\lambda}\left(\frac{\partial^2}{\partial s^2} + \frac{\partial^2}{\partial u^2}\right)\rho. \qquad (7.3.13)$$

Like this, it provides a local picture of the mixing dynamics, with u representing the direction along and s the direction across the filaments of the distributions in Figs. 7.5b and 7.6.

For sufficiently long times, so that the extension of the distribution in the exponentially expanding direction is already much larger than the available phase space and substantial folding must have occurred, Eq. (7.3.13) no longer describes the dynamics in the unstable direction u, but continues to hold for the interplay of diffusion and contraction in s [ZP94]. For a Gaussian distribution,

$$\rho(s, u) = \frac{1}{2\pi\sigma_s\sigma_u}\exp\left(-\frac{1}{2}\left(\frac{s^2}{\sigma_s^2} + \frac{u^2}{\sigma_u^2}\right)\right), \qquad (7.3.14)$$

the full Fokker–Planck equation reads

$$\frac{\partial}{\partial t}\rho = \lambda\left(\frac{u^2}{\sigma_u^2} - \frac{s^2}{\sigma_s^2}\right)\rho + 2D\left(\frac{s^2 - \sigma_s^2}{\sigma_s^4} + \frac{u^2 - \sigma_u^2}{\sigma_u^4}\right)\rho. \qquad (7.3.15)$$

Focusing on the long-time asymptote where the terms in u play no role any longer, and seeking a stationary state with $\partial\rho/\partial t = 0$, leads to the condition

$$0 = \left(-\frac{\lambda}{\sigma_s^2} + \frac{2D}{m\lambda\sigma_s^4}\right)\rho. \qquad (7.3.16)$$

It translates into a definite value for the variance in the stable direction [ZP94], $\sigma_s^2 = 2D/m\lambda^2$, or equivalently

$$D = \frac{m\lambda^2\sigma_s^2}{2}. \tag{7.3.17}$$

Equation (7.3.17) establishes a universal scale of coarse-graining that occurs if mixing goes along with diffusion. As it stands, λ is the inverse timescale of a contracting dynamics as it occurs in dissipation as well, but it is inseparably tied by volume conservation to the expansion of phase space that characterizes chaotic stretching and folding.

In its structure, Eq. (7.3.17) closely resembles Eq. (7.2.8), in that it relates the diffusion constant to a microscopic quantity, the scale of classical coarse-graining, to a macroscopic observable, the Lyapunov exponent. In this sense, its significance is comparable to that of the Fluctuation–Dissipation Theorem, Eq. (7.2.12). In both cases, these relationships reveal how vertical, top-down, information currents are redistributed among degrees of freedom on the microscopic level, to become macroscopically manifest again as noise.

The dynamics of a single pair and of a chain of coupled harmonic oscillators, worked out in Sect. 5.6, provides some clues how such an exchange occurs. In the case of only two coupled freedoms, energy and information oscillate periodically between them, on a timescale larger than the periods of the individual systems. Owing to correlations, the initially individual information rapidly becomes common property. The total information content can then in general no longer be attributed uniquely to the nominal degrees of freedom, except for periodic zeros of the correlation function. Yet for a bipartite system, it always remains possible to distinguish shared information neatly from the remaining individual portions.

With increasing number of freedoms, however, expanding the pair to a chain of oscillators, these remainders of individuality are rapidly lost in the mass. The total system, as a closed system with a Hamiltonian time evolution, never violates the conservation of entropy, see Eq. (5.3.17). A set of linearly coupled harmonic oscillators, being separable into normal modes, even conserves energy and entropy of each of these collective modes, Eq. (5.6.38). However, the time to return to a probability density that again factorizes into normal modes, so that the information could be uniquely attributed to individual freedoms, diverges rapidly.

The situation becomes vastly more complicated if the many-particle system is no longer separable but strongly interacting, so that it inextricably couples all single-particle degrees of freedom, and their collective dynamics becomes chaotic and therefore mixing: Under this condition, anticipated by Boltzmann in his notion of molecular chaos [Bec67], energy and information are intensely exchanged between all degrees of freedom and in particular also distributed over all scales. However, the interactions coupling the particles, such as notably scattering processes, occur almost exclusively on atomic and molecular scales, that is, they are decidedly microscopic in nature.

To summarize, if mixing induced by chaotic dynamics combines with coarse-graining due to many-body interactions, it explains at least a fundamental asymmetry in the information transport among scales: descending and ascending currents communicate, predominantly at the smallest scales. This is not sufficient, however, to explain also the asymmetry in time, manifest in the inconsistency between the Second Law and time-reversal invariance. The remaining conflict appears most pronounced in the guise of Zermelo's *recurrence objection* [Rei56, Bre94, Zeh10]: Referring to the Poincaré recurrence theorem (see Sect. 5.5.1), it points out that any state of a many-body system, in particular also states far from equilibrium, will eventually recur, to arbitrarily close agreement with the initial state and infinitely often. Not unlike objection, Zermelo's argument pales in view of the expected actual numbers. According to crude estimates, the recurrence time in chaotic systems diverges rapidly with number of degrees of freedom [LL92, AK07] and reaches astronomic time scales. The global interaction implied by diffusion and other many-body phenomena increases the number of effectively coupled freedoms to thermodynamical system sizes and accordingly boosts the expected return times. Large deviations from equilibrium represent instances of Poincaré recurrence, in a broad sense. Figure 7.7 therefore gives us also a rough idea which orders of magnitude to expect also in the present context. Even if this sheds doubt on the practical relevance of the recurrence objection, the challenge remains to resolve it in a more rigorous argument.

With the theoretical tools currently available, analyzing any further how information redistributes in large multipartite systems is a formidable task. In order to make any progress, quantum mechanics has to be taken into account. Indeed, the roles of entanglement and information processing for thermalization in quantum many-body systems are currently an active subject of research, to be addressed more closely in Sect. 8.6. Thermalization and the global approach to equilibrium are central problems in the understanding and application of the Second Law. Questions around the dynamics of information accompanying these processes go much further. They concern particularly phenomena occurring in systems that are still far away from thermal equilibrium.

7.3.3   Grand Total: The Second Law of Thermodynamics

As a synthesis of the previous subsections, we can sketch a panorama of processes involving information flows relevant for a basic global principle, such as the Second Law:

- **Chaotic dynamics** in microscopic degrees of freedom induces vertical information currents between different scales inside an identifiable degree of freedom or a small group of freedoms. Specifically, for closed Hamiltonian systems, information is conserved, so that bottom-up currents due to expansion compensate top-down currents due to phase-space contraction.
- **Dissipation** sends information into small scales and redistributes it by many-body interactions among a large number of degrees of freedom, converting it into shared information that can no longer be attributed to individual subsystems.
- **Noise and fluctuations** in turn inject information into individual freedoms that has no localizable source in other specific subsystems within a many-body system.

Their internal dynamics lifts it to macroscopic observability, for example as diffusion.

Chaos and thermal noise constitute the two principal sources of unpredictability in classical physics and appear as random generators in an otherwise predictable world. However, within this common category, the nature of their stochasticity is quite different. Chaos can be modelled and demonstrated even in one-dimensional dynamical systems. The information surfacing as chaotic motion originates exclusively in details of the initial condition in this same dimension. The horizon of predictability, for example in weather forecast, can be shifted ahead into the future by improving the accuracy of the input data, in particular measurements of the initial conditions, and the capacity and precision of numerical simulation. Fluctuations, as a many-body phenomenon, are far less tractable. Besides providing precise initial conditions, predicting fluctuations would require tracing information back to its origin in a multitude of subsystems where information currents ramify rapidly and are lost in the mass. This calls for a more sophisticated theoretical approach, involving concepts of both quantum mechanics and many-body physics. Numerical simulations of the information flow in multipartite systems challenge even present-day computer technology.

In fact, the repertoire of physical phenomena that exemplify these processes is much larger than schematic scenarios allowing for a relatively straightforward quantitative analysis. Nonequilibrium statistical mechanics, in particular, is full of phenomena that consume and at the same time generate entropy. Exemplifying unpredictability, they are nevertheless highly correlated and thus represent order. These processes are behind the seemingly boundless creativity we observe as complex macroscopic patterns in nature. On planet Earth, we encounter a plethora of examples, in natural as well as in man-made systems. To name but a few,

- **Life**, as the most prominent case, based on particularly favourable conditions on this planet, has generated highly organized structures on a vast range of spatial and time scales. They all depend on the continuous injection of low-entropy solar energy and accelerate its down-conversion to thermal radiation [Lov00, Mic10, Kle10, Kle12, Kle16], as compared to a sterile planet.
  At the same time, there is no doubt that chance is an essential ingredient of life, and not only of its individual manifestations but possibly even of its very existence on this planet [Mon71, EW83, Kup90]. Genetic variability represents perhaps the most striking evidence of the creativity of processes far from equilibrium.
- **Turbulence** absorbs entropy (as well as energy) at macroscopic scales and cascades it, by the formation of eddies of many different sizes, down to almost molecular length scales [FF04]. It provides a particularly compelling example of a predominantly top-down entropy flow, accompanied by random pattern formation in space and time as a manifestation of simultaneous bottom-up information transport.
- **Weather**, specifically cloud formation, demonstrates very directly how solar radiation leads to the generation of spatial structures of a diversity that appears inexhaustible (Fig. 7.9a). Being a consequence of the special composition of Earth's

Fig. 7.9 Pattern formation in the Earth's atmosphere. Cloud formation (**a**, over tropical rain forest in South America) is an impressive example of the seemingly endless creativity of self-organized structure formation in nonequilibrium thermodynamics. In snowflakes (**b**), crystallization combines order (their hexagonal symmetry breaks the isotropy of the gaseous state) with random features in all other aspects. They literally epitomize "frozen randomness".

atmosphere as it evolved since the advent of life [Lov00], it can be considered as another manifestation of life on this planet [Mic10, Kle10, Kle12, Kle16]. Weather is an instance of macroscopic chaos, where a collective phenomenon in a huge many-body systems allows again for an approximate description in a state space of few dimensions [Lor63], albeit with strong dissipation, see Sect. 6.3.

- **Condensation and crystallization** reveal microscopic order on macroscopic scales at the cold side of the corresponding phase transitions. At the same time, they disturb the macroscopic homogeneity and isotropy that prevails on their hot side by preferring certain structures and directions in space (Fig. 7.9b), hence are examples of spontaneous symmetry breaking. On balance, however, condensation and crystallization release heat, thus remain perfectly consistent with the Second Law.

- **Chemical pattern formation** consumes chemical, hence low-entropy, energy to excite periodic oscillations in space and time, as for example in the Belousov-Zhabotinsky reaction [ZGP83]. They are a prototypical instance of *dissipative structures*, featured, for example, in Ilya Prigogine's seminal work on irreversibility and indeterminism in nonequilibrium thermodynamics [PN77].

- **Lasers** and masers generate electromagnetic radiation with a high degree of phase coherence, breaking the statistical symmetry of random phase correlations that characterizes natural light sources. Similar phenomena have not been observed outside the realm of man-made technology, indicating that lasing is evidence of life.

- In **digital computing**, electronic systems are maintained at a critical state between chaos and thermalization, where any deterministic sequence of states can be performed within a finite set of discrete macroscopic nominal states, depending on initial and boundary conditions. While this requires the computer to be supplied with low-entropy energy, which it dumps as heat into the environment, its nominal states are shielded against bottom-up entropy currents (noise) originating in the microscopic degrees of freedom. See Chap. 9.

All these nonequilibrium processes have in common that they generate order by consuming low-entropy energy, they comprise bottom-up as well as top-down information currents. However, they comply with the Second Law in that all in all, more coherent energy is converted into heat than vice versa.

Till here, the term *information* has been preferred in this book, as referring indiscriminately to structures on all scales, without distinguishing it clearly from that of *entropy* (as it appears, for example in the term Kolmogorov-Sinai entropy). In the context of the Second Law, it becomes indispensable to differentiate unambiguously between the two concepts, giving an explicit definition of entropy as it is used in thermodynamics. In the thermodynamical context, entropy refers exclusively to **information in microscopic degrees of freedom and on the smallest scales**. This becomes manifest in particular in the relation of Boltzmann's microscopically defined quantity $S = k\ln(\Omega)$ to temperature T and heat Q (Clausius' relation) [Rei65],

$$\Delta S = S_{\text{fin}} - S_{\text{ini}} = \int_{\text{ini}}^{\text{fin}} \mathrm{d}Q \frac{1}{T}, \tag{7.3.18}$$

where the notation $\mathrm{d}Q$ indicates an inexact differential [Rei65], and the integration extends from an initial to a final state of a thermodynamical process. It exclusively refers to information contained in microstates but discards all information embodied in macroscopic structures.

This microscopic interpretation is evidently assumed in a standard formulation of the Second Law of Thermodynamics [Rei65]:

1. **In any process in which a thermally *isolated* system goes from one macrostate to another, the entropy tends to increase, i.e.,**

$$\Delta S \geq 0. \tag{7.3.19}$$

2. **If the system is not isolated and undergoes a quasi-static infinitesimal process, in which it absorbs heat $\mathrm{d}Q$, then**

$$\mathrm{d}S = \frac{\mathrm{d}Q}{T}, \tag{7.3.20}$$

   where T is a quantity, called the "absolute temperature", characteristic of the macrostate of the system.

In typical physical systems as they are discussed in equilibrium thermodynamics, this distinction is unproblematic; it can be based on the large gap in the spectrum of spatial scales between atomic and molecular structures and the size of macroscopic samples, such as containers filled with a nearly homogeneous gas or liquid or samples of a pure metal. This can hardly be said about systems of a certain complexity and certainly does not apply under conditions far from equilibrium, in particular in biological systems. For instance, strange attractors are characterized by structures on all scales below their overall size. As soon as the gap in the spectrum of spatial scales gets filled in, the distinction between entropy and information is blurred. Without the restriction to microstates, entropy would be strictly conserved, as stated in Sect. 5.3. The separation of microscopic and macroscopic structure is characteristic also for other thermodynamic quantities and underlies, for example, the distinction between heat and free energy.

In the present context, it is therefore more appropriate to refer to information currents between scales, instead of information contents of different length regimes. This suggests rephrasing the first part of the Second Law, the relevant one here, as.

In any process in which a thermally isolated system
goes from one macrostate to another, more information
is transferred from large to small than from small to large scales.

Even if the solar system is not exactly isolated, the phenomena listed above provide illustrative examples for this principle. They generate macroscopic order, but depend on the condition far from equilibrium, maintained on the Earth's surface by solar radiation.

Causality, as explored in Sect. 3.1, constitutes striking general evidence for a preferred direction of information currents. The overwhelmingly superior frequency of causal processes over chance, which we perceive as a linear order in time, defines a causal arrow of time on its own. It coincides with the thermodynamical time arrow and represents a particular manifestation of it [Zeh10].

In the foregoing subsection, it was pointed out that the underlying microscopic processes are perfectly reversible, so that all information currents in a given sense can occur likewise in the opposite direction. However, they break another, less obvious symmetry: In interacting many-body systems, the exchange and redistribution of information among degrees of freedom occurs predominantly at the lower end of their spectrum of length scales.

This asymmetry in itself does not yet explain the time arrow inherent in the Second Law. It has to be combined with an another asymmetry, a bias in the initial conditions: The ubiquity of systems far from equilibrium in the universe, such as the solar system, implies that it must have been born in a condition of a surplus of information at the largest scales within the total length spectrum covered in the course of its development. In this interpretation, the thermodynamical time arrow is a consequence of a cosmological direction of time [Zeh10, Haw17], defined by an initial imbalance between the information contents of micro- and macro-scales in the initial state of the universe, in favour of the latter. We are still participating in, and at the same time witnesses of, the gradual relaxation of this vast primordial fluctuation.

With this preferred direction of information flow, attention focuses on the way information is processed at the opposite extreme, at the smallest scales. The preceding subsections have demonstrated that classical mechanics is not apt for this task in an essential respect: It lacks an inherent constant of nature, a minimum action scale that would limit the information content of phase space. It is only with quantum mechanics that an impenetrable bottom is inserted where descending vertical information currents must end. The consequences of this fundamental element of the physics of information will be explored in Chap. 8.

Wherever the total information is bounded, for example in the context of Fourier transformation (Sect. 2.3) and of discrete state spaces (Sects. 5.5 and 8.5.2), it appears that the largest scales, forming the top end of the spectrum, directly connect back to the opposite end and are as crucial for information processing as the smallest scales. In the present context, this suggests to expect some particular processes to occur at the largest scales in the universe, possibly inserting a ceiling to bottom-up information flows as quantum mechanics does for top-down currents reaching the bottom. The question touches upon how cosmology and general relativity treat information, and it remains completely open at least as long as a synthesis of general relativity with quantum mechanics has not yet been achieved [Haw17].

Chapter 8
Information and Quantum Mechanics

Working out a detailed picture in Chaps. 5 to 7 how information is processed in systems of classical physics has directed the view towards what is going on at the smallest scales. Microscopic degrees of freedom absorb information dumped from the macroscopic level, they generate the information that becomes manifest in thermal noise and emerges at large scales by chaotic dynamics, and they exchange and redistribute entropy massively among themselves. At the same time, it has become clear that classical mechanics does not offer us a satisfactory answer how information is represented, let alone processed, at these scales. It suffers from gaps and inconsistencies in basic questions: Is there any limit to the distinguishability of the states of classical systems? Does classical mechanics allow for an unbounded resolution, that is, are delta functions in phase space permissible states? Does any upper bound exist for the entropy? Any lower bound? Are fractal structures in phase space, self-similar down to arbitrarily small details, compatible with fundamental principles? Settling them requires to study how nature processes information at the bottom of scales where quantum mechanics supersedes classical physics. If we regard information processing in classical systems as nature's assembly language, this descent into the lowest basement now amounts to looking directly into the chips and analyzing their function on the level of machine code.

Quantum mechanics entered the stage promising conclusive answers to these questions. To be sure, it was not conceived precisely to this end. Other observations, not obviously related to information, shattered the fundaments of classical physics and provoked a radically new approach. Classically inexplicable phenomena addressed in introductory textbook chapters include the photoelectric effect, Compton radiation, the discrete Hydrogen spectrum, and related experimental evidence. A prominent role is definitely due to blackbody radiation and its theoretical analysis by Max Planck. The immediate objective guiding his work on an expression for the spectral density in black bodies, published in 1900, was to reproduce the then fresh measurement results by Lummer, Pringsheim, and Kurlbaum. They had found significant discrepancies from the Wien radiation law towards the infrared regime [Pla49]. Notwithstanding, reviews published later by Planck himself [Pla49] reveal that he saw very clearly the

© Springer Nature Switzerland AG 2022
T. Dittrich, *Information Dynamics*, The Frontiers Collection,
https://doi.org/10.1007/978-3-030-96745-1_8

consequences of his findings for the concept of entropy, notably concerning a lower bound, and for the continuous character of physical quantities in general.

He was also aware of the closeness of his work to that of Josiah Willard Gibbs. Gibbs' objection to entropy as it is calculated in classical thermodynamics refers directly to the problem of extensiveness. Indirectly it bears on the entropy of mixing processes, hence on the distinguishability of microscopic states [Gib02, Gib93, Rei65, Jay92]. Gibbs' paradox can only be resolved by introducing a positive but otherwise undetermined lower bound for the resolution of differences of physical states. Even Boltzmann, when he coined the concept of complexions as physically distinguishable states underlying his definition of entropy, knew that he had to assume a certain irreducible discreteness. Planck picked up this line of thought, introducing a finite resolution explicitly in his calculations and linking it, through black-body radiation, to measureable quantities. He thus attributed solid physical reality to this lower bound as a fundamental constant of nature.

It is not clear whether the first generation of pioneers of quantum mechanics, Bohr, de Broglie, Einstein, Heisenberg, Schrödinger and contemporaries, were conscious of the intimate relation of their work to the concept of information. They certainly could not yet even know its definition on the classical level, as refined by Shannon later in 1948, revealing its closeness to Boltzmann's entropy. The first researcher to explicitly link quantum mechanics with information theory was John von Neumann, in particular by introducing a definition of entropy consistent with the principles of quantum mechanics, and by other results connecting the two fields. Inspired by a seminal paper by Richard Feynman [Fey82], quantum information was complemented with quantum computation and grew into a discipline of its own right, stimulated by the perspective of building a quantum computer and the development of technologies that promised to apply the corresponding theoretical achievements in the laboratory.

At the same time, awareness is increasing that the concept of information not only has to be reconsidered from scratch in the quantum realm, but that its treatment plays the role of a constitutive principle of quantum mechanics: It becomes more and more evident that quantum physics, including its most bizarre and puzzling manifestations, has to be interpreted as a reformulation of physics that takes fundamental principles and limits of information processing in the microcosm systematically into account. In this sense, quantum mechanics has much in common with a similarly radical proposal, Special Relativity Theory. Originating in the intellectually fertile early twentieth century, both are prototypical instances of scientific revolutions according to Thomas Kuhn [Kuh70]. Special Relativity can be reinterpreted as a theory working out the consequences of a single information-theoretic postulate: Information cannot propagate faster than the velocity of light. Could also quantum mechanics be derived from a set of basic postulates concerning the quantity, representation, dynamics, etc. of information on the microscopic level? While this goal has not yet been achieved completely, there exist promising attempts suggesting that such an idea is within reach [Zei99, Har01, CBH03, Bub05, CAP11]. An information theoretic perspective of quantum mechanics might even guide the conception of future textbooks and curricula.

The present section should be read as an outline of such an approach. It is decidedly not intended as an introduction to quantum information (readers are referred to monographs such as, e.g., refs. [NC00, BEZ00, HI&15, BCS04]), let alone to quantum mechanics in general. A first general part, Sects. 8.1 to 8.3, develops basic concepts and features of quantum mechanics, its state space and evolution laws, from information-theoretic principles. It will turn out, notably, that quantum mechanics is not a discretization of the microcosm. It is in a way much less, in that it does not imply a rigid digitization of space or other quantities, an idea discussed as *digital physics* [Zus69, Fre03, Flo09]. At the same time, it is also much more: It provides a comprehensive solution to the problem how to construct what might be called a "finitist physics", where the amount of information in a finite physical system is also finite [Mar03, San21].

From there, the section enlarges on two subjects, each of which sheds a spotlight on a particular interface between quantum and classical mechanics: Quantum measurement is the one privileged channel through which, from our viewpoint as macroscopic observers, we have access to quantum systems. An account of quantum measurement, Sect. 8.4, should elucidate how something as elementary as facts, to which we can attribute probabilities and hence information, can arise in the first place in quantum mechanics, even before analyzing how information is transferred from the quantum to the classical level.

Chaotic systems, in turn, are microscopes peeking arbitrarily far into the smallest scales. They provide a natural source of microscopic information. Quantum chaos, outlined in Sect. 8.5, attempts reconstructing classical chaos as far as possible in the framework of quantum mechanics and reveals the differences between classical and quantum information dynamics in a particularly striking manner. The chapter concludes in Sect. 8.6, returning to the topic of Sect. 7.3, irreversibility and the Second Law, to reconsider thermalization and the approach to equilibrium from the advanced viewpoint of quantum mechanics.

Another feature quantum mechanics and Special Relativity have in common is that they supersede classical physics and generalize it, albeit into extreme regimes of different parameters, but at the same time still contain it as a limiting case. In the case of SRT, this is a relatively clear asymptotic relation. It reinterprets fundamental concepts, such as simultaneity, in a completely new way. Yet Lorentz transformations, the mathematical backbone of SRT, are a straightforward extrapolation of Galilei transformations, approaching them in the limit $v/c \ll 1$ of small relative velocities v as compared to c, the velocity of light (cf. Eqs. (3.1.30, 3.1.31)).

The case of quantum mechanics is by far more involved. To be sure, one aspect of the approach from quantum to classical physics can be reduced to a similar asymptotic limit, known as the short-wavelength approximation. In analogy to optics, where light rays emerge as trajectories of wave packets in the limit of short wavelengths (the *eikonal approximation* [PH69]), the equations of motion of classical mechanics, specifically Hamilton's Equation (5.1.2), are recovered as asymptotic solutions of the Schrödinger equation if all characteristic actions S of the system at hand are large compared to Planck's constant, $S/\hbar \gg 1$.

However, this *semiclassical limit*, while it gives rise to beautiful mathematics, only solves a minor part of the problem. It does not give us any hint as to a deeper question: How positive facts, the basic material of all our classical knowledge, crystalize in a substrate of quantum states and their superpositions. Meanwhile, decades after the introduction of semiclassical approximations in the late 1920s by pioneers like Wentzel, Kramers, Brillouin, and van Vleck [Bri26, Vle28], evidence is increasing that this facet of classicality is related to the size of systems, not in the sense of spatial extent but in terms of the number of degrees of freedom they comprise and the way they interact. Classical physics thus appears rather as an emergent phenomenon, similar to manifestations of macroscopic order such as crystal lattices [Lau05]. Also here, the interchange and redistribution of information among microscopic freedoms, described by quantum mechanics, is the key issue. It forms a recurrent topic throughout this section. As an active field of research, it presently enjoys increasing attention.

Quantum mechanics not only reorganizes physics along the way information is processed. It demonstrates how the familiar concept of information tends to dissolve as we look too deeply into the microscope, and conversely lets us adumbrate how the very facts, making up physical reality as we perceive it, emanate from processes in Hilbert space outside our reach. Daring a bold analogy to another idea sharing its roots in the early twentieth century's Vienna, we might say that quantum mechanics opens our view towards the unconscious of physical reality.

8.1   Information Theory Behind the Principles of Quantum Mechanics

On the face of it, quantum mechanics deals with the microcosm, that part of our world which, while being so close to us, is least accessible to our senses. Basic entities such as atoms were considered as mere hypothetical constructions over centuries, till their existence could be evidenced directly in the laboratory. It therefore does not surprise that, besides revealing new physics, it required reconstructing our relation to reality. Once the mathematical framework was established, notably by the work of Heisenberg and Schrödinger, it became clear that it implied an entire new ontology of its own. In classical physics, it is measurements that give us access to and, through their results, constitute the state of a system. This clear-cut and rigorous role of observation in empirical science is abandoned in quantum mechanics. It introduces a new basic level of reality, independent of our observation and only indirectly accessible for it. Only on this level, the world of quantum states, principles of content, representation, and processing of information can be established that overcome the inconsistencies of classical physics in this respect. They can be understood as defining a fundamental data structure, which is not specific for any particular representation but constitutes in the first place what is reality and what can be known about it.

The computer as a metaphor, alluded to in the introductory section, proves helpful also here [Mar03, Llo06]. The state of a computer (as an information processing device, not as a physical system, see Sect. 9) is given by its total memory content, including RAM, CPU, video card, and other components and accessories. In a digital computer, it is equivalent to one huge binary sequence. However, it can never be accessed simultaneously and in its entirety. Output devices allow to project out certain (very limited) facets of the total state, read them out as a symbol string, as visualize them as pixels of a printed image or on a screen, make them audible as speech or musical tunes ..., depending on the output device they are sent to.

Quantum mechanics is even more radical. A quantum state is a completely feature-less item, an element of an abstract space. The state space of quantum mechanics has nothing to do with any configuration space or phase space. Neither do its elements correspond to any position in physical space, nor has its dimension anything to do with spatial dimension. Again, comparing it to a memory module characterized by its capacity is more appropriate.

According to all evidence available to date, it has a well-defined mathematical structure: that of a vector space endowed with a norm and a metric, both based on a scalar product. In a more precise mathematical term, it is a Hilbert space. Specifically, for normalizable states, it takes the form of a high-dimensional hypersphere. This geometry determines much of the characteristic principles of quantum mechanics; it is completely different, on the one hand, from that of the continuous phase space of Hamiltonian mechanics, and on the other from a high-dimensional discrete lattice one might naïvely expect as appropriate for quantized variables, and which in fact underlies classical digital computing, see Subsect. 9.3.1. In this structure, a funda-mental feature becomes manifest that is decisive also for an information-theoretic interpretation of quantum mechanics: The space is closed under linear superposition. Any linear combination of quantum states, belonging to the same Hilbert space, is itself a legitimate element of this same Hilbert space. This feature is completely alien to classical physics; it cannot even be compared with a weighted average defining an ensemble in statistical physics. There is though a second, apparently unrelated field that shares this same mathematical structure with quantum mechanics: signal anal-ysis. Indeed, interpreting quantum states as signals, not as facts, and their evolution as signal transformation, is another helpful analogy. However, in quantum mechanics the signal does not represent a deeper reality underlying it. To the contrary, reality as we perceive it derives from this signal.

Owing to its Hilbert-space structure, the quantum mechanical state space enjoys perfect homogeneity and isotropy. In classical mechanics, symplectic symmetry, the central achievement of Hamilton's formulation, already surmounts the asymmetry between position and momentum, but still respects the distinguishability of different degrees of freedom. By contrast, the only invariant property characterizing Hilbert spaces is their dimension, analogous to the capacity of a memory module. Also in this respect, quantum mechanics is closely akin to signal processing. In Fourier transformation, the total number of independent real coefficients is an invariant parameter that quantifies the total information content of a signal in the time as well as in the frequency or any combined representation. Similarly, Hilbert space dimension is intimately related to information and provides the entry point for an information-theoretic interpretation.

The dimension of a Hilbert space can be finite, $D_{\mathcal{H}} \in \mathbb{N}$, or enumerably infinite. Only if the Hilbert space is continuous, it can be compared to classical config-uration or phase space (cf. the categories of Fourier transformation discussed in Subsect. 2.3.1, Table 2.1). A finite dimension defines directly the range of entropy values that can be attributed to the system: from 0 for a completely determined state to $\log(D_{\mathcal{H}})$ for a completely unknown state. The fact that this range can be delimited means that the memory content of a quantum system is not indefinite. Quantum mechanics thus inserts a bottom to the hierarchy of scales alluded to in Sect. 1.5. The total information, introduced in Subsect. 1.4 in a somewhat imprecise manner as the absolute reference point for both potential and actual information, $I_{tot} = I_{pot} + I_{act} = $ const, now gains a clear meaning and a specific value. Another basic aspect that remained open in the general framework set in Sect. 1 is the unit of information: In quantum mechanics, the lower bound defining this unit is a two-state system, such as an electron spin, with a two-dimensional Hilbert space, $D_{\mathcal{H}} = 2$. In quantum mechanics, the bit is therefore the natural choice for a unit of information. The term "qubit", though, has more connotations than just a unit, see Sect. 8.2.

The dimension of Hilbert space attains a fundamental significance for the quantity and representation of information in a quantum system as the *only* invariant parameter determining these aspects. Otherwise, quantum mechanics is extremely flexible: Comparing a quantum system with a pixelated computer screen or printing device is misleading. Quantization is not pixelated, it is not tantamount to discretizing a specific quantity. Where a quantum mechanical operator represents a classical observable, for example, its eigenstates can delimit any kind of partition of a corresponding phase space, not just a square grid but countless other structures as well, see Subsect. 8.1.3.

Quantum mechanics is not only open as to the way a finite information content translates into a discretization of specific quantities. It even ignores subdivisions considered as absolutely invariant in classical physics, such as in particular the distinction of different subsystems or degrees of freedom. **In quantum mechanics, information constraints are more fundamental than and supersede any partition into subsystems**. If a quantum system comprises only one bit but two or more degrees of freedom, this single bit cannot be attributed to one of these freedoms but is shared by all of them. This principle leads immediately to *entanglement*, see Subsect. 8.1.4, maybe the most emblematic and at the same time weirdest feature of *quantum mechanics*. Applied to spatially separated subsystems, it implies *non-locality*, a particularly striking manifestation of entanglement.

There is one point where classical mechanics, in the mature form it reached in Hamilton's formulation, already anticipates a crucial feature of quantum mechanics: In closed systems, the time evolution strictly conserves entropy. In this respect, canonical transformations induced by Hamilton's equations of motion, see Subsect. 5.3, coincide with unitary transformations solving Schrödinger's equation, see Subsect. 8.3.2. As in classical statistical mechanics, this invariance characterizing the microscopic time evolution has fundamental implications for quantum statistics.

Postulating a new ontology with quantum states organized in Hilbert spaces, quantum mechanics is not complete without defining an interface with classical mechanics: As macroscopic observers, living in a world of classical facts, how can we communicate with quantum systems, get access to their states and control them? How does quantum information translate to classical information and vice versa? If classical physics is indeed based on a deficient picture of reality, how come that it is so incredibly successful, describes nature to such high accuracy that highly sophisticated laboratory equipment is needed to observe any discrepancy? Quantum mechanics answers the first two questions systematically by integrating the measurement process, the default interface between the quantum and the classical world, as a key element into its theoretical construction, see Sect. 8.4. In order to resolve the third question, semiclassical theories and approximations treating classical mechanics as an asymptotic limit of quantum mechanics have to be complemented by a theory of decoherence, explaining classical behaviour as an emergent effect in macroscopic systems (Subsect. 8.3.3).

8.1.1 Postulates of Quantum Mechanics Related to Information

The following list summarizes these information theoretic principles underlying quantum mechanics [CDL77, NC00, BA&17]. The list is intended as an overview, a lose collection of basic tenets of quantum mechanics. It does not pretend the rigour of an "information theoretic axiomatization", which could emerge as a long-term objective of research on the foundations of quantum mechanics but is not yet available, let alone completeness or mutual independence of the postulates listed.

1. The state of a quantum system is uniquely and completely defined by its state vector $|\alpha\rangle$, an element of a linear vector space $\mathcal{H}$ with a norm and metric induced by a scalar product $\langle . | . \rangle$, that is, of a Hilbert space.
2. Two states are maximally different, distinguishable in the sense of entropy measures, if their state vectors $|\alpha\rangle$, $|\beta\rangle$ are orthogonal, $\langle \alpha | \beta \rangle = 0$.
3. If two systems are located at different positions but coincide in their internal states, they are treated as strictly identical. No entropy can be attributed to their distinction. Their internal state vectors $|\alpha\rangle$, $|\beta\rangle$ can differ at most in a phase. In particular, in systems with half-odd integer spin (fermions, e.g., electrons), the state vectors change sign under an interchange of coordinates, in systems with integer spin (bosons, e.g., hydrogen atoms) they remain the same.
4. The size of a quantum system, measured as its capacity to store and represent information, is determined by its dimension $D_\mathcal{H}$, the total number of linearly independent directions in its Hilbert space, through its logarithm $\log(D_\mathcal{H})$.
5. This imposes an upper bound on the information to be attributed to a state of this system. It can be accessed partially and in different representations, but no more can be known, no more information can be extracted than that.

6. The Hilbert space dimension cannot be less than 2, corresponding to a two-state system. One bit therefore is a lower bound and a natural unit of information.

7. A state vector does not need to be restricted to a Hilbert space direction representing an individual particle or degree of freedom. If it overlaps with two or more such subspaces, the particles or freedoms it encompasses are *entangled*, they share the information associated to the space vector.

8. This applies even if entangled subsystems or freedoms are spatially separated by any distance. In that case, the state is *nonlocal*.

9. If the Hilbert space of a system comprising f degrees of freedom can be associated to a space $\mathbb{R}^{2f}$, isomorphous to a classical phase space of dimension $2f$, an area or hypervolume Ω within that space cannot represent a Hilbert space of more than $\Omega/(2\pi\hbar)^f$ dimensions, i.e., each Hilbert-space dimension occupies an area or volume $(2\pi\hbar)^f$ (a Planck cell) in it. This induces a subdivision of phase space that depends on the system and generally is not a regular (e.g., square) tiling.

10. If quantum states are considered as signals, their processing amounts to operations on Hilbert space. Observable quantities are represented by *Hermitian operators*. It matters in which order operations are applied: For two operators $\hat{A}$, $\hat{B}$, the *commutator* $[\hat{A}, \hat{B}] = \hat{A}\hat{B} - \hat{B}\hat{A}$ in general does not vanish.

11. The time evolution of a closed quantum system, as a continuous sequence of operations in infinitesimal time, is determined by the *Schrödinger equation*. It results in *unitary transformations* of its Hilbert space. They leave its dimension invariant and in particular conserve the information content of a quantum state as it evolves in time. See Subsect. 8.3.

In the following subsections, some crucial consequences of these principles will be worked out in more detail. Three further postulates, referring to measurements as the interface between quantum mechanics and directly observable classical states, will be quoted in Sect. 8.4.

8.1.2 Hilbert Space Vectors as Basic Information Carriers

Quantum mechanics introduces a basic level of reality, independent of our experience. Quantum states are only indirectly related to the facts forming the object of classical physics and obey a mathematical structure distinct from that underlying logics and in particular classical information theory. The state of a quantum system, symbolized in the notation introduced by P.A.M. Dirac by "$|a\rangle$", is an abstract entity which comprises all that can be known about the state, but generally becomes observable in the sense of classical physics only partially. This mathematical structure coincides with that of a closely related field [BC&00, OZK01]: In signal analysis, it is an obvious requirement that the space $\mathcal{H}$ of a class of signals be closed under linear superposition. If this space comprises, say, signals given by ordered sets of $D_{\mathcal{H}}$ complex numbers, represented as vectors $|a\rangle \in \mathcal{H}$, this means that

$$\forall |a\rangle, |b\rangle \in \mathcal{H}, \alpha, \beta \in \mathbb{C} : \alpha |a\rangle + \beta |b\rangle \in \mathcal{H}. \tag{8.1.1}$$

As harmless and self-evident as this superposition principle may look, it is blatantly at variance with our classical conception of states of a system. To cite the drastic example conceived by Schrödinger to illustrate this discrepancy [Sch35]: A superposition $\frac{1}{\sqrt{2}}(|\text{alife}\rangle + |\text{dead}\rangle)$ of incompatible states of an organism appears unthinkable. In particular, it must not be understood as a probabilistic statement. The complex prefactors α and β do not express likelihoods but vector components, the quantum state they specify is uniquely determined and free of any uncertainty.

By the same token, quantum states represent qualities, the absolute magnitude of state vectors does not bear any significance of its own, only their direction matters. This suggests normalizing them, for example

$$\|a\|^2 := \langle a | a \rangle = 1. \tag{8.1.2}$$

As a consequence, Hilbert space attains the topology of a hypersphere embedded in $D_{\mathcal{H}}$ dimensions. The Definition (8.1.2) involves a scalar product $\langle a | b \rangle \in \mathbb{C}$, which is defined canonically on a linear vector space $\mathcal{H}$ through its dual space $\mathcal{H}^\dagger$ [MW70], represented in Dirac's notation by "bra" vectors $\langle a | = |a\rangle^\dagger \in \mathcal{H}^\dagger$, the Hermitian conjugates of "kets" $|a\rangle$.

In quantum mechanics, the scalar product has a special physical significance of its own: Unlike discrete lattices of $D_{\mathcal{H}}$ points, Hilbert spaces allow for gradual transitions between distinct states, for example by varying the coefficients α and β in Eq. (8.1.1). The closeness of two state vectors $|a\rangle$, $|b\rangle$ is quantified geometrically as their overlap, the scalar product $\langle a | b \rangle$. They are distinct in the sense of information theory if they are orthogonal, that is, if and only if

$$\langle a | b \rangle = 0. \tag{8.1.3}$$

Otherwise, the angle enclosed between $|a\rangle$ and $|b\rangle$,

$$\theta = \arccos(|\langle a | b \rangle| / \|a\| \|b\|), \tag{8.1.4}$$

can vary between 0 (parallel), $\pi/2$ (orthogonal) and π (antiparallel state vectors). Equation (8.1.3) becomes an absolute criterion for distinguishability. This prevents arguments such as Gibb's paradox from the outset to apply to quantum mechanics. More specifically, in many-body systems considered in quantum statistics and other fields, the common state vectors of pairs of particles at different positions but with identical internal states have to be invariant under exchange, up to a relative phase. It depends on internal quantum numbers such as in particular the spin and gives rise to the distinction of fermions and bosons, see postulate 7 above. This rigorous principle has profound consequences, reflected in qualitative differences between quantum and classical statistical mechanics [Rei65, CDL77].

The condition (8.1.3) leads directly to the question how many distinct states a Hilbert space $\mathcal{H}$ admits, which now becomes a natural number attributed as an invariant characteristic parameter to the Hilbert space. It is intimately related to the concept of a basis. A set of states $B = \{|a_1\rangle, |a_2\rangle, \ldots, |a_N\rangle\} \subseteq \mathcal{H}$ qualifies as an orthonormal basis of $\mathcal{H}$ if all its elements are orthonormalized,

$$\forall |a_m\rangle, |a_n\rangle \in B : \langle a_m | a_n \rangle = \delta_{n-m}, \tag{8.1.5}$$

so that all state vectors in $\mathcal{H}$ have a unique decomposition in B, and B is complete, that is, all states in $\mathcal{H}$ can be completely represented in B,

$$\forall |a\rangle \in \mathcal{H} : \exists \alpha_n \in \mathbb{C} : |a\rangle = \sum_{n=1}^{N} \alpha_n |a_n\rangle, \tag{8.1.6}$$

or equivalently, if (denoting the identity operator on $\mathcal{H}$ as $\hat{\mathbf{I}}_{\mathcal{H}}$)

$$\sum_{n=1}^{N} |a_n\rangle \langle a_n| = \hat{\mathbf{I}}_{\mathcal{H}}. \tag{8.1.7}$$

Uniqueness and completeness together require that the number of elements of B must not be smaller nor larger than $D_{\mathcal{H}}$, so that $N = D_{\mathcal{H}}$. Possible values of this integer, from 2 to infinity, characterize important categories of quantum systems, some of which deserve a particular attention:

$D_{\mathcal{H}} = 2$: two-state systems

The lowest dimension that allows for an evolution in time is $D_{\mathcal{H}} = 2$, corresponding to an information capacity $\mathrm{lb}(D_{\mathcal{H}}) = 1$. This defines a two-state system, epitomized for example in two-level atoms (ground and excited state $|g\rangle, |e\rangle$), in electron spins (spin down $|\downarrow\rangle$ and up $|\uparrow\rangle$), and in linearly (horizontal $|\leftrightarrow\rangle$ and vertical $|\updownarrow\rangle$) or circularly (left $|+\rangle$ and right $|-\rangle$) polarized photons [BEZ00]. Anticipating the context of quantum computation, Sect. 10, an appropriate general notation for the basis vectors is $|0\rangle$ and $|1\rangle$. The term "qubit" for this case is used ambiguously, sometimes referring to two-state system with all the quantum physics it implies, sometimes to a mere unit of information for quantum physics.

General state vectors in this space, denoted $\mathcal{H}^{(2)}$, are given by

$$|a\rangle = \alpha_0 |0\rangle + \alpha_1 |1\rangle, \ |\alpha_0|^2 + |\alpha_1|^2 = 1. \tag{8.1.8}$$

Unitary operators on this space can be decomposed in an operator basis comprising the unit operator $\hat{\sigma}_0$ and the Pauli operators $\hat{\sigma}_1 \equiv \hat{\sigma}_x$, $\hat{\sigma}_2 \equiv \hat{\sigma}_y$, and $\hat{\sigma}_3 \equiv \hat{\sigma}_z$ [CDL77],

$$\hat{U} = \sum_{n=0}^{3} a_n \hat{\sigma}_n, \quad \hat{\sigma}_0 = \begin{pmatrix} 1 & 0 \\ 0 & 1 \end{pmatrix}, \hat{\sigma}_1 = \begin{pmatrix} 0 & 1 \\ 1 & 0 \end{pmatrix}, \hat{\sigma}_2 = \begin{pmatrix} 0 & -i \\ i & 0 \end{pmatrix}, \hat{\sigma}_3 = \begin{pmatrix} 1 & 0 \\ 0 & -1 \end{pmatrix}.$$

$$(8.1.9)$$

They correspond to rotations of the unit sphere in $\mathbb{C}^2$ and play a fundamental role in quantum computation, see Subsects. 10.2.1 and 10.2.2.

$D_{\mathcal{H}} \in \mathbb{N}$: discrete systems

Hilbert spaces with a finite dimension $D_{\mathcal{H}} \in \mathbb{N}$ are also called discrete Hilbert spaces. They are ideal laboratories for the study of genuine quantum features in general and for quantum information in particular. Of special interest, notably in the context of quantum computation, are systems composed of n qubits, with Hilbert spaces $\mathcal{H}^{(2n)}$, see Subsect. 10.2.1.

$D_{\mathcal{H}} \to \infty$: continuous systems

Closest to classical physics, however, are infinite-dimensional Hilbert spaces, as they allow for operators with a continuous spectrum. By straightforward transformations, they can be represented in spaces that are in many respects analogous to classical phase space, for example as Wigner functions, cf. Eq. (8.1.15), and allow for a direct comparison with classical mechanics. At the same time, they also admit an unbounded information content and should be considered as a mathematical idealization of Hilbert spaces with a very high but finite dimension.

Since quantum states can be represented as (sets of) complex numbers, they bear, in particular, a phase. While the absolute value can be interpreted in terms of probabilities, the phase is a quantum feature without classical analogue. Absolute values of the phase have no physical significance at all. Relative phases between two states $|a\rangle$, $|b\rangle$ can be observed through inner products $\langle a|b\rangle \in \mathbb{C}$. They become manifest in interference phenomena and are a hallmark of quantum mechanics. However, representations of quantum mechanics, including its wave nature, in real numbers do exist, avoiding phases from the outset. In Wigner functions, for example, it is oscillatory patterns featuring also negative values that allow reproducing constructive as well as destructive interference (Figs. 8.1 and 8.2).

8.1.3 Heisenberg's Uncertainty Principle and Information in Phase Space

One of the most emblematic features of quantum mechanics is the non-zero value of Planck's constant, comparable in its fundamental significance to the finiteness of the velocity of light. In the above list of Hilbert space categories, the situation is quite clear for discrete Hilbert spaces. Their finite dimension translates directly into correspondingly finite information content. Going to infinite-dimensional Hilbert spaces however, the notion of Planck cells epitomizing the finite quantum of action becomes elusive.

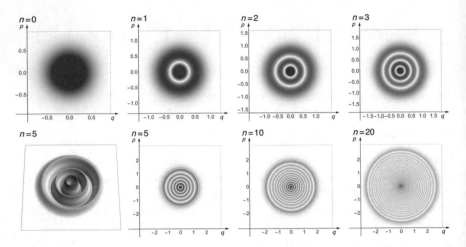

Fig. 8.1 Energy eigenfunctions of the harmonic oscillator, Eq. (8.1.21), represented as Wigner quasi-probability functions in phase space, Eq. (8.1.17), for $n = 0, 1, 2, 3$ (upper row) and $n = 5, 10, 20$ (lower row). Note the change in scale between the panels of the upper row. Colour code ranges from red (negative) through white (zero) through blue (positive). Parameter values are $m = 1$, $\omega = 1$, and $\hbar = 0.2$. In the lower row, energy contours at levels $H(p, q) = E_n = \hbar\omega(n + 1/2)$ are marked as black dotted lines inside Planck cells, defined in Eq. (8.1.15), highlighted as green rings. Positive maxima and negative minima along concentric rings inside the classical energy contours are mere quantum coherence effects.

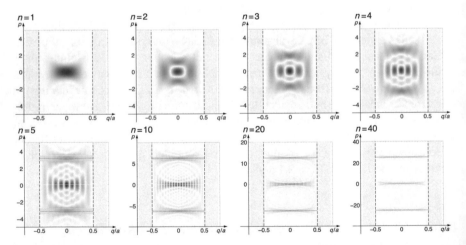

Fig. 8.2 Energy eigenfunctions of the infinitely deep, square potential well, Eq. (8.1.26) (forbidden areas $|q| > a/2$ indicated by grey hatching), represented as Wigner quasi-probability functions in phase space, Eq. (8.1.29), for $n = 1, 2, 3, 4$ (upper row) and $n = 5, 10, 20, 40$ (lower row). Colour code as in Fig. 8.1. Parameter values are $a = 1$, $m = 1$, and $\hbar = 0.2$. In the lower row, momentum levels $p_{n\pm} = \pm\sqrt{2mE_n} = \pm n\pi\hbar/a$ are marked as black dotted lines inside Planck cells, defined in Eq. (8.1.28), highlighted in green. Note how two well-defined maxima at $p_{n\pm}$ emerge from the smooth contours of the Wigner function and get sharper with increasing quantum number n. The oscillatory pattern appearing at $p \approx 0$ in all states, except for the ground state $n = 1$, is the signature of the coherent superposition of the two maxima at $p_{n\pm}$ and a pure quantum coherence effect.

A suitable starting point to understand its origin and its consequences regarding information are the purely mathematical uncertainty that affect combined time–frequency representations of signals, discussed in Subsect. 2.3.3. A signal given by a continuous function in time is equivalent to a continuous function in the frequency domain. Representing the same one-dimensional signal simultaneously in two dimensions, as a function of time *and* frequency, results in redundancy. It implies a restriction of the total information contained in such a representation, which in turn becomes manifest as a finite area in such a space, containing finite information. As a mathematical constant, the area $\Delta t \, \Delta \omega$ per dimension of the function space takes on an adimensional value, $\Delta t \, \Delta \omega = 1$ or $\Delta t \, \Delta \omega = 2\pi$, depending on the way Δt and $\Delta \omega$ are measured.

Time and frequency representations correspond to specific tilings of the time–frequency plane, namely into "vertical" and "horizontal" stripes, resp. (Fig. 2.8). Interpreting the uncertainty relation as a lower bound of the area of time–frequency cells suggests generalizing it to the theorem that an area A of time–frequency space cannot accommodate more than $A/2\pi$ such cells, whatever be their shape.

Time–frequency uncertainty as a mathematical fact carries over to quantum mechanics, applying to pairs of quantities analogous to time and frequency, that is, related by Fourier transform. Physics comes into play by mapping time–frequency space to phase space, where the area is not adimensional but is measured in units of action. This bold step was suggested by the crucial findings that led to quantum mechanics: Planck's analysis of blackbody radiation, Einstein's explanation of the photoelectric effect, the Bohr model of the Hydrogen atom, they all indicate that the quantization manifest in these cases applies to the same quantity, action, and requires a cell size of no less than $h = 2\pi \hbar$.

Quantum mechanics relates physical quantities to their frequency domains (e.g., position to wavenumber) through the concept of matter waves. This notion hovers in the background of all the examples mentioned but was proposed explicitly by pioneers such as Bohr and de Broglie, who suggested to relate the momentum $\mathbf{p} = m\mathbf{v}$ of a microscopic object to the wavenumber $\mathbf{k}$ of the matter waves associated to it through

$$\mathbf{p} = \hbar \mathbf{k}. \tag{8.1.10}$$

Similarly, Einstein's quantization of light requires to relate the energy E to the frequency $\omega = 2\pi \nu$ of a photon by

$$E = \hbar \omega. \tag{8.1.11}$$

Combining these identities with the time–frequency (position-wavenumber, etc.) uncertainty relation (2.3.42), we arrive at Heisenberg's time-energy, position-momentum, angle-action, ... uncertainty relations

$$\Delta t \Delta E \geq 2\pi \hbar,$$
$$\Delta q \Delta p \geq 2\pi \hbar, \qquad (8.1.12)$$
$$\Delta \phi \Delta I \geq 2\pi \hbar,$$

and so on. They involve pairs of classical quantities (u,v), related by *canonical conjugation* [JS98, GPS01]. In quantum mechanics, pairs of the corresponding operators, for example $\hat{p}$ and $\hat{q}$, are related by canonical commutation relations [CDL77] of the type

$$[\hat{q}, \hat{p}] = i\hbar, \qquad (8.1.13)$$

(note however that no operator $\hat{t}$ can be associated to time [SG64]).

As discussed in Subsect. 2.3.3, mathematical inequalities to quantities related by Fourier transformation can be expressed as lower bounds for the total information in the two domains, cf. Eq. (2.2.53). This applies also to the quantum–mechanical uncertainty relations (8.1.12). Based on a quantum version of the Shannon entropy (see Eq. (8.2.24) below), an inequality for the information in two representations related by conjugation, known as Beckner-Bialynicki-Birula-Mycielsky inequality, can be derived [BM75, Bia06, Bia07]. For position and momentum, it reads (defining entropy with the natural logarithm)

$$I_q + I_p \geq 1 - \ln\left(\frac{\delta q \delta p}{\pi \hbar}\right). \qquad (8.1.14)$$

Through the parameters δq, δp, this lower bound rises with improving resolutions of position and momentum measurements, resp. If instead of the Shannon entropy, a definition directly inspired by quantum mechanics is used, such as the von Neumann entropy S_{vN}, a rigorous lower bound $S_{vN} \geq 0$ can be stated, the identity being valid for pure states, cf. Eq. (8.2.22).

As in signal processing, the evident versatility of the uncertainty relations (8.1.12) generalizes further to the condition that within an area S, now of phase space, at most $S/2\pi \hbar$ quantum states can be accommodated, independently of the cell shape they occupy. Allowing for a number f of degrees of freedom, with a cell size $(2\pi \hbar)^f$, this becomes a condition for the dimension of the Hilbert space associated to a phase-space hypervolume S,

$$D_{\mathcal{H}} \leq \frac{S}{(2\pi \hbar)^f}. \qquad (8.1.15)$$

which restricts the memory capacity of this phase-space area, measured in bits, to

$$I_S = \mathrm{lb}(D_{\mathcal{H}}) \leq \mathrm{lb}\big(S/(2\pi \hbar)^f\big). \qquad (8.1.16)$$

Nearly all physical systems of practical interest are bounded in their phase-space extension, in position by binding potentials or ultimately by the present size of the expanding universe, in momentum by the total energy of the system. Equations (8.1.15, 8.1.16) thus imply that their Hilbert space dimension and with it their information content are finite, albeit possibly reaching astronomical values (some 10^{90} bits total information capacity of the universe [Llo02]).

Quantum operators with a continuous eigenvalue spectrum can be associated to classical observables and by their contour lines and gradients to structures in phase space. Yet it is not obvious how general phase-space cells should be defined in quantum mechanics. This question is evidently crucial for quantum statistical mechanics. Indeed, representing quantum mechanics in a setting as close as possible to classical phase space has been an objective in quantum statistics from early on. Numerous solutions have been proposed, but they all suffer in one way or other from the redundancy of a simultaneous representation in two non-commuting variables. The task of constructing a quantum analogue of a classical probability density function on phase space cannot be completely accomplished, not all the conditions a classical probability density fulfills can be simultaneously satisfied by phase-space representations of quantum mechanics.

One of the most influential proposals has been the Wigner function [Ber77a, HO&84]. It has the particular advantage of providing a complete representation of quantum mechanics in phase space, i.e., the transformation from pure, even from mixed quantum states, see Subsect. 8.2.1, to Wigner functions is invertible. Suffice it, for the purpose of visualizing phase-space cells, to outline its construction. The Wigner function associates a Hilbert space vector $|\psi\rangle$, given as a wave function $\psi(\mathbf{q}) = \langle \mathbf{q}| \psi \rangle, \mathbf{q} \in \mathbb{R}^f$, in the position representation, to a continuous distribution on $\mathbb{R}^{2f}$,

$$P_W(\mathbf{p}, \mathbf{q}) = \frac{1}{h^f} \int d^f q' \psi^* \left(\mathbf{q} + \frac{\mathbf{q}'}{2} \right) \psi \left(\mathbf{q} - \frac{\mathbf{q}'}{2} \right) \exp\left(i \frac{\mathbf{p} \cdot \mathbf{q}'}{\hbar} \right). \qquad (8.1.17)$$

It is real-valued and normalized on $\mathbb{R}^{2f}$, $\int d^f p \int d^f q P_W(\mathbf{p}, \mathbf{q}) = 1$, but not positive definite, hence cannot be interpreted as a classical probability density. Where negative values occur, they encode quantum coherence; they are even unavoidable to allow for destructive interference between quantum states represented as real-valued functions. As is evident from Eq. (8.1.17), the position and momentum dependences of the Wigner function are derived from the function $\psi^*(\mathbf{q}_2)\psi(\mathbf{q}_1)$ as its value along the diagonal $\mathbf{q}_+ = (\mathbf{q}_1 + \mathbf{q}_2)/2$ and its Fourier transform along the second diagonal $\mathbf{q}_- = \mathbf{q}_2 - \mathbf{q}_1$. In the following paragraphs, the structure of phase-space cells is illustrated by means of a few elementary examples.

- Returning to **discrete systems** with $D_{\mathcal{H}} \in \mathbb{N}$, they are represented in phase space by replacing the time–frequency coordinates of Subsect. 2.3.1 with position and momentum, appropriately scaled. If periodicity of the wave function is required in both domains,

$$\psi(q + Q) = \psi(q), \tilde{\psi}(p + P) = \tilde{\psi}(p), \qquad (8.1.18)$$

where $\tilde{\psi}(p) = \langle p | \psi \rangle = (2\pi\hbar)^{-1/2} \int_{-\infty}^{\infty} dq\, \psi(q)\, e^{-ipq/\hbar}$ is the momentum-space wave function, Q and P are analogous to the time and frequency periods T and $\omega_0 = 2\pi/T$, resp., of discrete Fourier transform, cf. Eqs. (2.3.24, 2.3.25). Taking the de Broglie relation (8.1.10) between wavenumber and momentum into account, their product must satisfy the condition $PQ = 2\pi\hbar$, that is, exactly N Planck cells fit in the total space PQ.

As in discrete Fourier transform, periodicity in p and q implies crosswise discretization in q and p,

$$p_\lambda = \lambda\Delta p, \ \Delta p = \frac{P}{N} = \frac{2\pi\hbar}{Q}, \ q_n = n\Delta q, \ \Delta q = \frac{Q}{N} = \frac{2\pi\hbar}{P}, \lambda, n = 1, \ldots, N,$$
$$(8.1.19)$$

and relations

$$\langle p_\mu | p_\lambda \rangle = \delta_{\mu - \lambda (\mathrm{mod}\, N)}, \ \langle q_n | q_m \rangle = \delta_{n - m (\mathrm{mod}\, N)}, \ \langle p_\lambda | q_n \rangle = N^{-1/2} e^{2\pi i \lambda n / N}$$
$$(8.1.20)$$

between the corresponding eigenstates, $\hat{p} | p_\lambda \rangle = p_\lambda | p_\lambda \rangle, \hat{q} | q_n \rangle = q_n | q_n \rangle$. The two representations thus amount to sets of N horizontal (vertical) rectangles of area $2\pi\hbar$ each, stacked one upon the other (side by side) in phase space, cf. Fig. 2.6b.

- The quantum mechanical **harmonic oscillator** is defined by the Hamiltonian operator (for a comprehensive exposé of the quantum mechanical harmonic oscillator, see, e.g., [CDL77])

$$\hat{H}(\hat{p}, \hat{q}) = \frac{\hat{p}^2}{2m} + \frac{m\omega^2}{2} \hat{q}^2. \qquad (8.1.21)$$

- The eigenfunctions $|n\rangle$, $\hat{H} |n\rangle = E_n |n\rangle$, pertaining to discrete energies $E_n = \left(n + \frac{1}{2}\right)\hbar\omega$, are in position representation [Sch01]

$$\psi_n(q) = \langle q | n \rangle = (m\omega/\hbar\pi)^{1/4} \left(2^n n!\right)^{-1/2} H_n(\xi) e^{-\xi^2/2}, \qquad (8.1.22)$$

where $H_n(\xi)$ is the nth Hermite polynomial of the argument $\xi = \sqrt{m\omega/\hbar}\, q$. Substituted in the Definition (8.1.17), these wave functions transform to Wigner functions [Sch01]

$$P_{\mathrm{Wn}}(p, q) = \frac{(-1)^n}{\pi\hbar} \exp(-2H(p, q)/\hbar\omega) L_n(4H(p, q)/\hbar\omega), \qquad (8.1.23)$$

denoting the real-valued function that corresponds to the Hamiltonian operator (8.1.21) as $H(p,q)$. $L_n(x)$ is the nth Laguerre polynomial. For examples of Wigner functions of excited eigenstates of the harmonic oscillator, see Fig. 8.1.

It is remarkable that the Wigner functions representing number states of the harmonic oscillator in phase space depend on position and momentum only through the classical energy $H(p,q)$, so that their contour lines coincide with the elliptic contours of the classical Hamiltonian. All these Wigner eigenfunctions extend over the entire phase space, but for high quantum numbers $n \gg 1$, they concentrate more and more sharply on the corresponding classical energy contours $H(p,q) = E_n$. Nevertheless it is possible to associate phase-space cells to them. They form nested concentric elliptic rings, delimited by neighbouring energy contours $H(p, q) = E_n \pm \hbar\omega/2 =: E_n^\pm$. The half-axes of each elliptic contour, given by $E_n^\pm = H(\Delta p_n^\pm, 0)$ and $E_n^\pm = H(0, \Delta q_n^\pm)$, resp., are $(\Delta p_n^\pm)^2 = 2m E_n^\pm$, $(\Delta q_n^\pm)^2 = 2E_n^\pm/m\omega^2$. They enclose a total phase space area $S_n^\pm = \pi \Delta p_n^\pm \Delta q_n^\pm = 2\pi E_n^\pm/\omega = 2\pi\hbar(n + \frac{1}{2} \pm \frac{1}{2})$, so that the size of a single cell is indeed

$$\Delta S_n = S_n^+ - S_n^- = 2\pi\hbar. \tag{8.1.24}$$

- A **particle in an infinitely deep, square potential well** of width a represents one of the most elementary textbook examples of quantization. It can be described by the Hamiltonian

$$\hat{H}(\hat{p}, \hat{q}) = \frac{\hat{p}^2}{2m} + V(\hat{q}), V(q) = \begin{cases} 0 & |q| < a/2, \\ \infty & |q| \geq a/2. \end{cases} \tag{8.1.25}$$

The hard-wall potential restricts the spatial range to the compact interval $-a/2 \leq q \leq a/2$ with Dirichlet boundary conditions $\psi(\pm a/2) = 0$. As a consequence, the momentum is quantized, $p_\lambda = \hbar k_\lambda = \lambda\pi\hbar/a$, $\lambda \in \mathbb{Z}\backslash\{0\}$, and the energy eigenfunctions, with energies $E_n = n^2\pi^2\hbar^2/2ma^2$, are

$$\psi_n(q) = \sqrt{\frac{2}{a}}\Theta(D(q))\begin{cases} \sin(k_n q) & n \text{ even}, \\ \cos(k_n q) & n \text{ odd}, \end{cases} \quad n \in \mathbb{N}, \tag{8.1.26}$$

with $D(q) = a - 2|q|$. The Heaviside function $\Theta(a - 2|q|)$ constrains the support of $\psi_n(q)$ to the interior of the box. Transformed to the momentum representation,

$$\tilde{\psi}_n(p) = \sqrt{\frac{a}{2\pi\hbar}}\text{sinc}\left(\frac{a}{2\hbar}(p - \hbar k_n)\right), \tag{8.1.27}$$

these eigenfunctions are localized with a width $\Delta p = \pi\hbar/a$ around the quantized momenta $\pm\hbar k_n$. Considering Planck cells associated to the energy eigenfunctions as composed of two rectangles of side lengths a and Δp each, centred at $\pm\hbar k_n$, then results in an estimate of their total size of

$$\Delta S = 2a\Delta p = 2\pi\hbar. \tag{8.1.28}$$

The Wigner functions resulting from the transformation (8.1.17) [Whe00],

$$P_{Wn}(\hbar k, q) = \frac{\Theta(D(q))}{2\pi\hbar a}\left[\frac{\sin(2k_n q)\sin(kD(q))}{k}\right.$$
$$\left.+\frac{k\sin(k_n D(q))\cos(kD(q)) - k_n\cos(k_n D(q))\sin(kD(q))}{k_n^2 - k^2}\right], n \text{ even,}$$

$$P_{Wn}(\hbar k, q) = \frac{\Theta(D(q))}{2\pi\hbar a}\left[\frac{\cos(2k_n q)\sin(kD(q))}{k}\right.$$
$$\left.+\frac{k_n\sin(k_n D(q))\cos(kD(q)) - k\cos(k_n D(q))\sin(kD(q))}{k_n^2 - k^2}\right], n \text{ odd,}$$
$$\tag{8.1.29}$$

are restricted in position to the same range as the underlying wave functions $\psi_n(q)$. Figure 8.2 shows some examples of these Wigner eigenfunctions. Features standing out increasingly sharply with higher quantum number n are the classical maxima at $p = \pm\hbar k_n$, described by the second terms in square brackets in Eq. (8.1.29), and the interference patterns around $p = 0$, given by the first terms.

– **Lattices of coherent states** provide the quantum mechanical structure closest to an intuitive notion of a pixelated classical phase space. Delta functions $\delta(\mathbf{p} - \mathbf{p}_0, \mathbf{q} - \mathbf{q}_0)$ in phase space epitomize the idea of classical mass points with precisely defined positions $\mathbf{q}_0$ and momenta $\mathbf{p}_0$, and in classical statistical mechanics, they are perfectly admissible as probability density functions, despite the diverging actual information they represent. In quantum mechanics, they violate uncertainty relations and are not allowed. Wavefunctions that at least come as close as possible to delta functions are Gaussians, since the uncertainty product with their Fourier transforms, likewise Gaussians, cf. Table 2.2, reaches the lower bound imposed by the mathematical uncertainty relation (2.3.43).

In quantum mechanics, Gaussian wave functions arise in particular as ground states of harmonic oscillators. The ground state of the system (8.1.21) in position representation reads

$$\psi_0(q) = \langle q|0\rangle = \left(\frac{m\omega}{\pi\hbar}\right)^{1/4}\exp\left(-\frac{m\omega}{2\hbar}q^2\right), \tag{8.1.30}$$

with widths $\Delta q = \sqrt{\hbar/m\omega}$ and (from the Fourier transform of $\psi_0(q)$) $\Delta p = \sqrt{m\hbar\omega}$. The corresponding Wigner function, Eq. (8.1.23) for $n = 0$,

$$P_{W0}(p, q) = \frac{1}{\pi \hbar} \exp\left[-\frac{1}{\hbar \omega}\left(\frac{p^2}{m} + m\omega^2 q^2\right)\right], \qquad (8.1.31)$$

is a Gaussian in phase space, with half axes by a factor $\sqrt{2}$ larger than $\Delta p/2$ and $\Delta q/2$.

A dense set of delta spikes cannot be construed from Gaussians of non-zero width such as $P_{W0}(p, q)$. It is tempting, however, to build a lattice of equidistant Gaussians, say on a discrete square lattice of spacing Q in position and P in momentum, so that each cell assumes a size of $P Q = 2\pi \hbar$ (Fig. 8.3). Wave functions can be moved by any amount q in position and p in momentum by translation operators [CDL77]

$$\hat{u}(q) = \exp\left(\frac{-i}{\hbar} q \hat{p}\right), \hat{v}(p) = \exp\left(\frac{i}{\hbar} p \hat{q}\right). \qquad (8.1.32)$$

Applying them to Eq. (8.1.26), the centroid of the ground state can be shifted from the origin to the point (p_0, q_0),

$$\psi_\gamma(q) = \langle q | \gamma \rangle = \left(\frac{m\omega}{\pi \hbar}\right)^{1/4} \exp\left[-\frac{m\omega}{2\hbar}(q - q_0)^2 + \frac{i}{\hbar} p_0(q - q_0)\right], \quad (8.1.33)$$

corresponding to a Wigner function,

$$P_{W\gamma} P_{Wn}(p, q) = \frac{1}{\pi \hbar} \exp\left[-\frac{1}{\hbar \omega}\left(\frac{(p - p_0)^2}{m} + m\omega^2(q - q_0)^2\right)\right]. \qquad (8.1.34)$$

Fig. 8.3 A von Neumann lattice places coherent (minimum uncertainty) states (green contour lines), Eq. (8.1.33), on the vertices of a square lattice (dashed back lines) with lattice constants P, Q, that coincide with the widths Δp, Δq, resp., of the coherent states in momentum and position (Eq. (8.1.34)). Each cell of the lattice occupies a Planck cell (solid green lines) of size $\Delta p \Delta q = 2\pi \hbar$. Contour lines and shading are fictitious.

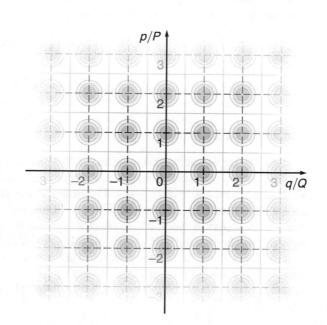

Shifted Gaussians of minimum uncertainty are called *coherent states* [Sch01], denoted $|\gamma\rangle$, where the parameter $\gamma = \sqrt{m\omega/2\hbar}q_0 + i\, p_0/\sqrt{2m\hbar\omega}$ determines the centroid in dimensionless complex coordinates.

The contour lines of $P_{W\gamma}(p,q)$ in phase space are ellipses. With elliptic tiles, a tessellation of phase space cannot be accomplished, but it suggests itself to replace them by rectangles with side lengths given by the axes of that elliptic contour that encloses a Planck cell, that is, choosing the lattice constants in p and q as, resp.,

$$P = \sqrt{\alpha\hbar m\omega}, \;\; Q = \sqrt{\alpha\hbar/m\omega}, \tag{8.1.35}$$

with $\alpha = 2\pi$, and placing their midpoints at

$$(p_\lambda, q_n) = (\lambda P, nQ) \text{ or } \gamma_{\lambda,n} = \sqrt{\frac{\alpha}{2}}(n + i\lambda), \lambda, n \in \mathbb{Z}, \tag{8.1.36}$$

to form what is called a *von Neumann lattice* [Neu18, BB&71, Per71, BGZ75, Zak89] (Fig. 8.3). If λ and n are restricted to finite ranges, say $\lambda = 1,..., \Lambda$ and $n = 1,..., N$, $\Lambda N = M$, this set represents a generalization of finite Hilbert spaces in the position and momentum representation, resp., see above.

Does this set of states provide a basis? Regarding completeness, it has been surmised by von Neumann [Neu18] and later proven rigorously [BB&71, Per71] that the value $\alpha = 2\pi$ marks the borderline: For $\alpha > 2\pi$, the set is not complete, for $\alpha \leq 2\pi$ it is. The problem with this strategy is, however, that coherent states are not orthogonal. Their squared overlap

$$|\langle\gamma|\gamma'\rangle|^2 = \exp\left(-|\gamma' - \gamma|^2\right) \tag{8.1.37}$$

vanishes only asymptotically for $|\gamma' - \gamma| \to \infty$. It remains non-zero also for elements of the discrete lattice defined in Eq. (8.1.36). Nevertheless, it can be said that again for the choice $\alpha = 2\pi$ in (8.1.35), the deviation from orthogonality is minimized without losing completeness.

Summarizing these elementary examples, it is already evident that "quantizing" a physical system by no means amounts to a simple pixelation of phase space. Quantum mechanics is not Lego in phase space. The patterns formed by energy eigenstates are most varied, already in cases where a closed outline can still be roughly attributed to them. We shall see that in more complex systems, in particular in the context of statistical mechanics, eigenstates are typically delocalized within larger manifolds, such as energy shells, so that the notion of Planck cells loses any direct geometrical meaning. Orthogonality is sustained instead by characteristic interference patterns on scales below the overall size of such manifolds. The question how to quantify information, based on phase-space structures associated to quantum states, will be discussed in Subsect. 8.2.1.

8.1.4 Entanglement and Non-Locality

The preceding arguments should have left the impression that quantum mechanics is absolutely strict and rigorous in postulating quantum bits as indivisible units, atoms of information space. At the same time, it comes with connotations of uncertainty, cloudiness, everything-related-to-everything. The key to resolve this apparent inconsistency is the concept of entanglement.

A rudimentary yet instructive example of entanglement is the superposition of classically incompatible states, such as the Schrödinger cat [Sch35],

$$|\psi\rangle = \frac{1}{\sqrt{2}}(|\text{alive}\rangle + |\text{dead}\rangle). \tag{8.1.38}$$

It is tempting but could not be further from the truth interpreting it as "hovering halfway between life and death". Instead, this is a state as uniquely defined as any other Hilbert space element. This becomes more evident if we introduce a "mixing angle" as a parameter,

$$|\psi(\phi)\rangle = \cos(\phi)|\text{alive}\rangle + \sin(\phi)|\text{dead}\rangle, \tag{8.1.39}$$

to rotate the state in its two-dimensional Hilbert space, thus crossing over smoothly between the classical alternatives. The state (8.1.38) is included as $|\psi(\pi/4)\rangle$. This shows that Schrödinger's cat is a state as unambiguous as the two classical alternatives. We could even define "white" and "black" Schrödinger cats as $|\psi(\pi/4)\rangle =:$ $|\text{white}\rangle$ and $|\psi(-\pi/4)\rangle =: |\text{black}\rangle$, say, so that the classical alternatives now appear as entangled states,

$$|\text{alive}\rangle = \frac{1}{\sqrt{2}}(|\text{white}\rangle + |\text{black}\rangle), \quad |\text{dead}\rangle = \frac{1}{\sqrt{2}}(|\text{white}\rangle - |\text{black}\rangle). \tag{8.1.40}$$

The direct physical consequences of superposition become more drastic if the role of intuitively incompatible alternatives, such as "dead" and "alive", is assumed by distinct degrees of freedom. It appears obvious to construct the Hilbert space of a multipartite system, with components 1, 2, ..., N, as a direct product of the Hilbert spaces of its subsystems,

$$\mathcal{H}^{(N)} = \mathcal{H}_1 \otimes \mathcal{H}_2 \otimes \cdots \otimes \mathcal{H}_N, \tag{8.1.41}$$

so that each degree of freedom n is associated to "its" Hilbert space $\mathcal{H}_n$. Situations can arise, however, where this straightforward construction has to be abandoned, in particular, where the total Hilbert space dimension is very low and its information capacity correspondingly scarce.

Consider a system comprising two spins with Hilbert spaces $\mathcal{H}_1, \mathcal{H}_2$, each spanned by a basis $\{|-\rangle_n, |+\rangle_n\}$, $n = 1, 2$ so that according to Eq. (8.1.41), the total Hilbert space has the basis

$$\{|-\rangle_1|-\rangle_2, \ |-\rangle_1|+\rangle_2, \ |+\rangle_1|-\rangle_2, \ |+\rangle_1|+\rangle_2\}. \tag{8.1.42}$$

Like this, the two bits of total information content refer separately to individual spin orientations. Alternatively, they might also refer to properties pertaining to both spins, in particular their *relative* orientation (aligned or opposed) [Zei99]. This becomes important for instance if a sum rule determines the value of the total angular momentum, as it may apply if the two spins emerged from a particle with zero total angular momentum, through a decay process that preserves this quantity (known as Einstein–Podolsky–Rosen pairs [EPR35], see Subsect. 8.4.3). In that case, the relevant Hilbert spaces are

$$\mathcal{H}_0 = \mathrm{span}(\{|-\rangle_1|+\rangle_2, \ |+\rangle_1|-\rangle_2\}), \mathcal{H}_{\pm 1} = \mathrm{span}(\{|-\rangle_1|-\rangle_2, \ |+\rangle_1|+\rangle_2\}), \tag{8.1.43}$$

corresponding to angular momentum quantum numbers $m = 0$ and $m = \pm 1$, resp. [CDL77]. State vectors in $\mathcal{H}_0$ and $\mathcal{H}_{\pm 1}$ can then be written in the general form, analogous to Eq. (8.1.39),

$$|\psi_0(\phi)\rangle = \cos(\phi)|+\rangle_1|-\rangle_2 + \sin(\phi)|-\rangle_1|+\rangle_2,$$
$$|\psi_{\pm 1}(\phi)\rangle = \cos(\phi)|+\rangle_1|+\rangle_2 + \sin(\phi)|-\rangle_1|-\rangle_2. \tag{8.1.44}$$

The angle ϕ determines the degree of entanglement of the two spins. While remaining aligned (in $\mathcal{H}_{\pm 1}$) or antiparallel (in $\mathcal{H}_0$), they retain their identity for $\phi = 0, \pi/2$ but are maximally entangled for $\phi = \pm\pi/4$. This continuum of states only exhausts one of the two available bits. However, the spins are embedded in three-dimensional space, with directions x, y, z, say. If the first bit has been used for the spin orientations in z, with spin states $|\pm_z\rangle_n$ and controlled by an angle ϕ_z, the second bit is free to be dedicated to the orientation along another axis, orthogonal to the first one, say x. There, the same transition from separate to entangled states as in Eq. (8.1.44) arises anew, now between spin states $|\pm_x\rangle_1|\pm_x\rangle_2$ controlled by a second angle ϕ_x.

With three orthogonal directions, there are six pairs $|-_u\rangle_n, |+_u\rangle_n, u = x, y, z,$ $n = 1, 2$, of spin eigenstates. As they all belong to the same two-dimensional Hilbert space, they cannot be linearly independent but are related, for example,

$$|+_x\rangle_n = 2^{-1/2}\big(|+_z\rangle_n + |-_z\rangle_n\big), \quad |-_x\rangle_n = 2^{-1/2}\big(|+_z\rangle_n - |-_z\rangle_n\big), \quad n = 1, 2.$$
$$\tag{8.1.45}$$

This implies that the two angles ϕ_x, ϕ_z can no longer be chosen independently. For the antiparallel sector $m = 0$, Eq. (8.1.44) combined with Eq. (8.1.45) gives

$$|\psi_0(\phi_x)\rangle = 2^{-1/2} \cos\left(\phi_x - \frac{\pi}{4}\right)\left(|+_z\rangle_1|+_z\rangle_2 - |-_z\rangle_1|-_z\rangle_2\right)$$
$$+ 2^{-1/2} \cos\left(\phi_x + \frac{\pi}{4}\right)\left(|-_z\rangle_1|+_z\rangle_2 - |+_z\rangle_1|-_z\rangle_1\right). \qquad (8.1.46)$$

While the first term represents parallel spins, only the second term corresponds to antiparallel orientation also in the z-direction. This choice therefore requires that $\cos(\phi_x + \pi/4) = \pm 1$, i.e., $\phi_x = -\pi/4 \pmod{\pi}$ and fixes ϕ_z at $\phi_z = 3\pi/4 \pmod{\pi}$. The "doubly antiparallel" state thus takes the definite form

$$|\psi_{00}\rangle = 2^{-1/2}\left(|-_z\rangle_1|+_z\rangle_2 - |+_z\rangle_1|-_z\rangle_2\right). \qquad (8.1.47)$$

Analogous states result if parallel orientation in x is combined with antiparallel orientation in z or vice versa, or for parallel spins along both axes.

This state (8.1.47) is as entangled as a Schrödinger cat state, but now between distinct particles! As counterintuitive as this result may appear, it has a straight-forward interpretation, based on the information theoretic principles of quantum mechanics [Zei99]. A pair of two-state systems has a total capacity of two bits, $lb(D_{\mathcal{H}^{(4)}}) = 2$. If they interact in such a way that their four-dimensional Hilbert space $\mathcal{H}^{(4)}$ decomposes into two sectors of one bit each, but now according to a common property of the two subsystems, for example a symmetry, then each of the two bits is equally shared by both parts. It means that the indivisibility of each of the two bits ignores the individual identity of the particles and supersedes even their spatial separation, revealing spatial localization as a mere classical phenomenon. Entanglement, manifest in information shared between two or more quantum systems, implies correlations, as will be worked out in the following subsection. They are to be distinguished from correlations resulting from a mere coincidence of states, as in classical systems, and indeed can be stronger in a direct quantitative and measurable sense, see Subsect. 8.4.3.

The situation turns downright bizarre if "spatial separation" is driven to the extreme, to macroscopic dimensions on geographic scales [TB&98, MH&12, YC&17]. Entanglement implies non-locality, another hallmark of quantum mechanics, irrespective of the spatial scale. Not surprisingly, quantum non-locality is suspect of violating Special Relativity. It can be absolved, however, recurring again to an information-theoretic analysis (Sect. 8.4).

8.2 Quantum Information

Quantum mechanics, as it has been presented in the foregoing subsections, appears as a self-contained construction, consistent but separated from the world as we know it, a scientific Laputa. What is clearly missing is a well-defined interface with classical physics and, in a wider view, with our human perception. How do quantum states and their evolution in time translate into facts, into trajectories—and indeed, into suitable input to calculate information?

Quantum mechanics appears very clear in this respect: Once the state vector is known, everything is known that can be known on a system, its information content is exhausted. Here, quantum mechanics matches the classical mechanics of discrete phase-space points. There is no space for partial knowledge, as it gives rise to the classical concept of statistical ensembles. It comes into play only when we ask how quantum systems interact with a macroscopic observer, go the laboratory and take a closer look at measurements. Much of the construction of this quantum–classical interface goes back to the seminal work of John von Neumann. He proposed both, a concept that allows combining quantum mechanics in Hilbert space with classical probability, and, based on it, a definition of information that applies to quantum states that are not completely defined.

8.2.1 The Density Operator and Von-Neumann Entropy

Why not construct quantum ensembles by assigning probabilities directly to Hilbert space vectors? This is not appropriate, because state vectors do not represent facts. Combining states multiplying them with complex coefficients, as in Eq. (8.1.1), amounts to linear superposition, not to statistical weighting. Partial knowledge can only be implemented on a higher ontological level.

In geometrical terms, extracting information on a Hilbert-space vector corresponds to a projection: The information, its quantity and precision, depends not only on the state measured, but also on the direction into which it is projected, that is, on what is observed. For example, Fourier analyzing a sinusoidal signal over a period that does not match the period of the signal can give rise to a broad spectrum. In particular, if the detector is insensitive to phases, the output will be a continuous distribution of intensities, even if the input signal is monochromatic.

In the same way, classical statistics comes into play if a quantum system is measured "in an improper way", also and in particular if no stochastic process is involved. Associating an operator $\hat{A}$ to a classical observable A, the result of a measurement of this quantity can be condensed into the mean of all expected measured values (the *expectation value*),

$$\left\langle \hat{A} \right\rangle = \sum_n p_n A_n, \tag{8.2.1}$$

denoting A_n, the nth eigenvalue in the proper basis of $\hat{A}$, $\hat{A}|n\rangle = A_n|n\rangle$, thus

$$A_n = \langle n|\hat{A}|n\rangle, \tag{8.2.2}$$

and p_n, the probability to find it. If before the measurement, the system was in a state $|\psi\rangle$ that did not coincide with one of those eigenstates $|n\rangle$ of $\hat{A}$, these probabilities are the result of projecting $|\psi\rangle$ onto the eigenstates $|n\rangle$, $p_n = |\langle n|\psi\rangle^2|$. In this case, they depend directly on the coefficients of a linear superposition, as in Eq. (8.1.1). However, lacking the phases, they generally do not allow to reconstruct the underlying state vector. Therefore, these probabilities are quantities of a completely different nature, incomparable with the complex coefficients of linear superpositions.

Irrespective of their possible origin in a measurement, sets of probabilities $\{\ldots, p_n, \ldots\}$ represent the classical information available on a quantum system for a macroscopic observer. They form a state space of their own right, hybrids complementing the quantum features of Hilbert-space elements with classical uncertainty that expresses partial knowledge. Mathematically, they can be cast in the form of *density operators*,

$$\hat{\rho} = \sum_n p_n |n\rangle\langle n|. \tag{8.2.3}$$

They are composed of projectors $|n\rangle\langle n|$ onto state vectors $|n\rangle$, so that the probabilities can be extracted in turn as

$$p_n = \langle n|\hat{\rho}|n\rangle, \tag{8.2.4}$$

and the expectation value (8.2.1) takes the form

$$\left\langle \hat{A} \right\rangle = \text{tr}\left[\hat{\rho}\hat{A}\right]. \tag{8.2.5}$$

With the Definition (8.2.4), all phase information is erased. Accordingly, the space $\mathcal{P}_{\mathcal{H}}$ comprising all density operators associated to a Hilbert space $\mathcal{H}$ is referred to as the *projective Hilbert space*.

Elements of $\mathcal{P}_{\mathcal{H}}$ share a number of general properties, implied by the quantum features of the projectors and by probability calculus [Fel68]:

1. Normalization of the probabilities p_n induces the **normalization** of the density operators

$$\text{tr}[\hat{\rho}] = \sum_n \langle n|\hat{\rho}|n\rangle = \sum_n p_n = 1. \tag{8.2.6}$$

2. Owing to the inherent symmetry of their Definition (8.2.3), density operators are **Hermitian**,

$$\hat{\rho}^\dagger = \left(\sum_n p_n |n\rangle\langle n|\right)^\dagger = \sum_n p_n^*(|n\rangle\langle n|)^\dagger = \sum_n p_n |n\rangle\langle n| = \hat{\rho}. \tag{8.2.7}$$

3. The condition $p_n \geq 0$, implies that the density operator is **non-negative**: For any normalized state $|\alpha\rangle$,

$$\langle\alpha|\hat{\rho}|\alpha\rangle = \sum_n p_n \langle\alpha|n\rangle\langle n|\alpha\rangle = \sum_n p_n |\langle\alpha|n\rangle|^2 \geq 0. \tag{8.2.8}$$

4. These properties apply likewise to $\hat{I} - \hat{\rho}$, the **complement** of $\hat{\rho}$. In particular, it is also positive definite. By the Schwarz inequality, $\langle\alpha|\alpha\rangle \geq \langle\alpha|n\rangle\langle n|\alpha\rangle = |\langle\alpha|n\rangle|^2$,

$$\langle\alpha|(\hat{I} - \hat{\rho})|\alpha\rangle = \langle\alpha|\alpha\rangle - \sum_n p_n |\langle\alpha|n\rangle|^2 \geq \langle\alpha|\alpha\rangle - \sum_n p_n \langle\alpha|\alpha\rangle = 1 - \sum_n p_n = 0. \tag{8.2.9}$$

5. Conversely, any operator satisfying properties 1 to 3 can be written in the form

$$\hat{\rho} = \sum_n w_n \hat{P}_n, \tag{8.2.10}$$

where the operators $\hat{P}_n$ are projectors, $\hat{P}_n^2 = \hat{P}_n$, and the coefficients w_n can be interpreted as probabilities, $w_n \in [0, 1]$, $\sum_n w_n = 1$.

Density operators can be represented in any orthogonal basis of $\mathcal{P}_\mathcal{H}$ like state vectors, cf. Eq. (8.1.5). If the N-dimensional discrete basis $\{|a_1\rangle, |a_2\rangle, \ldots, |a_N\rangle\}$ spans the underlying Hilbert space $\mathcal{H}$, a density operator is decomposed as

$$\hat{\rho} = \sum_{n',n''=1}^{N} |a_{n'}\rangle\langle a_{n'}|\hat{\rho}|a_{n''}\rangle\langle a_{n''}| = \sum_{n',n''=1}^{N} \rho_{n',n''} |a_{n'}\rangle\langle a_{n''}|, \ \rho_{n',n''} := \langle a_{n'}|\hat{\rho}|a_{n''}\rangle, \tag{8.2.11}$$

the coefficients $\rho_{n',n''}$ forming an $N \times N$ *density matrix*. The diagonal elements $\rho_{n,n} = \langle a_n|\hat{\rho}|a_n\rangle =: p_n$ represent classical probabilities. For an infinite-dimensional Hilbert space with a continuous basis, e.g., the set of position states $\{|\mathbf{x}\rangle | \mathbf{x} \in \mathbb{R}^f\}$, the diagonal elements $\rho(\mathbf{x}) = \langle\mathbf{x}|\hat{\rho}|\mathbf{x}\rangle$ give the probability *density* to be in $\mathbf{x}$. In particular, if $\hat{\rho} = |\psi\rangle\langle\psi|$, $\rho(\mathbf{x}) = |\langle\mathbf{x}|\psi\rangle|^2 = |\psi(\mathbf{x})|^2$.

The off-diagonal elements $\rho_{n',n''}$ with $n' \neq n''$, called *coherences*, encode quantum information. This distinction is ambiguous and depends on the basis used to represent $\hat{\rho}$; coherences can be concealed, transforming the density matrix to the basis given by its eigenstates; they can conversely be created representing a diagonal density matrix in a different basis.

Based on the expectation value (8.2.5) and the probability density $\rho(\mathbf{x})$, a probability current in position space is readily construed [CDL77]: Define a local velocity as the operator

$$\hat{\mathbf{j}}(\mathbf{x}) := \frac{1}{2}\left(|\mathbf{x}\rangle\langle\mathbf{x}|\frac{\hat{p}}{m} + \frac{\hat{p}}{m}|\mathbf{x}\rangle\langle\mathbf{x}|\right), \qquad (8.2.12)$$

symmetrizing the product of velocity $\hat{\mathbf{p}}/m$ and projection $|\mathbf{x}\rangle\langle\mathbf{x}|$ onto $\mathbf{x}$. Its expectation value, cf. Eq. (8.2.5), is then the current

$$\mathbf{j}(\mathbf{x}) := \left\langle\hat{\mathbf{j}}(\mathbf{x})\right\rangle = \mathrm{tr}[\hat{\mathbf{j}}(\mathbf{x})\hat{\rho}], \qquad (8.2.13)$$

i.e., for pure states $\hat{\rho} = |\psi\rangle\langle\psi|$,

$$\mathbf{j}(\mathbf{x}) = \frac{\hbar}{2mi}\left(\psi^*(\mathbf{x})\frac{\partial}{\partial\mathbf{x}}\psi(\mathbf{x}) - \psi(\mathbf{x})\frac{\partial}{\partial\mathbf{x}}\psi^*(\mathbf{x})\right), \qquad (8.2.14)$$

using the representations $\psi(\mathbf{x}) = \langle\mathbf{x}|\psi\rangle$ and $\langle\mathbf{x}|\hat{\mathbf{p}}|\mathbf{x}'\rangle = -i\hbar\delta(\mathbf{x}' - \mathbf{x})\partial/\partial\mathbf{x}$ of the wave function and the momentum operator, resp., in the position state basis. Comparing the divergence $\mathrm{div}\,\mathbf{j}(\mathbf{x}) = \nabla \cdot \mathbf{j}(\mathbf{x})$ with the time derivative $\partial\rho/\partial t$, according to the Schrödinger Eq. (8.3.1), validates a quantum continuity equation, analogous to the classical Eq. (5.2.21),

$$\frac{\partial}{\partial t}\rho(\mathbf{x}, t) + \nabla \cdot \mathbf{j}(\mathbf{x}, t) = 0. \qquad (8.2.15)$$

Whether $\hat{\rho}$ represents a single state vector (a *pure state*) or a probabilistic mixture of two or more states (a *mixed state*) can be concluded directly from a particular diagnostic, the *purity* or *participation ratio*. It is defined as

$$\xi_\rho = \mathrm{tr}[\hat{\rho}^2] = \sum_n p_n^2 \qquad (8.2.16)$$

and ranges between 0 and 1. The *inverse participation ratio* (IPR), in turn,

$$\xi_\rho^{-1} = \frac{1}{\mathrm{tr}[\hat{\rho}^2]} = \left(\sum_n p_n^2\right)^{-1}, \qquad (8.2.17)$$

can be interpreted as the number of states that contribute effectively to the density operator. It is akin to the Hilbert-space dimension but weights sectors of the Hilbert space according to their population. Accordingly, possible values of the IPR range from 1 to $D_{\mathcal{H}}$, the Hilbert space dimension, which also mark the limiting cases of uncertainty manifest in $\hat{\rho}$:

– **Pure states** with $\xi_\rho^{-1} = 1$ are possible only for all-or-nothing probabilities

$$p_n = \delta_{n-n_0} = \begin{cases} 1 \; n = n_0, \\ 0 \; \text{else}, \end{cases} \tag{8.2.18}$$

which implies $\hat{\rho}^2 = \hat{\rho}$ and thus $\mathrm{tr}[\hat{\rho}^2] = 1$, so that the density operator takes the form $\hat{\rho} = |\psi\rangle\langle\psi|$.

– Complete lack of knowledge of the system state corresponds to a **homogeneous probability distribution** over the entire Hilbert space, i.e., $p_n = \text{const} = 1/D_{\mathcal{H}}$ for $n = 1, \ldots, D_{\mathcal{H}}$, so that

$$\mathrm{tr}[\hat{\rho}^2] = \sum_{n=1}^{D_{\mathcal{H}}} \langle n| \left(D_{\mathcal{H}}^{-1} \sum_{n'=1}^{D_{\mathcal{H}}} |n'\rangle\langle n'| \right)^2 = \frac{1}{D_{\mathcal{H}}^2} \sum_{n,n'=1}^{D_{\mathcal{H}}} |\langle n|n'\rangle|^2 = \frac{1}{D_{\mathcal{H}}} \tag{8.2.19}$$

and therefore $\xi_\rho^{-1} = D_{\mathcal{H}}$. Evidently, considering infinite-dimensional Hilbert spaces, the inverse participation ratio has no upper bound, either.

These two cases are exactly analogous to the extremes of the classical probability, Eqs. (1.3.10) and (1.3.12). This suggests basing a definition of entropy, compatible with the postulates of quantum mechanics, on the density operator, a program proposed and accomplished by von Neumann [Neu18]. However, it would not be adequate to refer directly to the set of probabilities p_n, as they correspond to the diagonal elements of the density operator, see Eq. (8.2.3), and thus depend on the representation. A more promising strategy is to define the quantum entropy as expectation value, Eq. (8.2.5), of the logarithm of the probabilities $\ln(\hat{\rho})$, as in Eq. (5.3.2) [Weh78], which leads to the *von Neumann entropy* $S_{\mathrm{vN}} = \langle -c \ln(\hat{\rho}) \rangle = -c\mathrm{tr}[\hat{\rho} \ln(\hat{\rho})]$. As in Shannon's Definition (1.3.9), c fixes units. In quantum mechanics, it is customary to measure information in bits, cf. Subsect. 8.1.1, postulate 5, that is, to take the binary logarithm or set $c = 1/\ln(2)$, so that

$$S_{\mathrm{vN}} = \langle -\mathrm{lb}(\hat{\rho}) \rangle = -\mathrm{tr}[\hat{\rho}\,\mathrm{lb}(\hat{\rho})]. \tag{8.2.20}$$

Comparing Eq. (8.2.20) to the classical entropy, Eq. (5.3.2), highlights the close analogy as well as the crucial differences between the two concepts. The most conspicuous discrepancy is that in (8.2.20), there is no more prefactor d_r^f (f denoting the number of degrees of freedom) under the logarithm, which is included in the classical definition to determine the resolution in phase space and to compensate for the dimension $1/(\text{action})^f$ of the phase-space density. In view of the natural coarse-graining of quantum structures in classical phase space, discussed in Subsect. 8.1.3, one might expect to see d_r^f replaced by $(2\pi\hbar)^f$, the size of a Planck cell in f dimension. Indeed, in phase-space representations of quantum entropy, this is exactly what happens, see Eq. (8.2.28). However, in the basic Definition (8.2.3), the intrinsic properties of the density operator already take care of the finiteness of quantum information. In essence, it measures the dimension of that sector of the total Hilbert space that is effectively populated by the state specified by $\hat{\rho}$.

The two extreme cases of the purity, expressed in Eqs. (8.2.18) and (8.2.19), are directly reflected in corresponding limits of von Neumann's entropy:

- For **pure states**, a power expansion of the operator function $\hat{\rho}\ln(\hat{\rho})$ around $\hat{\rho}-\hat{I}$ takes a particularly simple form, since with $\hat{\rho}^2=\hat{\rho}$, $(\hat{I}-\hat{\rho})^n=\hat{I}-\hat{\rho}$, hence

$$\hat{\rho}\ln(\hat{\rho})=\sum_{n=0}^{\infty}\frac{1}{n!}\left(\frac{d^n}{dx^n}x\ln(x)\bigg|_{x=1}\right)(\hat{\rho}-\hat{I})^n$$

$$=(\hat{\rho}-\hat{I})\left(1-\sum_{n=2}^{\infty}\frac{1}{n(n-1)}\right)=0. \qquad (8.2.21)$$

The lower bound of von Neumann's entropy, reached for pure states, is therefore $S_{\text{vN}}=0$. This definite answer says in essence that "once you know in which pure state a quantum system is, nothing more can be learned about it". It contrasts with the classical view that more accurate measurements can always reveal more details, which does not allow for any lower bound of the entropy.

- Conversely, for a **homogeneous distribution** over the Hilbert space, $p_n=\text{const}=1/D_{\mathcal{H}}$ for $n=1,\ldots,D_{\mathcal{H}}$, so that

$$\text{tr}[\hat{\rho}\text{lb}(\hat{\rho})]=\sum_{n=1}^{D_{\mathcal{H}}}\frac{1}{D_{\mathcal{H}}}\text{lb}\left(\frac{1}{D_{\mathcal{H}}}\right)=-\text{lb}(D_{\mathcal{H}}), \qquad (8.2.22)$$

resulting in an entropy $S_{\text{vN}}=\text{lb}(D_{\mathcal{H}})$. Individually for each system, this is its upper bound. However, since the Hilbert-space dimension can be infinite, also von Neumann's entropy is not bounded from above if all kinds of quantum systems are compared.

Equation (8.2.20) is arguably also the definition of quantum entropy that comes as close as possible to the spirit of Shannon's original expression (1.3.9). This becomes most evident if the density operator is represented in the basis of its eigenstates, $\hat{\rho}=\sum_n p_n|n\rangle\langle n|$, where von Neumann's entropy takes the form

$$S_{\text{vN}}=-\sum_n p_n\text{lb}(p_n). \qquad (8.2.23)$$

In all other bases, however, the density operator is no longer diagonal, and Eq. (8.2.20) involves a double sum over basis states.

Enforcing a diagonal form as in (8.2.23) also for bases other than $\{|n\rangle\}$, say $\{|\alpha\rangle\}$, leads to a definition

$$S_{\{|\alpha\rangle\}} = -\sum_{\alpha} p_\alpha \mathrm{lb}(p_\alpha) \text{ or } S_{\{|\alpha\rangle\}} = -\int \mathrm{d}\alpha\, p(\alpha)\mathrm{lb}(p(\alpha)). \qquad (8.2.24)$$

sometimes referred to as Shannon quantum entropy (the second expression is valid for a continuous spectrum). Even for a pure state $\hat{\rho} = |\psi\rangle\langle\psi|$, expanding $|\psi\rangle = \sum_{\alpha} c_\alpha |\alpha\rangle$ and denoting $p_\alpha := |c_\alpha|^2$, the two entropies do not coincide but are related by

$$S_{\mathrm{vN}} = -\sum_{\alpha} |c_\alpha|^2 \mathrm{lb}(|c_\alpha|^2) - \sum_{\alpha \neq \alpha'} c_{\alpha'} \cdot c_\alpha^* \langle\alpha'|\mathrm{lb}(\hat{\rho})|\alpha\rangle = -\sum_{\alpha} p_\alpha \mathrm{lb}(p_\alpha) - \Delta S$$

$$= S_{\{|\alpha\rangle\}} - \Delta S. \qquad (8.2.25)$$

Since at the same time, for a pure state $S_{\mathrm{vN}} = 0$ and by definition, $S_{\{|\alpha\rangle\}} \geq 0$, it follows that the non-diagonal term ΔS must be positive. In general, the Shannon quantum entropy will give different results in different representations and is larger than von Neumann's entropy.

A particularly revealing choice for the basis $\{|\alpha\rangle\}$ is the set of coherent states $\{|\gamma\rangle\}$, parameterized by the complex dimensionless phase-space coordinates $\gamma = \sqrt{m\omega/2\hbar}\,\mathbf{q} + i\mathbf{p}/\sqrt{2m\hbar\omega} \in \mathbb{C}^f$, cf. Eq. (8.1.33). In this case, the Shannon entropy becomes

$$S_{\{|\gamma\rangle\}} = -\frac{1}{\pi^f}\int \mathrm{d}^{2f}\gamma\, \langle\gamma|\hat{\rho}|\gamma\rangle \mathrm{lb}\big(\pi^{-f}\langle\gamma|\hat{\rho}|\gamma\rangle\big) = -\int \mathrm{d}^{2f}\gamma\, P_{\mathrm{H}}(\gamma)\mathrm{lb}(P_{\mathrm{H}}(\gamma)). \qquad (8.2.26)$$

It can be interpreted as a quantum mechanical analogue of the classical entropy (5.3.1), now replacing the classical probability density by the *Husimi function*,

$$P_{\mathrm{H}}(\gamma) = \frac{1}{\pi^f}\langle\gamma|\hat{\rho}|\gamma\rangle. \qquad (8.2.27)$$

Unlike the Wigner function, which provides a one-to-one (invertible) representation of the density operator, the Husimi function is positive definite and can be understood as the probability for a system to be found in the coherent state $|\gamma\rangle$, normalized over the f-fold complex plane, $\int \mathrm{d}^{2f}\gamma\, P_{\mathrm{H}}(\gamma) = 1$. By definition, it involves a filtering with a minimum uncertainty Gaussian, a coarse-graining of phase-space structures on the scale $2\pi\hbar$, and thereby loses information. The prefactor π^{-f} balances the overcompleteness of the set of coherent states, see Subsect. 8.1.3.

Rescaled as $p_H(\mathbf{p}, \mathbf{q}) := (2\hbar)^{-f} P_H(\sqrt{m\omega/2\hbar}\mathbf{q} + i\mathbf{p}/\sqrt{2m\hbar\omega})$, the Husimi function becomes a phase-space probability density normalized over phase space, $\int d^f p\, d^f q\, p_H(\mathbf{p}, \mathbf{q}) = 1$, so that the Shannon entropy takes the form

$$S_H = -\int d^f p\, d^f q\, p_H(\mathbf{p}, \mathbf{q})\, \mathrm{lb}\big((2\hbar)^f\, p_H(\mathbf{p}, \mathbf{q})\big), \qquad (8.2.28)$$

known as *Wehrl entropy* [Weh78, Weh79], The factor $(2\hbar)^f$ now compensates for the dimension $(\text{action})^{-f}$ of the phase-space density and marks the resolution, the size of a Planck cell, of the Husimi function. Equation (8.2.28) is the most direct implementation of classical phase-space information, accounting for the finite resolution implied by quantum mechanics. However, the loss of actual information, manifest in the overcompleteness of coherent states, implies that the minimum value of the Wehrl entropy, reached if $\hat{\rho} = |\gamma\rangle\langle\gamma|$ is a pure coherent state, is $f\,\mathrm{lb}(\pi)/\pi^f$ [Weh79], not zero as for the von Neumann entropy.

Instead of deriving a quantum phase-space entropy from the Shannon entropy for coherent states, it can be implemented directly as the von Neumann entropy, representing pure and mixed states by the Wigner function. Expectation values as used for the definition of S_{vN}, see Eq. (8.2.20), can be written in terms of Wigner functions in the same way as classical phase-space averages (which has been Wigner's original motivation),

$$\left\langle \hat{A} \right\rangle = \mathrm{tr}[\hat{A}\hat{\rho}] = \int d^f p\, d^f q\, A_W(\mathbf{p}, \mathbf{q})\, P_W(\mathbf{p}, \mathbf{q}), \qquad (8.2.29)$$

where $A_W(\mathbf{p}, \mathbf{q})$ is the phase-space representation (*Weyl transform*) of the operator $\hat{A}$,

$$A_W(\mathbf{p}, \mathbf{q}) = \frac{1}{h^f} \int d^f q' \left\langle \mathbf{q} - \frac{\mathbf{q}'}{2}\Big| \hat{A} \Big| \mathbf{q} + \frac{\mathbf{q}'}{2}\right\rangle \exp\left(\frac{i\mathbf{p}\cdot\mathbf{q}'}{\hbar}\right). \qquad (8.2.30)$$

and $P_W(\mathbf{p}, \mathbf{q})$ that of the density operator, i.e., the Wigner function, generalizing Eq. (8.1.17) to mixed states.

Applied to the von Neumann entropy, this means

$$S_{vN} = \left\langle -\mathrm{lb}(\hat{\rho}) \right\rangle = -\int d^f p\, d^f q\, P_W(\mathbf{p}, \mathbf{q})\, (\mathrm{lb}(\hat{\rho}))_w(\mathbf{p}, \mathbf{q}). \qquad (8.2.31)$$

The Weyl transform $(\mathrm{lb}(\hat{\rho}))_W(\mathbf{p}, \mathbf{q})$ of $\mathrm{lb}(\hat{\rho})$ is *not* simply $\mathrm{lb}(P_W(\mathbf{p}, \mathbf{q}))$, i.e., the Weyl transform does not commute with the logarithm. The logarithm can be reduced to powers of its argument by Taylor expansion, but the Weyl transform of operator products $\hat{A}\hat{B}$ is in general not given by the product of the Weyl transforms of its factors, $(\hat{A}\hat{B})_W(\mathbf{p}, \mathbf{q}) \neq A_W(\mathbf{p}, \mathbf{q})B_W(\mathbf{p}, \mathbf{q})$, unless they commute.

A non-zero commutator results in correction terms. For the operator function $\log(\hat{\rho})$, the Weyl transform takes the form of a power series

$$(\ln(\hat{\rho}))_{\rm W}({\bf p}, {\bf q}) = \sum_{n=1}^{\infty} \frac{(-1)^{n-1}}{n}\left((\hat{\rho} - \hat{\rm I})^n\right)_{\rm W}({\bf p}, {\bf q}), \qquad (8.2.32)$$

where each power $\left((\hat{\rho} - \hat{\rm I})^{n-1}\right)_{\rm W}$ requires multiple phase-space integrations over the Wigner function $P_{\rm W}({\bf p}, {\bf q})$. The leading term $n = 1$ is simply $P_{\rm W}({\bf p}, {\bf q}) - 1$. The next order is already quite involved (abbreviating ${\bf r} := ({\bf p}, {\bf q})$ etc.),

$$\left((\hat{\rho} - \hat{\rm I})^2\right)_{\rm W}({\bf r}) = \left(\frac{2}{h}\right)^{2f} \int d^{2f}r' d^{2f}r'' e^{\frac{2i}{\hbar}({\bf r}({\bf r}' - {\bf r}' - {\bf r}' \wedge {\bf r}''))}\left(P_{\rm W}({\bf r}') - 1\right)\left(P_{\rm W}({\bf r}'') - 1\right).$$
$$(8.2.33)$$

The twofold integration over all of phase space in Eq. (8.2.33) shows that the entropy of a quantum state can no longer be understood as a sum or an integral over local contributions of points or compact regions in phase space, as for the classical information, Eq. (5.3.1). This is another manifestation of the redundancy of combined position-momentum representations. The Wehrl entropy (8.2.28) is local at least on the finite scale $2\pi\hbar$ but indeed, it already loses some of the specifically quantum information contained in the density operator.

In position space, by contrast, von Neumann's entropy can be written as the average of a local information density. Reading Eq. (8.2.20) as an expectation value $S_{\rm vN} = \langle \hat{i}_{\rm vN} \rangle = {\rm tr}[\hat{\rho} \hat{i}_{\rm vN}]$ suggests defining an operator-valued information density as

$$\hat{i}_{\rm vN} := -{\rm lb}(\hat{\rho}). \qquad (8.2.34)$$

Along the same lines as in 5.3, we can combine it with the current operator $\hat{\bf j}({\bf x})$, Eq. (8.2.12), to construct an entropy current operator

$$\hat{\bf i}_{\rm vN}({\bf x}) := -{\rm lb}(\hat{\rho})\left(|{\bf x}\rangle\langle{\bf x}|\frac{\hat{\bf p}}{2m} + \frac{\hat{\bf p}}{2m}|{\bf x}\rangle\langle{\bf x}|\right) = -{\rm lb}(\hat{\rho})\hat{\bf j}({\bf x}), \qquad (8.2.35)$$

and in analogy to Eq. (8.2.13), an entropy current in position space

$$\mathbf{i}_{\rm vN}({\bf x}) := \langle \hat{\bf i}_{\rm vN}({\bf x}) \rangle = {\rm tr}\left[\hat{\rho} \hat{\bf i}_{\rm vN}({\bf x})\right]. \qquad (8.2.36)$$

While the total entropy is conserved by unitary transformations, as will be shown in Subsect. 8.3, in general the information density is not conserved locally under the probability flow ${\bf j}({\bf x})$, that is, a continuity equation, similar to Eq. (8.2.15) for the probability current, does not apply here. A quantum quasi-probability flow in phase space, based on the Wigner function, and a corresponding entropy flow will be defined in Subsect. (8.3.1).

In conclusion, von Neumann's definition of quantum entropy has significant virtues as an adequate translation of Shannon's information into the language of the quantum world. It combines astonishing simplicity with strong mathematical properties, indispensable for its interpretation as an information measure. A host of alternative definitions have been proposed since its publication in 1932. They modify the original definition in order to adapt it to more special situations, often with the intention to avoid or mitigate peculiarities of quantum mechanics that appear inconvenient from a classical point of view. However, such advantages are achieved on the expense of universal validity and consistency with the principles of quantum mechanics, which characterize the von Neumann entropy. It will therefore serve as the central tool of all subsequent analyses of information in quantum mechanics.

Classical and quantum information are both intimately related to the dimension of the state space of the system at hand: the number of degrees of freedom in the classical, Hilbert-space dimension in the quantum case. In this point, however, the difference between the two concepts becomes particularly evident. In spatially extended systems, there is still a common basis for comparisons. On the classical level, the sheer size of the system (in configuration space, in phase space, ...) is of minor relevance. As an extensive quantity, classical entropy increases linearly with the number of spatial dimensions, of degrees of freedom, of subsystems. Within each dimension, it depends on the resolution of measurements (of position, velocity, etc.), which is a contingent parameter. In quantum mechanics, the entropy is determined by the Hilbert-space dimension, to be strictly distinguished from the spatial dimension. Extended systems of whatever spatial dimension are characterized by continuous spectra, hence by Hilbert spaces of infinite dimension. Unlike classical information, however, the finite size of a Planck cell restricts the number of dimensions per unit (hyper)volume. Under all practical circumstances, this implies a finite total Hilbert-space dimension increasing linearly with the system size, which results in a logarithmic dependence of the entropy on system size. Quantum systems with a discrete Hilbert space of finite dimension have no classical analogue at all. Their Hilbert-space dimension grows exponentially with the number of subsystems, so that the entropy scales linearly with it, it "counts qubits". These relationships are summarized in the following table (Table 8.1):

8.2.2 Entanglement and Quantum Information

Most of the ramifications of the classical concept of information, discussed in Subsect. 1.7, extend to quantum entropy identically or without major modification. A facet of information that will prove of fundamental relevance in the context of entanglement and non-locality is the question in how far the information content of a total system can be broken down to a sum over partial entropies of its subsystems, even if they interact with each other, a crucial issue in particular in current research of entanglement in strongly interacting many-body systems. Adopting the strategy applied on the classical level, define the partial or reduced density of a subsystem as the result of projecting out all other subsystems. For a bipartite system with subsystems A and B and total density operator $\hat{\rho}_{AB}$, the partial densities then read,

Table 8.1 In classical and quantum mechanics, information measures, such as the dimension of the state space accessible to a system and the information, depend in different ways on global size parameters. The table compares the functional dependences of the state-space dimension and the Shannon information, on the classical level, and of the Hilbert-space dimension and the von-Neumann entropy, on the quantum level, on the spatial system size (length, volume, phase-space area, etc.) and the spatial dimension and the number of degrees of freedom or subsystems.

	classical		quantum	
	state space dimension	Shannon information	Hilbert space dimension	von Neumann entropy
spatial dimension, number of degs. of freedom	linear	linear	exponential	linear
spatial extension, phase-space area	independent	independent	linear	logarithmic

$$\bar{\rho}_A = \mathrm{tr}_B[\hat{\rho}_{AB}], \quad \bar{\rho}_B = \mathrm{tr}_A[\hat{\rho}_{AB}]. \tag{8.2.39}$$

and the partial entropies are, according to Eq. (8.2.20),

$$S_A = -\mathrm{tr}_A\left[\bar{\rho}_A \mathrm{lb}(\bar{\rho}_A)\right] = -\mathrm{tr}_A\left[\mathrm{tr}_B[\hat{\rho}_{AB}]\mathrm{lb}(\mathrm{tr}_B[\hat{\rho}_{AB}])\right],$$
$$S_B = -\mathrm{tr}_B\left[\bar{\rho}_B \mathrm{lb}(\bar{\rho}_B)\right] = -\mathrm{tr}_B\left[\mathrm{tr}_A[\hat{\rho}_{AB}]\mathrm{lb}(\mathrm{tr}_A[\hat{\rho}_{AB}])\right]. \tag{8.2.40}$$

As in the classical case, the two subsystems become independent if the total density is assumed to factorize, $\hat{\rho}_{AB} = \hat{\rho}_A \otimes \hat{\rho}_B$, so that $\bar{\rho}_A = \hat{\rho}_A$, $\bar{\rho}_B = \hat{\rho}_B$, and the partial entropies reduce to the full von-Neumann entropies of these systems (likewise for S_B),

$$S_A = -\mathrm{tr}_A\left[\bar{\rho}_A \mathrm{lb}(\bar{\rho}_A)\right] = -\mathrm{tr}[\hat{\rho}_A \mathrm{lb}(\hat{\rho}_A)]. \tag{8.2.41}$$

Even if the total system is in a pure state (i.e., as a whole, it is completely isolated), $\hat{\rho}_{AB} = |\Psi\rangle\langle\Psi|$ so that $S_{AB} = 0$, the subsystems can be in mixed states. In that case, the total system cannot factorize, it is entangled, see below. Only if $S_{AB} = 0$ *and* the subsystems are uncorrelated, $\hat{\rho}_{AB} = \hat{\rho}_A \otimes \hat{\rho}_B$, the system becomes separable and the trace of $\mathrm{lb}(\hat{\rho}_{AB})$ over the total Hilbert space decays into a sum of partial traces. Taking normalization of the partial density operators into account, $\mathrm{tr}_A[\hat{\rho}_A] = \mathrm{tr}_B[\hat{\rho}_B] = 1$, one finds that in this case, partial entropies just add, in analogy to the classical relation (1.7.8),

$$S_{AB} = -\mathrm{tr}[\hat{\rho}_A \otimes \hat{\rho}_B \mathrm{lb}(\hat{\rho}_A \otimes \hat{\rho}_B)]$$

$$= -\text{tr}_B \lfloor \hat{\rho}_B \rfloor \text{tr}_A \lfloor \hat{\rho}_A \text{lb}(\hat{\rho}_A) \rfloor - \text{tr}_A \lfloor \hat{\rho}_A \rfloor \text{tr}_B \lfloor \hat{\rho}_B \text{lb}(\hat{\rho}_B) \rfloor$$
$$= S_A + S_B, \qquad\qquad\qquad (8.2.42)$$

Since quantum entropies are non-negative, it follows immediately that Eq. (8.2.43) can only be satisfied if $S_A = S_B = 0$, that is, if also the subsystems are in pure states.

Conversely, if the two subsystems are physically coupled so that their combination is then not only a mere conceptual union, the total system is entangled. In this case, quantum mechanics can forge much tighter links between them than would be possible with classical correlations:

Consider a bipartite system in a correlated state such that the total density operator does *not* factorize,

$$\hat{\rho}_{AB} \neq \hat{\rho}_A \otimes \hat{\rho}_B. \qquad\qquad (8.2.43)$$

In this case, the subsystems will no longer be in pure states,

$$\text{tr}_A\left[\left(\bar{\hat{\rho}}_A\right)^2\right] = \text{tr}_A\left[\left(\text{tr}_B[\hat{\rho}_{AB}]\right)^2\right] < 1, \quad \text{tr}_B\left[\left(\bar{\hat{\rho}}_B\right)^2\right] = \text{tr}_B\left[\left(\text{tr}_A[\hat{\rho}_{AB}]\right)^2\right] < 1,$$
$$(8.2.44)$$

and the identity $(\hat{I} - \hat{\rho})^n = \hat{I} - \hat{\rho}$, valid if $\hat{\rho}^2 = \hat{\rho}$, and with it Eq. (8.2.22), no longer apply to the partial entropies, so that

$$S_A = -\text{tr}_A\left[\bar{\hat{\rho}}_A \text{lb}\left(\bar{\hat{\rho}}_A\right)\right] > 0, \quad S_B = -\text{tr}_B\left[\bar{\hat{\rho}}_B \text{lb}\left(\bar{\hat{\rho}}_B\right)\right] > 0. \qquad (8.2.45)$$

As a consequence, the total entropy can no longer be the sum of partial entropies, rather $S_{AB} > S_A + S_B$. in fact, this generalizes to the convexity of quantum entropy [Weh91, NC00]: If a density operator is composed as $\hat{\rho}_{AB}(\lambda) = \lambda\hat{\rho}_A + (1 - \lambda)\hat{\rho}_B$ with $0 \leq \lambda \leq 1$, then $S_{AB}(\lambda) \geq \lambda S_A + (1 - \lambda)S_B$.

The definitions (8.2.40) of partial entropies S_A and S_B for a bipartite system look clearly distinct for the two subsystems. Indeed, from a classical point of view, see Eq. (1.7.5), there is no reason for them to agree (for example, one subsystem could be in a definite state, the state of the other unknown). Under the present assumption of the total system being in a pure state, $\hat{\rho}_{AB} = |\Psi\rangle\langle\Psi|$, however, quantum mechanics requires them to coincide. According to Eq. (8.2.40), the decisive feature distinguishing the two quantities is the order of the respective traces, $\text{tr}_A\left[f\left(\text{tr}_B[\hat{\rho}_{AB}]\right)\right]$ vs. $\text{tr}_B\left[f\left(\text{tr}_A[\hat{\rho}_{AB}]\right)\right]$, with $f(x) = x\,\text{lb}(x)$. That in fact they commute can be shown invoking *Schmidt decomposition* [EK95, NC00, GK05], a corollary of singular value decomposition [NC00, HJ13] applied to the density matrix of a bipartite system. In a composite Hilbert space $\mathcal{H} = \mathcal{H}_A \otimes \mathcal{H}_B$ with bases $\{|\alpha\rangle_A | \alpha = 1, \ldots, D_A\}$ and $\{|\beta\rangle_B | \beta = 1, \ldots, D_B\}$ for $\mathcal{H}_A$ and $\mathcal{H}_B$, resp., the total density operator can always be represented as a fourth-rank tensor $\rho_{\alpha,\alpha',\beta,\beta'}$,

$$\hat{\rho} = \sum_{\alpha,\alpha'=1}^{D_A} \sum_{\beta,\beta'=1}^{D_B} \rho_{\alpha,\alpha',\beta,\beta'} |\alpha'\rangle_{AA} \langle\alpha| |\beta'\rangle_{B\ B} \langle\beta|. \tag{8.2.46}$$

Schmidt decomposition then assures that alternative bases $\{|\psi_\mu\rangle_A | \mu = 1, \ldots, D_0\}$ and $\{|\phi_\nu\rangle_B | \nu = 1, \ldots, D_0\}$, resp., can be found, with the common dimension $D_0 = \min(D_A, D_B)$. This reduces the total density operator to a double sum

$$\hat{\rho} = \sum_{\nu,\nu'=1}^{D_0} c_\nu^* c_{\nu'} |\psi_{\nu'}\rangle_{AA} \langle\psi_\nu| |\phi_{\nu'}\rangle_{B\ B} \langle\phi_\nu|, \tag{8.2.47}$$

so that the two partial densities diagonalize simultaneously,

$$\bar{\hat{\rho}}_A = \mathrm{tr}_B \left[\sum_{\nu,\nu'=1}^{D_0} c_\nu^* c_{\nu'} |\psi_{\nu'}\rangle_{A\ A} \langle\psi_\nu| |\phi_{\nu'}\rangle_{B\ B} \langle\phi_\nu| \right] = \sum_{\nu=1}^{D_0} |c_\nu|^2 |\psi_\nu\rangle_{A\ A} \langle\psi_\nu|,$$

$$\bar{\hat{\rho}}_B = \mathrm{tr}_A \left[\sum_{\nu,\nu'=1}^{D_0} c_\nu^* c_{\nu'} |\psi_{\nu'}\rangle_{A\ A} \langle\psi_\nu| |\phi_{\nu'}\rangle \langle\phi_\nu| \right] = \sum_{\nu=1}^{D_0} |c_\nu|^2 |\phi_\nu\rangle_{B\ B} \langle\phi_\nu|. \tag{8.2.48}$$

In this form, the two traces indeed commute, and

$$S_A = -\mathrm{tr}_A \left[\bar{\hat{\rho}}_A \mathrm{lb}\left(\bar{\hat{\rho}}_A\right) \right] = -\sum_{\nu=1}^{D_0} |c_\nu|^2 \mathrm{lb}\left(|c_\nu|^2\right) = -\mathrm{tr}_B \left[\bar{\hat{\rho}}_B \mathrm{lb}\left(\bar{\hat{\rho}}_B\right) \right] = S_B. \tag{8.2.49}$$

On the classical level, a surplus of entropy in the subsystems indicates that they are correlated, reflected in a non-zero mutual entropy, Eq. (1.7.10). Transferring the classical definition to quantum entropy, the *quantum mutual entropy* (also known as *entanglement entropy*) between subsystems A and B is introduced as [CA97, NC00] (Fig. 8.4a)

$$S_{A\cap B} := S_A + S_B - S_{AB} = -\mathrm{tr}_A \left[\bar{\hat{\rho}}_A \mathrm{lb}\left(\bar{\hat{\rho}}_A\right) \right] - \mathrm{tr}_B \left[\bar{\hat{\rho}}_B \mathrm{lb}\left(\bar{\hat{\rho}}_B\right) \right] + \mathrm{tr}\left[\hat{\rho}\mathrm{lb}(\hat{\rho})\right]. \tag{8.2.50}$$

If the total system is separable, $S_{A\cap B} = 0$, which represents a lower bound for the mutual entropy (Fig. 8.4b). If it is in a pure state, $S_{AB} = 0$, Eq. (8.2.47) implies in particular that

$$S_{A\cap B} = 2S_A = 2S_B. \tag{8.2.51}$$

This is a remarkable result, because it clearly violates the classical upper bound (1.7.11) for the mutual entropy. Even based on the weaker estimate [CA97]

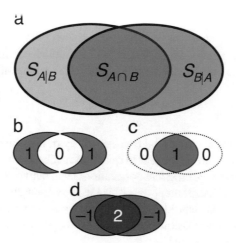

Fig. 8.4 Quantum mutual and conditional entropies, Eqs. (8.2.50, 8.2.54), for a bipartite system, represented (**a**) in a Venn diagram, analogous to Fig. 1.10. Panels (**b, c, d**) visualize these entropies for different degrees of correlation, for two spins with $S_A = S_B = 1$ bit: (**b**) uncorrelated systems (same probabilities 0.5 for $|+\rangle$ (up) and $|-\rangle$ (down) in both spins), (**c**) classically (anti)correlated spins (same probabilities for $|+\rangle_A|-\rangle_B$ and $|-\rangle_A|+\rangle_B$), (**d**) maximally entangled Bell states such as $\hat{\rho}_3$ or $\hat{\rho}_4$ in Eq. (8.2.63) (negative quantum mutual probabilities, see Eq. (8.2.56), cannot be adequately visualized in Venn diagrams). Designed following Fig. 1 in [CA97].

$$S_{A\cap B} \leq 2\min(S_A, S_B), \tag{8.2.52}$$

there remains a regime of extreme quantum correlations, $\min(S_A, S_B) < S_{A\cap B} \leq 2\min(S_A, S_B)$, that exceed all classical bounds.

The definition of quantum mutual entropy, Eq. (8.2.50), can be systematically extended to quantum mutual information for multipartite systems [Kum17]

$$S_{A\cap B\cap C} := S_A + S_B + S_C - S_{AB} - S_{BC} - S_{CA} + S_{ABC} \tag{8.2.53}$$

The signs of mono-, bi-, and tripartite terms in Eq. (8.2.53) can even be interpreted in terms of the mutual overlaps in a corresponding classical Venn diagram, see Figs. 1.11a and 8.4.

A closely related quantity measuring correlation is conditional information. Again following the classical Definition (1.7.14), the quantum conditional entropy of subsystem B, conditioned on the state of A, reads

$$S_{B|A} = -\text{tr}\left[\hat{\rho}\,\text{lb}(\hat{\rho}) - \bar{\hat{\rho}}_A\,\text{lb}(\bar{\hat{\rho}}_A)\right] = -\text{tr}[\hat{\rho}\,\text{lb}(\hat{\rho})] + \text{tr}_A\left[\bar{\hat{\rho}}_A\,\text{lb}(\bar{\hat{\rho}}_A)\right] = S_{AB} - S_A \tag{8.2.54}$$

and $S_{A|B} = S_{AB} - S_B$ (Fig. 8.4a). Moreover, with Eq. (8.2.50),

$$S_{B|A} = S_B - S_{A \cap B}, \quad S_{A|B} = S_A - S_{A \cap B}. \tag{8.2.55}$$

Unlike classical conditional information, quantum conditional and mutual entropies can become negative. If the total system is in a pure state, cf. Eq. (8.2.54),

$$S_{B|A} = -S_B = -S_A = S_{A|B}, \tag{8.2.56}$$

violating again a classical bound, the non-negativity of conditional entropy [CA97] (Fig. 8.4d). Correlations are as present in quantum mechanics as they are in classical physics, but they can exceed classical correlations even between two identical copies of the same system: Entanglement can blur the very identity of spatially separate particles. Quoting Erwin Schrödinger, "The best possible knowledge of a whole does not include the best possible knowledge of its parts—and this is what keeps coming back to haunt us." [Sch35].

As elementary a role entanglement plays for quantum mechanics, as surprisingly difficult is distinguishing it quantitatively from mere classical correlations: A quantitative entanglement measure, easy to use and providing necessary and sufficient conditions for non-classical correlations, is not on hand. Prototypical instances of entanglement, see Subsect. 8.1.4, refer to pure states. However, in order to include mixtures, working with the density operator is indispensable, and it proves easier to start from the opposite condition, separability. A sufficient condition for separability, say of a bipartite system with parts A and B in pure states, would be a density operator factorizing into two blocks, $\hat{\rho}_A$ and $\hat{\rho}_B$, as in Eq. (8.2.42). In order to extend it to mixed states, replace the single product $\hat{\rho}_A \otimes \hat{\rho}_B$ by a weighted sum [Per96, Sim00].

$$\hat{\rho}_{AB} = \sum_{n=1}^{N} p_n \hat{\rho}_{A,n} \otimes \hat{\rho}_{B,n}, \tag{8.2.57}$$

weighted with probabilities p_n, $\sum_{n=1}^{N} p_n = 1$ (note that Eq. (8.2.57) does not coincide with Schmidt decomposition, Eq. (8.2.47)). Suitable conditions for a lack of separability refer directly to the quantum mutual entropy, Eq. (8.2.50): The two subsystems are entangled if their mutual entropy exceeds the classical upper bound $S_{A \cap B} \leq \min(S_A, S_B)$. If a system A comprises three or more subsystems k, $k = 1, \ldots, K$, the mutual entropy gives way to the *quantum redundancy* [HH94].

$$S_{\cap A} := \sum_{k=1}^{K} S_k - S_A \tag{8.2.58}$$

where S_A denotes the entropy of the entire system.

Criteria based on entropic quantities can be cumbersome. A condition addressing directly the block structure of the density operator is the Peres–Horodecki criterion [Per96, HH&96, Hor97, HH&09]. For a bipartite system, separability would require that an operation, transforming valid density operator into valid density operators, should also do so if applied only to one of the two subsystems. Specifically for transposition, this means that a partial transposition, applied only to one subsystem, must produce a density matrix with non-negative eigenvalues. If the two Hilbert spaces are spanned by discrete bases, see Eq. (8.2.46), the partial transposition within B would amount to the operation

$$
\begin{aligned}
T_B : \hat{\rho} &= \sum_{\alpha,\alpha'=1}^{D_A} \sum_{\beta,\beta'=1}^{D_B} \rho_{\alpha,\alpha',\beta,\beta'} |\beta\rangle_B \, {}_B\langle\beta'| \, |\alpha\rangle_A \, {}_A\langle\alpha'| \rightarrow \\
\hat{\rho}^{T_B} &= \sum_{\alpha,\alpha'=1}^{D_A} \sum_{\beta,\beta'=1}^{D_B} \rho_{\alpha,\alpha',\beta',\beta} |\beta\rangle_B \, {}_B\langle\beta'| \, |\alpha\rangle_A \, {}_A\langle\alpha'|.
\end{aligned}
\tag{8.2.59}
$$

The Peres–Horodecki criterion then states that non-negativity of the eigenvalues of the partially transposed density matrix is a necessary condition for its separability, or equivalently, at least one negative eigenvalue is sufficient for entanglement. It has been shown that for Hilbert spaces of 2×2 and 2×3 dimensions, the condition is even necessary and sufficient [HH&96]. A criterion for entanglement, developed specifically for the emblematic case of EPR experiments, Subsect. 8.4.3, is Bell's inequality, Eq. (8.4.29).

All the above criteria have been conceived for finite-dimensional Hilbert spaces. Entanglement in systems with continuous spectra is at least as physically relevant. For this category, separability conditions equivalent to the Peres–Horodecki criterion have been proposed, translating the transposition into a reflection of phase space in the Wigner representation [Sim00], or, specifically for Gaussian states, as bounds on the variances of mixed operators that affect both subspaces [DG&00].

An instructive example to discuss how entanglement becomes manifest in quantum entropies is the pair of two-state systems alluded to in Subsect. 8.1.4. Instead of fixing the directions of polarization of two spins individually, they can be defined through a common property, the relative orientation along two mutually orthogonal axes [Zei99]: Choosing the mixing angles in Eqs. (8.1.44) as $\phi_{\pm} \pm \pi/4$, four different states arise, known as *Bell states* [Bel64]. They share the two available bits evenly between the two subsystems,

$$
\begin{aligned}
|\psi_1\rangle &= 2^{-1/2}(|+\rangle_A|+\rangle_B - |-\rangle_A|-\rangle_B), & |\psi_2\rangle &= 2^{-1/2}(|+\rangle_A|+\rangle_B + |-\rangle_A|-\rangle_B), \\
|\psi_3\rangle &= 2^{-1/2}(|+\rangle_A|-\rangle_B - |-\rangle_A|+\rangle_B), & |\psi_4\rangle &= 2^{-1/2}(|+\rangle_A|-\rangle_B + |-\rangle_A|+\rangle_B),
\end{aligned}
\tag{8.2.60}
$$

(now suppressing subscripts x etc. that indicate the reference axis). Represented in the canonical basis $\{|+\rangle_A|+\rangle_B, |-\rangle_A|+\rangle_B, |+\rangle_A|-\rangle_B, |-\rangle_A|-\rangle_B\}$, they correspond to four-dimensional state vectors

$$|\psi_1\rangle = \frac{1}{\sqrt{2}}\begin{pmatrix} -1 \\ 0 \\ 0 \\ 1 \end{pmatrix}, \quad |\psi_2\rangle = \frac{1}{\sqrt{2}}\begin{pmatrix} 1 \\ 0 \\ 0 \\ 1 \end{pmatrix}, \quad |\psi_3\rangle = \frac{1}{\sqrt{2}}\begin{pmatrix} 0 \\ -1 \\ 1 \\ 0 \end{pmatrix}, \quad |\psi_4\rangle = \frac{1}{\sqrt{2}}\begin{pmatrix} 0 \\ 1 \\ 1 \\ 0 \end{pmatrix},$$

$$(8.2.61)$$

or total density matrices $\hat{\rho}_n = |\psi_n\rangle\langle\psi_n|$, $n = 1,\dots,4$,

$$\hat{\rho}_1 = \frac{1}{2}\begin{pmatrix} 1 & 0 & 0 & -1 \\ 0 & 0 & 0 & 0 \\ 0 & 0 & 0 & 0 \\ -1 & 0 & 0 & 1 \end{pmatrix}, \hat{\rho}_2 = \frac{1}{2}\begin{pmatrix} 1 & 0 & 0 & 1 \\ 0 & 0 & 0 & 0 \\ 0 & 0 & 0 & 0 \\ 1 & 0 & 0 & 1 \end{pmatrix},$$

$$\hat{\rho}_3 = \frac{1}{2}\begin{pmatrix} 0 & 0 & 0 & 0 \\ 0 & 1 & -1 & 0 \\ 0 & -1 & 1 & 0 \\ 0 & 0 & 0 & 0 \end{pmatrix}, \hat{\rho}_4 = \frac{1}{2}\begin{pmatrix} 0 & 0 & 0 & 0 \\ 0 & 1 & 1 & 0 \\ 0 & 1 & 1 & 0 \\ 0 & 0 & 0 & 0 \end{pmatrix}. \quad (8.2.62)$$

Obviously, they are not diagonal nor do they factorize. The corresponding partial density operators are obtained by tracing over subsystem B or subsystem A, resp. The result is the same in all cases,

$$\overline{\hat{\rho}}_{A1} = \overline{\hat{\rho}}_{A2} = \overline{\hat{\rho}}_{A3} = \overline{\hat{\rho}}_{A4} = \begin{pmatrix} 0.5 & 0 \\ 0 & 0.5 \end{pmatrix} = \overline{\hat{\rho}}_{B1} = \overline{\hat{\rho}}_{B2} = \overline{\hat{\rho}}_{B3} = \overline{\hat{\rho}}_{B4}, \quad (8.2.63)$$

which amounts to an inverse participation ratio of $\xi_\rho^{-1} = 2$ and maximum ignorance, $S_A = S_B = 1$bit. It can be reconciled with the fact that we are here dealing with a pair of spins in perfectly well-defined states by taking into account that the partial density matrices project these states onto a basis orthogonal to the subdivision of the two available bits into relative orientations. That is, the Bell states represent a natural subdivision of the system that cuts perpendicularly across the frontier separating the two particles (Fig. 8.5).

Entanglement owes itself here to the combination of two factors, the extreme scarcity of the available information, two subsystems contributing one bit each, and the possibility to create Bell states, which allocate these bits in a way that ignores the partition into subsystems and even their spatial separation. A cartoon by Charles Addams, Fig. 8.6, perfectly visualizes this "separation of the inseparable", even if it was not drawn intending this interpretation. It will become strikingly evident in the discussion of entanglement in pairs of two-state systems, in the context of the Einstein–Podolsky–Rosen paradox in Subsect. 8.4.3, that correlations between the results of simultaneous measurements can exceed the strongest correlation that could be explained by a classical account of the same experiment.

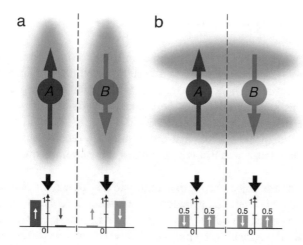

Fig. 8.5 Entanglement in a pair of two-state systems. The total quantum entropy of two bits available to a system comprising two spins A and B can be allocated in different ways to properties of this system. If the two bits (grey clouds) are assigned separately, each one to the individual orientation of one of the spins (panel **a**), the partial entropies likewise factorize into two pure states corresponding to well-defined spin polarizations (bar charts). If by contrast, the two bits refer to relative orientations (**b**), they split in a way that ignores the identity of the two subsystem, cf. Figure 8.6. This results in their maximal entanglement, reflected in complete uncertainty about their individual orientation (bar charts) and in partial entropies of 1 bit each, see Eq. (8.2.45). Two relative orientations in two mutually orthogonal directions, e.g., x and z, are necessary to fix the two bits as in Eq. (8.2.60). They are here symbolized by a single antiparallel polarization.

Fig. 8.6 Entanglement arises when quantum mechanics binds things into inseparable units (qubits), which from a classical point of view form separate entities, as depicted schematically in Fig. 8.5b. Cartoon by Charles Addams, published 1940 in *The New Yorker*. Reproduced with kind permission by The New Yorker Magazine Inc.

8.2.3   Decoherence and Quantum Information

Till here, it may well appear that the concept of information is disposable in quantum mechanics: Every quantum system is in a perfectly well defined state, not leaving any space for uncertainty, hence is assigned zero quantum entropy. Where non-zero values of this quantity arise, it is only because parts of the system are being ignored, despite their strong interaction with the rest. In terms of density operators, off-diagonal elements get lost that are required to reconstruct the underlying pure state, so that a non-zero entropy looks like a mere artefact, owing to improper treatment.

Can we attribute any objective physical reality to this loss? If not, our macroscopic world of classical facts would be revealed as a chimera. Decoherence, the decay of off-diagonal elements of the density, is in fact an indispensable process to connect quantum to classical physics. That it plays such an essential role for the interpretation of quantum mechanics has been recognized some decades ago, upon detailed studies that explained it physically and integrated it in the remaining canonical framework of quantum mechanics.

Also here, viewing microscopic systems as embedded in a macroscopic environment proves essential. In classical statistical mechanics, this is the decisive argument to reconcile time-reversal invariant microscopic equations of motion with macroscopic irreversibility. In quantum mechanics, its role is even more fundamental: Only the coupling to an environment, however weak, allows a microscopic system to be perceived in the first place, to be observed, measured, and manipulated. Indeed, a prototypical instance of decoherence had to be recognized already in the Copenhagen interpretation of quantum mechanics, as the interface connecting Hilbert-space vectors to classical observations: measurement.

As an elementary example, consider a quantum mechanical operator $\hat{A}$, associated to a classical observable A, say with discrete eigenstates, $\hat{A}|n\rangle = A_n|n\rangle$, n integer. An arbitrary normalized state $|\psi\rangle$ of the system, represented in this basis, $|\psi_{\text{ini}}\rangle = \sum_n c_n|n\rangle$ with $\sum_n |c_n|^2 = 1$, corresponds to a density matrix

$$\hat{\rho}_{\text{ini}} = |\psi_{\text{ini}}\rangle\langle\psi_{\text{ini}}| = \sum_{n,n'} c_n^* c_{n'} |n'\rangle\langle n|. \qquad (8.2.64)$$

Its quantum entropy, as a pure state, is $S_{\text{ini}} = 0$. However, the classical output of a measurement of $\hat{A}$ applied to this state should result in a set of probabilities p_n to find one of the eigenvalues A_n, as in Eq. (8.2.1). The corresponding density operator (8.2.3) is reached from the initial state (8.2.64) by the transition

$$\hat{\rho}_{\text{ini}} \to \hat{\rho}_{\text{col}} = \sum_n |c_n|^2|n\rangle\langle n| = \sum_n p_n|n\rangle\langle n|, \; p_n = |c_n|^2, \qquad (8.2.65)$$

that is, by "collapsing" towards diagonal form in the eigenbasis of $\hat{A}$. In this way, the quantum entropy increases to

$$S_{\text{col}} = -\sum_n p_n \text{lb}(p_n) = -\sum_n |c_n|^2 \text{lb}(|c_n|^2) > 0, \qquad (8.2.66)$$

if the system was not already in an eigenstate of $\hat{A}$ before the measurement. This so-called "collapse of the wave function" formed a central point in the Copenhagen interpretation, but remained an obscure process, an alien element incompatible with the otherwise rigorous and consistent mathematical framework of quantum mechanics. Since then, an almost comprehensive physical understanding has been accomplished, based on microscopic models that account explicitly for the degrees of freedom of the environment. In this way, the loss of coherence is resolved as a gradual process going on in finite time [Zur84], amenable to detailed analysis and simulation. Decoherence and quantum measurement will be discussed further in Subsect. 8.3.3 and Sect. 8.4, respectively.

8.3   Dynamics of Quantum Information

In order to understand how information processing in quantum systems goes on in time, it is appropriate to start with an overview over quantum mechanical time evolution in general. In quantum mechanics, as in classical mechanics, we find two very distinct levels of rigour: an exact microscopic account that obeys strong symmetries and conservation laws, and a rather more heuristic level which violates fundamental symmetries but allows integrating important phenomena such as dissipation and irreversibility. In this sense, Hamiltonian mechanics, see Sect. 5, is paralleled in quantum mechanics by unitary time evolution. The central consequence is again the strict conservation of entropy, to be elucidated in the sequel.

8.3.1   Unitary Time Evolution

The rigorous law determining the time evolution of quantum systems represented by Hilbert-space vectors, comparable to Hamilton's equations of motion for classical systems in phase space, is the Schrödinger equation,

$$i\hbar \frac{\mathrm{d}}{\mathrm{d}t} |\psi(t)\rangle = \hat{H} |\psi(t)\rangle. \qquad (8.3.1)$$

Its central element is the Hamiltonian operator $\hat{H}$, analogous to the classical Hamilton's function $H(\mathbf{p}, \mathbf{q})$. The Schrödinger equation has the character of a postulate, one of the cornerstones of quantum mechanics. As such, it cannot be derived from more fundamental principles. However, it can be motivated as a generalization of a field equation for the propagation of light waves, the Helmholtz equation. Like this, it has been conceived by Schrödinger, adapting it to the dispersion relation (energy as a function of wavenumber and frequency) for massive particles instead of massless photons. More important to lend significance and trustability to it are the general properties of quantum mechanical time evolution that it implies. They become more distinct by looking at the transformations of quantum states generated by integrating Eq. (8.3.1), a linear first-order differential equation, over time, which gives rise to a mapping from an "initial" quantum state at time $t = t'$ to a "final" state at $t = t''$,

$$|\psi(t')\rangle \rightarrow |\psi(t'')\rangle = \hat{U}(t'', t')|\psi(t')\rangle \tag{8.3.2}$$

As in Hamiltonian mechanics, time-reversal invariance, see below, implies that "initial" and "final" are interchangeable and do not indicate any time order. The transformation $\hat{U}(t', t'')$ is found to be,

$$\hat{U}(t'', t') = \exp\left(-\frac{i}{\hbar}\hat{H}(t'' - t')\right). \tag{8.3.3}$$

In fact, this integration can even be performed if the Hamiltonian depends on time, $\hat{H} = \hat{H}(t)$. In this case, a more sophisticated integration process (for details, see [Lou73, CDL77]) leads to the relation

$$\hat{U}(t'', t') = \hat{\mathcal{T}} \exp\left(-\frac{i}{\hbar}\int_{t'}^{t''} dt\, \hat{H}(t)\right). \tag{8.3.4}$$

The operator $\hat{\mathcal{T}}$ enforces "time ordering" of the integral in the exponent in the case that Hamiltonians at different times do not commute. It requires interpreting this integral such that the infinitesimal operators $\hat{H}(t)dt$ take effect in chronological order, i.e., $\exp\left(-i\hat{H}(t'')dt/\hbar\right)$ rightmost, $\exp\left(-i\hat{H}(t')dt/\hbar\right)$ leftmost. Equation (8.3.3) is retained if the Hamiltonian does not depend on time. In general, the time evolution is a function of both, the "initial" time t' and the "final" time t''. It only depends on their difference, $\hat{U}(t'', t') = \hat{U}(t'' - t') = \hat{U}(t)$, $t := t'' - t'$, if the Hamiltonian is constant, as in Eq. (8.3.3), and can then be expressed in terms of eigenstates $|E\rangle$ and eigenvalues E of $\hat{H}$, $\hat{H}|E\rangle = E|E\rangle$. If their spectrum is discrete, the time evolution operator can be decomposed,

$$\hat{U}(t) = \sum_n \exp\left(-\frac{i}{\hbar}E_n t\right)\hat{P}_n, \quad \hat{P}_n := |E_n\rangle\langle E_n|. \tag{8.3.5}$$

The time evolution depends on details of the spectrum $\mathrm{spec}(\hat{H}) = \{E_n|n \in \mathbb{N}\}$, but it will invariably be *quasiperiodic*, that is, composed as a sum of trigonometric functions. They are characterized by an alternation between episodes of very low values of the autocorrelation $\langle\psi(0)|\psi(t)\rangle$ ("collapses") and recoveries where

It returns to nearly unity ("revivals"), in a sequence that reflects the spectrum of eigenenergies.

The time-evolution operator $\hat{U}(t', t'')$, as defined in Eqs. (8.3.3) and (8.3.4), belongs to the category of *unitary operators*, which implies a number of constitutive properties of the time evolution generated by the Schrödinger equation:

1. The time-evolution operator obeys a Schrödinger equation of its own. Differentiation of Eqs. (8.3.3, 8.3.4) with respect to t', t'', yields, resp.,

$$\frac{\partial}{\partial t''}\hat{U}(t'', t') = -\frac{i}{\hbar}\hat{H}(t''), \quad \frac{\partial}{\partial t'}\hat{U}(t'', t') = \frac{i}{\hbar}\hat{H}(t'). \qquad (8.3.6)$$

2. The inverse of unitary operators is given by their Hermitian conjugate,

$$\hat{U}(t', t'') = \left(\hat{U}(t'', t')\right)^{-1} = \hat{U}^{\dagger}(t'', t'). \qquad (8.3.7)$$

Equation (8.3.7) embodies *time-reversal invariance* of unitary time evolution.

3. As an immediate consequence of Eq. (8.3.7), the time-evolution operator reduces to the unit operator for identical time arguments,

$$\hat{U}(t, t) = \hat{I}. \qquad (8.3.8)$$

4. Concatenating time steps amounts to multiplying the corresponding operators. Evolving a state in two subsequent stages, $|\psi(t')\rangle \rightarrow |\psi(t'')\rangle \rightarrow |\psi(t''')\rangle$, implies a total time-evolution operator from t' to t''',

$$\hat{U}(t''', t') = \widehat{T}\exp\left(\frac{-i}{\hbar}\int_{t'}^{t''}dt\,\hat{H}(t)\right)$$

$$= \widehat{T}\exp\left(\frac{-i}{\hbar}\int_{t'}^{t''}dt\,\hat{H}(t)\right)\widehat{T}\exp\left(\frac{-i}{\hbar}\int_{t'}^{t''}dt\,\hat{H}(t)\right)$$

$$= \hat{U}(t''', t'')\hat{U}(t'', t'). \qquad (8.3.9)$$

Note that for Eq. (8.3.9) to be valid, any order of t', t'', t''' is allowed, including inverse chronological order or t'' outside the interval $[t', t''']$.

5. Unitary time evolution conserves the scalar product and the norm

$$\left\langle\hat{U}a\middle|\hat{U}b\right\rangle = \langle a|\hat{U}^{\dagger}\hat{U}|b\rangle = \langle a|b\rangle, \quad |\hat{U}|a\rangle|^2 = |a|^2. \qquad (8.3.10)$$

6. As a consequence of the conservation of scalar products and norms, unitary transformations do not alter the dimension of Hilbert spaces and subspaces they are operating on, in analogy to classical canonical transformations conserving

phase-space volume. A corollary in turn is the *no-cloning theorem* [NC00]: Were it possible to copy an original state $|\psi\rangle$ and likewise another original $|\phi\rangle$ by a unitary transformation $\hat{U}$ from their Hilbert space $\mathcal{H}_0$ into an arbitrary state $|\alpha\rangle$ in a different Hilbert space $\mathcal{H}_1$,

$$\hat{U}\left(|\psi\rangle_{\mathcal{H}_0} \otimes |\alpha\rangle_{\mathcal{H}_1}\right) = |\psi\rangle_{\mathcal{H}_0} \otimes |\psi\rangle_{\mathcal{H}_1}, \ \hat{U}\left(|\phi\rangle_{\mathcal{H}_0} \otimes |\alpha\rangle_{\mathcal{H}_1}\right) = |\phi\rangle_{\mathcal{H}_0} \otimes |\phi\rangle_{\mathcal{H}_1},$$
$$(8.3.11)$$

the scalar product of the left-hand sides of Eq. (8.3.11), $\left(_{\mathcal{H}_1}\langle\alpha| \otimes_{\mathcal{H}_0}\langle\psi|\right)\hat{U}^\dagger\hat{U}\left(|\phi\rangle_{\mathcal{H}_0} \otimes |\alpha\rangle_{\mathcal{H}_1}\right) = \langle\psi|\phi\rangle$, could only coincide with the scalar product of the right-hand sides, $_{\mathcal{H}_1}\langle\psi| \otimes_{\mathcal{H}_0}\langle\psi||\phi\rangle_{\mathcal{H}_0} \otimes |\phi\rangle_{H_1} = |\langle\psi|\phi\rangle|^2$ if $\langle\psi|\phi\rangle = 1$ or $\langle\psi|\phi\rangle = 0$, i.e., if the two states to be copied are either identical or orthogonal. In terms of Hilbert space dimensions, $\hat{U}$ would constrict the total space $\mathcal{H}_0 \otimes \mathcal{H}_1$ to the dimension of $\mathcal{H}_0$.

7. Let an operator $\hat{A}$ transform states as $|a\rangle \mapsto |\bar{a}\rangle = \hat{A}|a\rangle$. In an alternative representation reached through a unitary operator $\hat{U}$, say $|a'\rangle = \hat{U}|a\rangle$ and $|\bar{a}'\rangle = \hat{U}|\bar{a}\rangle$, the equivalent transformation is $|\bar{a}'\rangle = \hat{U}|\bar{a}\rangle = \hat{U}\,\hat{A}|a\rangle = \hat{U}\hat{A}\,\hat{U}^\dagger\hat{U}|a\rangle = \hat{A}'\hat{U}|a\rangle =: \hat{A}'|a'\rangle$. Therefore, when Hilbert space vectors are transformed by $\hat{U}$, operators transform as

$$\hat{A} \mapsto \hat{A}' = \hat{U}\hat{A}\hat{U}^\dagger. \qquad (8.3.12)$$

8. It follows immediately that unitary transformations $\hat{U}$ conserve eigenvalues. Given the eigenvalue equation $\hat{A}|a\rangle = a|a\rangle$, we find for a transformed state $|a'\rangle = \hat{U}|a\rangle$,

$$\hat{A}'|a'\rangle = \hat{U}\hat{A}\hat{U}^\dagger\hat{U}|a\rangle = \hat{U}a|a\rangle = a\hat{U}|a\rangle = a|a'\rangle, \qquad (8.3.13)$$

that is, if a is an eigenvalue of $\hat{A}$, then so it is for $\hat{A}' = \hat{U}\hat{A}\hat{U}^\dagger$.

9. As a corollary of the conservation of eigenvalues, the total probability as the expectation value $\langle\hat{\rho}\rangle = 1$ of the density operator is also conserved. In particular, the probability current in position space (8.2.14) fulfills the continuity equation (8.2.15), which means that the probability is conserved even locally.

Resuming these features, they evidence a striking coincidence of unitary transformations in quantum mechanics with canonical transformations in Hamiltonian classical mechanics, see Subsect. 5.1:

– As a whole, unitary transformations form a **dynamical group** (property 2: inverse element, 3: neutral element, 4: closure under concatenation), parameterized by continuous time t.

The time evolution is **perfectly deterministic**, that is, to each state vector at some reference time t' evolves from or into exactly one state vector at a different time t''. States do not split nor merge, they cannot be created nor lost (properties 1, 2, 6). In particular, unitary time evolution never leads into a stable steady state, as this would violate conservation laws, such as properties 2 and 5, notably the conservation of information.

- It **preserves measures as well as distances** in the relevant state space, i.e., in Hilbert space (properties 5, 6, 8).
- As a consequence, also the **entropy is strictly conserved**, see below.

Equation (8.3.12) determines how operators transform under a unitary time evolution. It would be tempting to apply property 7 directly also to density operators. However, while formally, density operators pertain to the category of linear Hermitian operators, ontologically they play the role of descriptors of states, not of actions operating on them. Accordingly, the time evolution of density operators under unitary transformations can be deduced from the unitary evolution of states. For a pure state $\hat{\rho}(t') = |\psi(t')\rangle\langle\psi(t')|$, Eq. (8.3.2) implies that

$$\hat{\rho}(t') \rightarrow \hat{\rho}(t'') = |\psi(t'')\rangle\langle\psi(t'')| = \hat{U}(t'',t')|\psi(t')\rangle\langle\psi(t')|\hat{U}^\dagger(t'',t')$$
$$= \hat{U}(t'',t')\hat{\rho}(t')\hat{U}^\dagger(t'',t'). \tag{8.3.14}$$

This argument applies term by term also to mixed states, $\hat{\rho}(t') = \sum_n p_n |n(t')\rangle\langle n(t')|$, so that the evolution law

$$\hat{\rho}(t'') = \hat{U}(t'',t')\hat{\rho}(t')\hat{U}^\dagger(t'',t'), \tag{8.3.15}$$

as in Eq. (8.3.12), holds generally for density operators—if and only if the time evolution of all states involved is generated by the Schrödinger equation. It means in particular that mixed states may evolve unitarily as do pure states, but they cannot become pure states nor can pure states get mixed by unitary transformations.

Equation (8.3.15) derives from an equation of motion for Hilbert-space vectors, but it can itself be reduced again to a differential equation, now for infinitesimal time steps of elements of the projective Hilbert space. Differentiating Eq. (8.3.15) with respect to time, Eq. (8.3.4) implies

$$\frac{\partial\hat{\rho}}{\partial t}\hat{\rho}(t'') = \left(\frac{\partial}{\partial t''}\hat{U}(t'',t')\right)\hat{\rho}(t')\hat{U}^\dagger(t'',t') + \hat{U}(t'',t')\hat{\rho}(t')\left(\frac{\partial}{\partial t''}\hat{U}^\dagger(t'',t')\right)$$
$$= -\frac{i}{\hbar}\hat{H}(t'')\hat{\rho}(t'') + \frac{i}{\hbar}\hat{\rho}(t'')\hat{H}(t'') = \frac{i}{\hbar}\left[\hat{\rho}(t''),\hat{H}(t'')\right]. \tag{8.3.16}$$

Equation (8.3.18), also known as *von Neumann equation*, can be considered as a quantum mechanical analogue of Liouville's equation (5.2.12).

How does the density operator evolve in time if represented as a distribution on classical phase space, such as the Wigner function, Equation (8.1.17)? In order to transform von Neumann's equation to this representation, apply the Weyl transform (8.2.30) to both sides of Eq. (8.3.16): A corresponding evolution equation for the Wigner function,

$$\frac{\partial}{\partial t} P_W(\mathbf{p}, \mathbf{q}, t) = \left[H_W(\mathbf{p}, \mathbf{q}), P_W(\mathbf{p}, \mathbf{q}, t) \right]_{\text{Moyal}}, \qquad (8.3.17)$$

involves the Weyl transform $H_W(\mathbf{p}, \mathbf{q})$ of the Hamiltonian, cf. Eq. (8.2.30), and the *Moyal bracket*,

$$\left[A_W(\mathbf{p}, \mathbf{q}), B_W(\mathbf{p}, \mathbf{q}) \right]_{\text{Moyal}} := \sum_{n=0}^{\infty} \frac{(i/\hbar)^{2n}}{(2n+1)!} \left(\frac{\partial}{\partial \mathbf{p}_B} \cdot \frac{\partial}{\partial \mathbf{q}_A} - \frac{\partial}{\partial \mathbf{p}_A} \cdot \frac{\partial}{\partial \mathbf{q}_B} \right)$$

$$A_W(\mathbf{p}_A, \mathbf{q}_A) B_W(\mathbf{p}_B, \mathbf{q}_B)|_{\substack{\mathbf{p}_A = \mathbf{p}_B = \mathbf{p} \\ \mathbf{q}_A = \mathbf{q}_B = \mathbf{q}}}. \qquad (8.3.18)$$

In the limit $\hbar \to 0$, it reduces to the leading term $n = 0$, which coincides with the Poisson bracket of classical mechanics, Eq. (5.2.10) [LS98, GPS01]. Equation (8.3.18) can also be written in the form [BI11, SKR13, KOS17]

$$\frac{\partial}{\partial t} P_W(\mathbf{p}, \mathbf{q}, t) = - \left(\frac{\partial/\partial \mathbf{p}}{\partial/\partial \mathbf{q}} \right) \mathbf{j}_W(\mathbf{p}, \mathbf{q}, t), \qquad (8.3.19)$$

defining a quantum quasi-probability current in phase space $\mathbf{j}_W(\mathbf{p}, \mathbf{q}, t)$ as a differential operator acting on the Wigner function $P_W(\mathbf{p}, \mathbf{q}, t)$,

$$\mathbf{j}_W(\mathbf{p}, \mathbf{q}, t) := \left(\begin{array}{c} -\partial/\partial \mathbf{q}_H \\ \partial/\partial \mathbf{p}_H \end{array} \right) \sum_{n=0}^{\infty} \frac{(i\hbar/2)^{2n}}{(2n+1)!} \left(\frac{\partial}{\partial \mathbf{p}_P} \frac{\partial}{\partial \mathbf{q}_H} - \frac{\partial}{\partial \mathbf{p}_H} \frac{\partial}{\partial \mathbf{q}_P} \right)^{2n}$$

$$H_W(\mathbf{p}_H, \mathbf{q}_H) P_W(\mathbf{p}_P, \mathbf{q}_P, t)|_{\substack{\mathbf{p}_H = \mathbf{p}_P = \mathbf{p} \\ \mathbf{q}_H = \mathbf{q}_P = \mathbf{q}}}. \qquad (8.3.20)$$

In general, this current, and even its direction, need not coincide with the corresponding classical quantities. In particular, where the Wigner function takes negative values, the quasi-probability current will go in a direction opposite to the classical flow, see Fig. 8.7.

Evidently, Eq. (8.3.18) is equivalent to a continuity equation for the Wigner function and its current,

$$\frac{\partial}{\partial t} P_W(\mathbf{p}, \mathbf{q}, t) + \text{div}(\mathbf{j}_W(\mathbf{p}, \mathbf{q}, t)) = 0. \qquad (8.3.21)$$

The summation in Eq. (8.3.20) amounts to an expansion in even powers of $\hbar$. Its leading term $\sim \hbar^0$, to be interpreted as the classical limit of the current, is

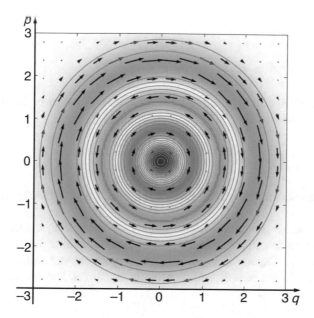

Fig. 8.7 Wigner function (colour code ranging from red (negative) through white (zero) through blue (positive) and grey contour lines) and quasiprobability current (8.3.20) (black arrows) for the third excited state of a harmonic oscillator, Eqs. (8.1.22, 8.1.23) with $n = 3$, $m = 1$, and $\hbar \omega = 1$ (note that the Wigner function of this same state, shown in Fig. 8.1, has been calculated for different values of m and $\hbar \omega$). Reproduced with permission from [BI11]. This representation of a quantum quasiprobability current is analogous to the classical probability density flow in harmonic oscillators, depicted in Fig. 5.18.

$$\lim_{h \to 0} \mathbf{j}_W(\mathbf{p}, \mathbf{q}, t) = \begin{pmatrix} -\partial H_W(\mathbf{p}, \mathbf{q})/\partial \mathbf{q} \\ \partial H_W(\mathbf{p}, \mathbf{q})/\partial \mathbf{p} \end{pmatrix} P_W(\mathbf{p}, \mathbf{q}) = \begin{pmatrix} \dot{\mathbf{p}} \\ \dot{\mathbf{q}} \end{pmatrix} P_W(\mathbf{p}, \mathbf{q})$$

$$= \dot{\mathbf{r}} \, P_W(\mathbf{p}, \mathbf{q}). \tag{8.3.22}$$

and coincides with the classical probability density flow (5.2.19), i.e., the phase-space velocity $\dot{\mathbf{r}} = (\dot{\mathbf{p}}, \dot{\mathbf{q}})$, now weighted with the Wigner function $P_W(\mathbf{p}, \mathbf{q}, t)$ instead of the classical probability density $\rho(\mathbf{r})$.

From here, we can construct an information current in phase space that transports a quantum phase-space entropy density, analogous to the classical information density (5.3.3). The Definition (8.2.31) of quantum entropy as a phase-space average suggests defining a quantum entropy density as

$$s_W(\mathbf{p}, \mathbf{q}, t) := P_W(\mathbf{p}, \mathbf{q}, t)(\mathrm{lb}(\hat{\rho}))_W(\mathbf{p}, \mathbf{q}, t), \tag{8.3.23}$$

where $(\mathrm{lb}(\hat{\rho}))_W(\mathbf{p}, \mathbf{q}, t)$ denotes the Weyl symbol of $\mathrm{lb}(\hat{\rho}(t))$. A corresponding current, measuring the flow of this quantity, is obtained replacing the Wigner function in Eq. (8.3.20) by the entropy density (8.3.23),

$$\mathbf{i}_W(\mathbf{p}, \mathbf{q}, t) := \begin{pmatrix} -\partial/\partial \mathbf{q}_H \\ \partial/\partial \mathbf{p}_H \end{pmatrix} \sum_{n=0}^{\infty} \frac{(i\hbar/2)^{2n}}{(2n+1)!} \left(\frac{\partial}{\partial \mathbf{p}_P} \cdot \frac{\partial}{\partial \mathbf{q}_H} - \frac{\partial}{\partial \mathbf{p}_H} \cdot \frac{\partial}{\partial \mathbf{q}_P} \right)^{2n}$$

$$H_W(\mathbf{p}_H, \mathbf{q}_H) s_W(\mathbf{p}_P, \mathbf{q}_P, t)|_{\substack{\mathbf{p}_H=\mathbf{p}_P=\mathbf{p} \\ \mathbf{q}_H=\mathbf{q}_P=\mathbf{q}}}. \tag{8.3.24}$$

In the classical limit, as in the case of the Wigner quasi-probability current, Eq. (8.3.22), it reduces to

$$\lim_{\hbar \to 0} \mathbf{i}_W(\mathbf{p}, \mathbf{q}, t) = \dot{\mathbf{r}}\, s_W(\mathbf{p}, \mathbf{q}), \tag{8.3.25}$$

to be compared with the classical definition (5.3.6) of an information current in phase space.

Quantum entropy is conserved under unitary time evolution, as will become clear in the next subsection, in the same way as canonical transformations conserve classical information. Caution is in order, though, with respect to both information currents: They do not comply with continuity equations, hence conserve entropy globally but not locally.

8.3.2   Unitary Transformations Conserve Information

The general properties of unitary time evolution comprise a set of symmetries and conservation laws, which considerably restrict quantum dynamics, in particular that of density operators. These conditions are reflected in the time evolution of the quantum entropy. Of special significance is the conclusion that under unitary time evolution, pure states cannot evolve into mixed states (and vice versa). That they are distinguished principally by their respective entropy suggests there might be a more general conservation law behind this verdict. Likewise, the comparison with the dynamics of classical phase-space information and its conservation under canonical transformations (Subsect. 5.3) let's us expect that a similar invariance should apply to quantum entropy under unitary transformations.

An invariance law for the von Neumann entropy, with comparable scope and valid under similar conditions as classical conservation of information, is:

Von Neumann entropy is conserved under unitary transformations, including in particular the time evolution of isolated systems.

As a proof, it would be sufficient to refer to property 8 of unitary transformations (conservation of eigenvalues), since quantum entropy is determined completely by the eigenvalues of the operator function $f(\hat{\rho}) = \hat{\rho} \ln(\hat{\rho})$, and its argument $\hat{\rho}(t)$ evolves unitarily in time [Weh91].

It is more instructive to derive the invariance of $\hat{\rho}(t) \ln(\hat{\rho}(t))$ from its unitary time evolution: Substitute Eq. (8.3.15) in Eq. (8.2.20),

$$S_{vN}(t'') = \text{tr}\left[\rho(t'')\,\text{lb}(\hat{\rho}(t''))\right]$$
$$= -\text{tr}\left[\hat{U}(t'',t')\hat{\rho}(t')\hat{U}^\dagger(t'',t')\,\text{lb}\left(\hat{U}(t'',t')\hat{\rho}(t')\hat{U}^\dagger(t'',t')\right)\right]. \quad (8.3.26)$$

The logarithm can be evaluated further expanding it into a power series around the unit operator $\hat{I}$, as in Eq. (8.2.32),

$$S_{vN}(t'') = -c\,\text{tr}\left[\hat{U}(t'',t')\hat{\rho}(t')\hat{U}^\dagger(t'',t')\sum_{n=1}^{\infty}I_n\left(\hat{U}(t'',t')\hat{\rho}(t')\hat{U}^\dagger(t'',t')-\hat{I}\right)^n\right],$$
$$\quad (8.3.27)$$

with $c:=1/\ln(2)$ and $I_n:=(-1)^{n-1}/n$. Rewriting $\hat{U}(t'',t')\hat{\rho}(t')\hat{U}^\dagger(t'',t')-\hat{I}=\hat{U}(t'',t')\left(\hat{\rho}(t')-\hat{I}\right)\hat{U}^\dagger(t'',t')(t'',t')$, the power series can be simplified to recompose the original logarithm,

$$\sum_{n=1}^{\infty}I_n\left(\hat{U}(t'',t')\hat{\rho}(t')\hat{U}^\dagger(t'',t')-\hat{I}\right)^n = \hat{U}(t'',t')\left(\sum_{n=1}^{\infty}I_n\left(\hat{\rho}(t')-\hat{I}\right)^n\right)\hat{U}^\dagger(t'',t')$$
$$= \hat{U}(t'',t')\ln(\hat{\rho}(t'))\hat{U}^\dagger(t'',t'). \quad (8.3.28)$$

Reinserting it in Eq. (8.3.27) and permuting operators results in

$$S_{vN}(t'') = -c\,\text{tr}\left[\hat{U}(t'',t')\hat{\rho}(t')\hat{U}^\dagger(t'',t')\hat{U}(t'',t')\ln(\hat{\rho}(t'))\hat{U}^\dagger(t'',t')\right]$$
$$= -c\,\text{tr}\left[\hat{\rho}(t')\ln(\hat{\rho}(t'))\right] = S_{vN}(t'). \quad (8.3.29)$$

The conservation of quantum entropy under unitary transformations, Eq. (8.3.29), as the corresponding classical law, deserves a few comments:

- The steps of the proof closely resemble the classical reasoning, Eqs. (5.3.14) to (5.3.17). The trace over the Hilbert space replaces the phase-space integral, and the invariance of the eigenvalues of the density operator under unitary transformations plays the same crucial role as the conservation of phase-space volume under canonical transformations.
- The range of validity of this conservation law is in fact much wider than the category of isolated autonomous systems. The more general expression (8.3.4) for unitary transformations generated by time-dependent Hamilton operators demonstrates that entropy conservation applies likewise to Hamiltonians with well-defined time dependence. This includes not only periodically driven systems (which in general do not conserve energy), but even systems where the time dependence is arbitrary, e.g., a specific realization of a random process.
- Incompatibility with the Second Law is an immediate consequence of classical entropy conservation. Naïvely, one might surmise that it could be resolved by

recurring to quantum mechanics, as it tends to remove other inconsistencies afflicting classical mechanics at the smallest scales. For instance, quantum uncertainty might provide exactly the coarse-graining of fine phase-space structures that is missing in classical mechanics to explain irreversibility and thermalization (Subsect. 7.3.2). The conservation of quantum entropy under unitary transformations frustrates this hope, it represents an even more compelling challenge of the Second Law. See Sect. 8.6.

– Evidently, a time evolution that leads from pure to mixed states (or back) does not conserve information. These processes are not consistent with unitary transformations, notably decoherence. The following subsection gives a brief overview how decoherence can still be accounted for within the framework of quantum mechanics.

8.3.3   Incoherent Processes and Classicality

Decoherence already enters the stage with the Copenhagen interpretation, at least implicitly, as an indispensable step to cross the borderline from Quantum to classical physics. It appears, however, as a black box, an abrupt and obscure transition. Where processes appear abrupt in physics, without being resolved as a gradual change, this usually indicates that our understanding is incomplete. In the case of decoherence, a deeper understanding was inspired from another side, the quantum theory of dissipation developed by Richard Feynman and contemporaries [FV63, CL84].

In both contexts, the challenge is reconciling irreversibility with time-reversal invariant unitary time evolution. The discrepancy can be traced back to what might be seen as an incomplete description: The macroscopic physical account ignores one part of the system that strongly affects its behaviour but is as not as visible as the other ("central") part, the "environment" formed by ambient and internal microscopic degrees of freedom. This deficiency can be corrected for in a straightforward manner: Include the disregarded freedoms in a more comprehensive account, then return to the original perspective by masking out the extra degrees of freedom.

For this strategy, various approaches at different levels of rigour are available. A host of models exist for the environment. The only condition, which however proves crucial, is that it comprises a large number of degrees of freedom, giving rise to a nearly continuous energy spectrum. The most direct way to limit the view to one part of a total system is tracing out the Hilbert-space dimensions of the other part, as in the definition (8.2.30) of reduced density operators. Alternative methods such as projections also exist. Whatever approach is chosen, the result is a time evolution that no longer takes the form of unitary operators but for example that of a master equation for the density operator, generating a semi-group instead of the unitary dynamical group.

It is this step that deliberately sacrifices part of the information available on the total system and thus violates entropy conservation. What appears as an artificial technical measure in the theoretical construction, in fact reflects objective physical railway"' It is analogous to the classical loss of information into microscales, manifest in the coarse-graining of phase-space densities discussed in Subsect. 7.3.2. In quantum mechanics, the consequences are even more drastic. Besides sacrificing information, above all this step erases entanglement with the environment and thus the essence of the quantum mechanical nature of the total system: quantum superpositions become classical alternatives, and classical behaviour of the system emerges.

For the present discussion, suffice it to sketch the outlines of the general reasoning, returning to the example of measurements. The starting point is a Hamiltonian operator that incorporates the minimum ingredients of a measurement on a quantum system. It comprises three terms,

$$\hat{H} = \hat{H}_S + \hat{H}_E + \hat{H}_{SE}, \tag{8.3.30}$$

describing, respectively, the (central) system, the environment, and the interaction between the two. It is plausible to assume that the two subsystems did not interact before the measurement, so that the initial state of the total system factorizes,

$$\lim_{t \to -\infty} |\Psi(t)\rangle = |\psi_{S,\text{ini}}\rangle \otimes |\psi_{E,\text{ini}}\rangle, \tag{8.3.31}$$

$|\psi_{S,\text{ini}}\rangle$ and $|\psi_{E,\text{ini}}\rangle$ denoting the initial states of system and environment, resp. In order to focus on the effect of the interaction, transform to the *interaction picture* [Lou73], that is, to a reference frame that already rotates in Hilbert space according to the time evolution generated by $\hat{H}_S$ and $\hat{H}_E$ alone, $|\Psi(t)\rangle = \hat{U}_0(t)|\Psi^I(t)\rangle$, $\hat{U}_0(t) = \exp\left(-i\left(\hat{H}_S + \hat{H}_E\right)t/\hbar\right)$, cf. Eq. (8.3.3). In this frame, the total state evolves under the additional action of $\hat{H}_{SE}$ as

$$|\Psi^I(t)\rangle = \hat{T} \exp\left(-\frac{i}{\hbar} \int_0^t dt' \hat{H}_{SE}^I(t')\right)|\Psi^I(0)\rangle, \tag{8.3.32}$$

see Eq. (8.3.4), with $\hat{H}_{SE}^I(t) = \hat{U}_0^\dagger(t)\hat{H}_{SE}\hat{U}_0(t)$ [Lou73].

As an elementary form of interaction, consider the two systems to couple through the product of two operators, $\hat{x}_S$ and $\hat{x}_E$, possibly vectors, acting within the Hilbert spaces of system and environment, resp.,

$$\hat{H}_{SE} = \hat{x}_S \cdot \hat{x}_E. \tag{8.3.33}$$

In a basis $\{|x_{S\alpha}\rangle |\hat{x}_S|x_{S\alpha}\rangle = x_{S\alpha}|x_{S\alpha}\rangle\}$ of eigenstates of $\hat{x}_S$, the initial state of the system can be represented as

$$|\psi_{S,\text{ini}}\rangle = \sum_\alpha C_\alpha |x_{S\alpha}\rangle. \tag{8.3.34}$$

In this way, Eq. (8.3.32) already becomes more explicit,

$$|\Psi^{I}(t)\rangle = \sum_{\alpha} C_{\alpha} |\mathbf{x}_{S\alpha}\rangle \exp\left(\frac{-i}{\hbar} \mathbf{x}_{S\alpha} \hat{\mathbf{x}}^{I}_{E\alpha}(t)\right) |\psi_{E,ini}\rangle = \sum_{\alpha} C_{\alpha} |\mathbf{x}_{S\alpha}\rangle |\psi_{E\alpha}(t)\rangle,$$

(8.3.35)

where $|\psi_{E\alpha}(t)\rangle := \exp\left(-i x_{S\alpha} \hat{\mathbf{x}}'_{E\alpha}(t)/\hbar\right) |\psi_{E}(0)\rangle$ is the state of the environment as it evolves in time, in line with the system state $|X_{S\alpha}\rangle$. The reduced density operator for the system is extracted as a trace over the environment,

$$\bar{\rho}^{I}_{S}(t) = tr_{E}\left[|\Psi^{I}(t)\rangle\langle\Psi^{I}(t)|\right] = tr_{E}\left[\sum_{\alpha,\alpha'} c^{*}_{\alpha} c_{\alpha'} |\mathbf{x}_{S\alpha'}\rangle |\psi_{E\alpha'}(t)\rangle \langle\psi_{E\alpha}(t)| \langle\mathbf{x}_{S\alpha}|\right]$$

$$= \sum_{\alpha,\alpha'} C^{*}_{\alpha} c_{\alpha'} \chi_{\alpha,\alpha'}(t) |\mathbf{x}_{S\alpha'}\rangle \langle\mathbf{x}_{S\alpha}|$$

(8.3.36)

The overlaps

$$\chi_{\alpha,\alpha'}(t) = tr_{E}\left[|\psi_{E\alpha'}(t)\rangle\langle\psi_{E\alpha}(t)|\right] \tag{8.3.37}$$

between time-evolved environment states play a decisive role. Suppose that these states, if not identical from the outset, tend to drift further apart in Hilbert space and eventually become orthogonal,

$$\chi_{\alpha,\alpha'}(t) \xrightarrow[t\to\infty]{} \delta_{\alpha'-\alpha}. \tag{8.3.38}$$

For the long-time limit of the reduced density operator, cf. Eq. (8.3.36), this means

$$\bar{\rho}^{I}_{S}(t) \xrightarrow[t\to\infty]{} \sum_{\alpha} |c_{\alpha}|^{2} |\mathbf{x}_{S\alpha}\rangle \langle\mathbf{x}_{S\alpha}|, \tag{8.3.39}$$

that is, it becomes diagonal in the eigenbasis of the coupling operator.

That states of the environment, evolving with different eigenstates of $\hat{\mathbf{x}}_{S}$, approach orthogonality, Eq. (8.3.38), is key for decoherence. To see how it comes about, we have to be more specific about the nature of the central system and the environment. As a simple example, consider a rudimentary model of a spin measurement. The system Hamiltonian

$$\hat{H}_{S} = \frac{1}{2}\hbar\omega_{0}\hat{\sigma}_{z} \tag{8.3.40}$$

contains the spin operator $\hat{\sigma}_z$, the same as the Pauli matrix $\hat{\sigma}_3$ defined in Eq. (8.1.9), and describes a system with two energy eigenstates, $|+_S\rangle$ and $|-_S\rangle$ (spin up or spin down), separated by an energy gap $\Delta E = E_+ - E_- = \hbar\omega_0$.

A similarly elementary model of the environment is a set of N harmonic oscillators as introduced in Eq. (8.1.21), with eigenfrequencies ω_n, described by the Hamiltonian

$$\hat{H}_E = \sum_{n=1}^{N} \hbar\omega_n \left(\hat{a}_n^\dagger \hat{a}_n + \frac{1}{2}\right). \qquad (8.3.41)$$

The operators $\hat{a}_n{}^\dagger$ and $\hat{a}_n$ switch up and down, resp., between the eigenstates of the nth oscillator, cf. Eq. (8.1.22) (*creation* and *annihilation operators* [Lou73, CDL77]). They relate to position and momentum operators of the nth oscillator as

$$\begin{aligned} \hat{q}_{En} &= \sqrt{\hbar/2m\omega_n}(\hat{a}_n^\dagger + \hat{a}_n), \\ \hat{p}_{En} &= i\sqrt{\hbar m\omega_n/2}(\hat{a}_n^\dagger - \hat{a}_n), \end{aligned} \quad \Leftrightarrow \quad \begin{aligned} \hat{a}_n &= \sqrt{m\omega_n/2\hbar}\hat{q}_{En} + i\hat{p}_{En}/\sqrt{2\hbar m\omega_n}, \\ \hat{a}_n^\dagger &= \sqrt{m\omega_n/2\hbar}\hat{q}_{En} - i\hat{p}_{En}/\sqrt{2\hbar m\omega_n}, \end{aligned}$$
$$(8.3.42)$$

confirming the identity of the Hamiltonians (8.3.41) and (8.1.21). For simplicity's sake, choose the system operator $\hat{X}_S$ in the interaction Hamiltonian also as $\hat{\sigma}_z$ and the environment operators as $\hat{x}_{En} = x_{En}\hat{a}_n^\dagger + x_{En}^*\hat{a}_n$, with coupling energies x_{En}, so that

$$\hat{H}_{SE} = \sum_{n=1}^{N} \hat{\sigma}_z \left(x_{En}\hat{a}_n^\dagger + x_{En}^*\hat{a}_n\right). \qquad (8.3.43)$$

With these ingredients, the action of the coupling Hamiltonian on the two subsystems is readily found. The central system remains unaffected since both $\hat{H}_{SE}$ and $\hat{H}_S$ contain $\hat{\sigma}_z$, hence commute with one another. Prepared in a superposition $|\psi_S(0)\rangle = c_-|-_S\rangle + c_+|+_S\rangle$, it will remain there. In the interaction picture, the coupling Hamiltonian acquires time dependence,

$$\hat{H}_{SE}^I(t) = \hat{U}_0^\dagger(t)\hat{H}_{SE}\hat{U}_0(t) = \sum_{n=1}^{N} \hat{\sigma}_z \left(x_{En}(t)\hat{a}_n^\dagger + x_{En}^*(t)\hat{a}_n\right). \qquad (8.3.44)$$

with $x_{En}(t) = x_{En}e^{i\omega_n t}$, $x_{En}^*(t) = x_{En}^*e^{-i\omega_n t}$: Creation and annihilation operators rotate in opposite directions in the complex plane. The effect on environment states becomes clearer if the factor $x_{En}(t)\hat{a}_n^\dagger + x_{En}^*(t)\hat{a}_n$ is rewritten in terms of position and momentum operators of the nth oscillator, $\hat{x}_{En}(t) = \sqrt{2\hbar m\omega_n^3}\hat{q}_{En}(t) + \sqrt{2\hbar\omega_n/m}\hat{p}_{En}(t)$, see Eq. (8.3.42), which shift its ground state to a phase-space point outside the origin, so that

$$H_{SE}^{I}(t) = \hat{\sigma}_z \sum_{n=1}^{N} \left(\sqrt{2\hbar m \omega_n^3} \hat{q}_{En}(t) + \sqrt{\frac{2\hbar \omega_n}{m\omega_n}} \hat{p}_{En}(t) \right), \tag{8.3.45}$$

with $\hat{q}_{En}(t) = (\text{Re}\, x_{En}(t)/\hbar\omega_n)\hat{q}, \hat{p}_{En}(t) = (\text{Im}\, x_{En}(t)/\hbar\omega_n)\hat{p}$. The coupling operator (8.3.45) pushes the oscillators, initially in their ground states $|\psi_{En}(0)\rangle = |0_{En}\rangle$ (coherent states centred at $q_0 = 0$ and $p_0 = 0$ as in Eq. (8.1.33)), out of the origin, to form pairs of coherent states

$$|\gamma_n^{\pm}(t)\rangle = \exp\left[\mp \frac{i}{\hbar} \sum_{n=1}^{N} \left(\sqrt{\frac{2m\omega_n}{\hbar}} \hat{q}_{En}(t) + \sqrt{\frac{2}{\hbar m \omega_n}} \hat{p}_{En}(t) \right) \right] |0_{En}\rangle, \tag{8.3.46}$$

centred at $\mathbf{r}_{0n\pm}(t) = \pm \bar{\mathbf{r}}_{0n} \mp (\tilde{p}_{0n}(t), \tilde{q}_{0n}(t))$,
$\bar{\mathbf{r}}_{0n} = \left(\sqrt{2/\hbar m \omega_n^3} \text{Re}\, x_{En}, \sqrt{2m/\hbar\omega_n} \text{Im}\, x_{En} \right)$ (see Fig. 8.8),

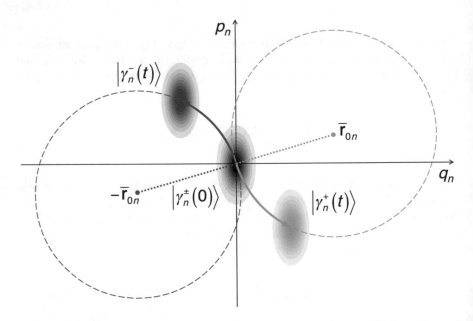

Fig. 8.8 Decoherence results if a system of interest is coupled to an environment that is not observed directly. As the central system gets entangled with the environment, the degrees of freedom of the environment evolve in directions depending on the coupling and on the state of the central system. For a spin coupled to a set of harmonic oscillators, Eqs. (8.3.40–8.3.43), if the harmonic oscillators are initially in their ground states (grey), their time evolved states will depart in opposite directions in phase space, depending on the polarization of the spin, up / + (green) vs. down / − (crimson), and on modulus and phase of the coupling constant x_n. For a coupling as in Eq. (8.3.43), the time evolved states move as coherent states $|\gamma_{n\pm}(t)\rangle$ on eccentric circles (dashed lines), returning periodically to their initial states.

$$\tilde{q}_{0n}(t) = \sqrt{\frac{2}{\hbar m \omega_n^3}} (\mathrm{Re}\, x_{En} \cos(\omega_n t) - \mathrm{Im}\, x_{En} \sin(\omega_n t)),$$

$$\tilde{p}_{0n}(t) = \sqrt{\frac{2m}{\hbar \omega_n}} (\mathrm{Re}\, x_{En} \sin(\omega_n t) + \mathrm{Im}\, x_{En} \cos(\omega_n t)). \tag{8.3.47}$$

Initially, these states depart in opposite directions. For the overlaps defined in Eq. (8.3.37), that means

$$\chi_{+-}(t) = \chi_{-+}(t) = \mathrm{tr}_E \left[\prod_n |\psi_{En}^+(t)\rangle\langle\psi_{En}^-(t)| \right]$$

$$= \exp\left(-\frac{2}{\hbar^2} \sum_n \frac{|x_{Sn}|^2}{\omega_n^2} \left(\sin\left(\frac{\omega_n t}{2}\right) \right)^2 \right). \tag{8.3.48}$$

For short times $t \ll T_n$, $T_n := 2\pi/\omega_n$, the contribution of each oscillator decays exponentially. This is an instance of the *collapses*, alluded to above. However, following a circular trajectory in phase space around a centre $\bar{r}_{0n}$ as sketched in Fig. 8.8, they return periodically, with period T_n, pairwise to the origin, so that the overlap $\langle 0_{n+}(t)|0_{n-}(t)\rangle$ rises to the initial value 1 again, representing *revivals* and evidencing the unitary time evolution underlying this process. The exact time dependence of the reduced density operator $\bar{\rho}_S(t)$ is a complicated function of all the parameters involved and may depend in particular on delicate number-theoretical properties of the spectrum of the environment, comprising all the frequencies ω_n of the harmonic oscillators. However, if their number takes large values $N \gg 1$, we expect the eventual near coincidences of all the individual revivals of the N bath modes to become very rare. Under plausible assumptions concerning the spectrum of the bath and the strength of its coupling to the central system, general conditions for an irreversible time evolution of the reduced density can be obtained.

In particular, in the limit $N \to \infty$ and for a sufficiently smooth spectral density of the environment, its response to the dynamics of the central system becomes *Markovian*, that is, it reacts instantaneously, without delay or memory. For the autocorrelation of the coupling operators, this means, cf. Eq. (7.2.2),

$$\langle \hat{x}_E(t'')\hat{x}_E(t') \rangle = 2T_E \langle \hat{x}_E^2 \rangle \delta(t'' - t'). \tag{8.3.49}$$

The parameter T_E controls the weight of the delta function. The analogy to thermal fluctuations, see Eq. (7.2.2), suggests interpreting it as an autocorrelation time. If it is short compared to all characteristic time scales of the central system, the response of the bath is Markovian. The time evolution of $\bar{\rho}_S(t)$ can then be written as a *master equation*, a differential equation that in general violates the conditions for unitarity, notably the conservation of entropy. For the present example of a spin measurement with Hamiltonians (8.3.40, 8.3.41, 8.3.43) for system, environment, and coupling (replacing individual coupling operators $\hat{x}_{En}$ by a global term $g\hat{x}_E$, $\hat{H}_{SE} = g\hat{x}_S\hat{x}_E$), it reads

$$\frac{\partial}{\partial t}\overline{\tilde{\rho}}_S(t) = \frac{-i}{\hbar}\left[\hat{H}_S, \overline{\tilde{\rho}}_S\right] + g^2 T_E\langle\hat{x}_E^2\rangle\left[\hat{x}_S, \left[\overline{\tilde{\rho}}_S, \hat{x}_S\right]\right]$$
$$+ \frac{1}{2}g^2\gamma\left(\left[\hat{x}_S\overline{\tilde{\rho}}_S, \left[\hat{H}_S, \overline{\tilde{\rho}}_S\right]\right] - \left[\left[\hat{H}_S, \overline{\tilde{\rho}}_S\right], \overline{\tilde{\rho}}_S\hat{x}_S\right]\right). \tag{8.3.50}$$

The first term on the r.h.s. generates unitary evolution of $\overline{\tilde{\rho}}_S(t)$; till here, the master equation coincides with the von Neumann equation (8.3.16). By contrast, the remaining terms induce an irreversible non-unitary time evolution. The second term generates the decay of coherences, hence entails a loss of information. Besides coupling g and response time T_E, it depends on the mean-square fluctuations $\langle\hat{x}_E^2\rangle$ of the environment. The third term describes processes of energy loss towards the bath, such as dissipation. The control parameter γ, related to the imaginary part of the bath response function, is to be interpreted as a friction constant. In the present context, however, only the second term on the r.h.s. is relevant.

If the central system is a spin, Eq. (8.3.50) amounts to an evolution equation for (2 × 2)-matrices and can be solved by expanding both sides in the basis $\{\hat{\sigma}_0, \hat{\sigma}_1, \hat{\sigma}_2, \hat{\sigma}_3\}$ given by the unit operator and the three Pauli matrices, cf. Eq. (8.1.9): $\overline{\tilde{\rho}}_S(t) = \sum_{n=0}^{3} a_n(t)\hat{\sigma}_n$. The master equation thus takes the form

$$\sum_{n=0}^{3} \dot{a}_n(t)\hat{\sigma}_n = \omega_0\left(a_1(t)\hat{\sigma}_2 - a_2(t)\hat{\sigma}_1\right) - 4g^2 T_E\langle\hat{x}_E^2\rangle\left(a_1(t)\hat{\sigma}_1 + a_2(t)\hat{\sigma}_2\right), \tag{8.3.51}$$

which results in differential equations for the individual coefficients $a_n(t)$,

$$\dot{a}_0 = 0, \ \dot{a}_1 = -\lambda a_1 - \omega_0 a_2,$$
$$\dot{a}_3 = 0, \ \dot{a}_2 = -\lambda a_2 + \omega_0 a_1. \tag{8.3.52}$$

abbreviating $\lambda := 4g^2 T_E\langle\hat{x}_E^2\rangle$. The second terms on the r.h.s. of the equations for $\dot{a}_1$ and $\dot{a}_2$ couple them crosswise, generating unitary rotation in the two-dimensional Hilbert space. The first terms induce the decay of the off-diagonal terms of the reduced density. With the solutions for $a_1(t)$ and $a_2(t)$, the time dependence of the off-diagonal elements of the reduced density matrix is found to be

$$\chi_{+-}(t) = \chi_{+-}(0)\exp(-(\lambda + i\omega_0)t) = \chi_{-+}^*(t). \tag{8.3.53}$$

It shows that diagonalization proceeds gradually on a finite timescale $\lambda^{-1} = 1/4g^2 T_E\langle\hat{x}_E^2\rangle$, which reduces with increasing strength of the environment coupling [Zur84]: It takes time, and we can calculate how long it takes. In the long-time limit $t \gg \lambda^{-1}$, the off-diagonal terms vanish.

This process describes a transition of the central system from an initially pure state, Eq. (8.3.31), to a mixture, and a corresponding increase of the entropy in the reduced density. For long times, it approaches the form

$$\overline{\rho}_S(t) = \frac{1}{2}\begin{pmatrix} 1 + a_3 & \sqrt{1 - a_3^2}\exp(-(\lambda + i\omega_0)t) \\ \sqrt{1 - a_3^2}\exp(-(\lambda - i\omega_0)t) & 1 + a_3 \end{pmatrix}, \quad (8.3.54)$$

where a_3, $-1 \le a_3 \le 1$, defines the initial polarization of the spin. From the eigenvalues $\overline{\rho}_{S-+}(t) = \frac{1}{2}\sqrt{1 - a_3^2}(1 \pm e^{(i\omega_0 - \lambda)t}) = \overline{\rho}^*_{S+-}(t)$, the purity, cf. Eq. (8.2.16), is readily found,

$$\xi_{\rho_S}(t) = \mathrm{tr}\left[\left(\overline{\rho}_S(t)\right)^2\right] = \frac{1}{2}\left(1 + a_3^2 + (1 - a_3^2)e^{-2\lambda t}\right) \rightarrow \begin{cases} 1 & t = 0, \\ \frac{1}{2}(1 + a_3^2) & t \gg \lambda^{-1}. \end{cases}$$
$$(8.3.55)$$

Of particular interest are Schrödinger cat states, i.e. here, states without initial polarization, $a_3 = 0$. In this case, the purity reduces to $\frac{1}{2}$ for $t \gg \lambda^{-1}$, and the von Neumann entropy has the asymptotes

$$S_{vN}(t) = \mathrm{tr}\left[\overline{\rho}_S(t)\mathrm{lb}\left(\overline{\rho}_S(t)\right)\right]$$
$$= \frac{1}{2}\left(1 + e^{-\lambda t}\right)\left(1 - \mathrm{lb}\left(1 + e^{-\lambda t}\right)\right) + \frac{1}{2}\left(1 - e^{-\lambda t}\right)\left(1 - \mathrm{lb}\left(1 - e^{-\lambda t}\right)\right)$$
$$\rightarrow \begin{cases} 0 & t = 0 \\ 1 & t \gg \lambda^{-1}. \end{cases} \quad (8.3.56)$$

Equations (8.3.55) and (8.3.56) clearly show the expected transition from a pure state (purity 1, entropy 0) to a state of maximum uncertainty (purity 1/2, entropy 1 bit), illustrated in Fig. 8.9.

If the system operator $\hat{x}_S$ is a spatial coordinate with an infinite discrete or continuous spectrum, Eq. (8.3.50) compares directly to a Fokker–Planck equation, such as Eq. (7.2.18), with the second term on the r.h.s. playing the role of the diffusion term and the third term that of the drift. To substantiate this interpretation, transform the reduced density operator to the interaction picture, cf. Eq. (8.3.36). In the representation of the observable $\hat{p}_S$, canonically conjugate to $\hat{x}_S$, it evolves according to

$$\frac{\partial}{\partial t}\overline{\rho}^I_S(p_S, t) = \langle p_S|\overline{\rho}^I_S(p_S, t)|p_S\rangle = D\frac{\partial^2}{\partial p_S^2}\overline{\rho}^I_S(p_S, t). \quad (8.3.57)$$

Equation (8.3.57) coincides with the diffusion Eq. (7.1.15), indicating that the double commutator $\left[\hat{x}_S, \left[\overline{\rho}_S, \hat{x}_S\right]\right]$ in Eq. (8.3.50) induces diffusion in p_S with a

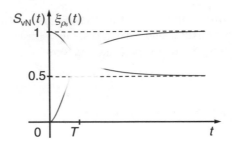

Fig. 8.9 Decoherence evidenced by two quantities, the purity $\xi_{\rho_S}(t)$ (blue) and the entropy $S_{vN}(t)$ (red), Eqs. (8.2.16, 8.2.20). For a spin coupled to a bath of harmonic oscillators, Eqs. (8.3.40–8.43), an initial decay away from a pure state with $\xi_{\rho_S}(0) = 1$ and $S_{vN}(0) = 0$ ends in a complete loss of coherence, approaching $\xi_{\rho_S}(t) \to 0.5$ and $S_{vN}(t) \to 1$ exponentially, see Eqs. (8.3.55, 8.3.56). The figure only covers these asymptotes; the intermediate regime is left blank.

diffusion constant $D = g^2 T_E\langle x_E^2\rangle$. It confers fluctuations in the environment operator $\hat{x}_E$, an unavoidable phenomenon in macroscopic many-body systems, to the central system.

In the representation of $\hat{x}_S$, the master equation takes the form [Zur91]

$$\left\langle x_S'' \left| \frac{d}{dt}\overline{\hat{\rho}}\,_S^{\,I}(t) \right| x_S' \right\rangle = -D(x_S'' - x_S')^2 \left\langle x_S'' \left| \overline{\hat{\rho}}\,_S^{\,I}(t) \right| x_S' \right\rangle \tag{8.3.58}$$

and is solved by

$$\left\langle x_S'' \left| \overline{\hat{\rho}}\,_S^{\,I}(t) \right| x_S' \right\rangle = \exp\left(-D(x_S'' - x_S')^2 t\right)\left\langle x_S'' \left| \overline{\hat{\rho}}\,_S^{\,I}(0) \right| x_S' \right\rangle. \tag{8.3.59}$$

This solution reduces decoherence to the essential: contraction of the density matrix towards the diagonal. It is illustrated in Fig. 8.10 for the collapse of a Schrödinger cat to a classical alternative.

Decoherence, manifest in the contraction of the reduced density matrix to the diagonal described by Eq. (8.3.59), finds a particular application in a context of vital importance for quantum mechanics: measurement. Also the collapse of the wave function during a quantum measurement is an instance of decoherence, caused by the interaction with a macroscopic environment, as will be elucidated in Subsect. 8.4.2.

In summary, the preceding analysis shows that decoherence does not actually destroy entanglement within the components of a small isolated quantum system, but rather aligns it with the macroscopically observable states of the larger system, the environment. As a consequence, the state of the central system no longer behaves as a coherent superposition but attains the character of a classical distribution of probabilities. Classicality thus appears to emerge from the embedding of a quantum system, however small, in a macroscopic environment.

We also see that the coupling to an environment redirects entanglement on the side of the central system very specifically into the eigenspaces of the coupling operator:

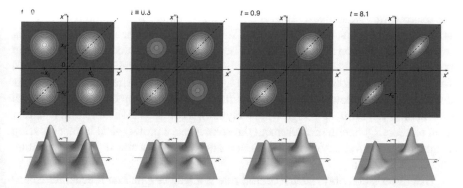

Fig. 8.10 Decoherence visualized as the contraction of the density matrix towards its diagonal $x' = x''$ (upper row: contour plots, lower row: three-dimensional renderings of $\rho(x', x'', t) = \langle x''|\hat{\rho}(t)|x'\rangle$ at times $t = 0, 0.3, 0.9, 8.1$). The initial state ($t = 0$), here a Schrödinger cat state composed of two Gaussian peaks centred at $\pm x_0$ (with $x_0 = 1$), encodes quantum coherence as two off-diagonal peaks of the density matrix at $x' = -x'' = \pm x_0$. Evolving in time according to Eq. (8.3.59), here with $D = 1$, these off-diagonal peaks decay ($t = 0.3, 0.9$) before also the "classical maxima" along the diagonal (dashed lines in the upper row) start contracting ($t = 8.1$).

States initially entangled across these eigenspaces lose coherence, but in this way also become immunized against further decoherence. Eigenspaces of the coupling operator are therefore called *decoherence-free subspaces*, which play a prominent role for example in the context of quantum computation [MSE96, PSE06, LCW98, LW03, Lid13], see Sect. 10.4.3. On the side of the environment, this induces a time evolution that depends on the state of the central system as well. In this way, information is exchanged between the central system and the environment.

This is a two-way process: The central system is altered by and thus receives information on the environment. Originating in a many-body system, it takes the form of noise. Conversely, it also leaves a trace in the macroscopic environment, less elusive than its own state: Decoherence can therefore be seen as a rudimentary instance of observation, even lacking a conscious observer. Quantum measurement, as a process where information in the classical sense is formed in the first place, will be scrutinized in the sequel.

8.4 Quantum Measurement

Decoherence and observation are intimately related: The environment observes the central system. Essential aspects of quantum measurement are anticipated by decoherence. Yet measurement involves a number of additional features that play a fundamental role in quantum mechanics. It may appear as a process occurring only in a very special kind of man-made equipment, used in some high-tech laboratories on planet Earth, hence to be of limited importance for a general physical account of the microworld. Were measurements a constitutive element of quantum mechanics, how could we understand in which way, for example, macroscopic phenomena our telescopes observe on Alpha Centauri arise from quantum processes out there? Yet quantum measurement provides a paradigm of how classical information, "hard facts", take shape in entangled quantum systems that do not represent a uniquely defined reality. It allows us to analyse in depth how bottom-up currents of classical information can form, at a stage where anything like classical information is hardly even seizable. Quantum measurement thus represents a prototypical border crossing from quantum information into classical reality, in some respects akin to quantum chaos, see Sect. 8.5.

That observation plays such a central role is a feature quantum mechanics shares with its near-contemporary companion, Special Relativity Theory: Never before had physical theory reflected so profoundly on the conditions of its own validation. It is the sheer force of experimental facts, incompatible with intuitive concepts of reality, which required reconstructing theory, based on undeniable epistemological insights. In its construction, Special Relativity had to incorporate the fact that information needs a finite time to propagate from event to observer. In quantum mechanics, the very concept of information has to be questioned in its familiar classical meaning and reconsidered from scratch.

This subsection traces quantum measurement through all its stages as a structured process in time, highlighting prominent features that occur at each step. The main stations of a quantum measurement, as outlined originally in Bohr's and Heisenberg's Copenhagen interpretation and worked out in further detail by von Neumann and more recent researchers, are sketched in Subsect. 8.4.1. The central step where the measured system irreversibly interacts with the measurement apparatus, known as the "collapse of the wave function", is analyzed (8.4.2), pointing out particular consequences such as entanglement and non-locality (8.4.3) and the quantum Zeno effect (8.4.4). The final stage is no less important. While it generates classical information in the form of a definite measurement result, it generates at the same time an opposite feature of quantum mechanics that till now has escaped all attempts to integrate it in the formalism of unitary quantum mechanics: Quantum randomness, a kind of unpredictability, unique to quantum processes and to be contrasted with classical randomness, will be the topic of Subsect. 8.4.5. Each of these specific facets of quantum mechanics challenges in one way or other the classical notion of causality. In how far they still allow constructing a concept of causality, compatible with the principles of quantum mechanics, will be considered in Subsect. 8.4.6.

8.4.1 Overview

For the pioneers, among them notably Niels Bohr, it was already evident that measurements are more than a footnote to the main body of quantum mechanics. They define how, as active observers, we can interact with quantum systems, coming from and

equipped with the concepts and tools of classical physics. The principles specifying this interface therefore enjoy the same rank as the other postulates consti-
tuting quantum mechanics. Complementing the list given in Subsect. 8.1.1, they
read [CDL77, NC00, BA&17]

12. A measurement of a physical quantity A can only result in one of the eigenvalues a of the operator $\hat{A}$, the corresponding quantum mechanical observable:
$\hat{A}|a\rangle = a|a\rangle$.

13. If the measured system has been in the state $|\psi_{ini}\rangle$ immediately before a
measurement of the observable $\hat{A}$, the probability to obtain the eigenvalue a
of this operator as a result of the measurement is $p_a = \left|\left\langle \psi \left| \hat{P}_a \right| \psi \right\rangle\right|^2$, denoting
projection operators $\hat{P}_a = |a\rangle\langle a|$.

14. Immediately after a measurement of an observable $\hat{A}$ that resulted in the
eigenvalue a, the measured system will be in a pure state, the corresponding
eigenstate, $|\psi_{fin}\rangle = |a\rangle$.

As a corollary, postulates 12 to 14 imply the repeatability of quantum measurements:
Entering with the final state $|\psi_{fin}\rangle = |a\rangle$ of a previous measurement as initial state
into a repetition of the same measurement, they assert that the measured value and
the state on exit will be the same as previously.

Inherent in these postulates is a sequence of necessary steps and conditions that
together constitute a broad notion of quantum measurement, encompassing everything from phototransduction in rhodopsin in the retina to high-tech particle detectors
at CERN's Large Hadron Collider, the following principal stages emerge, shared by
all quantum measurements:

1. **Preparation**: The input to a quantum measurement should be an initial condition, to be prepared in the laboratory using macroscopic devices, described as
classical objects. In particular, to exclude an uncontrolled bias on the measuring
instrument, it should not be correlated with the measured object ahead of the
measurement.

2. **Interaction**: Observation of the object requires it to interact with the instrument.
The total system comprising object and apparatus (or meter) now evolves in time
as a whole, still following the unitary laws of quantum mechanics. In this way,
their initial separation is lifted, they get entangled. The state of the apparatus is
diverted in different directions, according to the state of the object.

3. **Detection and collapse**: While the interaction of the meter with the object
defines the kind of observable that is measured, it does not yet amplify the
received signal to macroscopic magnitude. The detection proper requires a
macroscopic environment that registers the signal as an irreversible change of
its state. The reduced density operator describing the object (with the meter)
approaches diagonal form, specifically in the basis of eigenstates of the measured
observable. In this partial description, an initial quantum superposition reduces
to an incoherent sum of classical outcomes. This so-called (first) *collapse of the
wave function* is a non-unitary process that develops in the broader framework of

time evolution in the projective Hilbert space (see Subsect. 8.3.3). It epitomizes the above postulates 12 and 13, known together as the *projection postulate*, and is a central tenet in particular of von Neumann's account of quantum measurement. More recent developments relaxed the projection postulate in some directions.

4. **Disentanglement**: To complete the quantum–classical interface, besides leaving a lasting trace in the apparatus, the measurement should release the object in a state compatible again with classical physics. Specifically, the object should exit in the state indicated by the actual value displayed by the instrument, that is, in one of the eigenstates of the measured observable. Moreover, this guarantees that subsequent measurements on the same object will have the same result with probability 1 and thus warrants the reproducibility of the measurement.

This "second collapse", alluded to in step 4, appears to involve solely issues of classical probability, but it is here where a genuine quantum phenomenon, quantum randomness, arises as observable stochasticity.

In the same guise as the phases 1 to 4 of quantum measurement, the structural borders separating object, meter, and environment are not precisely defined. A sharp cut between object and meter as claimed by Heisenberg, hardly reflects reality. Which degrees of freedom of the apparatus are to be associated with the meter proper, which of them with the more passive role of the environment, is a matter of convention. Further ahead, the environment is most appropriately seen with an onion-shell structure. Including a conscious observer as the outermost shell is needless in a merely physical context, but it may be relevant for epistemology.

With these main phases in time, it also becomes apparent which structural components are essential in a quantum measurement:

- The typical **object** of observation in a quantum measurement is a microscopic system with a Hilbert space of low dimension. Systems often used in theoretical modelling are spins and photons.
- The subsystem on the side of the observer to enter into direct interaction with the object is the measurement apparatus or **meter**. Its principal feature is the measured observable, represented by some Hermitian operator, the **pointer** operator, which couples to its counterpart on the side of the object.
- While the meter need not comprise more than a single degree of freedom, it is indispensable for the behaviour of the observer as a classical system that it include an **environment**, macroscopic in the sense that the number of its degrees of freedom be large. Otherwise, no particular internal structure is required, a reservoir of harmonic oscillators is a standard model for it. In some contexts, it is even sufficient to consider the entire apparatus as one macroscopic object, lumping meter and environment into a single subsystem that interacts with the object.
- To achieve its purpose, the result of a measurement should be read by an **observer**, linking quantum measurement to its epistemological context. The further fate of the generated information, as it is perceived, processed, and communicated by an observer immersed in a society, may be highly complex and entail consequences

at disproportionate scales, as indicated in Chaps. 3 and 4. It has been surmised, for example by von Neumann [Neu18], that perception by a conscious mind is even required to complete the measurement or to achieve the collapse of the wave function. To comply with its role as quantum–classical interface, though, such a condition is not necessary. As will be shown below, a macroscopic environment lacking all technical particularities of an apparatus, let alone the complexity of a central nervous system, is sufficient.

In their absolute rigour, all these postulates and distinctions are evidently ideal-izations of physical reality. Quantum measurements are natural physical processes evolving in continuous time, amenable to comprehensive modelling and analysis in the framework of quantum mechanics proper. As an ordered sequence in time, the above steps are not disjoined and may well overlap in time. The collapse of the wave function is not instantaneous but proceeds gradually in finite time, to be detailed below, and practically never results in a complete elimination of quantum coherence. It can hardly be expected, either, that the second collapse would produce a perfectly pure state on output.

Also the postulate of repeatability of measurements invites to being released. A less restrictive approach to quantum measurement, generalizing von Neumann's theory in this sense, has been developed around the concept of *Positive operator-valued measurement* (POVM) [Per95, NC00, BA&17]. It replaces the projection operators $\hat{P}_a = |a\rangle\langle a|$ of postulate 13 by a set of positive Hermitian operators $\hat{F}_n = \hat{M}_n^{\dagger}\hat{M}_n$, which are complete, $\sum_n \hat{F}_n = \hat{I}$, cf. Eq. (8.1.7), but need not satisfy the condition $\hat{F}_n^2 = \hat{F}_n$ defining projectors [CDL77]. If the outgoing state $|\psi_{\mathrm{fin}}\rangle = \hat{M}|\psi\rangle / \sqrt{\langle \psi | \hat{M}^{\dagger}\hat{M} | \psi\rangle}$ of a previous measurement goes through the same procedure again, it will then not necessarily be reproduced identically. POVMs are a useful mathematical tool to describe, for example, measurements that destroy the observed system, such as in photon detection.

8.4.2 Von-Neumann Theory of Quantum Measurement

The three-component structure sketched above is incorporated in a total Hamiltonian comprising three self-energies and three interaction terms,

$$\hat{H} = \hat{H}_{\mathrm{O}} + \hat{H}_{\mathrm{M}} + \hat{H}_{\mathrm{E}} + \hat{H}_{\mathrm{OM}} + \hat{H}_{\mathrm{ME}} + \hat{H}_{\mathrm{EO}}, \tag{8.4.1}$$

(with obvious notation), or simply, if no distinction is required between meter and environment,

$$\hat{H} = \hat{H}_{\mathrm{O}} + \hat{H}_{\mathrm{OM}} + \hat{H}_{\mathrm{M}}, \tag{8.4.2}$$

analogous to Eq. (8.3.30). Before the measurement (stage 1), the observer—Iustitia with her blindfold—approaches the object without preliminary knowledge or bias: Object and meter should be initially uncorrelated, which suggests adopting also the factorization of the total initial state, as in Eq. (8.3.31),

$$|\Psi_{\text{ini}}\rangle = |\Psi_{\text{O,ini}}\rangle \otimes |\Psi_{\text{M,ini}}\rangle \otimes |\Psi_{\text{E,ini}}\rangle. \tag{8.4.3}$$

Switching on the interaction, information on the object is transferred to the meter, given that the time evolution of the meter depends on the state of the object. A suitable general form of the object-meter interaction is a product, as in Eq. (8.3.33), $\hat{H}_{\text{OM}} \sim \hat{\mathbf{x}}_{\text{O}} \cdot \hat{\mathbf{x}}_{\text{M}}$, of a pointer operator $\hat{\mathbf{x}}_{\text{O}}$ on the object side and a meter operator $\hat{\mathbf{x}}_{\text{M}}$. They can be vectors, to model measurements of vectorial quantities. Measurements do not last forever, they take place within a finite time segment, to allow for some resolution in time of the observation. Unlike Eq. (8.3.33), the coupling by $\hat{H}_{\text{OM}}$ is therefore time dependent,

$$\hat{H}_{\text{OM}}(t) = \hat{\mathbf{x}}_{\text{O}} \cdot \hat{\mathbf{x}}_{\text{M}} g(t). \tag{8.4.4}$$

The modulation $g(t)$, e.g., a box function with sharp or smooth shoulders, vanishes for times t outside a window $[t_{\text{ini}}, t_{\text{fin}}]$. Its duration is typically short on characteristic timescales of the object state: The measurement takes a snapshot of it.

The subsequent time evolution correlates object and meter, giving way to a coupled evolution in their common Hilbert space $\mathcal{H}_{\text{OM}} = \mathcal{H}_{\text{O}} \otimes \mathcal{H}_{\text{M}}$. Decomposing the common state in the canonical product basis $|\gamma_{\text{OM}}\rangle = |\alpha_{\text{O}}\rangle \otimes |\beta_{\text{M}}\rangle$, formed by states $|\alpha_{\text{O}}\rangle$ and $|\beta_{\text{M}}\rangle$ of object and meter, resp., this step amounts to the transition

$$|\Psi_{\text{O,ini}}\rangle \otimes |\Psi_{\text{M,ini}}\rangle \rightarrow |\Psi_{\text{OM,corr}}(t)\rangle = \sum_{\gamma} c_{\gamma}(t)|\gamma_{\text{OM}}\rangle, \tag{8.4.5}$$

where the time-dependent coefficients $c_{\gamma}(t)$ no longer factorize. In this way, the object state is imparted to the meter. In particular, it now shares entanglement on the side of the object. The common evolution depends on the coupling Hamiltonian $\hat{H}_{\text{OM}}$. If, for example, the object is a two-state system as in Eq. (8.3.40), with a Hilbert space spanned by $\{|-_{\text{O}}\rangle, |+_{\text{O}}\rangle\}$, prepared in a Schrödinger-cat state

$$|\Psi_{\text{O,ini}}\rangle = 2^{-1/2}(|-_{\text{O}}\rangle \pm |+_{\text{O}}\rangle), \tag{8.4.6}$$

this step takes the form

$$
\begin{aligned}
|\Psi_{\text{O,ini}}\rangle &= 2^{-1/2}(|-_{\text{O}}\rangle \pm |+_{\text{O}}\rangle) \otimes |\Psi_{\text{M,ini}}\rangle \\
&\rightarrow 2^{-1/2}(|-_{\text{O}}\rangle \otimes |-_{\text{M}}\rangle \pm |+_{\text{O}}\rangle \otimes |+_{\text{M}}\rangle).
\end{aligned} \tag{8.4.7}
$$

The object state is entangled with the meter, but a macroscopic record does not yet exist. Enters the environment. While the dimension of the Hilbert space of the meter may still be comparable to that of the object, it is indispensable that the environment be a many-particle system, $D_{\mathcal{H}_E} \gg 1$. The coupling of the object-meter system to the environment extends correlations, as in Eq. (8.4.5), to the entire system,

$$\left| \Psi_{\text{OM,corr}}(t) \right\rangle \otimes \left| \Psi_{\text{E,ini}} \right\rangle \rightarrow \left| \Psi_{\text{OME,corr}}(t) \right\rangle = \sum_{\Gamma} c_{\Gamma}(t) | \Gamma_{\text{OME}} \rangle, \qquad (8.4.8)$$

with states $| \Gamma_{\text{OME}} \rangle$ spanning the high-dimensional Hilbert space $\mathcal{H}_{\text{OME}} = \mathcal{H}_O \otimes \mathcal{H}_M \otimes \mathcal{H}_E$. For the measurement process as a whole, the decisive difference is that the state of a macroscopic apparatus now bears a trace the state of the object. Like this, the record becomes by far more stable than the quantum state of the object itself [BZ06]. Based on the consensus of multiple concordant records in the degrees of freedom of the environment, this phenomenon has much in common with the strategy of *distributed ledgers*, employed in network databases to increase their reliability and forming the nucleus in particular of the block chain technology [Wal16]. Like this, the object state leaves a fingerprint representing classical information, in the form of a multitude of redundant copies, in the state of the environment.

This stabilizing effect in turn retroacts on the object itself, in two ways: It privileges the representation of the object state in the basis of eigenstates $| \alpha_O \rangle$ of the object-meter coupling, the *pointer basis* [Zur81]. While entanglement within the object across this basis is suppressed, it is enhanced with the environment through the meter. For the object, this results in decoherence, as pointed out in Subsect. 8.3.3: If the number N of its degrees of freedom takes macroscopic values, $N \gg 1$, and its spectrum is sufficiently smooth, the reduced density operator of the object-meter system undergoes an irreversible time evolution

$$\hat{\rho}_{\text{OM}} = \text{tr}_E \left[\hat{\rho}_{\text{OME}} \right] = \text{tr}_E \left[\sum_{\Gamma, \Gamma'} c_{\Gamma}^*(t) c_{\Gamma'}(t) | \Gamma_{\text{OME}} \rangle \langle \Gamma'_{\text{OME}} | \right] \sum_{\alpha} p_{\alpha} | \alpha_{\text{OM}} \rangle \langle \alpha_{\text{OM}} |.$$

$$(8.4.9)$$

As detailed in Subsect. 8.3.3, a microscopic model of the environment allows to resolve this process as a continuous contraction of the reduced density matrix, represented in the pointer basis, to its diagonal elements, see Eqs. (8.3.58, 8.3.59). It is even possible to specify a non-zero timescale of this contraction in terms of the parameters, such as the correlation time of random fluctuations, characterizing the environment. This settles a question that remained unthinkable in the Copenhagen interpretation [Zur84]: How long takes the collapse of the wave packet?

The diagonal elements p_{α}, $\sum_{\alpha} p_{\alpha} = 1$, specify probabilities for the measurement to result in an eigenstate $\hat{\mathbf{x}}_O | \alpha_O \rangle = x_O | \alpha_O \rangle$ of the pointer operator. According to postulate 13 of quantum measurement, this probability must be given by the overlap of the system state $| \Psi_{\text{O,ini}} \rangle$ before the measurement with the pointer state $| \alpha_O \rangle$,

$$p_\alpha = \left| \langle \alpha_O | \Psi_{O,\text{ini}} \rangle \right|^2. \tag{8.4.10}$$

For the object prepared in the superposition (8.4.6), this means

$$\hat{\rho}_{OM} \rightarrow \frac{1}{2} (|-_O\rangle|-_M\rangle\langle-_M|\langle-_O| + |+_O\rangle|+_M\rangle\langle+_M|\langle+_O|). \tag{8.4.11}$$

This manifestly non-unitary step constitutes the collapse of the wave function. In the specific form given in Eqs. (8.4.9–8.4.11), resulting a weighted sum over projection operators, it has been introduced by Lüders [Lud50]. For the emergence of classical information, the collapse of the wave function is essential. It explains how a set of alternative macroscopic classical states weighted by probabilities, hence representing a positive amount of information, can arise from a microscopic pure quantum state with zero quantum entropy, blatantly violating the conservation of information, Eq. (8.3.29).

While on the face of it, the effect of the collapse is to *disrupt* entanglement, it also *enhances* it, namely within the subspaces of Hilbert space spanned by each state of the pointer basis. Where they comprise spatially separated subsystems, they can give rise to bizarre quantum effects such as non-locality, to be addressed in Subsect. 8.4.3. At the same time, the augmented stability of the pointer states tends to prevent the evolution of the measured object away from these states, an effect known as *watchdog effect* or *quantum Zeno paradox*, elucidated further in Subsect. 8.4.4.

8.4.3 Entanglement and Non-Locality in Quantum Measurement

Albert Einstein's sceptical attitude towards quantum mechanics, expressed in popular sayings and quotes such as "spooky action at a distance" or "God does not throw dice", is well known. As early as 1935, Einstein, together with Boris Podolsky and Nathan Rosen, cast his most urgent doubts in a publication [EPR35] that was to elicit a discussion that lasts to the present day and is by no means settled. Surprisingly, this 1935 paper only addresses a mild part of Einstein's criticisms, concerning the effect a measurement on one system may have on the state of another, apparently not interacting with the first, and it contains only a rudimentary version of what later, under the title of EPR experiment, should become a paradigm of the paradoxes related to quantum mechanical measurement. Of the three issues summarized as EPR paradox,

– restrictions of simultaneous measurements of **non-commuting observables**,
– **instantaneous correlations** between spatially separated measurements (quantum non-locality, "spooky action at a distance"),
– **absolute unpredictability** of measurement outcomes if the incoming state is a Schrödinger cat state (quantum randomness, "God does not throw dice"),

the EPR paper explicitly discusses only the first one. In what follows, non-locality
will be analyzed in the light of quantum information. Subsection 8.4.5 will focus on
quantum randomness.

In their paper, Einstein, Podolsky, and Rosen sketch the outlines of a gedankenexperiment, intended to bring out the consequences of measuring one observable on
the possible outcomes of measurements of another, not commuting with the first one.
They point out an interdependence of the two observations, a situation unprecedented
in classical physics and downright inacceptable for the authors. measurements of
different observables constitute independent elements of reality, they must not affect
each other.

Let a pair of free particles be described by the (fictitious) common wave function

$$\Psi(x_1, x_2) = \delta(x_2 - x_1 - x_0) = \frac{1}{2\pi\hbar} \int_{-\infty}^{\infty} dp \exp\left(\frac{2\pi i}{\hbar} p(x_2 - x_1 - x_0)\right).$$

$$(8.4.12)$$

It fixes the distance separating their positions at the value x_0 and obviously does not
factorize. Measuring the momentum of particle 1, say with the result p, however,
picks out this value from the flat continuous momentum spectrum of this state and
thus forces the common wave function to factorize,

$$\Psi(x_1, x_2) = \psi_{1,p}(x_1)\psi_{2,p}(x_2),$$
$$\psi_{1,p}(x_1) = \exp\left(\frac{-2\pi i}{\hbar} px_1\right), \quad \psi_{2,p}(x_2) = \exp\left(\frac{2\pi i}{\hbar} p(x_2 - x_0)\right), \quad (8.4.13)$$

imposing a likewise sharp momentum $p_2 = -p$ also on particle 2. On the other hand,
a position measurement on particle 1 with result x enforces a similar factorization,
now fixing positions, not momenta,

$$\Psi(x_1, x_2) = \psi_{1,x}(x_1)\psi_{2,x}(x_2),$$
$$\psi_{1,x}(x_1) = \delta(x_1 - x), \quad \psi_{2,x}(x_2) = \delta(x_2 - x - x_0).$$

$$(8.4.14)$$

That is, measurement on EPR pairs force their common wave function into
a product of independent eigenstates of the measured observable, simultaneously
for both subsystems. The authors conclude that a theory which either attributes
one feature (momentum) or the other (position) to reality, but not both, as does
quantum mechanics, cannot provide a complete description of reality. They are not
concerned with the question of correlations between the two particles implied by
their simultaneous observation, an issue that only later comes into focus.

A tightened version of the EPR experiment that should provoke countless studies
and publications around quantum measurement is proposed in 1957 by Bohm and
Aharonov [BA57]. They apply the argument put forward by Einstein, Podolsky, and
Rosen to a two-state system, a spin-$\frac{1}{2}$ particle, to construct a gedankenexperiment
with even more intriguing consequences than evoked by the original version.

The central motif of all EPR experiments is the dependence of the state of the measured system and the result of the measurement on what is measured and how, an idea that defies the classical conception that an observation should display but not alter reality (for purely classical exceptions to this rule, such as self-fulfilling prophecy, see Subsect. 3.2.2). In order to bring out this effect as drastically as possible, it is reasonable to choose the smallest quantum system as an example that meets all requirements, a correlated pair of two-state systems. Decades before the advent of quantum information, Bohm and Aharonov conceive a scheme for the observation of qubits (Fig. 8.11) that amplifies the consequences of the back-action on the object to the extreme.

Consider a decay process creating pairs of particles that inherit a conserved quantity from the parent particle, for example the disintegration of a two-atom molecule with total spin 0 (a singlet) into two atoms, A and B, with spin $\frac{1}{2}$ each, or a radioactive decay or electron–positron annihilation process that emits a pair of photons, polarized along mutually normal directions. They are created in a singlet state, a maximally entangled state of the same form as Eq. (8.1.44),

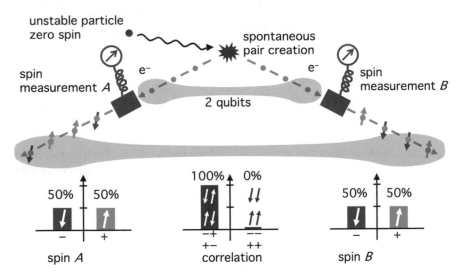

Fig. 8.11 The Einstein–Podolsky–Rosen experiment in the version proposed in 1957 by Bohm and Aharonov [BA57]. A pair of particles is created in a decay process that conserves the value of a quantity inherited from the parent particle, such as the total spin or the total polarization. If these particles are in an eigenstate of that quantity, measurements of another observable, not commuting with it, will have statistically distributed outcomes, for example equidistributed between two alternative possible values (such as spin up vs. spin down). Notwithstanding, the sum rule inherited from the parent particle implies that the results of simultaneous measurements of this observable on both particles are perfectly correlated (for example, anticorrelated between spin up and spin down), even if the two measurements are spatially separated in the sense of Special Relativity. This extreme correlation defies classical concepts, such as the independence of non-interacting particles. For details, see text.

$$|\psi_0\rangle = 2^{-1/2}(|-\rangle_1|+\rangle_2 - |+\rangle_1|-\rangle_2), \tag{8.4.15}$$

where the signs $+$, $-$ may refer to the spin orientation in any spatial direction of reference, to the sense of circular polarization (right / left), or one of a pair of orthogonal directions of linear polarization. They move apart in different directions and, once sufficiently separated to exclude all possible direct interactions, undergo simultaneous measurements of this same observable.

On the face of it, Eq. (8.4.15) implies that the outcomes of such measurements must be maximally anticorrelated. This is the EPR paradox in a nutshell: Entanglement leads to strong correlations between two events, even if they cannot interact with one another. What is more, while entanglement becomes manifest in anticorrelated outcomes, it is at the same time lifted by the measurement: Afterwards, the EPR pair is in a product state

$$|\Psi_{\text{fin}}\rangle = \begin{cases} |-\rangle_1|+\rangle_2 \text{ with probability } 0.5, \\ |+\rangle_1|-\rangle_2 \text{ with probability } 0.5. \end{cases} \tag{8.4.16}$$

The physics of such a process is much richer than these alternatives, however. Measurements of spins with a Stern-Gerlach magnet, of photon polarization with a polarizer and a counter, etc., comprise one or more control parameters to be set by the experimenter, defining the spatial orientation of the magnetic field, the direction of the polarizer in a plane perpendicular to the direction of propagation, etc. Denote the probability of the outcomes $A, B \in \{+,-\}$ for particles 1, 2 at their respective apparatus with settings $\mathbf{a}$, $\mathbf{b}$, resp., as $P(A,B|\mathbf{a},\mathbf{b})$. For a singlet state (8.4.14), quantum mechanics predicts the expectation value [Bel64]

$$S_{12}(\mathbf{a}, \mathbf{b}) := \langle(\sigma_1 \cdot \mathbf{a})(\sigma_2 \cdot \mathbf{b})\rangle = -\mathbf{a} \cdot \mathbf{b}, \tag{8.4.17}$$

averaging over the spin orientations σ_1 and σ_2 in three dimensions. If, for example, $\mathbf{a}$ and $\mathbf{b}$ are reduced to the angles α, β, with respect to some reference direction, of two Stern-Gerlach magnets with their fields perpendicular to the lines of flight of the two spins, Eq. (8.4.17) implies

$$P(+, +|\alpha, \beta) = P(-, -|\alpha, \beta) = \frac{1}{2}\left(\sin\left(\frac{\alpha-\beta}{2}\right)\right)^2 = \frac{1}{4}(1 - \cos(\alpha - \beta)),$$

$$P(+, -|\alpha, \beta) = P(-, +|\alpha, \beta = \frac{1}{2} - \frac{1}{2}\left(\sin\left(\frac{\alpha-\beta}{2}\right)\right)^2 = \frac{1}{4}(1 + \cos(\alpha - \beta)).$$
$$\tag{8.4.18}$$

that is, perfect anticorrelation (correlation) for parallel (antiparallel) fields, $|\alpha{-}\beta| = 0$ (π) and absence of correlations for $|\alpha{-}\beta| = \pi/2$. Indeed, in this case, the settings are two-dimensional unit vectors, $|\mathbf{a}| = |\mathbf{b}| = 1$, and [Bel64]

$$S_{12}(\mathbf{a}, \mathbf{b}) = P(+, +|\alpha, \beta) + P(-, -|\alpha, \beta) - P(+, -|\alpha, \beta) - P(-, +|\alpha, \beta)$$
$$= -\cos(\alpha - \beta) = -\mathbf{a} \cdot \mathbf{b}. \tag{8.4.19}$$

This far, the correlations implied by entanglement are surprising, but do not yet appear incompatible with classicality. After all, correlation does not imply causation (see Sect. 3). Instead of measurement 1 causing the outcome of measurement 2 or vice versa, their anticorrelation might be explained by a not directly observable common cause contained in the decay process, other than the mere sum rule requiring opposite polarization. Such hypotheses, studied as *hidden-variable theories* since the advent of the EPR paradox, have fuelled much of the debate around it. Thanks to the conception of experiments [CH&69] testing quantum mechanics against classical reconstructions and their realization since the 1980ies, using state-of-the-art quantum optical equipment [AGR81, AGR82, ADR82], hidden-variable theories are now ruled out with overwhelming evidence.

Even as this question appears to be settled, it is instructive to analyze the reasoning that finally led to this conclusion. It reveals the subtle points where quantum mechanics is indeed fundamentally distinct from classical physics, and suggests how to interpret them in terms of information (Figs. 8.12 and 8.13).

A common cause explaining the correlations in question can be included as a parameter, say λ (possibly a vector), which is inherited from the parent decay to both product particles and to be added as a common element to the individual settings **a** and **b**, but inaccessibility to direct observation. This implies that only averages

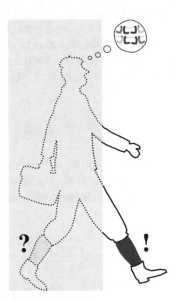

Fig. 8.12 A "Les chaussettes de M. Bertlmann et la nature de la réalité" [Bel81]. Bertlmann's socks are an imaginative metaphor for an interpretation of quantum correlations in EPR experiments in terms of classical causation: Bertlmann the epistemologist, a fictitious (but inspired by an existing) personality, is a genuine madcap. He sports the spleeny habit of never wearing socks of the same design—worse even, of consistently combining socks of complementary colour. The tint of his socks is absolutely unpredictable, but as soon as one sock is seen, the colour of the other is invariably determined.

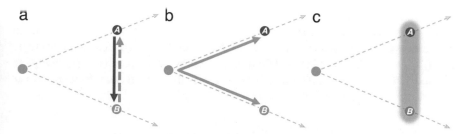

Fig. 8.13 Three alternative options to interpret quantum correlations in EPR experiments in information theoretic terms. Direct causation from measurement A to B or from B to A (panel **a**) is excluded by the correlation time that is shorter than the lower bounds implied by Special Relativity. Simultaneous causation by a concealed common cause in the past ("hidden-variable theories", panel **b**), epitomized in Bertlmann's socks, Fig. 8.12, is ruled out because quantum correlations violate Bell's inequalities, Eq. (8.4.29). The only option compatible with experimental data is to consider the two measurements as probing one and the same entangled state, involving only one indivisible qubit.

weighted by a probability density $\rho(\lambda)$ are physically relevant. The intended causal structure, relating measurements at A and at B as common effects to a decay specified by λ, requires the probabilities $P(A,B|a,b)$ to factorize as soon as their dependence on λ is taken into account, $P(A, B|\mathbf{a}, \mathbf{b}, \lambda) = P_1(A|\mathbf{a}, \lambda)P_2(B|\mathbf{b}, \lambda)$ (but only then). With the common probability

$$P(A, B) = \int_\Lambda d\lambda\, P_1(A|\mathbf{a}, \lambda)P_2(B|\mathbf{b}, \lambda), \tag{8.4.20}$$

averaged over an appropriate range Λ of the parameter λ, the expectation value (8.4.19) takes the form

$$S_{12}(\mathbf{a}, \mathbf{b}) = \int_\Lambda d\lambda \rho(\lambda)\Big(P_1(+|\mathbf{a}, \lambda) - P_1(-|\mathbf{a}, \lambda)\Big)\Big(P_2(+|\mathbf{b}, \lambda) - P_2(-|\mathbf{b}, \lambda)\Big). \tag{8.4.21}$$

The parenthesized factors under this integral can be considered as average outcomes of measurements at A and B, resp., given λ, $\langle A(\mathbf{a}, \lambda)\rangle = (+1)P_1(+|\mathbf{a}, \lambda) + (-1)P_1(-|\mathbf{a}, \lambda)$, $\langle B(\mathbf{b}, \lambda)\rangle = (-1)P_2(+|\mathbf{b}, \lambda) + (+1)P_2(-|\mathbf{b}, \lambda)$, which suggests writing Eq. (8.4.21) as

$$S_{12}(\mathbf{a}, \mathbf{b}) = \int_\Lambda d\lambda\rho(\lambda)\langle A(\mathbf{a}, \lambda)\rangle\langle B(\mathbf{b}, \lambda)\rangle. \tag{8.4.22}$$

Since the probabilities entering the averages $\langle A(\mathbf{a}, \lambda)\rangle$ and $\langle B(\mathbf{b}, \lambda)\rangle$ are restricted to the interval $[0,1]$, the moduli of these averages are likewise bounded

$$|\langle A(\mathbf{a}, \lambda)\rangle| \le 1, \quad |\langle B(\mathbf{b}, \lambda)\rangle| \le 1. \tag{8.4.23}$$

As a patent example, take the unit vector in the common direction of polarization of the two spins, say $\lambda = (\cos\theta \cos\phi, \cos\theta \sin\phi, \sin\theta)$ with $|\lambda| = 1$, as the hidden classical variable and assume the expected outcomes of the two measurements to be $\langle A(\mathbf{a}, \lambda)\rangle = \text{sgn}(\mathbf{a} \cdot \lambda)$ (+1 if the initial projection of the spin onto $\mathbf{a}$ is positive, else -1) and $\langle B(\mathbf{b}, \lambda)\rangle = -\text{sgn}(\mathbf{b} \cdot \lambda)$ [Bel64]. If λ is distributed homogeneously over the unit sphere, $P(\lambda) = P(\theta, \phi) = 1/4\pi$,

$$S_{12}(\mathbf{a}, \mathbf{b}) = \frac{-1}{4\pi} \int_0^\pi d\theta \sin\theta \int_0^{2\pi} d\phi\, \text{sgn}(\mathbf{a} \cdot \lambda)\text{sgn}(\mathbf{b} \cdot \lambda)$$

$$= 2\frac{|\alpha - \beta|}{\pi} - 1, \qquad (8.4.24)$$

where α and β are the individual angles of inclination of the two Stern-Gerlach magnets. This result coincides with the quantum mechanical prediction (8.4.19) at $|\alpha-\beta| = 0$, $\pi/2$, π (Fig. 8.14a). Therefore, in order to rule out a hidden-variable explanation, it is indispensable to choose different angles.

Indeed, at all other angles, the two predictions differ. This is not a consequence of an unfortunate construction of this particular example but a systematic deviation, related to the fact that classically, the expectation (8.4.24) shows first-order cusps at $|\alpha-\beta| = 0$, $\pm\pi$, where the quantum result has smooth quadratic extrema. In order to bring out a general discrepancy, independent of specific choices of the hidden-variable model, it is necessary to compare, besides the two values $\mathbf{a}$ and $\mathbf{b}$, additional settings $\mathbf{a}'$, $\mathbf{b}'$ for the measurements of particles 1 and 2, resp. [Bel81]. For the sum

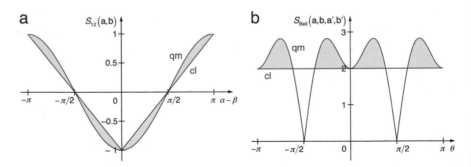

Fig. 8.14 The correlations between simultaneous measurements in an EPR experiment (Fig. 8.11), as implied by quantum mechanics, cannot be explained by a classical theory hypothesizing a "hidden variable" as common cause of the correlated results. The analysis proposed by John S. Bell revealed that the expectation value $S_{12}(\mathbf{a},\mathbf{b})$ of the product of the two outcomes (panel **a**), both restricted to the alternative $+1$ (spin up) and -1 (spin down), for the classical (blue, Eq. (8.4.24)) and the quantum result (red, Eq. (8.4.19)) coincide at standard configurations of the measurements (parallel, antiparallel, at right angles) but show markedly different functional shapes. A systematic deviation can be proven for a correlation function involving four different settings, $S_{\text{Bell}}(\mathbf{a}, \mathbf{b}, \mathbf{a}', \mathbf{b}')$ (**b**), cf. Figure 8.15b, where the quantum correlations, Eq. (8.4.32) (red) exceed their classical bound, Eq. (8.4.29) (blue), by a large margin. The excess of quantum correlations is marked in light red in both panels.

of two expectation values $S_{12}(\mathbf{a}, \mathbf{b})$ and $S_{12}(\mathbf{a}, \mathbf{b}')$, Eq. (8.4.22) implies

$$S_{12}(\mathbf{a}, \mathbf{b}) + S_{12}(\mathbf{a}, \mathbf{b}') = \int_\Lambda d\lambda \rho(\lambda) \langle A(\mathbf{a}, \lambda) \rangle (\langle B(\mathbf{b}, \lambda) \rangle + \langle B(\mathbf{b}', \lambda) \rangle), \quad (8.4.25)$$

and for its modulus, invoking the inequality (8.4.23) for $\langle A(\mathbf{a}, \lambda) \rangle$,

$$\left| S_{12}(\mathbf{a}, \mathbf{b}) + S_{12}(\mathbf{a}, \mathbf{b}') \right| \leq \left| \int_\Lambda d\lambda \rho(\lambda) (\langle B(\mathbf{b}, \lambda) \rangle + \langle B(\mathbf{b}', \lambda) \rangle) \right|$$
$$= \int_\Lambda d\lambda \rho(\lambda) \left| \langle B(\mathbf{b}, \lambda) \rangle + \langle B(\mathbf{b}', \lambda) \rangle \right|. \quad (8.4.26)$$

Similarly, for the difference, at another setting $\mathbf{a}'$ of measurement A,

$$\left| S_{12}(\mathbf{a}', \mathbf{b}) - S_{12}(\mathbf{a}', \mathbf{b}') \right| \leq \int_\Lambda d\lambda \rho(\lambda) \left| \langle B(\mathbf{b}, \lambda) \rangle - \langle B(\mathbf{b}', \lambda) \rangle \right|. \quad (8.4.27)$$

For the integrands in Eqs. (8.4.26, 8.4.27), the inequality

$$\left| \langle B(\mathbf{b}, \lambda) \rangle + \langle B(\mathbf{b}', \lambda) \rangle \right| + \left| \langle B(\mathbf{b}, \lambda) \rangle - \langle B(\mathbf{b}', \lambda) \rangle \right|$$
$$\leq 2 \max (\langle B(\mathbf{b}, \lambda) \rangle, \langle B(\mathbf{b}', \lambda) \rangle) \leq 2 \quad (8.4.28)$$

holds, using Eq. (8.4.23). This allows us to state an upper bound for the sum of Eqs. (8.4.26) and (8.4.27), assuming normalization, $\int_\Lambda d\lambda \rho(\lambda) = 1$

$$S_{\text{Bell}}(\mathbf{a}, \mathbf{b}, \mathbf{a}', \mathbf{b}') := \left| S_{12}(\mathbf{a}, \mathbf{b}) + S_{12}(\mathbf{a}, \mathbf{b}') + S_{12}(\mathbf{a}', \mathbf{b}) - S_{12}(\mathbf{a}', \mathbf{b}') \right|$$
$$\leq \left| S_{12}(\mathbf{a}, \mathbf{b}) + S_{12}(\mathbf{a}, \mathbf{b}') \right| + \left| S_{12}(\mathbf{a}', \mathbf{b}) - S_{12}(\mathbf{a}', \mathbf{b}') \right|$$
$$= \int_\Lambda d\lambda \rho(\lambda) 2 \left| \langle B(\mathbf{b}, \lambda) \rangle - \langle B(\mathbf{b}', \lambda) \rangle \right| \leq 2 \int_\Lambda d\lambda \rho(\lambda) = 2$$
$$(8.4.29)$$

Equation (8.4.28) constitutes *Bell's inequality* [Bel64]. It is this upper bound that prevents hidden-variable theories from reproducing the extreme correlations predicted by quantum mechanics for EPR experiments.

As a standard choice, take $\mathbf{a}, \mathbf{a}', \mathbf{b}$, and $\mathbf{b}'$ to lie in a common plane, forming angles of 0, 90°, 135°, and 225°, resp., with some reference direction (Fig. 8.15a) [Bel81]. Equation (8.4.19), the quantum mechanical expectation, then gives

$$S_{12}(\mathbf{a}, \mathbf{b}) = S_{12}(\mathbf{a}, \mathbf{b}') = S_{12}(\mathbf{a}', \mathbf{b}) = \cos(3\pi/4) = -1/\sqrt{2},$$
$$S_{12}(\mathbf{a}', \mathbf{b}') = \cos(\pi/4) = 1/\sqrt{2}, \quad (8.4.30)$$

so that

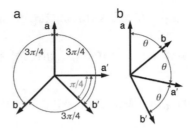

Fig. 8.15 Experiments demonstrating a violation of Bell's inequalities by quantum mechanical correlations require angles to be set at special values that define the orientation of the measurement apparatus, e.g., of a polarizer or a Stern-Gerlach magnet if photons or spins are measured. Shown are the configurations underlying Eq. (8.4.31) (**a**) and Eq. (8.4.33) and Fig. 8.14b (**b**).

$$S_{\text{Bell}}\left(\mathbf{a}, \mathbf{b}, \mathbf{a}', \mathbf{b}'\right) = |4\cos(3\pi/4)| = 2^{3/2} > 2, \qquad (8.4.31)$$

in clear contradiction to Eq. (8.4.28).

Another instructive configuration is to spread the four vectors as a fan, separated by equal angels θ, Fig. 8.15b. In that case, quantum mechanics predicts that

$$S_{12}(\mathbf{a}, \mathbf{b}) = S_{12}\left(\mathbf{a}', \mathbf{b}'\right) = S_{12}\left(\mathbf{a}', \mathbf{b}\right) = \cos(\theta), \quad S_{12}\left(\mathbf{a}, \mathbf{b}'\right) = \cos(3\theta), \quad (8.4.32)$$

and hence,

$$S_{\text{Bell}}\left(\mathbf{a}, \mathbf{b}, \mathbf{a}', \mathbf{b}'\right) = |3\cos(\theta) - \cos(3\theta)|, \qquad (8.4.33)$$

which exceeds the classical upper bound $S_{\text{Bell}}(\mathbf{a},\mathbf{b}) \leq 2$, Eq. (8.4.29), significantly in large intervals around odd-integer multiples of $\pi/4$ (Fig. 8.13b).

Bell's inequalities have served ever since their conception as a blueprint for *experimenta crucis* testing quantum mechanics against classical hidden-variable theories. They have been considerably refined in their theoretical content and significance, notably by the work of Clauser, Horne, Shimony, and Holt [CH&69], who cast them in a form directly applicable to quantum optical experiments employing linearly polarized photons as EPR pairs and polarization filters and photon detectors as measuring devices.

This strategy has been implemented shortly later by Aspect, Grangier, Dalibart, and Roger [AGR81, AGR82] in experiments at the frontier of the quantum optics of their time. Their already convincing results were subsequently improved further [ADR82] to make them watertight against remaining doubts. So-called loopholes, not envisaged by Aspect et al., concern in particular the possibility of causal relations between the generation of the entangled EPR pair and the settings at A and B ("freedom of choice loophole") or directly between them ("locality loophole"), which could short correlations through the EPR pair itself [SU&10, AA&15]. Attempts to close these loopholes use the very unpredictability of quantum randomness to isolate the settings at A and B from all possible external and mutual causal impacts [GV&15, SS&15]. At the same time, the competition for the largest distances over which a violation of Bell's inequality can be demonstrated, has reached values of more than 1000 km [TB&98, MH&12, YC&17]. At these magnitudes, EPR experiments clearly violate also the limits set by Special Relativity (Subsect. 3.1.4) for communication between the measurements (in one case, the required transmission time would have exceeded the time between subsequent settings by four orders of magnitude [YC&13]).

Does that mean quantum mechanics not only permits, but even requires faster-than-light communication, Einstein's "spooky action at a distance", as a consequence of its postulates? Also here, the concept of information is the key. What Special Relativity specifically denies is the *transfer of information* faster than the velocity of light. However, this is not implied by quantum non-locality. Referring directly to EPR experiments, entanglement within EPR pairs does not mean that the apparatuses at A and B become senders and receivers. Even if the respective measurement outcomes are correlated, they cannot transmit a signal. The reason is that a possible input signal, say the setting at A, only restricts particle B to a subspace of its Hilbert space (e.g., eigenstates of $\hat{\sigma}_x$, $\hat{\sigma}_y$, or $\hat{\sigma}_z$), but does not determine *which* state in this subspace it assumes: The output signal at B is random. The likewise random outcome of the measurement already at A prevents its use as a transmission line.

In summary, the following properties distinguish quantum correlations as a unique phenomenon: They are

- **instantaneous**, thus faster than any signal transmission, in particular faster than light by orders of magnitude,
- **stronger** by a wide margin than all classical correlations attainable by a concealed event in the past ("hidden-variable theories"),
- independent of any direct interaction and **not attenuated by spatial separation** between the correlated sites, and hence incompatible with classical explanations, based either on
- **direct causal relations** between the correlated measurements, or on a **common cause** affecting both measurements.

The only interpretation consistent with these conditions is the radical assumption that EPR measurement pairs constitute single indivisible entities, not pairs of systems between which signals could be exchanged.

In more general terms, quantum non-locality is not the consequence of a transmission of information from one site to another, it amounts to information being literally shared between two sites and simultaneously present at both of them. In the case of the EPR experiment as designed by Bohm and Aharonov, it is a single bit that cannot be assigned uniquely to one particle or to the other but invariably includes the states of both. That is, the partition of their common Hilbert space of dimension $D_{\mathcal{H}}$, into two qubits referring to common properties overrides the spatial partition into two non-interacting particles (Figs. 8.6, 8.11, 8.13c). This might appear as a consequence of the extreme scarcity of the available information, no more than one bit per subsystem, which could be assigned either to individual properties or to relations between the two particles [Zei99]. The original setup conceived by Einstein, Podolsky, and Rosen, Eqs. (8.4.11–8.13), though, shows that this is not the case: They do not refer to single qubits but to free particles living in Hilbert spaces of uncountably infinite dimension. Even in this case, entanglement entails an organization of the common Hilbert space according to shared information, ignoring the individual identity of particles.

Further evidence for the non-local and instantaneous character of entanglement, is provided by another gedankenexperiment, conceived by John Archibald Wheeler in the 1980s [Whe83]. With the EPR experiment, it shares the basic scheme of an entangled quantum object propagating along two alternative paths. However, by letting the two paths converge and bringing them to interference it probes the integrity and thus checks the wave against the particle nature of the entangled state (Fig. 8.16).

The simplest version of Wheeler's *delayed-choice experiment* adopts the setup of a Mach–Zehnder interferometer. The interferometer divides the path of a photon by a beam splitter, e.g., a half-silvered mirror, into two mutually orthogonal rays. After deflecting each of them by 90° at a fully silvered mirror, they intersect perpendicularly again. If they cross without interacting, detectors, e.g., photomultipliers, register each photon in either one of the two paths (Fig. 8.16a). If another beam splitter is inserted at the intersection (Fig. 8.16b), the photons are brought to self-interference. It is

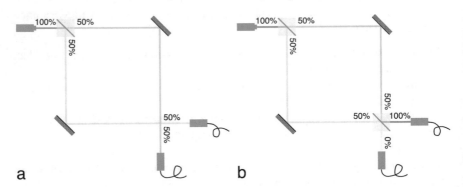

Fig. 8.16 Delayed-choice experiments modify the setup of a Mach–Zehnder interferometer to demonstrate the wave-particle ambivalence of photons. After passing a beam splitter (upper left in both panels), they propagate along two alternative paths (green vs. pink) that intersect anew at a crossing point (lower right). If they cross without interacting (**a**), they are registered with 50% probability in either path by detectors. If a second beam splitter is inserted at the intersection (**b**), they self-interfere constructively in one path (green) and destructively in the other (pink), resulting in 100% and 0% counting rates in the respective detectors. Inserting the second beam splitter only after a wave packet has entered the interferometer (delayed choice) allows testing questions concerning causality and non-locality.

constructive in the original direction of the incoming ray, so that the detector in this path registers photons, but destructive along the deflected path so that the other detector sees nil. An obvious interpretation is that in the first setup, photons behave as particles, going either one way or the other, while in the second, they behave as waves propagating simultaneously along both paths. Which alternative applies is controlled by the absence or presence of the second beam splitter.

Wheeler proposes to challenge this view by inserting or removing the second beam splitter only after the photon has passed the first one (that is, in terms of Special Relativity, in the future cone, see Sect. 3.1.5). If the second beam splitter is put in place only when the photon has already entered the split section of the interferometer but interference is still observed, this disputes the interpretation that it would have initially behaved as a particle, taking one route or the other but not both. It would require retroaction in time, to restore the wave nature of the photon from the moment on it passed the first branching, hence violate causality. The only alternative view is that in all cases, with or without delayed choice, the photon is actually present in both paths. For the first protocol, with the second beam splitter removed, it means in particular that the loss of coherence, resulting in the photon triggering one or the other detector, only occurs in the very moment of observation by that macroscopic device. Which one is triggered constitutes a quantum random process, see Subsect. 8.4.5, similar to that involved in EPR experiments.

Several alternative versions of the delayed-choice experiment have been proposed, based for instance on double-split interference, employing quantum-optics equipment or even gravitational lensing, on single- or two-photon polarization, and other substrates. It has been implemented in diverse state-of-the-art setups, providing strong experimental evidence for the behaviour detailed above [JW&06, PS&12, MK&13].

8.4.4 The Quantum Zeno Effect

Entanglement and non-locality, so drastically manifest in EPR experiments, provide convincing evidence of the inextricable interdependence of observed and observing system in quantum measurement. It results in a final state of the total system that bears information not only about the measured system but also about the apparatus. It amplifies the state of the measured system to macroscopic observability and at the same time stabilizes this state by anchoring it in a many-body system, the apparatus. These characteristics of quantum measurement amount to two antagonistic tendencies: the indeterminate branching from a state anywhere in the Hilbert space of the measured system towards one of the eigenstates of the measured observable (pointer states), addressed in the next subsection, and the stabilizing effect of the approach to the nearest pointer state, to be discussed in the sequel.

Repeatability of the measurement privileges pointer states as a particularly stable subset, forming attracting points within the otherwise homogeneous Hilbert space of the measured system. In classical symplectic (Sect. 5.3) as well as in quantum unitary dynamics (Sect. 8.3.2), the conservation of entropy prevents the existence of attractors, as an irreversible contraction of state space implies a loss of memory. In the case of quantum measurement, it comes about in two blatantly non-unitary steps: The first collapse of the wave function reduces the Hilbert space of the measured system to its projective Hilbert space. From there, the second collapse leads back to one of the pointer states, typically a discrete subset of lower dimension of the original Hilbert space.

During the pioneering years of quantum mechanics, attention focussed on the non-unitary character of the first collapse. Only much later, in the 1980ies, it turned towards the stabilizing effect of the second collapse. In a context not directly related to measurement, it was asked: Could decay processes be retarded or even entirely suppressed if the decaying system is continuously or periodically observed?

Feedback from observation to the observed system resembles self-fulfilling prophecy (Subsect. 3.2.2), an effect that requires the complexity of intelligent macroscopic subjects. It is all the more surprising that a similar effect should occur on atomic and subatomic scales. In the context of quantum mechanics, it owes its name to a paradox pointed out by the Greek philosopher Zeno of Elea (fifth century BC), where in fact he addresses the issue of continuity of motion: Referring to an arrow observed in flight, Zeno argues that no single observation could perceive the motion as such, it merely takes a snapshot of the arrow, thus freezes it in its instantaneous position. A sequence of discrete snapshots cannot establish continuous motion, either. The paradox was only resolved when differential calculus introduced the concept of infinitesimal steps in time and in space. Indeed, the original reasoning leading to the quantum Zeno effect [MS77, IH&90] does not assume continuous observation but the limit of a dense sequence of periodic measurements.

Within the framework of quantum measurement and decoherence developed above, a plausible quantitative argument substantiates the quantum Zeno effect (Fig. 8.17): Returning to the spin coupled to a macroscopic environment considered in Subsect. 8.3.3, prepare the spin in an eigenstate of, say, the Pauli spin operator $\hat{\sigma}_x$. As long as its time evolution is induced only by the Hamiltonian $\hat{H}_S = \frac{1}{2}\hbar\omega_0\hat{\sigma}_z$, Eq. (8.3.40), it consists of precession around the z-axis. the coefficients $a_n(t)$ of the corresponding density operator $\overline{\hat{\rho}}_S(t)$, from an initial state $\mathbf{a}(0) = (1, 0, 0)$, aligned with the x-direction, evolve as, cf. Eq. (8.3.52),

$$a_1(t) = \cos(\omega_0 t), a_2(t) = -\sin(\omega_0 t), a_3(t) = \text{const} = 0. \tag{8.4.34}$$

The probability to find the spin in its initial state again is then

$$P_1(t) = |\langle +_x|\psi(t)\rangle|^2 = |a_1(t)|^2 = \frac{1}{2}(1 + \cos(2\omega_0 t)), \tag{8.4.35}$$

while the probability to find it polarized in the y-direction increases as

$$P_2(t) = \left|\langle +_y|\psi(t)\rangle\right|^2 = |a_2(t)|^2 = \frac{1}{2}(1 - \cos(2\omega_0 t)). \tag{8.4.36}$$

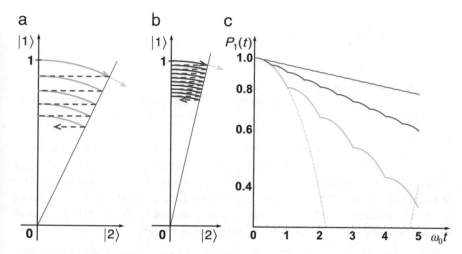

Fig. 8.17 The quantum Zeno effect can be reduced to the geometry of a particular path in Hilbert space: A metastable two-state system with a Hilbert space spanned by an excited state $|1\rangle$ and a ground state $|2\rangle$, prepared in $|1\rangle$, is periodically probed whether it is still in $|1\rangle$. It will alternately follow two types of time evolution (**a**): Free unitary evolution, pushing it out of the initial state (turquoise circle sections), and measurement projecting it back there (red dashed). If it is measured sufficiently frequently, the evolution out of the initial state can be considerably delayed (panel **b**, twice as many observations per unit time, as compared to panel **a**). In the extreme, continuous observation completely freezes the system. The decay of the probability $P_1(t)$ to remain in the initial state slows down with increasing frequency of observation, as shown by logarithmic plots of $P_1(t)$, given by Eq. (8.4.35) (turquoise, violet, orange: measurement frequency $1 \times$, $2 \times$, $4 \times$ as frequent, resp., as in (**a**), grey dashed: unobserved unitary evolution).

The expectation to see the spin in its initial state can be tested by measuring its x-component at some time t_1, for example by a short but intense coupling to a meter, strong enough to achieve a complete collapse of the state. As demonstrated in Subsect. 8.3.3, this projects the spin onto an eigenstate of the measured operator, here $\hat{\sigma}_3$, so that after the observation, it is in a mixed state with coefficients

$$a_1(t_1) = \cos(\omega_0 t_1), \quad a_2(t_1) = 0, \tag{8.4.37}$$

from where it evolves unitarily again, see Fig. 8.17.

If the measurement is repeated periodically at times $t_n = n\Delta T$, $\Delta T = T/N$, N times within a total time interval T, and sufficiently frequent, so that $\Delta T \ll T_0$, $T_0 = 2\pi/\omega_0$, the cosines in Eqs. (8.4.34–8.4.37) can be approximated by a quadratic decay, $\cos(x) = 1 - \frac{1}{2}x^2 + O(x^4)$. The probability to find the spin in its initial state after the full series of N measurements thus becomes

$$P_1(T) = \lim_{N \to \infty} \left[\frac{1}{2} \left(1 + \cos\left(\frac{2\omega_0 T}{N} \right) \right) \right]^N$$

$$= \lim_{N \to \infty} \left[1 - \left(\frac{\omega_0 T}{N} \right)^2 \right]^N = \lim_{N \to \infty} \exp\left(-\frac{\omega_0^2 T^2}{N} \right) = 1, \qquad (8.4.38)$$

i.e., the spin is frozen in its initial state.

Equation (8.4.38) represents the essence of the quantum Zeno effect. However, the concept is so general and versatile that various comments are in order to even approximately embrace it:

What the microscopic model of measurement and decoherence underlying Eqs. (8.4.34–8.4.38) achieves does not exactly come up to the postulate of repeatability of quantum measurements. Upon exiting the apparatus, the measured system has no preference for any state other than the pointer state corresponding to the result of the measurement, but it is generally not in a pure state. A microscopic theory substantiating the return to purity is presently not available.

- In the limit $N \to \infty$, Eq. (8.4.38) actually describes continuous observation, sometimes distinguished as "watchdog effect" from the quantum Zeno effect proper [Kra81]. A more realistic description of continuous measurement, however, would have to take an incomplete projection into account and result in a slow exponential decay of $P_1(t)$. Continuous measurement generalizes immediately to "observation without human observer", that is, to every metastable quantum system coupled to a macroscopic environment. This would imply that the mean lifetime we observe in particular in the decay of unstable nuclei and elementary particles should be considered as a "renormalized lifetime". The environment thus assumes a double role: According to the watchdog effect, it stabilizes observed systems. However, could these systems be isolated completely from, e.g., the electromagnetic vacuum field, they would not decay in the first place but evolve unitarily forever.
- The quantum Zeno effect hinges on the projection postulate as embodied, for example, in Eq. (8.4.9). This means that, if the time between two subsequent observations does not reduce to zero but remains a positive delay $0 \ll \Delta T \ll 2\pi/\omega_0$, the result of the previous measurement is not completely restored to a pure state. Instead of Eq. (8.4.37), the probability to be in the undecayed state

$$P_1(n\Delta T) = \left[1 - (\omega_0 \Delta T)^2 \right]^n \approx \exp\left(-n(\omega_0 \Delta T)^2 \right) \qquad (8.4.39)$$

then approximately amounts to a stepwise exponential decay, resulting in the long-time asymptote $\lim_{n \to \infty} P_1(n\Delta T) = 0$.

- The quantum Zeno effect provides direct evidence for the special role pointer states play in Hilbert space: They become attractors, in the sense that the system approaches the pointer state closest to its initial state. Unlike dissipation, where the ground state is the only attractor, the quantum Zeno effect can also stabilize states with a non-zero energy. If the spectrum of the pointer operator is discrete, up to an infinite number of them can coexist in Hilbert space, rendering the system multistable. With two or more pointer states, instabilities separating their basins of attraction are unavoidable. The non-unitary approach from an entangled initial state to a definite pointer state then also involves randomness, as is evident in the case of spin measurement discussed in the following subsection.

- The pointer operator, i.e., the observable that couples the measured system to the meter, may even depend on time. If it does not vary too rapidly, it can be used to "drag" the measured system along a prescribed target path in Hilbert space, inducing time evolution instead of freezing it [AV80, AS93]. The *inverse quantum Zeno effect* uses frequent or continuous observation to enforce an evolution that follows the sought protocol at every instant of time. It provides a valid option of quantum control.

8.4.5 Quantum Randomness

The generation of uncertainty by amplifiers resembles the generation of vacuousness by news organizations when there is no news. In politics things are often not "real" until they are widely discussed, so news media effectively make small events real by amplifying them. If the reported event is already fairly large, such as a troop movement or a cut in the discount rate, the amplified version is a reasonably faithful copy of the original. But if the event is small, e.g., a pork-barrel amendment or an unintentional but inflammatory misstatement, the amplified version can vary significantly from one report to the next and in this sense become uncertain. The limit of this process is reached when there is nothing at all to report, at which point the reporters begin interviewing each other and filling up air time with each other's opinions. Thus we have Paula Zahn asking Wolf Blitzer what he thinks the President's position will be on the upcoming tax cut fight, and so forth. In the news business this is called a slow day. In physics it is called quantum noise.

Robert B. Laughlin on quantum noise, from [Lau95], Chap. 5, p. 54f.

While in the quantum Zeno effect, the stabilizing action of pointer states becomes manifest, quantum randomness represents the other side of the coin, the instability owing to the presence of two or more outcomes of a measurement. It is in itself a unique quantum phenomenon, distinct from all classical manifestations of randomness as outlined in foregoing sections.

Quantum measurement is not complete with the first collapse, generating a mixed state given by a set of probabilities for the measured system to be in one of the eigenstates of the measured observable. With good reason, the Copenhagen interpretation postulates that the measured system should exit in a pure state, selected among the pointer states with a probability given by the projection postulate. From the mixed state, Eq. (8.4.9), this step takes the form

$$\hat{\rho}_0 = \sum_\alpha p_\alpha |\alpha\rangle\langle\alpha| \rightarrow \begin{cases} \vdots & \vdots \\ |\alpha\rangle\langle\alpha| \text{ with probability } p_\alpha, \\ \vdots & \vdots \end{cases} \qquad (8.4.40)$$

Calling it "the second collapse of the wave function" is not really adequate: While the "first" collapse leads from a pure to a mixed state, the "second" restores a pure state and in this sense "uncollapses" the wave function.

Regarding the final state $|\alpha\rangle$ reached after the measurement, only two conditions apply:

- on average over many repetitions of the measurement, the distribution of outcomes must satisfy the projection postulate, Eq. (8.4.10),
- otherwise, the sequence of outcomes is absolutely unpredictable.

It might appear that under these conditions, the second collapse reduces to an issue of classical statistics, that is, to a mere urn problem if the pointer basis is discrete. However, it also involves a process of manifestly quantum mechanical nature: the disentanglement of the measured object from the meter, a step that undoes the precedent entanglement but is at least as theoretically challenging. Even treating the second collapse as succeeding the first one neatly separated in time is an idealization. In fact, they overlap during a real measurement.

All instances of classical randomness discussed in Chaps. 5 to 7 have one essential feature in common: It is possible to trace their entropy production back to a source that is not directly accessible but can at least be identified. In the case of deterministic chaos, it is the potentially inexhaustible information content of the initial condition that fuels the entropy production, in the case of thermal fluctuations, it is in addition the huge number of degrees of freedom and the complexity of the many-body interactions that render any prediction hopeless. That is, their unpredictability is of a merely practical nature, their randomness is therefore apparent or *epistemic* [BA&17].

By contrast, a constitutive feature of quantum randomness is that such a hidden source is not only impossible to identify, but is pretended not to exist in the first place. Quantum randomness is a paradigm of intrinsic or *ontic randomness* [BA&17], maybe the only instance existing among all known physical phenomena. It is therefore a cornerstone of the theoretical conception of quantum mechanics, tied as a necessary complement to the probabilistic interpretation of quantum states. The guaranteed absence of any source that would allow reproducing or predicting it places quantum randomness in a higher category than any classical analogue. Random number generation from quantum randomness serves as benchmark for other random number generators, based, e.g., on classical noise or on deterministic pseudorandom algorithms simulating chaotic dynamics. As such, it is an indispensable resource and even attains a significant commercial value, for example in undecipherable cryptography and to generate randomized samples as input in causality studies based on screening, see Subsect. 3.1.2. Where apparent randomness recursively extends causal chains further and further back into the past, intrinsic randomness truncates the recourse, cf. Fig. 3.5.

This prominent role raises the question of how to check the authenticity of quantum randomness. In fact, there exists a host of quantum processes with a stochastic character. Well-known examples are quantum noise and radioactive decay:

- **Quantum noise,** alluded to in the introductory quote of Robert Laughlin, denotes the fluctuations observed in quantum systems coupled to the modes of the electromagnetic field. Its quantum character is manifest in particular in the fact that even in the zero-energy ground state of the field, the quantum vacuum, fluctuations persist [GZ09].
- **Radioactive decay** is another random phenomenon closely related to quantum vacuum fluctuations. They provide the energy required to surmount the potential barrier separating a metastable state, for example an atomic nucleus, from its decay products. The decay is an abrupt and irreversible non-unitary transition (but see, e.g., Ref. [MM&19]) that occurs unpredictably, but on average follows an exponential decay law characterized by a definite half life.

Quantum randomness in a more specific sense refers to random outcomes of EPR-type experiments and thus comes furnished with an unerring quality label: the non-local nature of the correlations between two measurements with space-like separation. Binary random sequences can be generated already in measurements on single two-state systems (Fig. 8.18) and thus far are not tied to non-locality in an EPR setup. However, the origin in correlations between space-like separated sites, verified as violation of the Bell inequality (8.4.29), can serve as a test criterion, referring to its source, to certify quantum randomness [PA&10]. At the same time, it has become a standard strategy to employ certified quantum random number generation in turn to

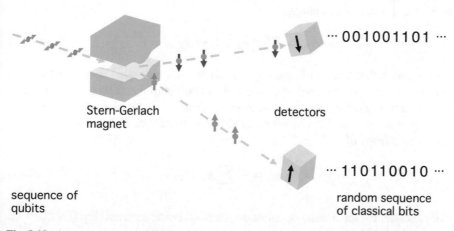

$\cdots 001001101 \cdots$

Stern-Gerlach
magnet

detectors

$\cdots 110110010 \cdots$

sequence of
qubits

random sequence
of classical bits

Fig. 8.18 A prototypical example of quantum randomness is spin measurement. If spin-$\frac{1}{2}$ particles are prepared in, say, an eigenstate of $\hat{\sigma}_x$, that is, as qubits without preferred orientation in the vertical direction, subsequent measurements of $\hat{\sigma}_z$ will result with equal probability 0.5 each in eigenstates of $\hat{\sigma}_z$, spin up or spin down, cf. Eq. (8.4.10). The sequence of actual outcomes constitutes a binary random process that is considered as fundamentally unpredictable: an instance of intrinsic randomness.

achieve "loophole-free Bell tests" [AA&15, HB&15, GV&15, SS&15]. One of the remaining loopholes in EPR experiments is a possible common classical cause that determines simultaneously the settings at the two arms of the setup. In order to close it, the setting should occur at random. Since any residual predictability could invalidate the cause for quantum correlations, this requires random numbers of undoubtful quality. In this way, quantum random numbers are fed back into experiments intended to corroborate their very quantum nature.

Concerning criteria for quantum randomness based on the structure of data sets, standard conditions reduce to the absence of correlations. More sophisticated criteria, as they are considered for classical random processes, such as algorithmic complexity (see Subsect. 4.2), are not usually applied in the quantum context, with few exceptions such as [Zur89].

With these characteristics, quantum randomness occupies a special position in quantum theory, as an essential element that however resists all attempts to integrate it consistently in the rest of the theoretical framework. It is disturbing that there should exist one particular phenomenon, randomness in quantum measurement, where this otherwise astonishingly successful framework of unitary quantum time evolution fails. The situation resembles the closely related case of the collapse of the wave function, which in the Copenhagen interpretation still appeared as a gap in the description. Nevertheless, it could finally be integrated, explaining it on the basis of a microscopic model as a consequence of the coupling to an apparatus. Could we expect a similar solution for quantum randomness?

Insight in this respect is gained by tracking the information balance of object and meter throughout the measurement. If we assume the states of object and meter to factorize before the measurement, as in Eq. (8.4.3), both systems enter the process in pure states with zero entropy,

$$S_{O,\text{ini}} = -\langle \text{lb}(\hat{\rho}_{O,\text{ini}}) \rangle = S_{M,\text{ini}} = 0. \tag{8.4.41}$$

As soon as with the onset of coupling, the object gets entangled with the meter and further with the environment, it shares information with them. Its partial entropies must increase, to reach a maximum after the (first) collapse of the wave function. Tracing over meter and environment results in a mixed state as in Eq. (8.4.9) with a non-zero entropy of

$$S_{O,\text{clps}} = -\sum_{\alpha} p_{\alpha} \text{lb}(p_{\alpha}). \tag{8.4.42}$$

For example, for the density operator during a spin measurement, Eq. (8.4.10) with $p_- = p_+ = 0.5$, the value is

$$S_{O,\text{clps}} = 1 \text{ bit}. \tag{8.4.43}$$

It is perfectly consistent with the complete ambiguity we have at this stage of the measurement about its outcome, spin up or spin down.

With the second collapse, however, things become problematic. On the face of it, the situation appears clear: The object returns to a pure state, i.e., in the above example, $\hat{\rho}_O \to |-_O\rangle\langle-_O|$ or $\hat{\rho}_O \to |+_O\rangle\langle+_O|$, and thus

$$S_{O,\text{fin}} = 0. \tag{8.4.44}$$

Indeed, with the object again in a uniquely defined state, no uncertainty is left. Comparing only the final to the initial state, quantum measurement even conserves entropy. However, this result challenges the intuitive idea that the random choice among several alternative outcomes α with non-zero probabilities p_α constitutes a surprise, hence generates entropy, one bit per spin measurement. Von Neumann entropy does not capture this unpredictability.

It is instructive to take a wider view, including not only the measured object but also the meter and the environment in the entropy balance. For this entire system, closed as a whole, the conclusion is evidently correct that no information can have entered nor leaked out. Between its components, however, entropy is exchanged during the coupled phase when they are entangled, thus share mutual information. There is no reason why on exit, each component should not have left a trace in the states of the other two. On the contrary, the purpose of the measurement requires meter and apparatus to attain information on the initial state of the object. By reciprocity of entanglement, it is then plausible that they also leave a trace in the object in turn, conferring information on their initial state to it (Fig. 8.19). This would suggest that the randomness generated in the second collapse reflects in fact the contingency in the initial states of many-body systems, magnifying it to macroscopic observability. That is, in the simplest setting, it is amplified quantum noise. Using vacuum fluctuations as input to generate quantum random numbers is indeed an approved technology if the source is a macroscopic phenomenon of quantum optics, e.g., laser phase diffusion [JC&11, AA&15]. Even two-state systems such as single photons have been proposed as a source, in setups closely resembling delayed-choice experiments, see Subsect. 8.4.3 [JA&00].

8.4.6 Quantum Causality

The concept of causality, as discussed in Subsect. 3.1, not only is a cornerstone of epistemology, in particular of classical physics, but epitomizes the notion of a directed flow of information along well-defined world lines in an ordered space–time. This ideal world is questioned from different sides by the quantum effects outlined in the preceding subsections. The challenge of reconstructing causality within the framework of quantum mechanics thus concentrates all those puzzles in a nutshell.

Quantitative criteria for causal relations can be specified in terms of information transfer between events occurring at definite places and times, measured as functional relations between conditional probabilities. None of these basic concepts transfers

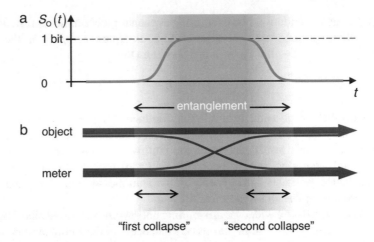

Fig. 8.19 Quantum randomness generated in quantum measurements becomes manifest in terms of its entropy balance. The von-Neumann entropy S_O associated to the reduced density operator of the measured object (**a**) starts at 0, corresponding to the initially pure state of the object. Upon entanglement with the meter and the concomitant collapse of the wave function, it increases to a maximum that reflects the uncertainty of the outcome (e.g., 1 bit in the case of a measurement of unbiased spins). The final return to a pure state, consistent with the actual measurement outcome, reduces the entropy to its initial value 0 again. This process can be interpreted as an exchange of information between object and meter (**b**), as a result of their correlation during the phase of coupling and thus of entanglement of the two subsystems.

unaltered to quantum mechanics. Specific issues where quantum theory gets into conflict with the classical concept of causality are:

- The directed information flow implicit in a particular cause–effect relation can only be defined relative to an independent background time arrow. Collectively, they determine in turn a causal structure that coincides with the global directedness of time underlying the Second Law of Thermodynamics. **Unitary quantum time evolution**, by contrast, is time-reversal invariant and thus lacks a preferred direction of time. In that, it agrees with the time evolution induced by canonical transformations in classical mechanics. That is, in closed quantum systems, effects and causes are indistinguishable.
- In **entangled states** of two or more particles, the idea of processes going on in defined positions at defined times dissolves, and with it the general causal structure of space–time, based on the possibility (or impossibility) of sending signals from an event A to an event B. In the case of the EPR experiment, for example, neither one of the classical alternatives (Fig. 8.13)—A causes B, B causes A, or both share a common cause C—explains the observed correlations. In particular, local hidden variables, a kind of common cause that could reconcile quantum mechanics with classical causality, are ruled out.

- **Non-locality**, a spatial manifestation of entanglement, scrambles the spatial order required to distinguish space-like from time-like relations. A symmetric superposition of (*A here, B there*) and (*A there, B here*) could turn into a corresponding superposition of (*first A, then B*) and (*first B, then A*) if space–time allowed for such rotations, in particular for closed time-like curves (CTCs, see Fig. 3.1). They become possible, e.g., in the context of quantum gravity, as a feature of exotic solutions of Einstein's field equations of gravitation [Goe49]. In this case, also the direction of signals exchanged between *A* and *B*, hence the basis of causal order, becomes undefined, giving rise to correlations with indefinite causal order [OCB12, Bru14].

- **Delayed-choice experiments**, Subsect. 8.4.3, go even further than "spooky action at a distance" in that they tempt us to interpret them as outright action backwards in time. In this case, an inversion of the order of cause and effect is not owed to an ambiguity of the global time arrow, but occurs on the background of a definite causal time order. It only appears inescapable, though, if it is taken for granted that the path a particle takes must be defined at least up to the resolution allowed by the uncertainty relations. Causal time order can be rescued at the price of accepting that even a single particle, e.g., a photon, can be present simultaneously at places separated by macroscopic distances.

- Finally, **quantum randomness** challenges causality from the opposite side. It generates sequences of random events, in particular binary alternatives, which lack any correlation and cannot be traced back to any even hidden origin to be considered as a cause. They are therefore "more random" than chaotic time series or algorithmic pseudorandom number generators and serve as security standard in cryptography. Injecting entropy without any source, quantum randomness appears to violate even the principles of unitary quantum time evolution, since that would imply conservation of entropy. Unlike decoherence, which could be convincingly integrated, quantum randomness is still an alien element in quantum mechanics and also challenges causality in general.

Several attempts have been proposed how to construct models of quantum causality, generalizing the formalism in such a way that at least part of the conflicts between quantum mechanics and causal order can be resolved. For example, spatial order could be entangled with time order by expanding the concept of quantum states so as to include entire actions and events extended in space *and* time, so-called quantum processes.

Such models contribute to the ongoing discussion how to apply quantum mechanics to the theory of gravity, more generally, how to construct a common fundament to unify quantum theory and General Relativity. Since both theories deal with information-theoretical postulates—quantum mechanics with the quantity and density of information, Relativity with signalling, it seems evident that they can only be reconciled on basis of a more general account of how physical systems process information in space and time.

In view of what we know by now, in particular about decoherence and measurement as many-body phenomena, definite events in space and time, the elements of information, take shape only for a macroscopic observer under certain conditions on size and interactions of the systems involved, crossing the border from the quantum to the classical world [Zur91]. In this sense, also causal order should be considered as an emergent phenomenon. More insight is expected from ongoing research on the dynamics of entanglement in strongly interacting quantum many-body systems.

8.5   Quantum Death and Resurrection of Chaos

The foregoing sections give diverse evidence that information is a crucial concept to understand quantum mechanics and contrast it with classical physics. Quantum mechanics not only defines principles concerning the content and the processing of information that are lacking in classical physics, it even allows us to see how information can emerge in the first place from quantum states and processes. Phenomena that connect the quantum with the classical level, involving quantities anchored in either realm, are particularly relevant here. Quantum measurement is a prime example, where entities as elusive as the quantum state of a spin or a photon crystallize to hard classical facts. A second phenomenon amplifying microscopic states to macroscopic visibility is chaos. However, unlike quantum measurement, it does not just take a snapshot but is dynamical in itself. Studying the manifestations of chaos in quantum systems therefore highlights other facets of how information is processed in quantum systems. The kinship of chaotic dynamics to measurement might insinuate that quantum mechanics would render the stochastic character of deterministic chaos even more pronounced. Surprisingly, the opposite is the case.

8.5.1   Quantum Chaos: A Deep Probe into Quantum Information Processing

By their evolution in time, chaotic systems generate a continuous flow of information which, as we have seen (Sects. 5.3, 6.3, and 7.3), can be traced back to microscopic initial and boundary conditions as its source [Sha80]. It is precisely the permanent entropy production, a defining property of chaos, that seriously challenges the way classical mechanics handles information. In simple models defined on discrete spaces, such as the discrete Bernoulli and baker maps (Sect. 5.5), it already became evident that restricting the total information capacity of their state space to finite values drastically alters their chaotic dynamics and on long time scales, prevents chaos entirely.

In fact, classical chaos is incompatible with the information-theoretic postulates of quantum mechanics, regarding static as well as dynamical properties. As concerns the dynamics, the conservation of entropy in unitary time evolution, Eq. (8.3.29), combined with the finite information supply a closed quantum system disposes of, prevents a permanent production of entropy. In a static sense, since fractal phase-space structures—the self-similar interlocking of chaotic with regular regions in Hamiltonian systems (Fig. 6.9), strange attractors in dissipative systems (Sects. 6.3, 6.4)—are not allowed by quantum uncertainty.

The desire to resolve these fundamental inconsistencies has spurred research on chaotic behaviour in quantum systems and given rise to an area and a research community of their own: quantum chaos. It is often understood as the quest for phenomena in molecular, in atomic, and down to nuclear systems that exhibit the one or other symptom of classical chaos. In fact, however, it has a much wider scope: It asks in how far the very concept of classical chaos, with its pivotal role for example in the foundation of statistical mechanics, is compatible with the postulates of quantum mechanics, viewing ergodicity and mixing, turbulence and strange attractors, with the eyes of Schrödinger and Heisenberg. Indeed, the trailblazing work by Casati, Chirikov, Izrailev, and Ford [CC&83], which kicked off the field, had actually been intended to demonstrate that stochastic behaviour in simple Hamiltonian systems, see Subsect. 5.5, survives quantization. The big surprise was that exactly this is *not* the case (Table 8.2).

The purpose of this subsection is presenting the main results of this active field, to illuminate the way quantum systems can imitate chaotic entropy production and reconcile the classical with the quantum account of chaotic dynamics. It is not intended as a comprehensive review of quantum chaos. Various issues of central importance in the field, such as semiclassical approximations and spectral statistics,

Table 8.2 The unsatisfactory way classical mechanics treats aspects related to information content, density, and production in dynamical systems becomes particularly manifest in chaotic systems, in terms of divergences of these quantities, in static structures such as strange attractors as well as in dynamical scenarios. Quantum mechanics regularizes these singularities by imposing fundamental bounds, but at the expense of altering drastically the character of chaotic behaviour.

Divergences of information density and production in classical chaos ...	are regularized in quantum mechanics.
Static structures:	
Fractal patterns (strange attractors, self similar phase space structures) store an infinite amount of information, therefore require an unbounded information density.	Quantum uncertainty restricts the resolution of phase space structures, imposing an upper bound on the information density and smoothing self-similar patterns.
Dynamical scenario:	
Chaotic time evolution permanently generates entropy, lifting information from invisibly small to large scales.	In systems with a finite Hilbert space dimension, unitary quantum time evolution eventually becomes quasiperiodic.

cannot be included. They are covered in comprehensive monographs [Ozo88, Gut90, Rei92, Sto99, Haa10], anthologies [GVZ91, GC95], and reviews [Eck88].

To set the stage, the following Subsection 8.5.2 will summarize some results obtained for models of classical chaos on discrete spaces, as precursors of quantum chaos. Subsection 8.5.3 proceeds to the real thing: quantizing classically chaotic systems to see how and why their behaviour deviates so drastically from that of their classical counterparts. It will draw our attention to the specific mechanisms, such as dynamical localization, behind that general phenomenon summarized as the quantum suppression of chaos.

The conclusion that the quantum death of classical chaos owes itself to the finite information content of closed quantum systems suggests that opening them towards a macroscopic environment might again alter the situation. This idea is pursued further in Subsect. 8.5.4, in two related directions: Periodic measurements on classically chaotic quantum systems as well as incorporating dissipation require interaction with environment. Indeed, the results of simulations of models augmented accordingly corroborate that in both cases, chaotic behaviour indicated by a lasting entropy production is at least partially recovered. They therefore provide prototypical examples how entropy is produced at the smallest scales, to emerge at large scales as randomness.

8.5.2   Discretizing Classical Chaos

Quantum mechanics cannot be reduced to a mere discretization of space, a "pixelation of phase space", as its implications are of a much more fundamental nature. Notwithstanding, particularly in the context of complex dynamical processes, precisely this kind of "poor man's quantization" can provide first insights what to expect for, and how to interpret, the results of a more rigorous quantum mechanical treatment.

In Sect. 5.5, it has been demonstrated how discretizing the Bernoulli and the baker map not only impedes chaotic behaviour, but even leads to periodic recurrences of the initial state. In what follows, a similar model is construed, which however already allows for an interpretation as a mechanical system in continuous time and thus for a direct quantization.

A free particle subject to a time-dependent external force, a periodic train of delta pulses given by a parabolic potential barrier (a "kicked inverted oscillator"), is described by the Hamiltonian [FMR91]

$$H(p, q, t) = \frac{p^2}{2m} - \frac{m \lambda^2 q^2}{2} \sum_{j=-\infty}^{\infty} \delta\left(\frac{t}{T} - j\right), \tag{8.5.1}$$

with parameters m (mass), λ (escape rate from the potential barrier), and T (period of the delta kicks). Integrating the equations of motion

$$\dot{p} = m\lambda^2 q \sum_{j=-\infty}^{\infty} \delta\left(\frac{t}{T} - j\right), \quad \dot{q} = \frac{p}{m}, \tag{8.5.2}$$

from a moment immediately after the jth kick, $t_j = jT + \varepsilon$, $\varepsilon \to 0^+$, to immediately after the next kick leads to a map from $(p_j, q_j) = (p(t_j), q(t_j))$ to $(p_{j+1}, q_{j+1}) = (p(t_{j+1}), q(t_{j+1}))$,

$$p_{j+1} = (1 + \lambda^2 T^2) p_j + m\lambda^2 T q_j, \quad q_{j+1} = q_j + \frac{T}{m} p_j. \tag{8.5.3}$$

Choosing parameters such that $m = 1$, $\lambda = 1$, and $T = 1$ (time measured in units of the inverse escape rate), Eq. (8.5.3) can be written as

$$p_{j+1} = 2p_j + q_j, \quad q_{j+1} = q_j + p_j. \tag{8.5.4}$$

Assuming moreover periodic boundary conditions with period 1, both for p and for q, results in a map from the unit square onto itself. It can be represented as a matrix,

$$[0, 1[\times [0, 1[\to [0, 1[\times [0, 1[, \ \begin{pmatrix} p \\ q \end{pmatrix} \mapsto \begin{pmatrix} p' \\ q' \end{pmatrix} = T \begin{pmatrix} p \\ q \end{pmatrix} (\mathrm{mod}\ 1), \quad T = \begin{pmatrix} 2 & 1 \\ 1 & 1 \end{pmatrix}. \tag{8.5.5}$$

It factorizes into a matrix generating free motion and another generating the kick,

$$T = T_{\mathrm{kick}} T_{\mathrm{free}}, \ T_{\mathrm{free}} = \begin{pmatrix} 1 & 0 \\ 1 & 1 \end{pmatrix}, \ T_{\mathrm{kick}} = \begin{pmatrix} 1 & 1 \\ 0 & 1 \end{pmatrix}. \tag{8.5.6}$$

At first sight, the *Arnol'd cat map* (8.5.5) [AA68, Sch84, FMR91, LL92] appears to be just a linear shear. As in the case of the Bernoulli and baker maps, though, its nonlinearity is hidden in the *modulo* operation. How nonlinear in fact it is reveals its action on a simple pattern, which becomes disrupted even after a single application of the map (Fig. 8.20). The cat map is an example of strong chaos: A positive Lyapunov exponent of $\lambda = \ln\left(\frac{1}{2}(3 + \sqrt{5})\right)$, constant all over its phase space, ranks it in the category of K-systems (systems with a positive Kolmogorov-Sinai entropy, cf. Eq. (3.1.14)) in the hierarchy of classical chaos [Sch84, LL92].

Interpreted as an invertible operator, the coordinate transformation (8.5.4) also propagates phase-space probability density distributions $\rho(\mathbf{r})$, as defined in Eq. (5.2.2),

$$\rho_{n+1}(\mathbf{r}) = \rho_n(T^{-1}\mathbf{r}), \mathbf{r} := \begin{pmatrix} p \\ q \end{pmatrix}. \tag{8.5.7}$$

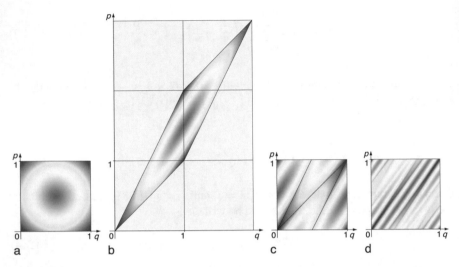

Fig. 8.20 The Arnol'd cat map (8.5.5) combines a skew linear shear with a *modulo* operation. An initial distribution on the definition space, a unit square, such as the simple pattern in (**a**), remains perfectly intelligible after the shear alone (**b**), but gets significantly disrupted by the *modulo* (**c**). Already after two applications of the map (**d**), the original pattern can no longer be discerned.

In order to "pixelate" a smooth density distribution, its arguments have to be discretized, for example

$$\rho_{n,m} := \rho(p_n, q_m), \quad q_m = \frac{m}{N}, \quad p_n = \frac{n}{N}, \quad m, n \in \{1, \ldots, N\}. \tag{8.5.8}$$

Applying the transformation T to the discrete coordinates $\mathbf{r}_{n,m} = (p_n, q_m)$ then amounts to a permutation of the elements of the $N \times N$-matrix $\rho_{n,m}$, generated by a permutation tensor **C**,

$$\rho \mapsto \rho', \rho'_{n,m} = (\mathbf{C}\rho)_{n,m}. \tag{8.5.9}$$

It is analogous to the combination of matrices **B** and $\mathbf{B}^{-1}$ permuting rows and columns of $\rho_{n,m}$, introduced in Subsect. 5.5 to implement the baker map on a discrete phase space. Since in the Arnol'd cat map, transformations of rows and columns are coupled, the two permutation matrices now combine into a single fourth-rank tensor **C**. Instead of specifying it explicitly, suffice it to state the underlying index permutation, for instance (due to the inevitable rounding involved, this choice is not unique)

$$\begin{pmatrix} n \\ m \end{pmatrix} \mapsto \begin{pmatrix} n' \\ m' \end{pmatrix} = \begin{pmatrix} (2n + m - 2)(\bmod N) + 2 \\ (n + m - 2)(\bmod N) + 2 \end{pmatrix}. \tag{8.5.10}$$

Figure 8.21a shows the results of applying the discrete Arnol'd cat map with $N = 16$ to the pixelated pattern shown in the upper left panel. The most conspicuous

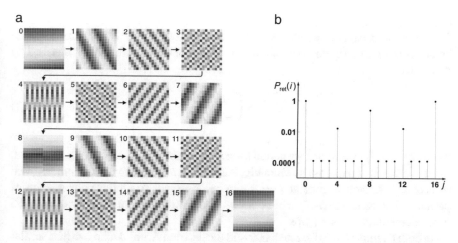

Fig. 8.21 Partial and complete recurrences in the discrete Arnol'd cat map, Eqs. (8.5.7–8.5.10), for a phase space with 16×16 cells. Panel (**a**) shows the result of 16 consecutive applications of the map to the initial pattern (upper left corner), identifying each pixel with a characteristic colour, as in the analogous Fig. 5.12. The initial pattern is recovered identically after 16 time steps. The probability to return, defined in Eqs. (8.5.11, 8.5.12) as the overlap of the initial with the final density distribution, is depicted as a logarithmic plot in panel (**b**). Note the symmetry of the return probability with respect to time reversal at $j = 8$, half the recurrence time, which is apparent also in the sequence of the detailed patterns in part (**a**).

feature is the identical recurrence of the initial pattern is after N applications of the map. This confirms what has been found already for the discretized baker map, cf. Figure 5.12: Operating on a state space of finite dimension D, a deterministic time evolution must repeat itself after at most D time steps. In the present case, the state space comprises $D = N \times N$ phase-space cells. The fact that the recurrence time 16 is even much shorter than the upper bound $D = 16^2 = 256$ is explained by the symmetries of the map and the particular parameter choice $N = 2^4$.

A global measure of the coincidence of the final with the initial state is the *probability to return*. For a probability density defined on a $2f$-dimensional phase space, it is defined as the overlap of the final with the initial density,

$$P_{\text{ret}} \left(j'', j' \right) = \int \mathrm{d}^{2f} r \rho \left(\mathbf{r}, j'' \right) \rho \left(\mathbf{r}, j' \right). \tag{8.5.11}$$

For a discrete phase space as in Eq. (8.5.8), it reduces to a double sum

$$P_{\text{ret}} \left(j'', j' \right) = \sum_{n,m=1}^{N} \rho_{n,m} \left(j'' \right) \rho_{n,m} \left(j' \right). \tag{8.5.12}$$

For the same discretized Arnol'd cat map, the time evolution of $P_{\text{ret}}(j) := P_{\text{ret}}(j, 0)$ is recorded in Fig. 8.21b. The figure not only confirms the expected return to unity after 16 steps, but also an additional time reversal invariance of the sequence of patterns with respect to the time $j = 8$, half the recurrence time: It reflects the special symmetries of the map, in particular the similarity of the transformation matrix T to its inverse

$$T^{-1} = \begin{pmatrix} 1 & -1 \\ -1 & 2 \end{pmatrix},$$
(8.5.13)

which up to a rigid shift and a rotation by $\pi/2$ generates the same shear as T. Indeed, in the sequence of patterns shown in Fig. 8.21a, a corresponding near symmetry with respect to inversion of time at $j = 8$ can be observed. A similar phenomenon of partial and complete revivals will be found in the full quantum versions of this and other elementary chaotic maps.

In both examples, the discrete baker and the discrete Arnol'd map, we find similar information cycles, different in turn from those in chaotic systems on continuous spaces: In the latter, we see a permanent flow of information passing through a window of observability; it enters at the smallest resolvable scale, undergoes a process of mere expansion or of stretching and folding, to leave that window again towards smaller scales. By contrast, in discrete maps, information reaching the largest scale by chaotic expansion is "recycled" to reemerge at the smallest scale (see Figs. 5.12 and 8.21a) and undergo the same process anew.

Discretized maps represent but the simplest instances of a much more general situation: simulating chaotic dynamics on digital computers. In particular during the heydays of chaos theory when the available computer capacity was so limited that unavoidable discretization was still an issue, various studies have been dedicated to the question how chaotic dynamics develops (or not) on granular spaces [CP82, HW85, WH86, BR87]. They consider different discretization techniques (truncation, round off, finite measurement precision with random errors) and levels of abstraction from specific computer platforms but coincide in their main conclusions:

- the time evolution on discrete spaces terminates in loops (periodic orbits) of finite length,
- the entropy production stalls after a finite time,
- orbit lengths and crossover times to periodic behaviour depend on the bin size, thus increase with the magnitude of the total discrete state space.

In this context, the analogy of dynamical systems with computers is particularly striking and substantiates its value as an analytical tool.

8.5.3   Quantum Death of Classical Chaos

In order to find out about the consequences quantum mechanics implies for chaos, it suggests itself to recast models of classical chaos in the language of quantum theory.

How to do that, however, is neither obvious nor unique. Since quantum mechanics amounts to a more precise and consistent description, ascending to this level from a classical model is not a reduction but a creative step. While the so-called canonical quantization rules for classical mechanical systems (replace position q, momentum p, and functions of these by operators $\hat{p}, \hat{q}$ and corresponding operator-valued functions, obeying the canonical commutation relation $[\hat{p}, \hat{q}] = -i\hbar$) provide a framework, there always remains some freedom to be filled with suitable decisions, which follow heuristic rather than rigorous criteria. In the sequel, a selection of chaotic systems, the baker map, the Arnol'd cat map, and the kicked rotor, are "quantized" in this manner, to confront their dynamics with the chaotic behaviour of their classical counterparts.

Quantum baker map

Elementary models of classical chaos, such as the baker map and the cat map, are defined as maps of the unit square $[0,1[\times [0,1[\subset \mathbb{R}^2$ onto itself. This restriction to a bounded space already has far-reaching consequences for the quantum mechanical treatment. It requires the Hilbert-space dimension of these systems to be finite, $D_{\mathcal{H}} = N \in \mathbb{N}$. Unlike the classical case, this is already sufficient to discretize their state space. A suitable implementation replaces the unit square by a rectangle in phase space, say of size P in momentum and Q in position. Periodic boundary conditions $\psi(q + Q) = \psi(q)$ of wave functions $\psi(q) \sim \exp(-ipq/\hbar)$, imply quantization of the momentum eigenvalues. Periodicity in p, $\tilde{\psi}(p+P) = \tilde{\psi}(p)$, entails quantization of q, cf. Eq. (8.1.19),

$$\hat{p}|\lambda\rangle = p_\lambda|\lambda\rangle, \, p_\lambda = \lambda\frac{2\pi\hbar}{Q} = \lambda\frac{P}{N}, \, P = \frac{2\pi\hbar N}{Q},$$
$$\hat{q}|n\rangle = q_n|n\rangle, \, q_n = n\frac{2\pi\hbar}{P} = n\frac{Q}{N}, \, Q = \frac{2\pi\hbar N}{P}, \quad \lambda, n \in \{0, \dots, N-1\}. \quad (8.5.14)$$

These definitions are consistent with the finite number N of Planck cells, $2\pi\hbar N = PQ$. Defining dimensionless momentum eigenfunctions as

$$\langle n|\lambda\rangle = \psi_\lambda(q_n) = \tilde{\psi}_n^*(p_\lambda) = \sqrt{\frac{2\pi\hbar}{PQ}} \exp\left(i\frac{p_\lambda q_n}{\hbar}\right) = \frac{1}{\sqrt{N}} \exp\left(2\pi i\frac{n\lambda}{N}\right)$$

$$(8.5.15)$$

then leads to a discrete Hilbert space structure that coincides exactly with the space of discrete Fourier transform, cf. Eqs. (2.3.24, 2.3.25): States are represented by N-dimensional complex coefficient vectors $\psi \in \mathbb{C}^N$, $\tilde{\psi} \in \mathbb{C}^N$, in position or momentum representation, resp., and are transformed between these representations by N-dimensional discrete Fourier transforms, cf. Eq. (8.1.19). Adopting the notation of Eq. (2.3.55),

$$\tilde{\psi} = \hat{\mathbf{F}}_N \, \psi, \quad \psi_n = \langle n|\psi\rangle = \sum_{\lambda=0}^{N-1} \left(\hat{\mathbf{F}}_N\right)_{n\lambda} \tilde{\psi}_\lambda, \quad \left(\hat{\mathbf{F}}_N\right)_{n\lambda} = \frac{1}{\sqrt{N}} \exp\left(2\pi i\frac{n\lambda}{N}\right),$$

$$\tilde{\boldsymbol{\psi}} = \widehat{\mathbf{F}}_N^{-1}\,\boldsymbol{\psi}, \quad \tilde{\psi}_\lambda = \langle \lambda | \psi \rangle = \sum_{\lambda=0}^{N-1} \left(\widehat{\mathbf{F}}_N^{-1}\right)_{\lambda n} \psi_n, \quad \left(\widehat{\mathbf{F}}_N^{-1}\right)_{\lambda n} = \frac{1}{\sqrt{N}}\exp\left(-2\pi i\frac{\lambda n}{N}\right).$$

$$(8.5.16)$$

On this discrete space, it is now straightforward to implement the classical maps on the unit square. For example, rephrase the baker map as a sequence of four elementary geometric operations (Fig. 5.10):

1. expand the unit square $[0, 1[\times[0, 1[$ by a factor 2 in q,
2. divide the expanded q-interval $[0,2[$ into two equal sections, $[0,1[$ and $[1,2[$,
3. shift the right one of the two rectangles, $(q, p) \in [1, 2[\times[0, 1[$, by 1 to the left in q and by 1 up in p, $[1, 2[\times[0, 1[\to [0, 1[\times[1, 2[$,
4. contract the full rectangle $[0, 1[\times[0, 1[\cap[0, 1[\times[1, 2[= [0, 1[\times[0, 2[$ by a factor 2 in p to recover the original square $[0, 1[\times[0, 1[$.

Quantum mechanically, operations on state vectors in a given representation of the discrete Hilbert space, such as shifts, expansions, and contractions, are performed in this same representation, switching between representations by Fourier transform where necessary. The classical operations 1 to 4 thus translate into quantum operations as follows (assuming N to be even):

1. In the q-representation, divide the coefficient vector $\boldsymbol{\psi} = (\psi_0,\dots, \psi_{N-1})$ into two halves of dimension $N/2$, $\boldsymbol{\psi}_- = (\psi_0,\dots, \psi_{N/2-1})$ and $\boldsymbol{\psi}_+ = (\psi_{N/2},\dots, \psi_{N-1})$,
2. expand $\boldsymbol{\psi}_-$ and $\boldsymbol{\psi}_+$ both to size N by adding a 0 behind each coefficient,
3. Fourier transform both partial vectors separately to the p-representation, which due to their inflated structure amounts to applying an inverse $(N/2 \times N/2)$-Fourier transformation to the original $\boldsymbol{\psi}_-$ and $\boldsymbol{\psi}_+$ of size $N/2$,

$$\tilde{\boldsymbol{\psi}}_\mp = \widehat{\mathbf{F}}_{N/2}^{-1}\boldsymbol{\psi}_\mp, \quad \left(\widehat{\mathbf{F}}_{N/2}^{-1}\right)_{\lambda n} = \frac{1}{\sqrt{N/2}}\exp\left(-4\pi i\frac{\lambda n}{N}\right), \qquad (8.5.17)$$

4. stack $\tilde{\boldsymbol{\psi}}_+$, the p-representation of the right half of $\boldsymbol{\psi}$, on top of $\tilde{\boldsymbol{\psi}}_-$, the p-representation of the left half, so as to represent the upper half of the spectrum of spatial frequencies,

5. transform the combined state vector $\tilde{\boldsymbol{\psi}} = \left(\tilde{\boldsymbol{\psi}}_-, \tilde{\boldsymbol{\psi}}_+\right)$ from the N-dimensional p-representation back to the q-representation, applying an $(N \times N)$-Fourier transform, $\boldsymbol{\psi} = \widehat{\mathbf{F}}_N\tilde{\boldsymbol{\psi}}$.

Combining these steps into a single transformation, a quantized baker map is obtained in the compact form of a linear transformation of the coefficient vector $\boldsymbol{\psi}$ [BV87, BV89],

$$\boldsymbol{\psi} \mapsto \boldsymbol{\psi}' = \hat{\mathbf{B}}_N\boldsymbol{\psi}, \quad \hat{\mathbf{B}}_N := \widehat{\mathbf{F}}_N\begin{pmatrix} \widehat{\mathbf{F}}_{N/2}^{-1} & 0 \\ 0 & \widehat{\mathbf{F}}_{N/2}^{-1} \end{pmatrix}. \qquad (8.5.18)$$

Like this, however, it still suffers from a deficiency: Owing to the asymmetric way zeros have been intercalated in step 2 above, the transformation (8.5.18) breaks a symmetry of the classical baker map: The origin $(p_0, q_0) = (0, 0)$ of the quantum position-momentum index space coincides with the classical origin $(p, q) = (0, 0)$ of phase space, but the diagonally opposite corner $(p_{N-1}, q_{N-1}) = N(P, Q)/(N - 1)$, appropriately scaled, does not coincide with $(p, q) = (1, 1)$, thus breaks the symmetry of the classical map. By a slight modification, it can be recovered on the quantum side [Sar90]: Shift the two discrete grids of the transformations (8.5.16) between the two representations by ½, replacing $\hat{F}_N$ by $\hat{G}_N$,

$$\left(\hat{G}_N\right)_{n\lambda} = \frac{1}{\sqrt{N}} \exp\left(2\pi i \frac{\left(n + \frac{1}{2}\right)\left(\lambda + \frac{1}{2}\right)}{N}\right), \tag{8.5.19}$$

and similarly for $\hat{G}_{N/2}$, so that the quantum baker map takes the final form [Sar90]

$$\psi \mapsto \psi' = \hat{B}_N \psi, \ \hat{B}_N := \hat{G}_N \begin{pmatrix} \hat{G}_{N/2}^{-1} & 0 \\ 0 & \hat{G}_{N/2}^{-1} \end{pmatrix}. \tag{8.5.20}$$

Composed of unitary transformations, it is itself unitary, hence conserves entropy. Evidently, n subsequent time steps accumulate to the transformation

$$\psi_0 \mapsto \psi_j = \hat{B}_N(j)\psi_0, \ \hat{B}_N(j) := \left(\hat{B}_N\right)^j. \tag{8.5.21}$$

As an indicator of recurrences, define the quantum mechanical probability to return to a specific initial state, in analogy to the classical Eq. (8.5.11), as

$$P_{\mathrm{ret}|\psi\rangle}(t) = |\langle\psi(t)|\psi(0)\rangle|^2 = \left|\langle\psi(0)|\hat{U}(t)|\psi(0)\rangle\right|^2, \tag{8.5.22}$$

with the time-evolution operator $\hat{U}(t)$ defined in Eq. (8.3.3). On average over the entire Hilbert space, it simplifies to

$$P_{\mathrm{ret}}(t) = \left|\mathrm{tr}[\hat{U}(t)]\right|^2. \tag{8.5.23}$$

For discrete time j and a one-step evolution operator $\hat{U}$, $P_{\mathrm{ret}}(j) = \left|\mathrm{tr}\left[\hat{U}^j\right]\right|^2$. The probability to return has been evaluated in Fig. 8.22a for the quantum baker map with a Hilbert space of dimension $D_{\mathcal{H}} = N = 8$. Recurrences are not as periodic and not as complete as in the discretized version of the classical map, cf. Figure 5.12, but they do occur. As an example, the transformation matrix $\left(\hat{B}_N\right)^j$ at $j = 490$, where it comes particularly close to the unit matrix, is depicted as a grey-level plot

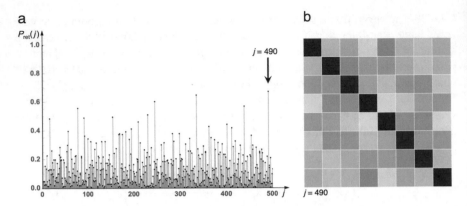

Fig. 8.22 Recurrences in the quantum baker map, Eq. (8.5.20), are neither as periodic nor as precise as in the discretized classical baker map, Fig. 5.12, or the discrete classical Arnol'd cat map, Eqs. (8.5.8–8.10) and Fig. 8.23, or the quantum cat map, Eq. (8.5.28) and Fig. 8.23. They occur as approximate revivals, identifiable as marked peaks of the return probability, Eq. (8.5.23) (**a**). For the strong peak at time step $j = 490$ (arrow in panel (**a**)), the accumulated transformation matrix in the position representation $\hat{B}_N(j)$ (**b**), cf. Eq. (8.5.21), here with $N = 8$, indeed comes close to a unit matrix. Grey-level code in (**b**) ranges from light grey (0) through black (1).

in Fig. 8.22b. It is evident that the system shows no tendency towards a steady state, equidistributed over its Hilbert space.

Quantum Arnol'd cat map

With the time-dependent Hamiltonian (8.5.1), the Arnol'd cat map is amenable to canonical quantization, replacing classical position and momentum by quantum operators. A suitable starting point is the factorization, as in Eq. (8.5.6), into an operator generating free motion and another generating the kick. In terms of unitary time evolution operators, the map thus takes the form

$$\hat{U}_{\text{cat}} = \hat{U}_{\text{kick}}\hat{U}_{\text{free}}, \tag{8.5.24}$$

with the two factors given by substituting the Hamiltonian (8.5.1) as an operator into Eq. (8.3.4), understanding the small parameter ε to approach zero from above,

$$\hat{U}_{\text{free}} = \exp\left(\frac{-i}{\hbar}\int_{jT+\varepsilon}^{(j+1)T-\varepsilon} dt\, \hat{H}(\hat{p},\hat{q},t)\right) = \exp\left(\frac{-i}{\hbar}\frac{\hat{p}^2 T}{2m}\right),$$

$$\hat{U}_{\text{kick}} = \exp\left(\frac{-i}{\hbar}\int_{(j+1)T-\varepsilon}^{(j+1)T+\varepsilon} dt\, \hat{H}(\hat{p},\hat{q},t)\right) = \exp\left(\frac{i}{\hbar}m\lambda^2\hat{q}^2 T\right). \tag{8.5.25}$$

The two integrations over time span, respectively, the phase of free motion between the kicks and the delta kick at $t_j = jT$. With the parameters chosen as in Eqs. (8.5.4–8.6), these operators reduce to

$$\hat{U}_{\text{free}} = \exp\left(\frac{-i}{\hbar}\frac{\hat{p}^2}{2}\right), \quad \hat{U}_{\text{kick}} = \exp\left(\frac{i}{\hbar}\hat{q}^2\right). \tag{8.5.26}$$

If the Arnol'd cat map is implemented on the finite-dimensional Hilbert space constructed in Eqs. (8.5.14–8.5.16), $\hat{U}_{\text{free}}$ and $\hat{U}_{\text{kick}}$ can be written as $N \times N$-square matrices, for example

$$\begin{aligned}\left\langle\lambda\left|\hat{U}_{\text{free}}\right|n\right\rangle &= \frac{1}{\sqrt{N}}\exp\left(i\pi\frac{\lambda(\lambda-2n)}{N}\right), \\ \left\langle\lambda\left|\hat{U}_{\text{kick}}\right|n\right\rangle &= \frac{1}{\sqrt{N}}\exp\left(i\pi\frac{n(n-2\lambda)}{N}\right),\end{aligned} \quad n, \lambda \in \{0, \ldots, N-1\}. \tag{8.5.27}$$

Recombined, they result in the cat map as a transformation matrix, say in position representation,

$$\begin{aligned}\left\langle n''\left|\hat{U}_{\text{cat}}\right|n'\right\rangle &= \sum_{\lambda=0}^{N-1}\left\langle n''\left|\hat{U}_{\text{kick}}\right|\lambda\right\rangle\left\langle\lambda\left|\hat{U}_{\text{free}}\right|n'\right\rangle \\ &= \frac{1}{N}\sum_{\lambda=0}^{N-1}\exp\left(\frac{i\pi}{N}\left(\lambda^2 + 2\lambda(n''-n') - n''^2\right)\right).\end{aligned} \tag{8.5.28}$$

The typical outcome of a numerical experiment with the quantum Arnol'd cat, here with $N = 12$, is presented in Fig. 8.23. Panel (b) shows the return probability, Eq. (8.5.23), as a function of discrete time j. Not only is it periodic with a period, in this case identical to the size N of the discrete state space, as the discrete classical map, that is, the accumulated transformation matrix returns to the identity after N

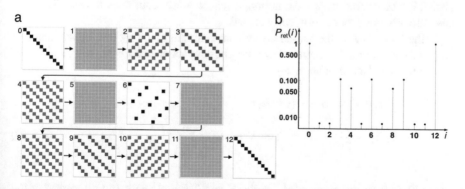

Fig. 8.23 The quantum Arnol'd cat map, Eq. (8.5.28), here with $N = 12$, shows a similar pattern of recurrences (**a**) as its counterpart, the discrete classical map (8.5.10), see Fig. 8.21. The return probability, Eq. (8.5.23), depicted as a semi-logarithmic plot in (**b**), indicates a complete revival at $j = N = 12$. This is confirmed by the evolution of the transformation matrix $\hat{U}_{\text{cat}}^j$, represented in (**a**) as the moduli of the matrix elements. The sequence of matrices also illustrates the symmetry of the time evolution with respect to time reversal at $j = N/2 = 6$.

time steps. Also the particular invariance of the time evolution with respect to time reversal at $j = N/2$, half the period of the map, reappears in the quantum version.

This symmetry appears even more striking if it is verified in terms of the transformation matrices, panel (a), here represented as the moduli of the matrix elements. The only patterns not repeated symmetrically during the period are those at 0 (identity) and at $N/2 = 6$. In those cases where this modulus is constant ($j = 1, 5, 7, 11$), the information on the state of the system is hidden in the phases of the matrix elements (not shown).

Quantum kicked rotor

The two models considered so far, the quantized baker and cat maps, corroborate what could have been anticipated from the behaviour of discretized classical chaotic maps: On a finite Hilbert space, their quantum mechanical time evolution is quasiperiodic, in extreme cases such as the Arnol'd cat map even exactly periodic, hence is incompatible with chaos. However, these are "toy models" in the sense that for the sake of mathematical transparence and tractability, they sacrifice all the intricacies characterizing even elementary mechanical systems. Finite-dimensional Hilbert spaces, in particular, can be realized only approximately and with highly sophisticated laboratory equipment, see Chap. 10 on quantum computation.

Does the hypothesis that chaos is suppressed in closed quantum systems also hold in systems with a Hilbert space of infinite dimension? A model that bears evidence as to this question, yet still permits a detailed analysis of its dynamics, is the kicked rotor [Chi79, CC&83, LL92]. Like the Arnol'd cat map, it derives from a Hamiltonian system with a single degree of freedom, rendered chaotic by a pulsed external force. Both systems therefore can absorb (or lose) an unbounded amount of energy. The decisive condition for the present context is though that their time evolution is unitary, so that the conservation of information, Subsect. 8.3.2, continues to hold: They are energetically open, but entropically closed.

In the kicked rotor, the spatial coordinate is an angle θ, $0 \le \theta < 2\pi$, canonically conjugate to an angular momentum l. Its phase space therefore has the topology of a cylinder. The Hamiltonian reads

$$H(l, \theta, t) = H_0(l) + V(\theta, t),$$

$$H_0(l) = \frac{l^2}{2\Theta}, \quad V(\theta, t) = K \cos(\theta) \sum_{j=-\infty}^{\infty} \delta\left(\frac{t}{T} - j\right), \tag{8.5.29}$$

where Θ denotes the moment of inertia, K and T the strength and the period of the kicks, resp. As in the steps from Eqs. (8.5.1) through (8.5.3), the equations of motion

$$\dot{l} = K \sin(\theta) \sum_{j=-\infty}^{\infty} \delta\left(\frac{t}{T} - j\right), \quad \dot{\theta} = \frac{l}{\Theta}, \tag{8.5.30}$$

integrated from immediately after the jth kick, $t_j = jT + \varepsilon$, $\varepsilon \to 0^+$, to immediately after the next kick leads to a map from (l_j, θ_j) to (l_{j+1}, θ_{j+1}),

$$l_{j+1} = l_j + KT \sin(\theta_{j+1}), \quad \theta_{j+1} = \theta_j + \frac{T}{\Theta} l_j (\mathrm{mod}\ 2\pi). \qquad (8.5.31)$$

Scaling parameters such that time is measured in units of T and inertia in units of Θ, it becomes the *standard* or *Chirikov map* [Chi79, LL92, Ott93, Haa10],

$$(l, \theta) \mapsto (l', \theta'), \quad l' = l + K \sin(\theta'), \quad \theta' = \theta + l(\mathrm{mod}\ 2\pi). \qquad (8.5.32)$$

That the image θ' appears on the r.h.s. of the first equation, not θ, is owed to the definition of time sections immediately *after* each kick. The standard map can also be motivated as the simplest case of a *perturbed circle map* or *twist map*, which describes nearly integrable motion subject to a weak nonintegrable perturbation ~ K [LL92, Ott93]. Besides the obvious periodicity in θ, this map is also invariant under shifts $l \to l + 2\pi$ of the angular momentum. The parameter K provides a continuous control, missing in the baker and the Arnol'd cat map, how far the kicked rotor deviates from a free rotor, hence from integrability. It allows us to scan through its entire dynamical scenario from integrable to fully chaotic, as predicted by the Kolmogorov-Arnol'd-Moser (KAM) and Poincaré-Birkhoff theorems [Sch84, LL92].

For the standard map, it encompasses the following major phases:

$K = 0$:

The system is integrable, trajectories are circles closing around the cylindrical phase space (Fig. 8.24a),

$$l' = l, \quad \theta' = \theta + l. \qquad (8.5.33)$$

$0 < K < K_c \approx 0.9716$:

Deformed toroidal trajectories of the type (8.5.34) are interspersed with mixed zones where chaotic motion and regular islands interpenetrate in an intricate, self-similar structure (Fig. 8.24b, c).

$K = K_c \approx 0.9716$:

At the critical perturbation strength, the last of the deformed toroidal trajectories that survived the perturbation turns into a *cantorus* [LL92], a porous fractal embracing the cylinder along the former torus. With gaps opening in the cantorus, it no longer forms a closed barrier but opens the way for trajectories to move along the cylinder without bound.

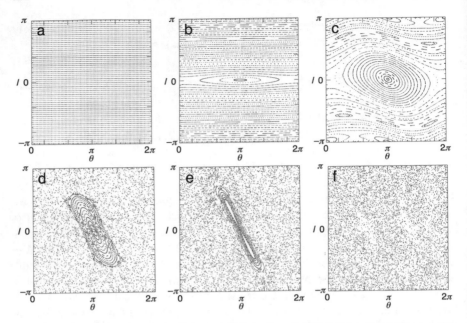

Fig. 8.24 The dynamical scenario of the kicked rotor, Eqs. (8.5.29–8.32), represented as strobo-scopic phase-space plots at different values of the kick strength K, from integrability through mixed behaviour where regular coexists with chaotic motion, through almost homogeneous chaos: (**a**) $K = 0$, integrable motion, as in Eq. (8.5.34), (**b**) $K = 0.02$, incipient chaos with chains of regular islands intruding into slightly deformed tori, (**c**) $K = 0.6$, typical mixed phase space according to the KAM scenario, with a few strongly deformed tori surviving, (**d**) $K = 3.0$, the regular island centred at $l = 0$, $\theta = \pi$, is surrounded by a chaotic sea, connected all along and around the cylin-drical phase space, (**e**) $K = 4.2$, this regular island has undergone a Hopf bifurcation, (**f**) $K = 7.0$, homogeneously chaotic phase space, up to invisibly small regular islands. Note how the complexity of the phase-space geometry increases initially from perfect homogeneity to intricate self-similar structures (panels (**a**) to (**c**)), but subsequently reduces again, down to almost homogeneous and isotropic chaos (panels (**d**) to (**f**)).

$K_c < K < 4$:

A chaotic sea extends all along the cylinder, but dotted with (chains of) regular islands where trajectories hop around the cylinder (Fig. 8.24d).

$K = 4$:

The largest and last of the regular islands, centred at $l = 0$, $\theta = \pi$, bifurcates, giving way to an unstable fixed point and a pair of stable islands (Fig. 8.24e).

$K \gg 4$:

The chaotic sea along and around the cylinder becomes asymptotically homoge-neous, interrupted only by tiny regular islands supporting so-called accelerator modes [Kar83, LL92] (Fig. 8.24f).

In this last regime of homogeneous chaos, the discrepancy between classical and quantum dynamics turns most striking: The stretching and folding of phase space by the standard map manifests itself globally as a spreading in the direction of the angular momentum, induced by the kicks, and as increasingly fine structures in the angle distribution, which compensate for the spreading in angular momentum. If this small-scale structure is erased by averaging over θ, the angle can instead be treated as a random variable ξ, equidistributed around the circle, resulting in a random walk in the angular momentum l,

$$l' = l + K \sin(\xi). \tag{8.5.34}$$

The theory of random walks, Eqs. (7.1.9, 7.1.10) with spatial steps $\Delta x = K$ and time steps $\Delta t = 1$, predicts a diffusive spreading of the angular momentum

$$\left\langle (l_j - \langle l_0 \rangle)^2 \right\rangle = j D(K), \quad D(K) = \frac{K^2}{2}. \tag{8.5.35}$$

The diffusion constant $D(K)$ is an asymptotic value, approached for $K \gg 1$ (note that in agreement with convention in the present context, Eq. (8.5.35) defines the diffusion constant D by a factor 2 larger than the canonical diffusion law, Eq. (7.1.10)). It ignores deviations from homogeneous chaos, such as accelerator modes [Kar83, LL92]. Angular momentum diffusion in the kicked rotor is a striking example of a manifestly irreversible process that arises from a deterministic and time-reversal invariant dynamics by coarse graining, hence is referred to as *deterministic diffusion* [Sch84, LL92, Ott93].

Like the baker and the Arnol'd cat maps, the kicked rotor conserves phase-space area and therefore entropy. However, averaging over the angle eliminates the top-down information current that compensates for the entropy production due to angular momentum diffusion. A Gaussian distribution

$$\rho(l_j, j) = \frac{1}{\sqrt{2\pi \left(\sigma_0^2 + j D(K)\right)}} \exp\left(\frac{-(l_j - \langle l_0 \rangle)^2}{2(\sigma_0^2 + j D(K))}\right), \tag{8.5.36}$$

in accord with Eq. (8.5.35) (for an initial variance $\sigma_0^2 = \langle (l_0 - \langle l_0 \rangle)^2 \rangle$), corresponds to an information content, cf. Eq. (7.1.21) with an angular-momentum unit d_l (undetermined till it will be given a definitive value in the context of the quantum map below),

$$I(j) = c\left(1 + \ln\left(\frac{1}{d_l}\sqrt{\sigma_0^2 + j D(K)}\right)\right) \tag{8.5.37}$$

and an asymptotic production rate per time step, for long times $j \gg \sigma_0^2/D(K)$,

$$\Delta I = I(j+1) - I(j) = \frac{c}{2j} \tag{8.5.38}$$

that decreases algebraically but remains positive forever.

Based on a periodically time-dependent Hamiltonian, the standard map is quantized along the same steps as the Arnol'd cat map. Also here, the unitary evolution operator over a single period of the driving factorizes into free motion and kick,

$$\hat{U}_{\mathrm{qkr}} = \hat{U}_{\mathrm{kick}} \hat{U}_{\mathrm{free}}, \tag{8.5.39}$$

where now (cf. Eq. (8.5.26))

$$\hat{U}_{\mathrm{free}} = \exp\left(\frac{-\mathrm{i}}{\hbar}\frac{\hat{l}^2 T}{2\Theta}\right), \quad \hat{U}_{\mathrm{kick}} = \exp\left(\frac{-\mathrm{i}}{\hbar} KT \cos(\hat{\theta})\right). \tag{8.5.40}$$

Scaling classical parameters as in Eq. (8.5.32) and the period of the kicks as $\tau :=
\hbar T/\Theta$, and introducing a quantum kick strength $k := KT/\hbar$, the one-step time-evolution operator simplifies to

$$\hat{U}_{\mathrm{qkr}} = \exp(-\mathrm{i}k\cos(\hat{\theta}))\exp\left(-\mathrm{i}\tau\frac{\hat{l}^2}{2\hbar^2}\right). \tag{8.5.41}$$

Unlike the cat map, the kicked rotor does not require to "artificially discretize" its Hilbert space by imposing periodic boundary conditions in position and momentum. Eigenvalues of the angular momentum are already discrete,

$$\hat{l}|\lambda\rangle = \hbar\lambda|\lambda\rangle, \lambda \in \mathbb{Z}, \langle\theta|\lambda\rangle = \frac{1}{\sqrt{2\pi}}\exp(\mathrm{i}\lambda\theta). \tag{8.5.42}$$

This means that the unperturbed Hamiltonian $\hat{H}_0(\hat{l})$, Eq. (8.5.29), read as an operator, has a discrete spectrum, but in contrast to the foregoing examples, its Hilbert space is of infinite dimension.

Efficient algorithms to simulate the time evolution generated by the quantized standard map can be achieved by alternating forward and backward Fourier transformations between angular momentum and position representation and taking advantage of the diagonal form of the free rotation and kick operators in the respective representation,

$$\left\langle\lambda'\left|\hat{U}_{\mathrm{free}}\right|\lambda\right\rangle = \exp(-\mathrm{i}\tau\lambda^2/2)\delta_{\lambda'-\lambda},$$
$$\left\langle\theta'\left|\hat{U}_{\mathrm{kick}}\right|\theta\right\rangle = \exp(-\mathrm{i}k\cos(\theta))\delta(\theta'-\theta) \tag{8.5.43}$$

A full time step thus reduces to a quantum map in angular momentum representation,

$$\langle \lambda''|\psi''\rangle = \langle \lambda''|\hat{U}_{qkr}|\psi'\rangle = \sum_{\lambda'=-\infty}^{\infty} \langle \lambda''|\hat{U}_{qkr}|\lambda'\rangle\langle \lambda'|\psi'\rangle,$$

$$\langle \lambda''|\hat{U}_{qkr}|\lambda'\rangle = \frac{1}{2\pi}\int_0^{2\pi} d\theta \langle \theta|\hat{U}_{kick}|\theta\rangle e^{i(\lambda''-\lambda')\theta}\langle \lambda'|\hat{U}_{free}|\lambda'\rangle$$

$$= b_{\lambda'-\lambda''}(k)\exp\left(-i\tau\lambda'^2/2\right), \tag{8.5.44}$$

where $b_n(x) := i^n J_n(x)$ and $J_n(x)$ denotes the Bessel function of order n. Initial states, in all numerical experiments documented in this subsection, have been chosen as $|\psi(0)\rangle = |0\rangle$, the angular-momentum ground state, or $\hat{\rho}(0) = |0\rangle\langle 0|$ in the quantum case and accordingly as $\rho(l, \theta, 0) = \delta(l)/2\pi$ in the classical case.

This was the strategy proposed and implemented for the first time by Casati et al. in their [CC&83]. As a criterion in how far the quantum evolution would reproduce the chaotic dynamics of the classical map, they compared the time dependence of the kinetic energy with the linear growth, Eq. (8.5.35), indicating angular-momentum diffusion in the classical kicked rotor. Given that the eigenvalues of the unperturbed Hamiltonian form a countably infinite set and that the rotator could absorb energy without bound from the driving, there is no obvious reason to expect a behaviour deviating qualitatively from the classical case. Yet numerical results showed the opposite. To be sure, for generic parameter choices, the energy starts to grow at the classical rate, but after a finite time the increase slows down and eventually stalls altogether, giving way to fluctuations of the energy around a constant mean. Evidently, the same applies to the entropy production. This crossover is even quite abrupt, as compared to the duration of the preceding classical phase (Fig. 8.25a).

More light is shed on the dramatic difference between the classical and the quantum time evolution of the kicked rotor by a numerical experiment simulating a time reversal on both systems (Fig. 8.25b). In the classical case, it amounts to the operation $T : l \rightarrow -l, \theta \rightarrow \theta, j \rightarrow -j$ on the classical trajectory, followed by propagation with the time-reversed mapping

$$l_{j+1} = l_j + K\sin(\theta_j), \quad \theta_{j+1} = \theta_j + l_{j+1}(\mathrm{mod}\ 2\pi). \tag{8.5.45}$$

In the quantum case, it consists in the single operation (see, e.g., [Haa10])

$$\langle \lambda|\psi_j\rangle \rightarrow \langle \lambda|\psi_{j+1}\rangle = \langle -\lambda|\psi_j\rangle^*, \tag{8.5.46}$$

on the state at time j and subsequent propagation with

$$\hat{U}_{qkr}^\dagger = \exp\left(i\tau\frac{\hat{l}^2}{2\hbar^2}\right)\exp(ik\cos(\hat{\theta})) \tag{8.5.47}$$

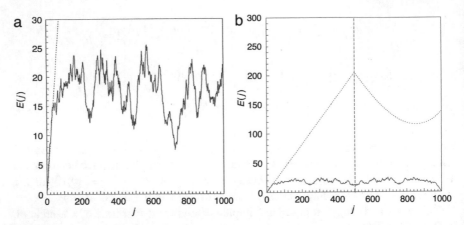

Fig. 8.25 Angular-momentum diffusion in the classical vs. the quantum kicked rotor. Panel (**a**) shows the linear increase of the kinetic energy (dotted), Eq. (8.5.35), absorbed by the classical kicked rotor, Eqs. (8.5.29–8.5.32), compared to the time evolution of its expectation value in the quantum kicked rotor (full line), Eq. (8.5.44), over the first 1000 time steps, starting from a zero angular momentum initial state. In panel (**b**), a time reversal is performed both on the classical and on the quantum system, see text, after time step $j = 500$ (dashed vertical line). While the quantum system returns to its initial state at $j = 1000$, the classical system recovers its diffusive spreading after a few hundred time steps. Parameter values are $K = 10$ and $\hbar = 0.6\pi/(\sqrt{5} - 1)$ for the quantum model. They result in an estimate, cf. Eq. (8.5.56), of $j^* \approx 33$ for the crossover time to quasiperiodic behaviour.

instead of $\hat{U}_{\text{qkr}}$. Performed with machine precision on the classical trajectory (l_j, θ_j) registered at $j = 500$, the operation achieves a reversal of angular-momentum diffusion, but only for the first few hundred subsequent time steps. Meanwhile, the chaotic time evolution amplifies the small but unavoidable numerical errors exponentially, wipes out the information on the reversed initial conditions at $j = 500$, and restores the diffusive energy growth. In the quantum simulation, by contrast, numerical errors in the time reversal increase only linearly. Within 500 steps, it returns to the initial state with good precision.

A valuable hint how to interpret this surprising result is that like this, the quantum kicked rotor behaves *as if* its Hilbert space had a finite dimension. The crossover from chaotic diffusion to quasiperiodic behaviour is reached at a critical time j^* when the kinetic energy approaches a value that, according to the classical diffusion law (8.5.35),

$$\left\langle \left(l_j - \langle l_0 \rangle\right)^2 \right\rangle = j^* \frac{K^2}{2} =: L^2, \tag{8.5.48}$$

corresponds to a variance L of the wave packet, hence to a finite effective Hilbert space dimension $D_{\mathcal{H}} \approx L/\hbar$: It remains localized around a centre $l_c = \langle l_0 \rangle$.

That the system does not explore the full angular momentum range, but remains restricted to a small section of it, calls for further analysis. The Hamiltonian of the unperturbed rotor is invariant under discrete shifts of the angular momentum. Any localization of the wave function requires a breaking of this symmetry. The hint explaining this phenomenon came from solid-state physics: While metals conduct electric currents more or less well, other substances like in particular glass are isolators. The decisive difference lies in their microscopic structure. The periodicity of the crystal lattice of bulk metals supports wave functions of their conduction electrons that extend all over the sample. Amorphous substances such as glass lack this symmetry, preventing extended energy eigenfunctions. This quantum effect, occurring in systems with static or "frozen" spatial disorder, is known as *Anderson localization* [And58, LR85], see Subsect. 4.1 (Fig. 8.26).

There is no spatial disorder in the quantum kicked rotor. Localization occurs, however, in angular momentum space, where the free-motion component $\hat{H}_0(\hat{l})$ of the Hamiltonian (8.5.29) is decisive. In itself, it is periodic in angular momentum, with a period $\Delta l_{\mathrm{cl}} = 2\pi\Theta/T$, cf. Eq. (8.5.31), but with quantum mechanics, an additional periodicity comes into play: that of angular momentum quantization, with period $\Delta l_{\mathrm{qm}} = \hbar$, cf. Eq. (8.5.42). The two periods are commensurate if and only if their ratio is a rational number,

$$\Delta l_{\mathrm{qm}}/\Delta l_{\mathrm{cl}} = \hbar T/2\pi\Theta = \tau/2\pi \in \mathbb{Q}. \qquad (8.5.49)$$

For all other values of the parameter $\tau/2\pi$, the superposition of the two periodic lattices in angular momentum space is aperiodic. This disorder of number-theoretical nature is sufficient to induce localization of the wave functions. *Dynamical localization* has been proposed from the side of solid-state physics, by Fishman, Grempel, and Prange, to explain the localized eigenfunctions in the quantum kicked rotor

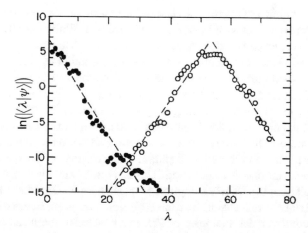

Fig. 8.26 Dynamical localization in the quantum kicked rotor. The figure shows two examples of energy eigenfunctions of a model closely akin to the quantum kicked rotor, Eqs. (8.5.29–8.31), as a semi-logarithmic plot in angular momentum representation (solid and open circles). Exponential localization, Eq. (8.5.48), is indicated by straight lines (dashed). Reproduced with permission from [FGP84].

[FGP82, FGP84]. For rational values of $\tau/2\pi$, an exceptional case of measure zero, it turns out that indeed, the kinetic energy increases indefinitely with time [IS80], giving rise to the term *quantum resonance*.

As a time-dependent system, the kicked rotor does not possess eigenvalues and eigenfunctions of the energy. Owing to its discrete time-translation symmetry, however, closely related concepts do apply, *quasienergies* and quasienergy eigenfunctions, referring to the eigenfunctions $|\psi(\varepsilon_n)\rangle$ and eigenvalues mod 2π of the one-step time evolution operator (8.5.41), $\hat{U}_{\mathrm{qkr}}|\psi(\varepsilon_n)\rangle = \exp(-i\varepsilon_n)|\psi(\varepsilon_n)\rangle$. As eigenphases of a unitary operator, the parameters $\phi(\varepsilon_n) = \varepsilon_n T/\hbar$ are defined mod 2π, which restricts the quasienergies to an interval, e.g., $0 \leq \varepsilon_n < 2\pi\hbar/T$. Dynamical localization amounts to an exponential decay of these quasienergy eigenfunctions around a centre $\lambda_c(\varepsilon_n)$ that depends on the eigenvalue ε_n,

$$\langle\lambda|\psi(\varepsilon_n)\rangle \sim \exp\left\{-\frac{\hbar|\lambda - \lambda_c(\varepsilon_n)|}{L}\right\}. \tag{8.5.50}$$

Since these quasienergy states are normalizable, their spectrum is discrete, indicated here by assigning a discrete index n to the quasienergy. The *localization length L* scales with the classical diffusion constant roughly as $L \approx D(K)/\hbar = K^2/2\hbar$.

How does this explain the crossover to periodicity? If in a numerical experiment, the initial state is chosen as an angular-momentum eigenstate, $\langle\lambda|\psi_0\rangle = \delta_{\lambda-\lambda_0}$, its expansion in quasienergy states,

$$|\psi_0\rangle = \sum_n c_n|\psi(\varepsilon_n)\rangle, \quad c_n = \langle\psi(\varepsilon_n)|\psi_0\rangle = \langle\psi(\varepsilon_n)|\lambda_0\rangle, \tag{8.5.51}$$

will comprise a number of roughly $2L/\hbar$ significant terms, corresponding to the finite effective Hilbert space dimension $D_{\mathcal{H}}$ alluded to above. While initially, moduli and phases of the expansion coefficients c_n must be orchestrated such that they superpose to an angular-momentum eigenstate, this "conspiration of phases" will be lost during the time evolution. On average, the wave function will assume a width $2L$ in angular momentum and exhibit the same exponential localization, Eq. (8.5.50), as the quasienergy eigenstates. Containing only a finite number $D_{\mathcal{H}}$ of discrete frequencies, however, the time evolution is quasiperiodic and in particular returns infinitely often, in an erratic fashion, to revivals, i.e., to states close to the initial state.

At the same time, during an initial phase, the behaviour of the quantum kicked rotor must imitate the classical spreading. This is required already by the energy-time uncertainty relation, Eq. (8.1.12), applied to quasienergies: The discreteness of the spectrum will only be reflected in the time evolution, viz. as quasiperiodicity, as soon as phase differences reach the order of 2π. If we estimate the mean separation of $L/\hbar$ quasienergies distributed around the circle as $\Delta\varepsilon \approx 2\pi\hbar^2/LT$, this will be the case after a critical number j^* of time steps given by

$$j^* = \frac{2\pi}{\Delta\phi(\varepsilon)} = \frac{2\pi\hbar}{\Delta\varepsilon T} \approx \frac{L}{\hbar}. \tag{8.5.52}$$

It is a particular instance of the *Heisenberg time*, the characteristic time span during which a classically chaotic quantum system is expected to reproduce classical chaos. For systems in continuous time, with a mean level separation ΔE, the energy-time uncertainty relation suggests defining the Heisenberg time as

$$t_H = \frac{2\pi\hbar}{\Delta E}, \tag{8.5.53}$$

in accordance with Eq. (8.5.52). It also indicates the maximum range in time over which semiclassical approximations to an exact quantum time evolution can be trusted.

A more general analysis refers directly to the entropy production by a system with finite information content I_{tot} : Quantum features become visible as soon as the total entropy produced exhausts this supply. If, for example, entropy is produced at a constant rate h_{KS}, the Kolmogorov-Sinai entropy, Eq. (3.1.10), and the initial information is defined by the total Hilbert space dimension, $I_0 = c\ln(D_\mathcal{H})$, the crossover to quantal behaviour is expected for

$$t_E = c\frac{\ln(D_\mathcal{H})}{h_{KS}}. \tag{8.5.54}$$

called the *Ehrenfest* or *log time*. It is the relevant timescale to describe the transition to a repetitive dynamics in quantum systems such as the quantized baker and Arnol'd cat maps, but applies as well to their discretized classical versions.

In the kicked rotor, the entropy production by angular momentum diffusion is not constant but given by Eqs. (8.5.37, 8.5.38). Comparing the increasing entropy of the diffusion cloud, Eq. (8.5.37) (with a quantum angular-momentum resolution $d_l = \hbar$) to a total information content $I_{tot} = c\ln(D_\mathcal{H}) \approx c(1 + \ln(L/\hbar))$,

$$c(1 + \ln(L/\hbar)) = I_{tot} = I(j^*) = c\left(1 + \ln\left(\frac{1}{\hbar}\sqrt{\sigma_0^2 + j^*D(K)}\right)\right), \tag{8.5.55}$$

allows to estimate the crossover time j^*. If the system is prepared in an angular momentum eigenstate, so that $\sigma_0^2 = \hbar^2$, this term can be neglected against the diffusion term $j^*D(K)$, giving a crossover time $j^* \approx L^2/D(K)$. With the approximate values $D(K) \approx K^2/2$, Eq. (8.5.35), and $L \approx D(K)/\hbar$, we arrive at

$$j^* \approx L \approx D(K)/\hbar = K^2/2\hbar, \tag{8.5.56}$$

which agrees well with the numerical data , see Fig. 8.25a.

It might appear as if the above examples for the quantum suppression of classical chaos represented rare exceptions. Systems with finite-dimensional Hilbert space are artificial constructions, hard to realize even in high-tech laboratories. Dynamical localization, the decisive mechanism preventing chaotic behaviour in the kicked rotor, appears to be an anomaly, not to be expected in more typical systems. In fact, however, it can be shown that all periodically forced systems, if their unperturbed Hamiltonian has a discrete energy spectrum and the driving is not at resonance with it, exhibit a quasiperiodic time evolution, incompatible with chaotic entropy production [HH82].

In view of the artificial nature of the quantum kicked rotor, an experimental verification does not look like an easy task, it is all the more surprising that it found its way into the laboratory quite rapidly. In particular, two techniques, recently developed in the quantum optics of its time, have been explored to realize it in experiments: highly excited states of Hydrogen [BS88, GS&88], Fig. 8.27a, or Rydberg atoms [BG&89, BB&91], Figs. 8.27b, c, irradiated by microwave fields in a cavity, and cold atoms in standing laser fields [MR&94, RB&95, MR&95] (Fig. 8.28). While in the former setups, the train of delta kicks is replaced by a sinusoidal driving and diffusion occurs in the energy space of excited atomic states, the latter experiment was closer to the theoretical model in that sharp laser flashes were used for the driving, and momentum diffusion was observed as increasing width of a cloud

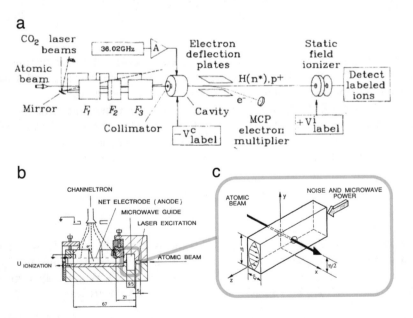

Fig. 8.27 Quantum optics experiments confirming dynamical localization. Schematic setup (**a**) of an experiment employing microwave irradiation to excite Hydrogen atoms. The central block (**b**) of a similar experiment based on highly excited Rubidium states comprises a laser irradiation to generate excited states, further microwave excitation (waveguide enlarged in (**c**)), and finally ionization in static electric fields. Reproduced from [GS&88] (**a**) and [BB&91] (**b, c**) with permission.

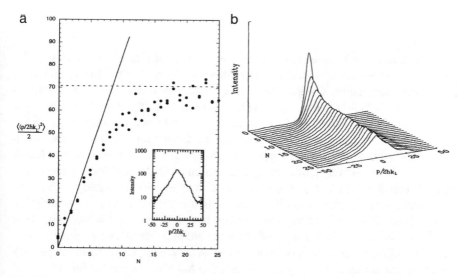

Fig. 8.28 Experiments with cold two-level atoms in an optical lattice provide direct evidence of dynamical localization. Measurements of the width of an expanding cloud of atoms as a function of time (**a**) demonstrate the transition from classical chaotic diffusion to quasiperiodic motion, to be compared with numerical simulations, see e.g., Fig. 8.27a. The inset is a logarithmic plot of the distribution over the lattice at the final time, to be compared with Fig. 8.26. The evolution of the spreading cloud from an initial narrow Gaussian shape to an exponential envelope as in Eq. (8.5.50) is visualized directly in a linear plot (**b**). Reproduced from [MR&95] with permission.

of atoms spreading in the optical lattice generated by standing laser fields. In both experiments, the measured signals provided clear evidence of dynamical localization (Fig. 8.28).

Symptoms of the quantum agony of classical chaos can be summarized in a general diagnostics as follows:

– Closed quantum systems with a finite effective Hilbert space dimension $D_{\mathcal{H}}$ only contain a finite amount of information, of the order of $\ln(D_{\mathcal{H}})$, which is conserved in a unitary time evolution.
– While in classical chaos, structures on arbitrarily small scales provide an inexhaustible supply of entropy, the finite information content of an isolated quantum system is recycled in a closed loop.
– With a limited supply of information, quantum time evolution cannot indefinitely produce novelty but eventually turns repetitive, giving way to quasiperiodic behaviour and zero entropy production.

A crucial point in this reasoning is the conservation of entropy in unitary quantum time evolution. It therefore applies not only to closed systems, but likewise to systems with an external forcing—provided it is deterministic, defined by a unique functional dependence on time: The quantum suppression of classical chaos is a coherence phenomenon. However, it loses its power as soon as external impacts of a stochastic nature disrupt coherence, as will become evident in the following subsection.

8.5.4 Resurrection of Chaos by Decoherence and Dissipation

Chaotic motion—in weather, in turbulence, even in celestial mechanics—is an undeniable fact. At the same time, these phenomena are in principle subject to the laws of quantum mechanics. The quantum death of classical chaos therefore constitutes a blatant contradiction. It cannot be the last word, at least if applied to macroscopic systems. In fact, similar doubts extend to other emblematic quantum phenomena, such as tunneling and non-locality. They are far from robust, demonstrating them in the laboratory requires an enormous technological effort. The key is coherence: As hard as it is protecting coherence against external perturbation for an appreciable time, as invariably is it lost by even the faintest contact with the world out there.

Therefore, "lifting the splendid isolation" in a more comprehensive description that includes the effects of an environment, it should be possible to elicit traces of chaotic behaviour in the type of quantum models introduced in the previous subsection. There are countless instances of decoherence in all fields of physics where quantum mechanics plays a role. Section 8.3 gives an idea of this wealth of phenomena and how to account for them. In the sequel, two prototypical manifestations of an environment will be illustrated: Measurements focus on decoherence as an exchange of entropy with the environment. Their particular effects on classically chaotic quantum systems complement the outline of quantum measurement given in Subsect. 8.4. Dissipation, in turn, never comes along without decoherence, but above all involves an irreversible loss of energy. In the context of quantum chaos, this allows us in addition to see how quantum mechanics modifies fractal structures in phase space, notably strange attractors.

Elements of a theory of quantum measurement and more generally of environment-induced decoherence have been introduced in Subsects. 8.4.2 and 8.3.3, resp. They serve as starting point to study quantum chaos under observation: Take a classically chaotic quantum system, such as the quantum kicked rotor, and substitute it for the term $\hat{H}_0$ representing the observed central system in the Hamiltonian, see, e.g., Eq. (8.4.2). Constructing the interaction with the meter is more involved. In the above analysis of quantum measurement, it has been assumed that the measurement is a single event, lasting only a finite time. In order to study the effect of observation on quantum chaos, however, it is more appropriate to consider repeated or continuous measurements, in synchrony with, for example, the external forcing that induces chaos.

For instance, choose the coupling Hamiltonian as (cf. Eq. (8.4.4))

$$\hat{H}_{\mathrm{OM}}(t) = g\hat{\mathbf{x}}_{\mathrm{O}} \cdot \hat{\mathbf{x}}_{\mathrm{M}}\Theta(t), \tag{8.5.57}$$

where g controls the coupling strength. Its time dependence is now explicitly given by a Heaviside step function $\Theta(t)$, switching the measurement on at $t = 0$, but never switching it off again. The role of angular-momentum diffusion as suitable diagnostic to verify chaotic behaviour and detect the effect of observations on it, suggests to choose angular momentum as the measured observable, setting, see Eq. (8.5.42),

$$\hat{\mathbf{x}}_O := \hat{l} = \hbar \sum_{\lambda=-\infty}^{\infty} \lambda |\lambda\rangle\langle\lambda|. \tag{8.5.58}$$

Following the procedure in Subsect. 8.3.3, a meter is modelled as a set of harmonic oscillators, see Eq. (8.3.41),

$$\hat{H}_M = \sum_n \hbar\omega_n \left(\hat{a}_n^\dagger \hat{a}_n + \frac{1}{2} \right). \tag{8.5.59}$$

As expounded in Subsect. 8.3.3, this model for a quantum system coupled to an environment can be evaluated, resulting under suitable assumptions for the parameters of coupling and bath in a master equation for the reduced density operator $\overline{\hat{\rho}}_O(t)$ of the observed system, Eq. (8.3.50). In the present context where only decoherence, i.e., the loss of information, but no energy loss comes into play, only the second term on the r.h.s. of the Eq. (8.3.50), representing random transitions, is relevant, reducing the master equation to [DG90b, DG90c, DG92]

$$\frac{d}{dt}\overline{\hat{\rho}}_O = -\frac{i}{\hbar}\left[\hat{H}_O, \overline{\hat{\rho}}_O \right] + \kappa \left[\hat{x}_O, \left[\overline{\hat{\rho}}_O, \hat{x}_O \right] \right], \tag{8.5.60}$$

where the parameter γ controls the decoherence rate. Proceeding in synchrony with the kicks, Eq. (8.5.60) is integrated over a single time step, from immediately after a kick to immediately after the next one, as in Eq. (8.5.44). The result is a mapping for the reduced density operator,

$$\langle \lambda'' | \overline{\hat{\rho}}_O'' | \mu'' \rangle = \sum_{\lambda',\mu'=-\infty}^{\infty} b_{\lambda'-\lambda''}(k) b_{\mu'-\mu''}^*(k) \exp\left(-\frac{i\tau}{2}(\lambda'^2 - \mu'^2) \right)$$
$$\exp\left(-\kappa(\lambda'-\mu')^2 \right) \langle \lambda' | \overline{\hat{\rho}}_O | \mu' \rangle. \tag{8.5.61}$$

While the factors in the first line on the r.h.s. incorporate the time evolution generated by the Hamiltonian $\hat{H}_O$ of the isolated kicked rotor, the exponential in the second line contains the effect of the measurements: The off-diagonal elements of the density matrix decay exponentially in time, with a rate that increases with their separation $(\lambda'-\mu')^2$ from the diagonal. These processes contract the density matrix towards its diagonal but leave the angular momentum, and with it the kinetic energy, untouched. Incoherent transitions towards the diagonal are a patent example of decoherence, and they illustrate the general image of quantum measurement drafted in Sect. 8.4: The measured system approaches eigenstates of the measured observable, here the angular momentum, playing the role of pointer states.

The mapping (8.5.61) serves to calculate the evolution of the measured system, e.g., by numerical iteration, and to compare it with that in the unmeasured case. Figure 8.29 shows the global increase of the kinetic energy for the classical kicked rotor (dotted), the unobserved quantum kicked rotor (dashed), as in Fig. 8.25a, compared to the quantum kicked rotor under continuous measurement (full line).

Even with very weak coupling (panel a, $v := 1 - \exp(-\kappa) = 10^{-3}$), the system still exhibits the abrupt stalling of classical diffusion, but then returns to a linear increase with time, albeit much slower than expected classically. With significantly stronger coupling (panel b, $v = 0.5$), chaotic diffusion in the measured quantum system reaches almost the classical level. Figure 8.30 looks closer into the angular momentum distribution, in a logarithmic plot, after 512 iterations of the map. Here, weak measurements (Fig. 8.30a) are reflected in a broadening of the distribution, as compared to the unobserved system, while it still maintains the characteristic irregular quantum profile around an approximately exponential envelope. With much stronger coupling (Fig. 8.30b), the distribution of the measured angular momentum becomes almost identical to the classical Gaussian diffusion cloud.

With this scenario, two independent time scales compete: the characteristic time j^*, see Eq. (8.5.56), of the crossover from classical diffusion to dynamical localization, and the decoherence time $j_{dec} \approx 1/\kappa$: For weak coupling, $j_{dec} \gg j^*$, the observed system will undergo the transition at j^* to dynamical localization with an exponential profile of the angular-momentum distribution and only much later approach the classical Gaussian distribution, while for sufficiently strong coupling, $j_{dec} \ll j^*$, classical diffusion occurs from the beginning, and symptoms of dynamical localization will never occur.

We can conclude that incoherent processes induced by a macroscopic environment destroy dynamical localization. They restore chaotic entropy production, at least partially, by supplementing the limited information reserve of the kicked rotor with random fluctuations imparted by the environment.

The loss of coherence is a pure quantum phenomenon, as is coherence. It affects all quantum systems interacting with a many-body environment. Dissipation, by contrast, refers more generally to a simultaneous loss of information *and* energy that occurs already on the classical level, as expounded in Sect. 6. In order to take dissipation into account in the model of the quantum kicked rotor coupled to an environment, transitions between different angular momentum states have to be included that tend to reduce the kinetic energy.

In order to compare the dissipative quantum dynamics with a corresponding classical behaviour, also the classical map has to be modified accordingly. Unlike the classical equations of motion (8.5.30) and map (8.5.32), it can no longer be derived from a Hamiltonian but has to be construed directly from a Newtonian equation of motion with friction, for example complementing Eq. (8.5.30) with an Ohmic friction term with friction constant γ,

$$\dot{l} = -\gamma l + K \sin(\theta) \sum_{j=-\infty}^{\infty} \delta\left(\frac{t}{T} - j\right), \quad \dot{\theta} = \frac{l}{\Theta}, \qquad (8.5.62)$$

Integrated as above over a single period from immediately after one kick till immediately after the next one, it leads to the map from (l_j, θ_j) to (l_{j+1}, θ_{j+1}),

$$l_{j+1} = e^{-\gamma}l_j + K\sin(\theta_{j+1}), \quad \theta_{j+1} = \theta_j + \frac{e^{-\gamma}-1}{\gamma}\frac{T}{\Theta}l_j(\bmod\,2\pi). \quad (8.5.63)$$

In order to achieve a form as simple as Eq. (8.5.32), define the angular momentum attenuation per time step, $\Lambda := e^{-\gamma}$, and rescale the angular momentum as well as the kick strength by a factor $|\ln(\Lambda)|/(1-\Lambda)$:

$$(l,\theta) \mapsto (l',\theta'), \quad l' = \Lambda l + K\sin(\theta'), \quad \theta' = \theta + l(\bmod\,2\pi). \quad (8.5.64)$$

In this form, the map is known as *Zaslavsky's map* [Zas79, SW85]. For $\Lambda = 1$, the standard map (8.5.32) is recovered. For $\Lambda < 1$, it exhibits attractors, phase-space manifolds approached in the long-time limit by its trajectories. Increasing the kick strength K from zero onwards, the map now no longer passes through the dynamical scenario sketched above but through a sequence of bifurcations and inverse bifurcations [SW85], beginning with a simple point attractor at $l = 0$, $\theta = \pi$ that replaces the stable island of the undamped map (Fig. 8.24b), and culminating in a single-piece strange attractor at a critical value $K_1(\Lambda)$. The attracting manifolds must be restricted in angular momentum to the strip

$$|l| \leq l_{\max} = \frac{K}{1-\Lambda}, \quad (8.5.65)$$

since for $|l| > l_{\max}$, friction reduces the angular momentum for all values of θ.

The strange attractors of the dissipative standard map are self-similar structures, comparable to the strange attractor of the dissipative baker map (Fig. 6.5) and likewise with a fractal dimension between 1 and 2. As argued in Sect. 6.4, cf. Eq. (6.4.9), the actual information contained in a self-similar set diverges and with it the Hilbert space dimension required to capture it in a quantum mechanical state, unless it is capped by a finite resolution. Unlike classical mechanics, where a maximum resolution can only be incorporated "by hand", it is intrinsic in quantum mechanics as a direct consequence of the Heisenberg uncertainty relation. It does not result in a rigid pixelation, as in computer graphics, however, but becomes manifest as a flexible coarse-graining that smoothes fractal objects in multiple ways, depending on their position and orientation in phase space.

Its effect can be studied by quantizing the dissipative standard map (8.5.64), along similar steps as in the modelling of the quantum kicked rotor under continuous observation. The crucial difference is that dissipation involves incoherent transitions that, besides contracting the reduced density operator towards its diagonal, also induce a drift towards lower energy. In terms of angular momentum eigenstates $|\lambda\rangle$, see Eq. (8.5.42), they take the form of annihilation operators of angular momentum quanta [DG90a],

$$\hat{\Gamma}_\lambda := \sqrt{|\lambda|+1}(|\lambda\rangle\langle\lambda+1| + |-\lambda\rangle\langle-\lambda-1|), \lambda \in \mathbb{N}_0 \quad (8.5.66)$$

(and their adjoints $\hat{\Gamma}_\lambda^\dagger$ as creation operators). Instead of the coupling to a meter, cf. Eqs. (8.5.57, 8.5.58), the interaction with a heat bath absorbing kinetic energy from the kicked rotor takes the form

$$\hat{H}_{OM} = \sum_{\lambda,n} g_{\lambda,n}\left(\hat{\Gamma}_\lambda^\dagger \hat{a}_n + \hat{\Gamma}_\lambda \hat{a}_n^\dagger\right).\tag{8.5.67}$$

The coupling constants $g_{\lambda,n}$ have to be specified so as to reproduce the dissipative classical map, assuming a bath composed of harmonic oscillators as in Eq. (8.5.59). Under similar assumptions as above, regarding the respective characteristic time scales of the kicked rotor and the heat bath, a master equation for the reduced density operator, analogous to Eq. (8.3.50), is derived,

$$\left\langle \mu \left| \frac{d}{dt}\bar{\bar{\rho}}_o \right| \lambda \right\rangle = -\frac{i}{\hbar}\left[\hat{H}_O, \bar{\bar{\rho}}_o\right] + |\ln(\Lambda)|\left\{-\frac{1}{2}(|\lambda| + |\mu|)\left\langle \mu \left| \bar{\bar{\rho}}_o \right| \lambda \right\rangle \right.$$
$$\left. + \Theta_{\lambda\mu}\sqrt{(|\lambda|+1)(|\mu|+1)}\left\langle \mu + \text{sgn}(\mu) \left| \bar{\bar{\rho}}_o \right| \lambda + \text{sgn}(\lambda) \right\rangle\right\}.$$
$$\tag{8.5.68}$$

In addition to the first and second terms generating unitary time evolution and decoherence, resp., as in Eq. (8.5.60), this master equation now comprises a third term causing transitions towards lower angular momentum, that is, it generates dissipation proper. During the kicks, only the unitary term contributes, while decoherence and dissipation can be neglected. Integrating Eq. (8.5.68) over a single time step, a map for the reduced density operator ensues in the form of a propagator $G\left(\lambda'', \mu''; \lambda', \mu'\right)$ in angular-momentum representation,

$$\left\langle \mu' \left| \bar{\bar{\rho}}_o \right| \lambda' \right\rangle \mapsto \left\langle \mu'' \left| \bar{\bar{\rho}}_o \right| \lambda'' \right\rangle = \sum_{\lambda',\mu'} G\left(\lambda'', \mu''; \lambda', \mu'\right)\left\langle \mu' \left| \bar{\bar{\rho}}_o \right| \lambda' \right\rangle\tag{8.5.69}$$

(for details, see Refs. [DG87, DG88, DG90a]).

The dissipative quantum map (8.5.69) can be used in particular to simulate numerically the time evolution under the quantized kicked rotor with dissipation. Besides the crossover time j^* and the decoherence time j_{dec}, a third characteristic time scale comes into play, the energy decay time $j_{diss} = 1/\gamma$, given by the inverse friction constant γ, see Eqs. (8.5.62, 8.5.63). In general, decoherence, as a true quantum phenomenon, is much faster than the loss of energy, $j_{dec} \ll j_{diss}$. Both processes can occur, though, before or after the crossover to quasiperiodic behaviour. Figure 8.31 shows a few such cases in terms of the time dependence of the kinetic energy, to be compared with Figs. 8.25a and 8.29: For very weak coupling to the heat bath, $\Lambda = 1 - 5 \times 10^{-6}$ (graph d), only decoherence becomes manifest as a slow increase of the kinetic energy over its quasiperiodic evolution in the unitary case (e), i.e. here, the order of time scales is $j^* < j_{dec} < j_{diss}$. With somewhat stronger friction,

$\Lambda - 1 - 10^{-1}$ (c), a remnant of quasiperiodic fluctuations is still visible before the kinetic energy approaches a smooth increase. At appreciable coupling, $\Lambda = 1-10^{-3}$ (b), there is still a a kink in the energy growth at j^*, but giving way directly to a marked slowing down of the increase, due to friction, as compared to the unbounded growth in the classical kicked rotor without friction (a).

For $j \gg j_{\mathrm{diss}}$, the system approaches a stationary state where chaotic diffusion is balanced by dissipation. For $K > K_1(\Lambda)$, the classical threshold to a one-piece strange attractor, this stationary state is not only characterized by a constant mean value of the energy but in particular also by a periodic time dependence of $\overline{\rho}_0(t)$. Its period coincides with that of the kicks, so that observed stroboscopically, it appears constant. This asymptotic density is depicted in Fig. 8.32b as snapshots of a Wigner function $P_W(l, \theta, t)$ immediately after each kick (as to how to adapt the basic definition (8.1.17) to a cylindrical phase space, see Refs. [Ber77a, DG90a]). It has support at the quantized angular momenta $l_\lambda = \hbar\lambda$ and is plotted for each value l_λ in a pseudo-three-dimensional rendering as the continuous graph $P_W(\theta)$ on that baseline.

For comparison, Fig. 8.32a shows the corresponding classical strange attractor. The finite value of the scaled Planck's constant, here $2\pi\tau = 0.01$ (filled circle in the upper left corner), leads to a smoothing that destroys the self-similarity of the classical strange attractor (evident in the enlarged detail in the upper inset in Fig. 8.32a). At the same time, it reveals features of the classical dynamics that are not visible in the classical attractor itself: The stationary quantum phase-space distribution extends

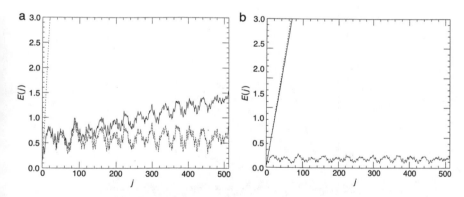

Fig. 8.29 Angular-momentum diffusion in the classical vs. the quantum kicked rotor, unobserved and under continuous measurement, over the first 512 time steps, starting from zero angular momentum initial states. With weak coupling to the meter (full line in (a), Eq. (8.5.63) with $\nu := 1-\exp(-\kappa) = 10^{-3}$), the linear increase of the kinetic energy is restored, but it still shows the crossover from classical diffusion (dotted) to irregular quantum fluctuations as in the unobserved case (dashed). With stronger coupling ($\nu = 0.5$, (**b**)), the rate of increase of the kinetic energy reaches almost its classical value. The data shown are based on a model slightly different from Eqs. (8.5.57–8.59), where instead of the expectation value of the angular momentum, its full distribution is measured [DG90b, DG90c, DG92]. Other parameter values are $K = 5$ and $\tau/2\pi = 0.1/G$ ($G := (\sqrt{5} - 1)/2$). Reproduced with permission from [Dit19].

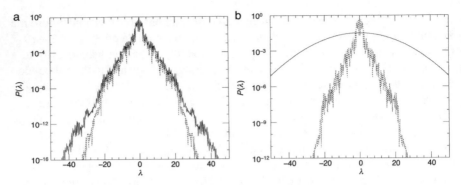

Fig. 8.30 Logarithmic plot of the angular-momentum distribution after 512 time steps in the quantum kicked rotor, unobserved and under continuous measurement. After spreading diffusively as in the classical case, the isolated quantum system crosses over to a quasiperiodic evolution where it fluctuates around an exponential envelope (dashed lines in both panels), cf. Eq. (8.5.50) and Fig. 8.28. Under weak continuous measurement, the observed system continues to spread diffusively even after the crossover, but slower than in the classical case (full line in (a), with $\nu := 1 - \exp(-\kappa) = 10^{-3}$). With stronger coupling, a Gaussian envelope, characteristic of classical diffusion, is approached even before the crossover to localization could occur (full line in (b), with $\nu = 0.5$). The data shown are based on the same model, slightly different from Eqs. (8.5.57–8.59), that underlies Fig. 8.29. Initial states and other parameter values as in Fig. 8.29. Reproduced with permission from [Dit19].

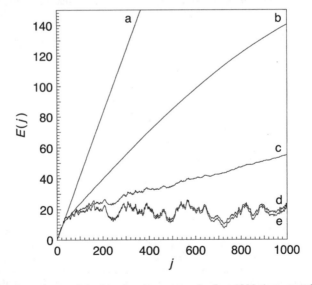

Fig. 8.31 Time dependence of the kinetic energy over the first 1000 time steps in the quantum kicked rotor with dissipation, Eqs. (8.5.68, 8.5.69), starting from zero angular momentum states. Graphs d, c, b show the increasingly faster growth of the kinetic energy with rising friction constant, $\Lambda = 1\text{--}5 \times 10^{-6}$ (d), $1\text{--}10^{-4}$ (c), $\Lambda = 1\text{--}10^{-3}$ (b), compared with the unitary quantum kicked rotor (e) and the classical standard map without friction (a). In cases c and d, a remainder of quasiperiodic fluctuations is still seen, superposed on the smooth energy growth, while in case b, the approach to a steady state sets in already. Reproduced with permission from [DG90a].

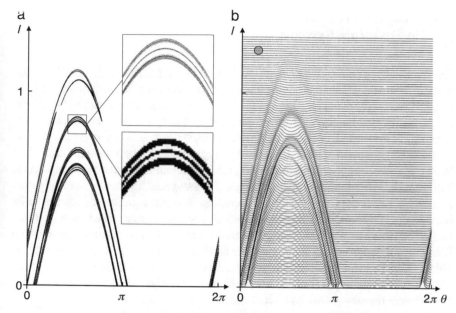

Fig. 8.32 Asymptotic density distributions of the classical (**a**) versus the quantum (**b**) kicked rotor with dissipation. (**a**) The classical strange attractor is represented as the support of the asymptotic phase-space distribution at time sections immediately after each kick. The upper inset shows an enlargement of one of the branches of the attractor to visualize the self-similar repetition of its laminated structure. The lower inset is a magnification of the same region in the large plot, revealing its Cartesian pixelation. (**b**) For the quantum phase-space distribution, the Wigner function $P_W(l,\theta)$ of the reduced density operator at the same time sections is plotted along the discrete support $l_\lambda = \hbar\lambda,\, \lambda \in \mathbb{Z},\quad \theta \in [0, 2\pi[$ it attains for a cylindrical phase space [Ber77, DG90a]. The filled circle in the upper left corner indicates the size of a Planck cell. Parameter values are $K = 5,\, \Lambda = 0.3,\, \tau/2\pi = 0.1/G,\, G := (\sqrt{5} - 1)/2$. Reproduced with permission from [DG90a].

beyond corresponding branches of the classical attractor along a *centre manifold*, a preferred direction of stretching of the classical dynamics that forms a backbone of the attractor. By contrast, the fine wavelets visible inside the sine-shaped main support lines of the distribution can be attributed to a remainder of quantum coherence, which is generated at each kick by the unitary evolution and erased again during the subsequent damped rotation.

This form of quantum mechanical smoothing of fractal geometries in classical phase space illustrates why quantization cannot be reduced to pixelation: The Cartesian grid underlying the pixelated .pdf image of the classical attractor can be directly discerned in the lower inset in Fig. 8.32a and compared with the quantum distribution in panel b, smoothed on a scale indicated by the filled circle of the size of a Planck cell. In this way, quantum mechanics complies with the bounds on information density in phase space, but without determining more specific structures than required by the uncertainty relations.

8.6   Mixing, Irreversibility, and Information Production in Quantum Systems

The discussion of entropy flows in classical physical systems, Sects. 5, 6, and 7, culminated in a global picture, interpreting irreversibility in terms of information dynamics: It is the result of the unbalanced exchange of entropy between macroscopic and microscopic scales, and its distribution and scrambling among the degrees of freedom of many-body systems [RUR18]. The question exactly how entropy is processed and redistributed at the smallest scales, notably in approaching thermal equilibrium, had to be postponed, relegated to a more comprehensive quantum mechanical treatment. Two closely related issues were left open: Coarse-graining, the erasure of information on tiny details, is an essential condition to attain equilibrium but has to be incorporated "by hand" in classical statistical mechanics. And the amount of information on the state of a microscopic system its state space can accommodate has no fundamental upper bound in classical mechanics, giving rise to paradoxes such as Gibbs'. The fact that in its postulates, quantum mechanics addresses exactly these problems nurtures hopes it would solve them definitively, and invites to reconsider the classical reasoning with its ad-hoc constructions from scratch.

The foregoing subsections, however, cast doubts on this hope. To be sure, quantum uncertainty and the concept of pure states strictly limit the amount of information to be obtained about a quantum system, thus regularizing divergences of classical mechanics related to entropy. At the same time, the unitary time evolution of closed quantum systems conserves entropy just as rigorously as the canonical transformations of classical mechanics. The information required to reconstruct an initial state of a quantum system is never erased, ruling out irreversibility. As a consequence of both conditions, deterministic chaos, a crucial ingredient of mixing and thus for thermalization, is suppressed in closed quantum systems. Therefore, also in quantum mechanics, a subtle interplay of several elements is necessary to explain the approach to thermal equilibrium, in particular a chaotic dynamics and the interaction of a large number of subsystems. Many details remain open: With quantum thermalization, this book has reached the frontier of ongoing research. Conclusive answers are not yet available, the present section merely provides a snapshot of the state of the art.

8.6.1   The Role of Chaos: Berry's Conjecture

In Chap. 7, mixing in classical systems was characterized by two simultaneous tendencies: With increasing time, the combination of stretching and folding in chaotic motion leads locally to near homogeneity down to smaller and smaller scales, and globally to a uniform spreading over the entire accessible phase space. The information required to reconstruct the initial from the final state is relocated to correlations on ever-shorter distances in phase space, see, e.g., Figs. 7.5b and 7.6. This tendency of classical chaos towards uniformity in phase space is reflected in various phenomena:

For example, unstable periodic orbits of chaotic systems approach a uniform distribution over phase space as their number proliferates with increasing length [HO84]. At the same time, spatial structures must approach delta-correlated stochastic processes.

In classical statistical mechanics, mixing emerges as an asymptotic tendency in the long-time limit of the chaotic dynamics of strongly interacting many-body systems. However, it cannot be deduced for an isolated trajectory evolving from a sharp initial condition. It is necessary to explore also its neighbourhood, be it by averaging over time or by including an ensemble of non-zero width of initial conditions. Quantum mechanics is fundamentally distinct in this respect: If a quantum system approaches an equilibrium state in the long-time limit, it should do so in particular if considered as isolated, prepared in a pure state and evolving unitarily into pure states. That is, the tendency to thermalize must be predisposed from the outset in the eigenfunctions of the system.

Are the eigenfunctions of Hamiltonians describing classically chaotic quantum systems distinguished by specific characteristics? This question has been left aside in Sect. 8.5. The general conditions mentioned above allow us to devise a plausible expectation. On the one hand, such wave functions should be approximately homogeneous on the phase-space manifold accessed by the system. For a closed system, this is the constant-energy surface corresponding to its eigenenergy. On the other hand, they should be random functions, with isotropic correlations decaying on a non-zero scale given by the local de Broglie wavelength (Figs. 8.33, 8.34). There exists a surprising exception to these rules: so-called "scars" [Hel84, Ber89], eigenfunctions of chaotic systems that tend to exhibit an enhanced amplitude along unstable periodic orbits of the corresponding classical system (Fig. 8.35). As conspicuous as

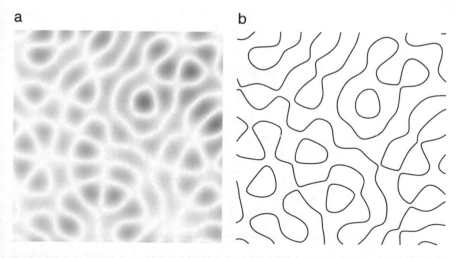

a   **b**

Fig. 8.33 Typical Gaussian random wave, Eq. (8.6.2) but for a purely real wave function, shown in a section of about 6×6 wavelengths, (**a**) as a density plot, colour code ranging from red (negative) through white (zero) through blue (positive), and (**b**) reduced to its nodal lines (contour lines at $\psi(\mathbf{q}) = 0$).

scars may appear in individual eigenfunctions, on average they do not modify the statistical properties of ensembles of eigenfunctions.

This suggests constructing them as *Gaussian random waves*: a superposition of plane waves, with wave vectors distributed isotropically, common wavenumber

$$k_n(\mathbf{q}) = \frac{|\mathbf{p}_n(\mathbf{q})|}{\hbar} = \frac{1}{\hbar}\sqrt{2m(E_n - V(\mathbf{q}))}, \tag{8.6.1}$$

for eigenvalues E_n of Hamiltonians $\hat{H}(\hat{\mathbf{p}}, \hat{\mathbf{q}}) = \hat{\mathbf{p}}^2/2m + \hat{V}(\hat{\mathbf{q}})$, random phases ϕ_n, distributed uniformly around the unit circle, and Gaussian random amplitudes $A_n(\mathbf{p})$. These assumptions, known as *Berry's conjecture* [Ber77b, Ber78, MK79, Ber91, Sre94], result in an ansatz for generic chaotic wave functions of the form

$$\psi_n(\mathbf{q}) = \langle \mathbf{q} | \psi_n \rangle$$
$$\sim \int d^f p \, A_n(\mathbf{p}) \delta\left(\mathbf{p}^2 - 2m(E_n - V(\mathbf{q}))\right) \exp(\mathrm{i}(\mathbf{p} \cdot \mathbf{q} - \phi_n)). \tag{8.6.2}$$

Equation (8.6.2) had been conceived originally for models of quantum chaos with few degrees of freedom, such as time-dependent systems in one dimension or two-dimensional billiards (Fig. 8.34). A crucial, if mathematically delicate, step is to extend it so as to include many-body systems. A prototypical model, in particular for thermalization, for systems with many interacting degrees of freedom, is the hard-sphere gas (Fig. 8.36b): an ensemble of N masses interacting only by a contact potential, as for hard spheres of radius a,

$$V(r) = \begin{cases} 0 & \text{for } r < a, \\ \infty & \text{else,} \end{cases} \tag{8.6.3}$$

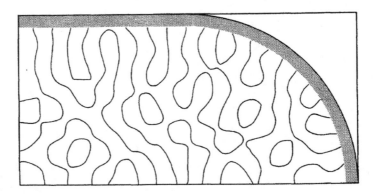

Fig. 8.34 Nodal lines of an energy eigenfunction with wavenumber $k = 50.158$ of one quadrant of the stadium billiard with odd-odd parity. The parameters of the billiard are $R = 0.665$ (radius of the semicircles) and $a = R$ (width of the central rectangle), resulting in an area $\pi/4$ of the quadrant. Reproduced from [MK79] with permission.

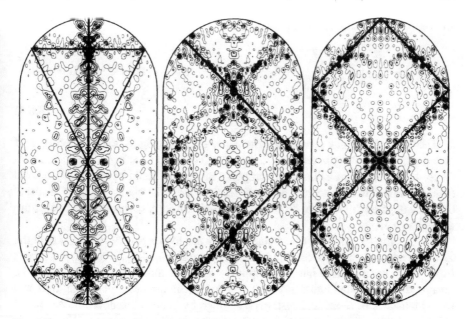

Fig. 8.35 Scarred eigenfunctions of the stadium billiard, depicted as contour lines. Dominant scarring orbits are indicated by bold lines. The aspect ratio of the billiard is as in Fig. 8.34. Reproduced from [Hel84] with permission.

a

b

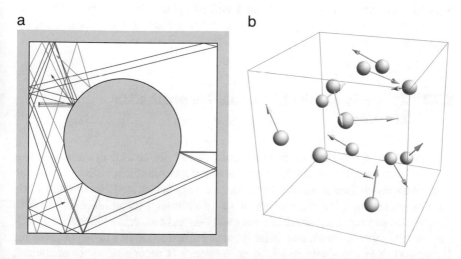

Fig. 8.36 The Sinai billiard (**a**), a free particle moving inside a square potential box with a circular obstacle in the centre, is a standard model for chaos in classical Hamiltonian systems and the first one for which the properties of ergodicity and mixing could be rigorously proven. Sensitive dependence on initial conditions is illustrated in the figure by three trajectories with nearly identical initial positions, which get uncorrelated after a few collisions with the central disk. Raising the dimension from 2 to 3, replacing the rigid disk by a spherically symmetric hard-core potential between particles, and the number of particles to $N \gg 1$, see Eqs. (8.6.3, 8.6.4), the hard-sphere gas (**b**) is obtained, a paradigm for thermalization in classical many-body systems.

and otherwise moving as free particles,

$$H(\mathbf{P}, \mathbf{Q}) = \sum_{n=1}^{N} \frac{\mathbf{p}_n^2}{2m} + \sum_{\substack{n,n'=1 \\ n' \neq n}}^{N} V(|\mathbf{q}_{n'} - \mathbf{q}_n|), \mathbf{P} = (\mathbf{p}_1, \ldots, \mathbf{p}_N), \mathbf{Q} = (\mathbf{q}_1, \ldots, \mathbf{q}_N).$$

(8.6.4)

The hard-sphere gas generalizes the *Sinai billiard* (Fig. 8.36a), a prototype of chaotic motion in two dimensions, to ensembles of N spheres in f dimensions, with $N \gg 1$. For the Sinai Billiard, the property of mixing could be rigorously proven [Sin70]. Raised to $2Nf$ degrees of freedom, the hard-sphere gas (Fig. 8.36b) shares the same degree of dynamical disorder as the Sinai billiard, so that Berry's conjecture applies as well.

For this model, surfaces of constant energy are hyperspheres $\mathbf{P}^2 = 2mE$ in the $2Nf$-dimensional phase space, with voids along the diagonals $\mathbf{q}_n = \mathbf{q}_{n'}$. Hyperspheres in $\mathbb{R}^{2Nf}$ do not factorize into hyperspheres in lower dimensions representing the energy shells of individual degrees of freedom. It is therefore plausible to expect that eigenstates of N-particle systems described by Gaussian random waves would exhibit the same mixing properties as chaotic single-particle systems. Whether this extrapolation also applies to the spreading of entanglement is however not obvious. Moreover, there is no point in referring to chaotic dynamics if the ensemble consists of single particles with low-dimensional Hilbert spaces, such as spins. The challenges many-particle systems pose to quantum thermalization will be sketched in the following.

8.6.2 Typicality and the Eigenstate Thermalization Hypothesis

The phase-space structures of eigenstates of classically chaotic systems, summarized in Berry's conjecture, constitute an important specific mechanism how quantum systems approach thermal equilibrium. As soon as quantum statistics of many-body systems comes into play, the concept of space assumes a second meaning besides phase-space geometry: the space of configurations and networks of interacting particles, possibly with a low-dimensional internal Hilbert space that does not allow for anything such as ergodicity and mixing. Evidence is accumulating, of analytical, numerical, and even experimental nature, that the relevant properties for quantum thermalization apply as well to the vast majority of generic many-body systems, independently of whether they can be considered as chaotic in any sense. This calls for a more general approach, not recurring to dynamics. It is found in the notion of *typicality* [GL&06], anticipated already by von Neumann in his *quantum ergodic theorem* [Neu29]. It rests on a mathematical result that does not refer to any dynamical features but to the mere geometry of very high-dimensional Hilbert spaces $\mathcal{H}$

with $D_\mathcal{H} \gg 1$. Owing to normalization, the relevant manifold is the $(2D_\mathcal{H} - 1)$-dimensional unit hypersphere $S^{(2D_\mathcal{H}-1)}$, embedded in $\mathbb{C}^{D_\mathcal{H}}$. With growing dimension, the surface measure of $S^{(2D_\mathcal{H}-1)}$ is increasingly depleted around the poles and concentrated along the equators, defined as the set of points where one, and only one, of the coordinates takes the value 0 (Fig. 8.37).

This geometric fact is the crucial ingredient of *Levy's lemma* [MS86]. It deals with averages of smooth functions $f(\mathbf{x})$ over ensembles of points $\mathbf{x}$ drawn at random from a set with a uniform measure on $S^{(2D_\mathcal{H}-1)}$. The geometry of the hypersphere relates its equators to the medians of $f(\mathbf{x})$, and it relates their neighbourhoods to the set of points $\mathbf{x}$ with $f(\mathbf{x})$ not deviating drastically from the mean value $\langle f \rangle$. Statistical outliers, deviating by more than ε from the mean value $\langle f \rangle$ of some smooth function $f(\mathbf{x})$, are thus restricted to the pole caps of $S^{(2D_\mathcal{H}-1)}$. Their measure $\sim \exp(-cD_\mathcal{H}\varepsilon^2)$, diminishes exponentially with $D_\mathcal{H}$, c denoting a positive constant. If states in a Hilbert space $\mathcal{H}$ of a many-body systems are represented by vectors $x = (x_1, \ldots, x_n, \ldots, x_{D_\mathcal{H}})$, with component x_n giving the participation of subsystem n, say a single particle, Levy's lemma implies that for $D_\mathcal{H} \gg 1$, the overwhelming majority of accumulates in the vicinity of the equators, where none of the subspaces dominates. They are typical in the sense that with high probability, they are maximally entangled. More precisely, if the total system is restricted by energy conservation, i.e., if it is in a microcanonical state, a single subsystem will be found almost certainly in a canonical state [PSW06]. Being maximally entangled, these states are at the same time of a decidedly non-classical nature.

Based on the notion of eigenstates, which, to attain orthogonality, possess a complex fine structure on small scales but are homogeneous on global manifolds such as energy shells, the *eigenstate thermalization hypothesis* [Deu91, Sre94, PSW06, RDO08] has been postulated as an alternative scenario of quantum thermalization that does not resort to ensemble or time averaging:

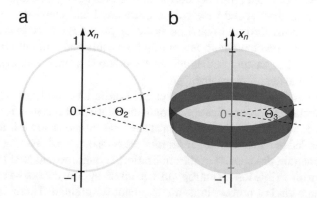

Fig. 8.37 With growing dimension, an increasing fraction of the total surface of (hyper)spheres is concentrated around the equators. If one sixth of the circumference of a unit circle occupies two sectors (red) of width $\Theta_2 = \pi/6 \approx 0.5$ each (panel **a**), the same fraction of the surface of a unit sphere (red belt in **b**) occupies a neighbourhood of the equator of minor width $\Theta_3 = \arcsin(1/3) \approx 1/3$. This tendency continues into higher dimensions, see text.

- Typical eigenstates within a small interval of eigenenergies (i.e., a microcanonical ensemble) of an isolated but internally interacting many-body system are locally distinct in their fine structure, but are globally approximately identical, covering their energy shell homogeneously.
- A localized initial state of a many-body system will overlap with a large number of such eigenstates. The "conspiracy of phases" required by the initial localization is lost with the unitary time evolution of the total system, entangling the initially excited subsystem with almost all others.
- The information on the initial state, encoded, e.g., in cross correlations, thereby spreads globally over the entire ensemble. It is not lost, but can no longer be retrieved by local measurements on small subsystems, only by observing the whole. In terms of locally accessible information, the process is therefore irreversible.

The eigenstate thermalization hypothesis, proposed on general analytical grounds, has been tested numerically, e.g., in sets of coupled bosons [RDO08], and corroborated by experimental studies using ultra-cold atoms in optical lattices [KT&16] or sets of coupled two-state systems realized on basis of SQUIDs (Superconducting Quantum Interference Devices) [NR&16].

8.6.3 Many-Body Localization: Threatening Thermalization?

So far, it might appear that quantum mechanics translates the classical image of the approach to thermal equilibrium to a completely different conceptual framework, but draws on the same essential ingredients: chaotic dynamics substantiating the conjecture of Gaussian random waves, as concerns the spreading over single-particle phase spaces, and typicality corroborating many-body entanglement. In this way, it underpins the Second Law with a more solid and consistent treatment of microscopic processes. At the same time, however, previous subsections have shown that in particular, classical chaos can hardly be established without restrictions in quantum mechanics, casting doubts whether the above scenario applies in fact so unconditionally to quantum mechanics.

The example of the kicked rotor, Subsect. 8.5.3, is indicative. Originally conceived exactly to substantiate thermalization in quantum statistical physics, it revealed the opposite. To be sure, chaotic behaviour is at least partially recovered if the quantum system is coupled to a macroscopic reservoir, see Subsect. 8.5.4. Falling back to this kind of argument in the context of thermalization, though, would lead to an infinite regress: The irreversible loss of entropy in a quantum system cannot be explained by taking the same kind of process in its environment for granted. Therefore, quantum thermalization must be understood in isolated systems.

In the case of the kicked rotor, it is a specific quantum coherence effect without any classical analogue that impedes chaotic behaviour: Dynamical localization prevents extended eigenstates in the absence of translational symmetries, in that

case in momentum space. Its counterpart in position space, Anderson localization, invokes quenched disorder, i.e., a random static potential, such as frustra-
tion in spin systems, to explain isolating properties of some materials. Could it be
that a similar phenomenon, occurring in many-body systems, challenges the above
quantum account of the Second Law, embodied in the eigenstate thermalization
hypothesis?

As a quantum coherence phenomenon, localization in turn does not tolerate
random interactions with an environment. In a many-body system, the coupling of a
local degree of freedom to the rest of a many-particle system has an effect comparable
to an environment: quantum localization and many-body interaction are antagonists.
This suggests a trade-off, depending on the balance between the degree of disorder
and the magnitude of a random static potential on the one hand and the strength of the
inter-particle interaction on the other. It should manifest itself in a phase transition
from a localized to a delocalized (ergodic) phase [BAA06, PH10], as the coupling
strength surpasses a critical value, the *mobility edge*. While on the delocalized side
of the transition (weak disorder, strong coupling), eigenstate thermalization occurs,
on the localized side (strong disorder, weak coupling), the system should never reach
equilibrium.

Results of numerical and experimental studies support this view. They indicate
that many-body localization impedes the spreading of information on the initial
state, encoded in cross correlations, over the entire system, as required by eigenstate
thermalization [Alt18]. The information remains confined instead to the localization
neighbourhood of the initial state, see, e.g., Figs. 8.26 and 8.28b. This phenomenon
can be evidenced by looking at the mutual entropy (called *entanglement entropy*
in this context), shared among a local subsystem and the rest of the system: In
the ergodic phase, it will increase from zero (pure initial state) as a linear function
of time to an asymptotic value, given in an N-particle system by $\log(N)$. In the
presence of localization, the entanglement entropy still increases with time, but only
logarithmically [ZPP08, BPM12], as a signature of the approach to a local equilibrium
within the localization neighbourhood. Besides evidence for many-body localization
based on numerical implementations of interacting quantum systems with a static
disordered background potential [ZPP08, BPM12, LLA15], a failure to relax to
equilibrium has been observed again in ultra-cold fermionic atoms, in optical lattices
lacking a discrete translation symmetry [SH&15], and in trapped ions [SL&16]. Even
substantiated like this, it is not clear in how far it is relevant for the host of equilibration
phenomena we observe in nature.

8.6.4   Perspectives: Equilibration and Entanglement

Reconstructing our view of the approach to thermal equilibrium within the framework of quantum mechanics has shifted perspectives and shed light on surprising
phenomena. The central role complex many-body dynamics assumes in classical
statistical mechanics is transferred to the structure of eigenstates of many-body

systems. Frozen disorder in background potentials can lead to the failure of thermalization where classically, no exceptions are to be expected. Above all, where entropy currents between different scales and among separate subsystems proved to be the key to understand thermalization in classical systems, see Chap. 7, they are complemented in quantum mechanics by the dynamics of entanglement.

Which role entanglement plays becomes particularly graphic, comparing two closely related irreversible phenomena: thermalization on the one hand and decoherence and dissipation on the other. In both cases, an asymptotic stationary state is approached through an asymmetric exchange of entropy and energy with a macroscopic many-body system, an environment. Yet they are distinct in a subtle aspect, the way the absorbed entropy and energy are redistributed among the degrees of freedom of the many-body system. This difference is reflected in the way the environment is modelled in the two cases: In decoherence and dissipation, see Subsect. 8.3.3, a heat bath composed of a large number of harmonic oscillators suffices, if only their total energy spectrum meets some smoothness conditions. Even if these oscillators are coupled, giving rise to the emergence of some collective modes (Subsect. 5.6.2), their identity is largely conserved in the uncoupled eigenmodes of the environment. In thermalization, by contrast, an exchange of energy between the subsystems through a mutual coupling is indispensable. According to the eigenstate thermalization hypothesis, typical eigenstates, representing the vast majority, are entangled among all individual particles of system.

These global conditions are reflected in the structure of entanglement in the two cases. In the case of decoherence, the "central system" is also the organizing centre of the links formed by entanglement. They extend from the central system into the environment, forming decoherence-free subspaces along these links (Subsect. 10.4.3). In eigenstates satisfying the eigenstate thermalization hypothesis, entanglement is rather more democratic, relating everyone to everyone, homogeneously and isotropically. Even in the case of many-body localization where entanglement is restricted to localization neighbourhoods, localization centres are scattered homogeneously over a sample.

The spreading of entanglement on the way to these long-time asymptotic structures can be observed in a number of diagnostics, in particular in entanglement entropies and in time-dependent correlations between different subsystems and sites [LB17]. It would be tempting to take a closer look into information currents resolved in space and time, and to compare them with the corresponding classical information flows, see Sect. 7.3. However, it is not easy to substantiate the distribution and exchange of information among small subsystems or even individual degrees of freedom, expected to accompany the spreading of entanglement. A crucial problem is that, while the total information is conserved, owing to correlations no such conservation law applies to subsystems, which impedes an unambiguous definition of entropy currents between them. The interdependence of entanglement spreading, information flows, and chaotic dynamics is presently an active frontier of research [RUR18]. For example, it has been possible to detect and describe the propagation front, the "light cone" that forms as the information spreads across a strongly interacting many-body system [LB17] (Fig. 8.38).

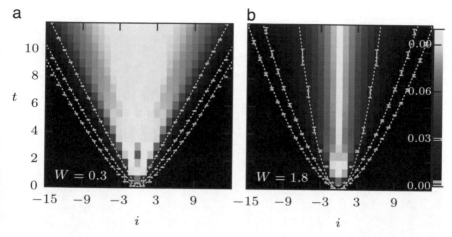

Fig. 8.38 In how far a quantum many-body system approaches thermal equilibrium is reflected in the way a local perturbation spreads over the system, to be observed as propagating correlations with the initially perturbed site. It is depicted here as the degree of correlation (colour code ranging from black (zero) through yellow for strong correlation) as a function of site position i relative to the initial perturbation (abscissa) and time t (ordinate), for a one-dimensional spin-$\frac{1}{2}$ chain with Heisenberg nearest-neighbour interaction and random on-site potential of mean magnitude W [LB17]. For weak random potential $W = 0.3$ (**a**), the system thermalizes and correlations spread linearly with time. With strong randomness $W = 1.8$ (**b**), many-body localization impedes the propagation of correlations, the correlated region expands only logarithmically with time. Reproduced from [LB17] with permission.

The discussion of chaotic classical systems embedded in an environment, Chaps. 6 and 7, focussed on the exchange of information in both directions, its loss into the environment as well as the randomizing leakage of entropy from the environment into an observable macroscopic system. This two-way flow proves decisive also in the context of decoherence, quantum measurement, and quantum chaos, and wherever a central system serves as a probe into a many-body quantum system. The entropy production, to become manifest in randomness if a large, classically chaotic system is coupled locally into a thermalizing quantum system, remains to be explored. It could provide valuable insights regarding quantum thermalization as well as quantum chaos.

Part II
Computers as Natural Systems

Chapter 9
Physical Aspects of Computing

Considering physical systems as information processors has proven enormously fruitful and illuminating: the computer as a metaphor. However, computers do not constitute an independent separate world. They are natural systems themselves, subject to the laws of physics and the limitations they entail for their functioning. This last part of the book is intended to integrate computing systems into their broader context in nature, to explain why on the one hand, they are "nothing but" physical dynamical systems, and on the other, to understand what makes them special—if anything. In the same sense that the first part, inspired by Wheeler's "It from bit" [Whe90a, Whe90b], interpreted natural systems as computers, the following sections complement it with the "Bit from it" [AFM15]. They provide an overview from a new perspective; they are not intended as an outline, even rudimentary, of the theory of computation. For comprehensive monographs and anthologies on this subject, see, e.g., Refs. [BEZ00, CT06]. Analyzing computers as dynamical physical systems can shed light on both sides, on physics and on computing.

In computers, we are confronted with the coexistence two distinct bodies of knowledge that both claim competence for this same subject: On the one hand, from the physical point of view (leaving genetic and neural "computing" aside), they are inanimate matter, thus are within the scope of statistical mechanics in general and are more specifically covered by Solid-state physics, electronics, and mechanics. In this context, digital computers in particular stand out by their specific set of discrete macroscopic states, established on top of the continuous space of microscopic states of classical physics. On the other hand, computer science, from the interested standpoint of their inventors and users, based on this space of "nominal" states, combines mathematics and engineering to develop a formal framework for the description and construction of computers.

Are these two approaches complementary or do they compete, even contradict one another? Challenging questions arise on either side: How can capabilities culminating in inference, even "artificial intelligence", emerge from mere physics? In how far do fundamental laws of physics affect even the high-level functionality of computing equipment? Conversely, is the deterministic course of computing

© Springer Nature Switzerland AG 2022
T. Dittrich, *Information Dynamics*, The Frontiers Collection,
https://doi.org/10.1007/978-3-030-96745-1_9

processes, controlled by software, compatible with the reductionist account of the underlying physical processes? Is the motion of an electron in a transistor that pertains to an integrated circuit determined by the Schrödinger equation or by the program that the chip is executing? Computers offer a prime opportunity to analyze how cognition can emerge from mere physics. After all, we designed and built them ourselves, thus have unrestricted access to the intentions and principles behind their construction. In physics, we are routinely using computers to simulate physical systems. We tend to forget that they are physical objects of their own right, which we ought to understand as thoroughly as those we are reproducing with their help.

An account of computing in terms of even microscopic classical physics is evidently not the last word. There exists a third, yet deeper, level of description: As has been expounded in the previous section, quantum mechanics is disparately more precise and consistent in the way it treats information and its dynamics in physical systems. To be sure, even considering extreme miniaturization, quantum effects do not play any functional role in classical computation (apart from electron transport in semiconductors, which forms the technological basis of present-day digital computing and evidently cannot be understood without quantum solid-state physics). Descending to the level of quantum mechanics, however, we encounter an entirely different category of computing systems, based on principles and strategies that reinvent computation from scratch. While classical computing enforces discretization as a high-level attribute arising on the basis of a continuous substrate, quantum computation takes advantage of the discretization nature provides for free on the quantum level (with all the caveats mentioned in Chap. 8).

Classical and quantum computing are rooted in two separate strands of historical development, to be presented here in chronological order, the present Chapter on classical information processing equipment, Chap. 10 on quantum computing. The subsequent sections, after integrating computers in a broader historical and technological context (Sect. 9.1), proceed "bottom up". Section 9.2 explains in detail how a discrete set of nominal states of a computing system can establish on a substrate of continuous classical variables. How computation proper as dynamical information processing emerges as directed navigation on that discrete space is expounded in Sect. 9.3. Section 9.4 finally ascends from the bottom level of gates and other basic elements of computers to higher tiers of the hierarchical organization of a computer.

9.1   What's so Special About Computing?

Ask a youngster, born in the twenty-first century, "Tell me, what is a computer?" As an answer, she will immediately point to her tablet or laptop, maybe with an expression of surprise about such a trivial question. "Computer" is an established concept by now, so elementary that it does not even need any explanation. Yet a hundred years ago it has been anything but obvious, indicating that the category is not as clear-cut as it may appear. In computers, various developments converge and culminate that have their roots in different scientific, technological, and cultural tendencies [BE16, Cop20].

As elaborated in the Introduction to this book, the advent of computers has left such a deep impact on intellectual life in the twentieth century that the concept became a universal metaphor, with a potential exceeding that of celestial mechanics or clockwork in previous epochs. Cells, brains, societies, the universe, "Everything is a computer" has become a viable claim. A computer pioneer as visionary as Konrad Zuse speculated about physical space as a whole to be considered as a computer [Zus69], quantum computer enthusiast Seth Lloyd sees computing extended to the entire universe: "What does the universe compute? It computes itself. The universe computes its own behaviour." [Llo06]. Contrast this all-embarrassing view, "every natural system evolves, thus calculates, hence is a computer", with the prosaic "nothing but" attitude that they are just another smart product of human inventiveness, "a computer has a CPU, an input and an output device, that's it", destined to be eventually superseded by the next major technological achievement. Computers are at the same time a special class of physical systems, products of high-tech engineering, and a cultural and historical phenomenon. The present section seeks a broad synthesis of these diverse facets.

9.1.1 Computers as Man-Made Tools

The overwhelming majority of tools and machines invented and made by humans in the course of history served to extend and surmount the natural limits of human action. Tools increase our force, our speed, our endurance, they replace physical labour. Others, in turn, complement and amplify our senses: telescopes let us look beyond the reach of our view, microscopes help us peeking into details not resolved by the naked eye. They provide access to information otherwise unattainable, yet deliver it as faithfully as possible, they do not really process it. Devices that actually allow us to overcome the limitations of our intellectual capacity, complementing or substituting mental work, reach far back into the historical past as well. In fact, the very onset of history, by definition, is marked by the advent of writing, an elementary tool to communicate data and protect them against the deficiencies of memory. Related, but even more elementary in nature are counting devices, starting with collections of similar objects such as pebbles or signs such as tallies, scores scratched on a bone or a wooden stick. Clocks in the widest sense count time in numbers of periodic events such as heartbeats, yardsticks measure distances in units of inches, etc. Other remarkable examples are mechanical locks, restricting access to protected spaces. They combine the functions of memory (encoded in the shape of the blade), recognition (of this code by the lock), and cryptography (preventing unauthorized copy), thus anticipate passwords and similar present-day authentication technology.

While the evolution of machinery serving physical purposes culminated in the nineteenth century in the advent of heat engines, computers mark the peak, for the time being, of the development of information processing tools. Heat engines consume energy to produce mechanical work. They generate excess entropy as an unintended but inevitable side effect. For the function of computers, by contrast, direct physical effects are irrelevant. Rather, the input, transformation, and output of information are decisive. Yet, also computing equipment consumes low-entropy energy and produces excess entropy in the form of heat. With these features, computers constitute a distinct and unique category of non-equilibrium thermodynamical systems.

Aside from the constant motion of clocks (which to be achieved required in fact the development of highly sophisticated technology [Sob95]), devices for information processing remained largely static for centuries. A first step into a functioning that involves a rudimentary variability has been the *abacus*, a calculation aid that transcends the counting of similar objects and already requires more specific material and construction. It already allows efficient addition and subtraction, but still requires substantial skill to handle the carry. The desire to mechanize even this task spurred the invention of mechanical calculators. Pioneers like Wilhelm Schickard, Blaise Pascal, and Gottfried Wilhelm Leibniz analyzed the mental operations involved to identify common patterns, that is, they extracted the underlying *algorithms*, and constructed mechanical machinery to simulate these operations. Mechanical calculators already master a considerable multiplicity of inputs and outputs. Typically driven by a simple hand crank as energy source, their purpose is however not the transformation of this mechanical energy into other forms of energy, but the diversity of the induced motion.

Mechanical calculators are able to mimic nontrivial mental tasks, multiplication and division, in the most advanced devices, besides addition and subtraction, but not more than these two to four operations. The initial step towards programmable computers, machines capable of performing a large repertoire of alternative tasks, was made by Ada Lovelace and Charles Babbage. Lovelace, inspired by the enormous versatility of the *Analytical Engine*, a mechanical computer conceived by Babbage, drew attention to the importance of performing a set of operations on a calculator in a predefined sequence and pointed out how even this task could be automated. In this way, she coined the concept of *programming* and anticipated the notions of *hardware* and *software*, a distinction that is not always obvious in artificial equipment. In the context of information processing in neural systems, in "wetware", it completely loses its meaning.

The huge potential of this idea was fully grasped only by Alan Turing. He elaborated a comprehensive theory of computation, which till today forms the nucleus of computer science. With his abstract general model for computing equipment, the *Turing machine*, he reduced it to its very essence and thus provided the conceptual framework for all discussions around computability, the fundamental conditions and ultimate limits of programmable computers. Turing's approach is finally defied only by the advent of quantum computation, as it depends on radically different principles, including a redefinition of the basic element of digital information processing, the bit.

Other technologies survive and coexist on the fringes of this mainstream development, of digital programmable computing. In particular, analogue computing, ranging from slide rules through orreries through analogue simulation, performed by mechanical, hydraulic, and electronic devices, still occupies an important niche (Sect. 3.2.1). A very particular class of tools supporting our brain are represented by artificial sources of randomness (Chap. 4), such as coin toss, dices, dreidels, roulette wheels, ... They serve to prevent foretelling by others, such as in gambling, or to augment our own creativity, such as in aleatoric music and art (see Fig. 4.1). In these devices, at least a subjective unpredictability is essential. In general, we expect computing equipment to be absolutely obedient and reproducible. It should strictly and reliably follow the rules we implemented in them, no more, no less: Computers should basically be deterministic systems (for pseudorandom number generation in deterministic computers, though, see Sect. 4.1).

An essential feature of computers is that they establish a well-defined set of states and of rules to switch between them, on a macroscopic level that is largely independent of the physical substrate chosen. Accordingly, only a minute fraction of the entropy we could associate to their physical microstates is actually relevant for their operation, namely the information communicated with their human users, entered as input, read off as output, and processed and stored internally to produce the intended function: abacus beads shifted up or down, keys pressed on a mechanical calculator or a keyboard, pixels lighted on a screen or printed on paper, voltage pulses entering or leaving gates, sections of magnetic tape polarized in one or other direction ... These *nominal states* of the machine gain their particular significance only by being intended as such by the designer and interpreted as such by the user. They are symbols in the widest sense, and thus are subject to conventions and other cultural and social conditions and their history. In short, the information processed by computers shares all the attributes of *semantic* information (see Sect. 1.2).

We can communicate with computers. Their ability to imitate our mental capacities has advanced so far that we can interact with them as with our equals, a feature that brings them into the vicinity of natural species such as cats and dogs, accompanying us as pets since early mankind. Another function computers have started to share with living organisms is foresight, see Sect. 3.2. Large-scale computer simulations have increased our ability to predict to an unprecedented level, literally advancing important aspects of our global future to the present, such as in climate simulation: They provide us with the control panel to manipulate the climate through the emission of greenhouse gases. In this sense, they contribute a substantial part to the role life and in particular intelligent beings play for the entropy flow on planet Earth. Computer simulations help us adding the dimension of time, past and in particular future, to the image of planet Earth that mankind is developing and refining. Section 3.3.3 introduced the concept of Information Gathering and Utilizing Systems (IGUS) as a class of thermodynamical systems far from equilibrium that play a special role as efficient catalyzers of entropy conversion. As tools invented by a particular instance of IGUS, human beings, to enhance their information processing capacities, computers themselves can be associated to the same category. While being deterministic, they challenge predictability in a much stronger sense than chaotic systems: Collecting information from a wide range of sources, they mix, combine, and transform it in multiple ways, and with their output, they reach targets on almost unlimited spatial and temporal scales. In order to anticipate the time evolution of a computer, that is, to emulate it, you need a computer of at least the same capacity, see the discussion in Sect. 4.3.1.

9.1.2   Computers and Computing in Natural Dynamical Systems

There is no doubt that state-of-the-art computing equipment already provides us with advanced cognitive capacities, of which "artificial intelligence" is only a spectacular instance. As such, computers pertain and should be compared to a huge class of natural and artificial manifestations of cognition. In order to bring some structure into this vast category, a generally applicable comparative quality would be very useful that allowed us at least to establish a hierarchical order. The concept of simulation, introduced in Sect. 3.2.1 as a generic term including anticipation and prediction,

System A is able to simulate (or emulate) system B: $A \succ B$,

meets this purpose: It constitutes an order relation. In order make it more explicit, we would have to refer to an isomorphism between dynamical features, i.e., the state space and the transition rules, of systems A and B. In the symmetric case that $A \succ B$ and $B \succ A$ (denoted $A \succsim B$), the two systems are equivalent in the sense that they can simulate one another. As an equivalence relation, $\succsim$ induces a hierarchy of equivalence classes within the set of dynamical systems. Like this, "simulation" not only applies to products of science and technology, but indeed also to natural systems: Surface waves on water simulate wave optics, water piping systems simulate electrical circuits, billiard ball trajectories simulate light rays, etc. With this universality, emphasizing the symmetry between simulating and simulated systems, the concept allows to classify computers on equal footing with systems not made for computing and thus supports the radical view paraphrased above as "everything is a computer".

In some cases, when the observable behaviour of a system is dominated by a specific phenomenon that suggests a concise and simple mathematical description, this system can be considered as special-purpose computer calculating exactly that number, function, or whatever other mathematical object. The following fictitious examples are not recommended to be put in operation:

- **Spaghetti-bundle number sorter**: To sort a set of N numbers in decreasing order, trim N uncooked spaghetti to lengths proportional to the numbers to be sorted. Slam the spaghetti as a compact bundle vertically onto a horizontal rigid surface. The tallest spaghetti, representing the largest number, sticks out on top of the bundle. Remove it and repeat the procedure with the rest of the bundle to find the second-largest number, etc. [Dew84, BE16].
- **The spider-web shortest-path finder**: A graph is a network of links (i, j) of lengths l_{ij} connecting nodes i and j. What is the shortest concatenation of links connecting node A to node B? To identify it, represent nodes by ringlets and links by inelastic strings of fixed lengths $\sim l_{ij}$ (Fig. 9.1a). Lift the resulting web of strings at nodes A and B and pull tight. The shortest path appears as a straight sequence of tense strings on top of the network, while all other strings hang down slag (Fig. 9.1b) [Dew84].

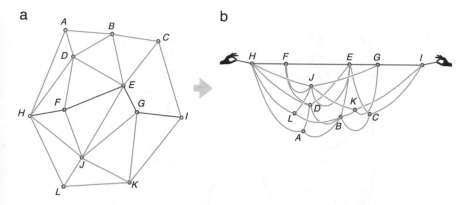

Fig. 9.1 The spider-web shortest-path finder. In order to identify the shortest path between two nodes (here H and I) in a graph (connecting pairs of nodes i, j with links (i, j) of length l_{ij}, panel **a**), represent the nodes by ringlets and the links by inelastic strings of corresponding size $\sim l_{ij}$, tied to the ringlets. Grasp the rings representing the extremes of the sought path and pull taut (**b**). The links forming the shortest path stand out under tension on top of the network (pink), all others hang down slag (green). Figure redesigned following [Dew84].

- **The wire-ring periodic-orbit calculator**: Closed orbits in a planar billiard of arbitrary shape [Sto99] are characterized by specular reflections at the walls, that is, by incoming and outgoing directions obeying the optical law of reflection. In order to identify a closed orbit with given topology (number of reflections), form a rigid ring of polished metal of the shape of the billiard circumference, mount as many freely sliding ringlets along the metal ring as reflections of the orbit with the wall are intended, and thread a string through one ring after the other, with loose ends hanging out of one of them (Fig. 9.2a). Pull taut. The ringlets will slide towards equilibrium positions such that at each ringlet, the strings form equal angles with the circumference (Fig. 9.2b), as in optical reflection.

- **Comb Fourier analyzer**: Crop the teeth of a comb so that the contour of their tips forms a ramp, with lengths decreasing linearly from the longest tooth on one end to the shortest on the other. Couple the comb mechanically or acoustically to a periodic signal. As some of the teeth oscillate in resonance, the amplitude of each tooth's vibration is a measure of the strength of its basic frequency in the signal, that is, the sequence of amplitudes represents its power spectrum. Length range, specific weight, and elasticity of the teeth delimit the frequency range covered.

- **Bed-of-nails detector of rationals**: Place a light source in the centre of an array of thin pegs, diameter d, arranged as a square lattice with lattice constant a, $d \ll a$ (Fig. 9.3). Observe the intensity of the light that permeates along straight channels between the pegs, shining on a screen around the array: Those directions in which light reaches the screen as a narrow bright strip, parallel to the pegs, correspond to rational angles $\phi_{p, q}$, $\tan(\phi_{p, q}) = p/q$, $p, q \in \mathbb{Z}$. The transmitted light decreases in intensity with increasing values of p and q. For $p^2 + q^2 > a^2/d^2$, the channel is blocked.

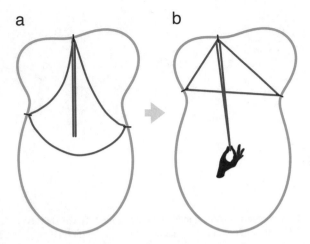

Fig. 9.2 The wire-ring periodic-orbit calculator. By Hamilton's principle, periodic orbits in a planar billiard [Sto99] mark a relative extremum of the action, that is, of their length. To find a minimum, form a rigid wire ring of the shape of the billsiard's circumference, mount as many ringlets on the ring as the orbit should have reflections on the wall, thread a string through these ringlets (**a**), and pull taut (**b**). Their positions equilibrate such that the directions of the string sections comply with the conditions of specular reflection at the wall.

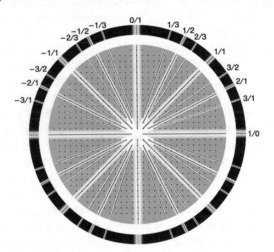

Fig. 9.3 The bed-of-nails detector of rationals. Fix cylindrical pins (nails) on a board, standing out perpendicularly at each vertex of a square grid. At the center of the board, replace one pin by a light source. Observe the light leaking out in the plane of the board onto a wall or screen. The directions, given by angles $\phi_{p,\,q}$ where light is transmitted along free channels between the pins, mark rational ratios, $\tan(\phi_{p,\,q}) = p/q$, $p,\,q \in \mathbb{Z}$. The values of p and q covered by the device are bounded from above by the ratio of pin diameter d and lattice constant a, $p^2 + q^2 < a^2/d^2$.

Analogue computers belong marginally to this class, but transcend it in that they can be "programmed" (rewired) to perform a larger, if still limited, repertoire of simulation tasks. At the opposite extreme, systems simulating other systems of vastly different nature and complexity, we find formal systems such as scientific theories condensing huge subject areas into a few lines of mathematical code, epitomized by, for instance, Maxwell's equations, Hamilton's equations, the Schrödinger equation, and extrapolated towards an imaginary "theory of everything" [LP00]. A theory formulated in mathematical terms obviously is not a dynamical system in itself. Yet, evaluated by hand or numerically, it defines an entire class of equivalent dynamical systems in the sense that they can all simulate each other (notably including their implementation on a digital computer).

Intermediate between simulation and conventional computers is "unconventional computing". The notion refers to the use of substrates other than semiconductor technology, but for computing tasks comparable to conventional computing. For example, employing inorganic substances in states far from equilibrium, logical gates can be implemented as interactions between reaction-wave fronts in reaction–diffusion systems [AC02] or crystallization fronts in supersaturated solutions [Ada09].

All these examples can be seen as natural systems that are reinterpreted by human observers as computers. At the same time, there exist other systems in nature, which, selected for this quality in a long evolution, achieved a highly optimized functionality for computation. It is illuminating to study how computation-like phenomena originated on Earth [Kup90]. An important root, exhibiting crucial capacities *in statu nascendi*, is catalysis by enzymes, which can be vastly more efficient and more specific than inorganic catalysis. It hinges on lock-and-key mechanisms where only a single catalyst can induce a particular reaction, e.g., opening an ion channel in the cell membrane. They already represent sophisticated chemical recognition, hence can be interpreted as cognition.

Even more precise in its selectivity is protein folding. The process leading from a linear sequence of amino acids, determined by the genetic code, to the final three-dimensional conformation of the protein, comprising several levels of structure (primary, secondary, tertiary, see Sect. 2.2.2) amounts to selecting a single choice among some 10^{140} options (referred to as *Levinthal's paradox*) [Kar97, DC97, Dur98, DO&08]. With very few examples of systematic malfunction, such as prions causing the Creutzfeld–Jacob and similar neuro-degenerative diseases, it works fail-safely on a nanosecond timescale. According to present knowledge, it is explained by a funnel-shaped profile of the energy landscape of protein conformations, allowing the protein to reach its native state by gradually lowering its free energy and reducing its entropy.

The starting point of protein synthesis, the genetic code, represents the biological phenomenon that comes closest to digital computation, see Sect. 2.2. The closeness of genetic processes, such as transcription and translation of DNA, to digital computation suggests to generalize them as "genetic computing" [KR08], culminating in the interpretation of the entire living cell as a computer [Dan09]. Genetic computing comprises static features, such as discretization and memory (of the order of several Gb in the case of the genome of mammals), as well as dynamical functions, such as copying (DNA replication and DNA-to-RNA transcription) and notably translation, the step from DNA code to linear amino acid sequence, from where protein folding proceeds. Working at ambient temperature, genetic information processing might be expected to suffer from a substantial error rate, owing to thermal noise affecting all its operations. However, exploiting its condition far from thermal equilibrium positively, it disposes of very efficient proof-reading and repair mechanisms that reduce the frequency of errors, for example in transcription, to extremely low values [BT&19].

In many species, a multicellular adult organism emerges from a subsequent development that creates an output of formidable complexity, a "biological 3d-printer"— from a linear discrete code, the DNA, to an intricate spatial structure. With computers, it shares the strategy of selective sequential activation or deactivation of genes in a well-orchestrated sequence, comparable to a master program calling subroutines, which call subroutines, etc. A patent objection against a too close analogy between computers and living organisms, that self-reproduction is a unique feature of life, has been refuted by serious speculations about self-reproducing machines [Neu66].

With the nervous system of higher animals, a new mode of information processing evolved that is based on completely different principles. Using entire cells (neurons) as basic processing units and electrochemical signals as basic information carriers, it is able to receive and combine environmental stimuli to elicit behavioural responses much more rapidly than the genetic system. On its lowest level, triggering and transmission of action potentials, neural information processing is discrete. Yet spikes are not comparable to bits, since on higher levels, this discretization is sacrificed again, encoding information in a way closer to analogue than to digital computing.

Central nervous systems (CNS) perform highly demanding cognition and inference tasks, including pattern recognition, learning, prediction, creativity, as sketched in Sects. 3.3 and 4.2, not to mention speech production and perception, and algebra. At the same time, it cannot be denied that the CNS forms an integral part of a living organism, as one of its organs. Besides nervous signalling proper, it permanently communicates with the rest of the body through chemical signals, notably hormones. The central nervous system is inextricably involved in metabolism and reproduction—hunger and sex. The requirements of information processing concerning energy supply, see Sect. 9.2.2, are amply reflected in the brain as the main consumer of chemical energy in the human organism. Conversely, variations in the available nutrients affect and modulate brain function on all levels.

Besides cognition, the brain generates and controls another type of emergent phenomena, an entire spectrum of passions and emotions, let alone consciousness and conscience, high-level capacities arising only in the CNS of humans and possibly some primates.

It is remarkable that on this analogue substrate, evolution created discreteness anew, on an even higher level, in various independent instances. Already in arthropods, we observe clear signs of discretization emerging in lasting products of their behaviour, such as spider webs and honeycombs. Several features of human cognition, in turn, amount to discretizing stimuli that originate in analogue signals:

- **Hearing** reduces the fundamental tones of perceived and generated sound to a discrete set of frequencies, as reflected in musical notation,
- **Vision** associates discrete colours (red, green, blue, …) to sections of the visible part of the continuous spectrum of light,
- **Counting** as perception of multiplicity employs pattern recognition to identify objects as individual members of the same equivalence class.

Intersubjective communicability is a plausible selection factor in favour of discretization:

- **Speech** has developed a well-defined vocabulary of words, discrete concepts encoded in sequences of discrete symbols, represented acoustically by frequencies (vowels) or sound elements (consonants),
- **Music** restricts a large part of sound production to a discrete spectrum of frequencies and a discrete set of durations, represented in musical notation,
- **Writing** encodes speech as a finite set of elementary signs (letters),
- **Trade** measures exchange value in discrete units, symbolized as money.

From very early on, information technology reflects the discreteness inherent in human cognition and communication:

- even since before Johannes Gutenberg, **typesetting** for printing devices uses movable sorts for letters,
- **typewriters** print letters on paper according to individual keys of the keyboard being touched,
- **musical instruments** such as organs, harpsichords, pianos, accordions, … introduce manual and pedal keyboards as digital input units,
- In many mechanical and electronic devices, the **control of continuous parameters** is restricted to a small set of discrete values.

In some cases, for example in the manufacturing of textiles from yarn, discretization is already inherent in the technology itself. Techniques such as weaving and knotting, knitting and crochet, have prepared minds since centuries for digital structures and pixelated images (Fig. 9.4) and even inspired one of the earliest devices with discrete automatic control, the Jacquard weaving loom (Fig. 1.1).

The idea of algorithms, beyond its tight mathematical meaning, is anticipated in daily life, in work routines and recipes, in fixed procedures in administration and law, even in religious ceremonies. It finds its most advanced manifestations in assembly lines and automated production in the industry. In all these diverse contexts, the objective of rationalizing processes and standardizing their output is the driving force behind the development of algorithms in this wide sense.

It is remarkable that Boolean algebra, the mathematical basis of all digital computation, has been proposed by George Boole, a son of the Victorian age, in the middle of the nineteenth century, way before the advent of computer technology. A child of the same age, Samuel Morse's code for telegraphy anticipates binary coding. In the 1920s, with quantum mechanics, discretization entered physics at the most fundamental level. Even in the arts, the same tendency became manifest: The Bauhaus, for

Fig. 9.4 Pixelation in weaving. The discretization intrinsic in traditional techniques of manual textile production, such as weaving, forces artists to render figurative motifs as discrete patterns, composed of a likewise discrete set of colours. The figure shows a traditional Andean woven design in the making on a hand loom. Patterns are created by threading coloured weft partially into the warp. Photo courtesy of Museo de Textiles Andinos Bolivianos, La Paz, Bolivia.

instance, pioneered the reduction of a boundless diversity of forms to a modular kit of a few basic shapes and colours, foreshadowing digital computer design. Discretization and modularity were already in the air when digital computing appeared on the scene. Besides technical efficiency and convenience, it is therefore the tendency to discretize inherent in human thinking itself that is reflected in digital computing. With this historical background, computers appear as a consequent, if the most striking, result of several long-standing tendencies, in particular of European / Western cultural development. They integrate in the evolutionary panorama as tools invented by a single species to enhance its cognitive competences [BE16], but quite specifically its performance in inference tasks.

Computation is an emergent phenomenon, exhibited to a varying degree by a special class of dynamical systems. It represents a particular way of processing information that does not quite fit in the general classification outlined in Chaps. 3– 7. Computing systems do not just convey information from large to small (dissipation) or from small to large scales (chaos) or distribute it among the parts of a composite system. To be sure, downward currents of entropy as well as energy are not directly involved in the operations of digital information processing, but they are in fact indispensable for classical computing, as will be explained below. Bottom-up entropy flow, by contrast, amplifying thermal and other microscopic fluctuations to macroscopic scales, poses a direct threat to deterministic computation (except for some unconventional modes of computation which positively exploit noise as a resource, see Sect. 9.2.2). It is therefore intentionally suppressed on the macroscopic level in the intended functioning of computers, a phenomenon termed *shielding* (e.g., logical gates composed of transistors are protected against electronic shot noise). It isolates the computation process proper from everything going on at atomic or even smaller scales that would interfere with logical operations. In particular, it suppresses all quantum effects. The same unique features of quantum mechanics that quantum computation attempts to take advantage of, such as superposition and interference, are completely eliminated in conventional computing as well as in genetic and neural information processing: They are decidedly classical phenomena, based on macroscopically manifest facts and incompatible with quantum superposition. While without doubt, quantum mechanics is the basis not only of semiconductor technology but also of the molecular interactions underlying DNA replication and transcription as well as synaptic signal transmission, the inherent randomness arising in these quantum many-body systems would threaten the strict determinism required for computation as well as for genetic control. In both contexts, it has to be reduced to a minimum.

An entire sequence of higher levels of integration, in turn, is built upon the bottom level of gates, buses, memory bits, ... (for example, machine language → assembler language → compiler language → operating system → application software, see Sect. 9.4.1). These higher levels are expected to be *platform independent*: for the functioning of level n, details of the implementation of the next lower level $n-1$ are irrelevant.

While unlike brains, computers are not based on excitable matter, they do constitute physical systems that reach a maximum of sensitivity and diversity in their response to input. The way computers process information places them in an intermediate position, a "critical point" of tamed instability, hovering between the extremes of chaotic expansion and dissipative contraction, between unbounded amplification and immediate attenuation of external stimuli. Arriving through privileged channels (input devices), input is processed along a multitude of possible paths equivalent to a vast class of dynamical systems, approaching the category of *universal Turing machine* (Sect. 9.3.1), and sends output, unlimited in physical scale, to a large variety of targets.

Computers do depend sensitively on their initial condition, program and initial input, as in deterministic chaos. Moreover, their subsequent time evolution must be precisely controllable by further input, comparable with time-dependent boundary conditions. Changing a single bit can send the computer on a completely different path through its state space. It appears as if this extreme controllability distinguishes digital computing from chaos. That is not quite the case. The very instability of chaotic systems, manifest in their sensitive dependence on initial conditions, can also be exploited to steer their time evolution onto any selected unstable periodic orbit embedded in a strange attractor and keep it there or send it to another orbit. The scheme for chaos control proposed by Ott, Grebogi, and Yorke [OGY90] uses very small control signals to prevent deviations from the intended orbit, owing to the intrinsic dynamics or to noise. The extreme sensitivity on initial as well as boundary conditions, resulting in particularly easy control, therefore is a feature digital computing does have in common with deterministic chaos.

The same versatility and flexibility of the time evolution locates computers also in the vicinity of brains again. An attractive hypothesis says that also brains maintain a critical state poised at a point of transition from strong damping to exponential growth of avalanches of neural activity, analogous to a second-order phase transition [SB95, CB99, Chi10, BT12]. Convincing evidence has been presented that such a condition of self-organized criticality [BTW87, Bak96] would be optimal for cognitive processes [SY&09].

Summarizing the above panorama, a number of characteristic properties of computing systems can be abstracted:

1. Computing is an emergent property of **systems far from thermal equilibrium**. Besides informational input and output, it requires a source of low-entropy energy as well as a sink for excess entropy, released as heat. This applies even to reversible computing (see below).
2. Computing requires **shielding**, that is, the blocking of the lowest level of information-bearing processes against the noise generated at even lower levels of the underlying physical structure.
3. Above this basic level, it comprises a **hierarchy of integration levels**, each of which is shielded against non-functional input from lower levels.

Digital computation, in particular, is characterized by the following additional features:

4. Among the full space of physical microstates of a computer, only a **finite discrete set, the nominal states**, are functionally relevant and interact via input and output signals with the human user.
5. Computers are expected to be identically reproducible in their function. Switching between nominal states in the process of computing follows a **fixed set of deterministic transition rules**.
6. The time evolution of a computer is clocked; it proceeds in **discrete time steps**.

These principles constitute a target rather than an actual state. Real computers as natural systems are subject to the fundamental laws of physics. Inhowfar do they allow to comply with the characteristics listed? Conditions and limits of their physical implementation will be outlined in the sequel.

9.2   Physical Conditions of Computation

Information can assume a multitude of forms, applies to every system that appears in at least two distinguishable states, and propagates along any channel associated to any physical interaction. Yet there exist a number of general limitations, implied by fundamental laws of physics, which restrict the way information is processed in natural systems, regarding the spatial as well as the time and energy scales involved. If we were to define a "fundamental computing capacity of matter", applying to every possible computer, it would depend on the following criteria:

Time scales:

– How does Special Relativity Theory limit the speed of propagation within and communication between components of a computer?

- How does thermodynamics limit the accuracy of information processing operations?
- How many operations can be performed before processing merges with noise?
- Is there a trade-off between accuracy and speed of computation?

Spatial scales:

- What is the maximum density of information storage, in (phase) space, according to classical physics, according to quantum mechanics?
- Which physical limits exist for the scaling-up of computers?

Energy scales:

- What is the minimum energy dissipated per operation?
- How does the total consumed/dissipated energy restrict the size of a computer?

Besides these physical conditions, there are of course **structural laws**, intrinsic conditions of the way information is represented and processed in a computer that limit its performance, such as:

- What are the restrictions digital computing imposes on the processes it is capable to simulate?

Each of these questions requires a detailed technical analysis to be answered, which in general is beyond the scope of this section. It will focus instead on those issues that are directly related to information flow and entropy production. To provide a solid basis for their discussion, the following two subsections are dedicated to the interface connecting analogue physics to digital computing: How can bits, the atoms of storage, be physically realized, how are they switched by an analogue control, and how can gates, the functional units of their processing, be implemented?

9.2.1   Implementing and Controlling a Single Bit: Macroscopic Discretization

Computers are macroscopic objects, built from solid (rarely liquid [AC02, Ada09], hardly ever gaseous) materials, with a correspondingly large number of degrees of freedom. This applies even to quantum computers, as will be discussed in Chap. 10. As such, they dispose of a huge space of internal microscopic physical states. This physical state space, in turn, has little to do with those states that are involved in their function, the input, the processing of information, and the output. In particular, as long as we do not take quantum effects into account, the physical states constitute a high-dimensional continuum, the total phase space of the computer, i.e., an analogue state space. The first question to be asked about the physics of computers is therefore: How does a discrete set of nominal states emerge from the physical continuum— whence the digital? We are here dealing with a "classical quantization" of admissible states that is independent of and occurs on spatial, time, and energy scales orders of magnitude above the atomic level where quantization proper dominates physics. In fact, the concept of discrete nominal states is by no means an acquisition of the computer age: Historical technology is full of them, see Sect. 9.1.2. Mechanical locks, railway points, and retractable ball pens have two of them, a typical manual transmission for cars has seven (reverse, idle, 1 to 5). If a driver does not switch carefully or the gear is out of order, it may well end in a state that no longer coincides with any nominal state of the transmission.

The discussion of Fourier transform in Sect. 2.3 led to the conclusion that analogue signals, if they encompass a finite bandwidth, can be encoded in the same number of discrete coefficients as if sampled with a finite rate. This dissolves an absolute distinction between analogue and digital: The decisive parameter is the number of independent real coefficients, which is finite and coincides in both representations. Digitizing signals, in the context of signal processing, though, goes an essential step beyond this basic bound on the information content (Fig. 9.5). It involves discretization not only of the independent variable, usually time, but also of the amplitude that depends on it. A continuous function of time $f(t)$, Fig. 9.1a, reduced to a sequence of values $f_n = f(t_n)$, Fig. 9.5b, sampled at discrete times $t_n = n\Delta t$, $\Delta t = T/N$ over a total time T, is discretized further, Fig. 9.5c, by passing it through a filter such as,

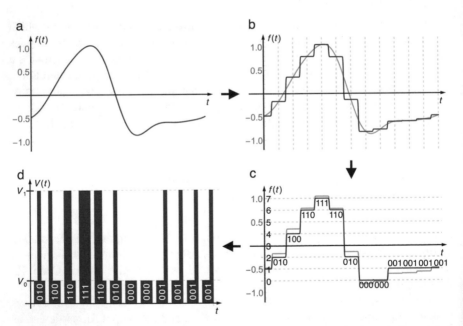

Fig. 9.5 Analog-to-digital converters transform continuous signals to bit streams. The original signal (**a** and light blue in **b**), is a continuous function of some parameter, say time. In a first step, it is reduced to a sequence of discrete real coefficients by sampling it at equidistant time sections (**b** and light blue in **c**). A second round-off or truncation procedure reduces the sequence of coefficients further to a staircase on a discrete grid, here of eight equidistant levels (**c**). Encoded as three-digit binary numbers, it can be transmitted as a sequence alternating between two voltage levels V_0 and V_1, forming three-digit blocks (**d**).

$$f_n \mapsto k_n \in \mathbb{Z}, \, k_n = \lfloor f_n/\Delta f \rfloor, \tag{9.2.1}$$

where $\lfloor x \rfloor \equiv \text{floor}(x)$ denotes the floor function (largest integer less than or equal to x). Instead of floor(x), $\lceil x \rceil \equiv \text{ceiling}(x)$ or a similar rounding or truncation will also do. The resolution Δf is determined by the maximum range $f_{max} - f_{min}$ of the sampled values f_n and the number of available levels K, typically $K = 2^M$, as $\Delta f = (f_{max} - f_{min})/K$. The output of this *analog-to-digital converter* (ADC) is a sequence of integers, usually in binary code encompassing M bits (Fig. 9.5d). All in all, this is a highly nonlinear filter that sacrifices part of the initial signal for the sake of transmitting the remaining, sharply delimited information content all the more accurately and securely.

Digitizing the signal in this way accomplishes a crucial advantage: A sequence of zeros and ones, for example encoded as two voltage levels V_0 and V_1, low and high, is much easier to protect against random contamination than the analogue original signal $f(t)$. Choosing the offset $\Delta V = V_1 - V_0$ sufficiently far above the noise level, the probability of transmission errors can be reduced to practically zero. Though analogue signals are more sensitive to weak noise, they are more robust against stronger disturbance disturbances. While their deterioration is gradual, even with a considerable amount of transmission error, a reduced (e.g., coarse-grained) message remains recognizable, digital signals tend to be irrevocably destroyed once the noise exceeds some threshold, their degradation is all-or-nothing. The reason is revealed in the spectral analysis of a digital signal (Fig. 9.6a). It consists of a convolution of the pure bit string, encoded, e.g., as a sequence of delta functions, with the shape of a single peak, typically a rounded box function $f_{box}(t)$.

$$f_{dig}(t) = \int_{-\infty}^{\infty} dt \, f_{box}(t) \sum_{n=1}^{N} a_n \delta(t - t_0 - n\Delta t)$$

$$= \sum_{n=1}^{N} a_n f_{box}(t - t_n), \, a_n \in \{0, 1\}, \, t_n := t_0 + n\Delta t. \tag{9.2.1}$$

According to the convolution theorem [Kap03], its spectrum is a product of the Fourier transform $\tilde{f}_{box}(\omega)$ of the box function (smooth white curve in Fig. 9.6b) with that of a discrete sequence of weighted delta functions, $\tilde{f}_{bits}(\omega)$, which in turn is periodic (dark blue dots in Fig. 9.6c).

$$\tilde{f}_{dig}(\omega) = \tilde{f}_{box}(\omega)\tilde{f}_{bits}(\omega), \quad \tilde{f}_{bits}(\omega + \Omega) = \tilde{f}_{bits}(\omega), \, \Omega = 2\pi/\Delta t. \tag{9.2.2}$$

Contaminating the signal with noise destroys this information-carrying structure at higher frequencies (light blue dots in Fig. 9.6b). However, as long as at least a single copy of this finite sequence of coefficients remains unspoilt, the original bit string can be retrieved (Fig. 9.6d). Analog-to-digital conversion is an elementary instance of shielding: By opening a gap between the signal-carrying frequencies and the noise, microscopic entropy is prevented from infiltrating into the relevant information.

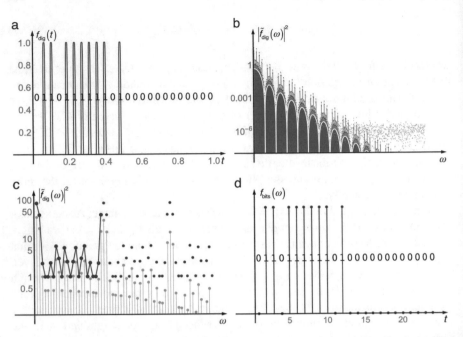

Fig. 9.6 Digitization of a continuous signal in the frequency domain. The digitized signal of Fig. 9.5d, truncated to the leading two binary digits, reduces to the 24-bit string 01-10-11-11-11-01-00-00-00-00-00-00 and could be encoded as the sequence of spikes depicted in (**a**). Fourier transformed as a periodic signal, it has the power spectrum shown in (**b**) (dark blue). A blow-up of the first 64 harmonics (**c**, light blue) reveals a repetitive small-scale structure comprising 24 frequencies. Folding out the spectrum of a single spike (continuous white curve in **b**), a periodic sequence of frequencies remains (dark blue in **c**), the first 24 of which encode the digital signal (joined dark blue dots in **c**). Even if Gaussian noise of variance $\sigma = 10^{-3}$ is added to the original clean signal (light blue dots in **c**), about nine periods of the relevant signal remain unaffected (not all shown). An inverse Fourier transform of the relevant 24 harmonics highlighted in (**c**) is sufficient to retrieve the original bit string (**d**).

Digital data are appropriately received, stored, and processed by devices that exhibit a similar separation of discrete nominal states from analogue microstates. It is a hallmark of nonlinear systems that the solutions of their evolution equations comprise a discrete set of stable states (see Sects. 3.1.3, 6.2, and 6.3). Typically it includes two or more alternative types that differ qualitatively in their behaviour, do not permit continuous transitions among themselves, and are accessed from disjoint subsets of the initial conditions: Coexisting *attractors* occupy subspaces of lower dimension than the embedding state space, surrounded by their *basins of attraction* [GH83, LL92, Str15]. On each attractor, the dynamics can still be very complex, ranging from point attractors through limit cycles through chaotic attractors. If two or more point attractors coexist, that is, in *bistable* or *multistable* systems (for an example from visual perception, see Fig. 9.7), we are dealing with a classical manifestation of "quantization", on macroscopic scales far above the realm of quantum mechanics.

Fig. 9.7 Bistability in visual perception. Ambiguous figures allow for two (or more) alternative interpretations of a two-dimensional image by our visual system. In the case shown, the grey-level pattern is perceived as projection of a three-dimensional surface comprising six cubes seen from above (illuminated from the right, black faces on top) or as seven cubes seen from below (illuminated from the left, black faces underneath). Switching between alternatives occurs unconsciously, but is amenable to mental control.

In the simplest cases, bistability or multistability results from a potential energy surface with two or more relative minima, such as in a quartic double well (Fig. 9.8a),

$$V_0(x) = -\frac{a}{2}x^2 + \frac{b}{4}x^4, \quad a, b > 0. \tag{9.2.3}$$

The symmetric pair of minima at $x_\pm = \pm\sqrt{a/b}$ is separated by a barrier of relative height $E_B = a^2/4b$ at $x_0 = 0$ (Fig. 9.8a). The contour lines of the Hamiltonian (Fig. 9.8b)

$$H(p, x) = -\frac{p^2}{2m} + V(x) \tag{9.2.4}$$

coincide with the trajectories of frictionless motion in this potential, comprising nearly harmonic oscillation with frequency $\Omega = \sqrt{2a/m}$ close to either one of the two stable equilibria, $H(p, x) \geq -E_B$, and strongly anharmonic oscillation between the wells above the barrier, $H(p, x) > 0$, separated by a figure-eight separatrix at $H(p, x) = 0$ (red line in Fig. 9.8b).

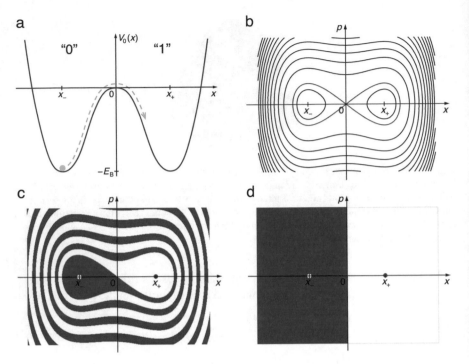

Fig. 9.8 Anatomy of a classical bit. A simple on–off switch is modelled as a bistable system, a symmetric double well (**a**), given by a quartic potential, Eq. (9.2.3), comprising two minima at x_- and x_+, separated by a central barrier of relative height E_B. The contour lines of the corresponding Hamiltonian (**b**), including the separatrix at $H(p, x) = 0$ (red), coincide with the trajectories of frictionless motion in this potential. With weak friction, i.e., in the underdamped regime (**c**), the minima of the Hamiltonian become point attractors, each one embedded in its basin of attraction (red/white for the attractors at x_-/x_+), winding around one another in a characteristic ying-and-yang shape. If friction is strong enough, in the overdamped regime (**d**), trajectories fall directly into the minimum closest to their initial point, reducing the basins of attraction to two contiguous half spaces.

In the presence of an Ohmic friction force $F(\dot{x}) = -2m\lambda\dot{x}$, the motion is given by Newton's equation, cf. Eq. (3.1.13),

$$m\ddot{x} = -2m\lambda\dot{x} + ax - bx^3. \tag{9.2.5}$$

It implies an exponential contraction of phase space with the rate 2λ towards a pair of point attractors at $x_\pm$. As long as friction is not too strong, in the underdamped regime $\lambda < \Omega$ [Kap03], contraction is still superposed with damped oscillations, with reduced frequency $\omega = \sqrt{\Omega^2 - \lambda^2}$, so that the two basins of attraction spiral into the respective potential well (Fig. 9.8c). For strong friction, however, in the overdamped regime $\lambda \gg \Omega$, the inertia term $m\ddot{x}$ on the l.h.s. of Eq. (9.2.5) can be neglected,

$$2m\lambda\dot{x} = -V'(x) = ax - bx^3, \tag{9.2.6}$$

trajectories just fall directly into the well closest to their initial point, and the attraction basins become identical with the left and the right half, $x < 0$ or $x > 0$, of phase space, resp. (Fig. 9.8d). This scenario constitutes a prototype of binary discretization in classical physics. The nominal states—call them "off" and "on", "0" and "1", "−" and "+", …,—correspond to the coordinate x being close to x_- or x_+, resp., within some tolerance.

Transitions between these states, to be addressed further below, inevitably occur in continuous time and pass through all intermediate values of x. In particular, they cross the unstable equilibrium at $x = 0$. For digital computing, these transition states remain outside consideration—*tertium non datur*, they do not participate in information processing proper. That is, the reduction of the continuous physical state space to two nominal states is by no means absolute, intermediate states do occur, but only on exceedingly short time scales, as long as the device is kept within the intended operating conditions (power supply, temperature, humidity, vibration, …).

Bistable elements of this type can be implemented in multiple forms. Every key of a computer keyboard, every computer mouse contains a mechanical switch that converts the analogue fingertip pressure to a binary output voltage signal. The simplest construction of a mechanical switch realizes a bistable potential with springs combined with metal profiles of appropriate shape. Pre-stressed metal or plastic sheets can often take two alternative shapes, separated by an unstable configuration, such as in twist-off caps. In magnetic material, the macroscopic magnetization typically exhibits two preferred orientations corresponding to opposite polarizations, an effect underlying all magnetic memory devices.

Operation in the overdamped regime is decisive. With only weak friction or in the absence of dissipation, the memory of the initial state would remain in the system for a long time or forever, e.g., in the form of repeated oscillations in either well or between them. For a proper function of the binary element it is indispensable, however, that the bit currently saved can be erased and overwritten at any time in a controlled unidirectional flow of information, bit by bit, from the input into the binary element and from there to other binary elements or to the trash, i.e., converted to heat. Being prepared for fresh input at every cycle is incompatible with a dynamics that conserves information strictly, as in a Hamiltonian system, or even approximately, as in a dissipative system in the underdamped regime.

Bistability does not require any static mechanical or electric potential with two or more stable equilibria, it can also arise as a purely dynamical phenomenon. An example of enormous technical importance for electronic engineering is the *flip-flop*, realized electromechanically with relays or electronically with transistors (Fig. 9.9). Its two stable states correspond to dynamical, not to static equilibria. While static bistable systems need opposing external forces to be switched from one minimum to the other or back, see below, flip-flops can be switched back and forth by the same input signal. In this way, they reduce periodic input signals to half of their frequency: They effectuate period doubling (Fig. 9.9d). At the same time, dynamical

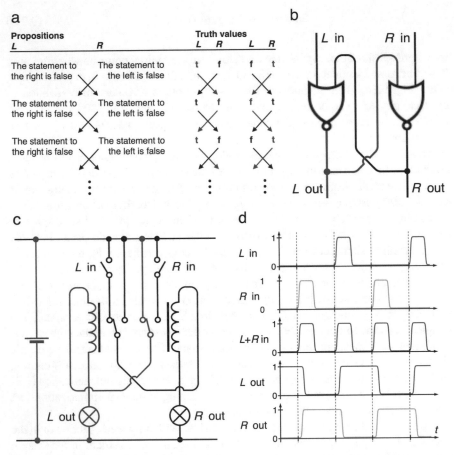

Fig. 9.9 Flip-flops are bistable elements realized as dynamical systems, instead of static double-well potentials, as in Fig. 9.8. Their function is based on a crosswise negative feedback between two binary elements. In logics, it can be found in a pair of propositions with negative mutual reference (**a**), cf. Fig. 4.5, and represented by a pair of cross-coupled NOR gates (**b**), cf. Sect. 2.1.1. It can be implemented as an electronic circuit combining two SPDT (Single-Pole Double-Throw) relays (**c**), each one requiring the other in the off state to be active itself. In all cases, the feedback dynamics allows for two stable solutions, which are symmetric under simultaneous exchange of the two binary elements $L \leftrightarrow R$ and of truth or state values $t \leftrightarrow f$, $on \leftrightarrow off$, $0 \leftrightarrow 1$, etc. The state $L = t$ (*on*), $R = f$ (*off*) is represented in panels (**a**–**c**) by highlighting the active element in red. Flip-flops are switched by externally triggering the element that is in the off state. A switching cycle is depicted in panel (**d**) as time-dependent input and output voltage levels, indicating causal order as small delays. Triggering both inputs simultaneously ($L = R = 1$) is not allowed; it would result in a race condition between the two inputs. For a similar mechanism, but with negative self-reference of a single binary element, see Fig. 4.4. Comparing the combined input (third curve in **c**) with the two outputs shows that flip-flops perform period doubling (one output pulse per two input pulses), so that concatenated flip-flops serve as binary counters.

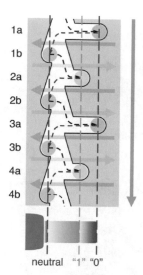

Fig. 9.10 Purely mechanical flip-flop. The retractable ball pen is a simple mechanical realization of a bistable system with memory. Starting from the "on" state ("1", writing, positions 1a and 3a), pressing the button (pink arrows) sends the ball pen into a neutral position (1b, 3b). From there, releasing the button (green arrows), the ball pen snaps into its "off" state ("0", retracted, positions 2a and 4a). Pressing the button anew leads to a neutral position again (2b), now preparing it for the writing state, from where the procedure repeats. Shown is a simplified scheme, based on a ball moving in channels. It assumes a static force towards the bottom (blue arrow) and alternating input forces in the transversal direction. The memory, which is indispensable for the flip-flop, requires strong friction in the overdamped regime.

bistability requires a constant supply of energy even to remain permanently in one of the alternative states. The same period doubling of a discrete input signal can be achieved again in a passive mechanical system, realized for example in retractable ball pens (Fig. 9.10). Finally, emblematic instances of bistability in neural information processing are ambiguous images, figures that our visual perception can interpret only in two alternative ways, but not in any intermediate form (Fig. 9.7).

In present computer technology, the standard switching element is the transistor, almost exclusively realized as MOSFET (Metal Oxide Semiconductor Field Effect Transistor) within very-large scale integrated circuits (Fig. 9.11). It uses the electric field of an isolated gate to control the type of majority carrier in a doped semiconductor bulk material. While in the "off"-state of the transistor, the bulk isolates two oppositely doped terminals, called "source" and "drain", a sufficiently strong potential on the gate creates a channel of the same type of majority carriers as in source and drain, thus forming a conducting bridge between them. Employed in the *switched mode*, the transistor has a characteristic (output current vs. input voltage) with a sharp increase from "off" to "on" (Fig. 9.11d). The sigmoid shape of this curve (see also Fig. 9.12c), reducing a continuous (analogue) input into a nearly discrete (binary) output, is at the heart of digital computing.

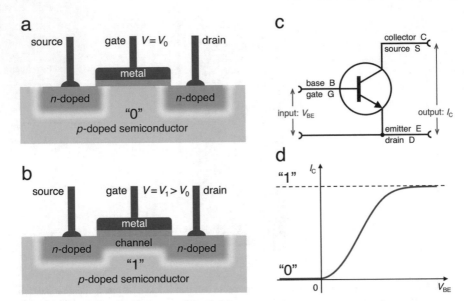

Fig. 9.11 Transistors are standard switching elements in digital computers. The most common type of transistors used is the Metal Oxide Semiconductor Field Effect Transistor (MOSFET), panels (**a**) and (**b**), typically realized in Very-Large-Scale Integrated (VLSI) circuits. It is controlled by the potential of a metal gate (blue), isolated by an oxide layer (dark grey) from the bulk, a doped semiconductor material (light green), thus forming a capacitor with it. If, e.g., it is p-doped, it isolates a pair of n-doped zones (*source* and *drain*, pink) on either side of the gate, corresponding to the "off" or "0" state of the transistor. With a sufficiently high positive voltage on the gate, however, its electric field will displace positive charge carriers (holes, i.e., missing electrons in the valence band) in its vicinity. A conducting channel forms between the two n-doped terminals where electrons are the majority carriers, even in the p-doped bulk. This is the "on" or "1" state (**b**). The onset of conductivity between source and drain (collector-emitter current I_C in Bipolar Junction Transistors, BJT), as output in response to a continuous increase of the gate-drain voltage (base-emitter voltage V_{BE} in BJT) as input signal, is sufficiently steep to qualify transistors as switches (**c**), similar to relays or vacuum tubes. This feature is reflected in the sigmoid shape of the characteristic (I_C vs. V_{BE}), with a sharp increase, almost a step function, from the nominal 0 (off) to the 1 (on) state (**d**). Note the similarity with Fig. 9.11c. Panels (**a**, **b**) redrawn after Ref. [ED19].

While components such as relays, vacuum tubes, or transistors essentially just amplify and discretize their input signal, switches in the proper sense include memory as an additional feature. It is of central relevance for the understanding of computation how such binary elements are controlled, i.e., switched and erased, by external signals. Since their two nominal states coincide with stable equilibria, some external agent investing energy is required to force the system out of one equilibrium and into the other, across the potential barrier (dashed green line in Fig. 9.8a). The most obvious option is to apply a spatially constant external bias $F_{con} = -V'_{con}(x) = \text{const}$, adding a control term $\sim x$ to the potential of Eq. (9.2.3),

$$V_c(x, t) = -\frac{a(t)}{2}x^2 + \frac{b}{4}x^4 - c(t)x, \ a(t), \ b \in \mathbb{R}^+, \ c(t) \in \mathbb{R}. \tag{9.2.7}$$

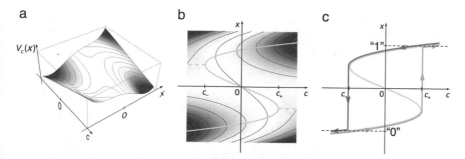

Fig. 9.12 Switching of a bistable system. A term cx, added to the symmetric bistable potential of Fig. 9.7, Eq. (9.2.3), generates a bias, which, if strong enough, can annihilate either one of the two potential minima, as is evident in the parameter dependence of the potential (**a**, colour code ranging from red for negative through white for zero through blue for positive potential). In the range $c_- < c < c_+$, three equilibrium points coexist, the central one unstable, the two lateral ones stable (**b**, light green line). Outside this region, only a single minimum remains. The unstable and one of the stable equilibria in the central range emerge by bifurcations at c_- and c_+, resp., from an inflection point (dashed light green). Assuming friction in the overdamped regime, varying c adiabatically from large negative values upwards (**c**, green curve) or from large positive values downwards (**c**, red curve) leads to hysteresis in the response to the parameter change.

It could be realized physically, e.g., as a bistable ferromagnet subject to a magnetic field $\sim c$ [Ben82]. Varying $c(t)$ from large negative values ($F_c(t) = -V_c'(x,t) = c(t) < 0$) through the neutral point $c(t) = 0$ to large positive values ($F_c(t) > 0$) drives the potential through qualitatively different shapes (Fig. 9.12a): For $c(t) < c_- = -\frac{2}{3}\sqrt{a^3/3b}$, there is only a single minimum at some coordinate $x_-(c) < 0$. At $c(t) = c_-$, a bifurcation occurs: In the range $c_- < c(t) < 0$, besides the stable equilibrium at $x_-(c(t)) < 0$, another stable equilibrium emerges at $x_+(c(t)) > 0$, together with an unstable point at $x_0(c(t))$ (Fig. 9.12b). Passing through the symmetric configuration at $c(t) = 0$, the scenario repeats in reverse order, replacing $-x \to x$ and $-c \to c$, with an inverse bifurcation at $c_+ = \frac{2}{3}\sqrt{a^3/3b}$. Switching therefore requires to apply a force $F_c(t) = c(t)$ with $|c(t)| > c_+$.

If the dynamics is strongly damped, as assumed for the binary element, this results in a response characterized by *hysteresis*: If, say, c is increased from large negative values all the way up to the upper bifurcation at c_+, the system will remain in equilibrium at the relative potential minimum at x_-, even in the presence of another minimum at x_+, for $c(t) < c_+$. At $c(t) = c_+$, however, the left equilibrium disappears, so that the system drifts into the well at x_+ and remains there upon increasing $c(t)$ further. If the procedure is repeated, now lowering c from above c_+, the switching does not occur at c_+, but only as soon as c_- is reached again, now from above (Fig. 9.12c). That is, the system retains a memory from where in parameter space it came from—it shows hysteresis, an effect known in particular from magnetic materials. This allows using bistable systems as memory elements. The options to maintain the recorded bit, protected against external perturbations $|c(t)| < c_+$, or to erase and rewrite it when required $|c(t)| > c_+$ is crucial for the function of the binary element.

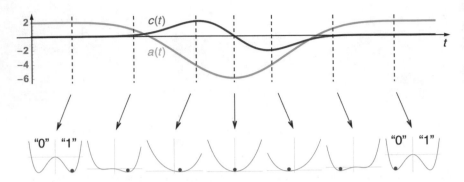

Fig. 9.13 A full switching cycle for a bistable system can be accomplished by driving two control parameters of a biased double-well potential, the barrier height $a(t)$ (green) as well as the bias force $c(t)$ (pink), cf. Eq. (9.2.7), through a time dependence as sketched. Starting from the system at rest in the right well, the protocol consists of lowering the barrier until the potential becomes monostable, imposing a bias towards the right, inverting the sign of the bias to send the system into the opposite well, and stabilizing it there by restoring the barrier. This process does not imply any lower bound of the dissipated energy, since it avoids a barrier crossing as in Fig. 9.8a, and the friction force can be reduced indefinitely by going sufficiently slowly through the cycle. Adopted from Refs. [Ben82, Lik82].

If we assume not only the bias force $c(t)$, but also the prefactor of the quadratic term ~$a(t)$ in the double-well potential to be controllable, as indicated in Eq. (9.2.7), the system can be driven through an entire switching cycle. With a protocol as sketched in Fig. 9.13, starting from the bistable configuration $a(t) > 0$, $c(t) = 0$, with the system in the right well, then lowering the barrier, $a(t) < 0$, and imposing a bias $c(t) > 0$ towards the left, the binary system moves through the neutral position with $a(t) < 0$, $c(t) = 0$, to the opposite side, where it is stabilized by restoring the barrier. In this way, the bistable potential returns to its initial state, but with the system in the other stable position. This switching cycle now does not require any minimum energy consumption, as the loss due to friction can be reduced indefinitely by slowing down the process [Lan61, BL85, Lan91, Ben03].

If unlike the switching cycle, the bias $c(t)$ is not controlled in such a way that it overshoots after passing through the symmetric configuration, but is kept at $c = 0$ while the barrier is restored, an erasure instead of a switching of the initially stored bit instead is achieved (Fig. 9.14). In this case, as soon as a symmetric potential is reached, the initial information is forgotten, as reflected in a broad symmetric probability density distribution under thermal equilibrium conditions at temperature T,

$$\rho(x,t) = \frac{1}{Z(T,t)} \exp\left(-\frac{V_c(x,t)}{k_B T}\right), \quad Z(T,t) = \int_{-\infty}^{\infty} dx \exp\left(-\frac{V_c(x,t)}{k_B T}\right),$$

$$(9.2.8)$$

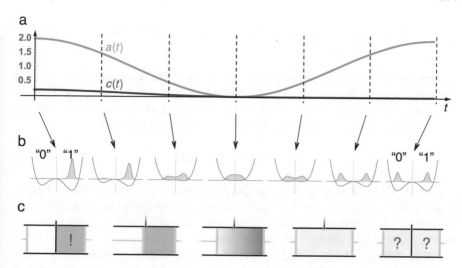

Fig. 9.14 Erasure of a bit can be accomplished with a protocol similar to that of a switching cycle (Fig. 9.13), varying the barrier $a(t)$ (green curve in **a**) and the bias $c(t)$ (pink) of a double-well potential as in Eq. (9.2.7) (**a**), but now terminating in a symmetric configuration, i.e., with the bias approaching zero. Starting with the system in a definite state (**b**, left to right) in one of the wells, removing barrier height and bias in parallel leads to a broad symmetric distribution centred at the origin. If the barrier is then restored to its original height, a symmetric bimodal distribution results that no longer contains any information as to the value of the bit. The pink curves in (**b**) indicate probability density distributions in thermal equilibrium for the potentials shown. The same process can be realized in a schematic Szilard engine [Szi29, Ben82] (**c**): Starting with the knowledge that a gas particle is in the right half of a two-compartment container, work can be gained by separating the two compartments with a dividing wall, entering a piston without doing any work into the empty (left) compartment, removing the wall to expand the space between the pistons adiabatically, extracting work, and entering the dividing wall again. The loss of knowledge on the particles' position, right or left, amounts to an entropy increase of $k_\mathrm{B}\ln(2)$, cf. Eq. (9.2.11). After Ref. [Ben82].

with the potential $V_c(x)$, Eq. (9.2.7), with $a(t) = c(t) = 0$. Restoring the barrier to its initial height then results in a bimodal distribution, concentrated in the two wells but without any preference for either one. If the biased distribution before the erasure was $\rho_c(x) = \rho_\mathrm{hump}(x - x_+)$ with an approximately symmetric shape $\rho_\mathrm{hump}(-x) \approx \rho_\mathrm{hump}(x)$, then the final distribution will be given approximately by

$$\rho_\mathrm{fin}(x) = \frac{1}{2}(\rho_+(x) + \rho_-(x)), \qquad (9.2.9)$$

comprising two peaks, $\rho_-(x) = \rho_\mathrm{hump}(x_- - x)$ at x_- and $\rho_+(x) = \rho_\mathrm{hump}(x - x_+)$ at x_+, of similar shape as $\rho_\mathrm{hump}(\pm x)$ and $\rho_+(0) = \rho_-(0) = 0$. The entropy in this bimodal distribution is (see Eq. (5.3.1), here in thermodynamic units with $c = k_\mathrm{B}$)

$$S_\mathrm{fin} = -k_\mathrm{B} \int_{-\infty}^{\infty} \mathrm{d}x \rho_0(x) \ln(d_x \rho_0(x)) = -k_\mathrm{B} \int_{-\infty}^{\infty} \mathrm{d}x \rho_\mathrm{fin}(x) \ln\left(\frac{d_x}{2} \rho_\mathrm{fin}(x)\right)$$

$$= k_\text{B} \ln(2) + S_\text{hump}, \quad S_\text{hump} = -k_\text{B} \int_{-\infty}^{\infty} \mathrm{d}x \rho_\text{hump}(x) \ln\big(d_x \rho_\text{hump}(x)\big), \quad (9.2.10)$$

irrespective of the specific form of $\rho_\text{hump}(x)$, denoting the corresponding entropy in the initial state as S_hump. The increment is therefore $\Delta S = S_\text{fin} - S_\text{hump} = k_\text{B} \ln(2)$, or 1 bit in logical units, over the well-defined initial state. This now does imply a minimum investment of energy for the erasure, dissipated into heat

$$\Delta Q = T \Delta S = k_\text{B} T \ln(2). \quad (9.2.11)$$

Equation (9.2.11) is widely known as *Landauer's principle* [Lan61, BL85, Lan91, Ben03]. The logically irreversible erasure requires a physically irreversible increase of entropy, while inversely (see below), logically reversible operations can nevertheless be performed in strongly dissipative gates.

It is switching and erasure where physical boundary conditions take their strongest effect on the operation of binary elements. In the overdamped regime considered, it is easy to estimate the minimum energy cost of a single switching: Lifting the system over the barrier consumes an energy E_B, the height of the barrier. Upon relaxation into the opposite well, it is not recovered as kinetic energy but dissipated into heat $Q = E_\text{B}$. When operated at an ambient temperature T, the system is subject to thermal noise of an average energy $k_\text{B} T$. In order to make sure that the switch is not triggered accidentally by a mere thermal fluctuation, the barrier must be chosen high enough, $E_\text{B} \gg k_\text{B} T$. This loss can be avoided by following the protocol indicated in Fig. 9.13 instead and slowing down the operation indefinitely, but at the expense of a diverging switching time. Unlike implementations of binary elements with static potentials, bits based on dynamical equilibria, such as flip-flops, consume energy permanently, not only when switched.

9.2.2 Implementing Gates: Reversible and Irreversible Operations

As soon as the operation of bits comes close enough to abrupt switches between binary alternatives, digital computing can be described adequately as navigation in discrete steps on a huge but finite space $\mathcal{L} = \otimes_{m=1}^{M} \mathcal{B}_m$ of $N = 2^M$ discrete states, represented by M bits (two-state systems) $\mathcal{B}_m$. Formally, each step then amounts to a map from the present state $\mathbf{C}_n \in \mathcal{L}$ to the subsequent state $\mathbf{C}_{n+1} \in \mathcal{L}$, depending on the present state as well as possibly on some contingent, but likewise discrete input $\mathbf{I}_n \in \mathcal{I}$,

$$\mathcal{L} \otimes \mathcal{I} \to \mathcal{L}, \mathbf{C}_\text{n}, \mathbf{I}_\text{n} \mapsto \mathbf{C}_{n+1}. \quad (9.2.12)$$

The dependence of each bit in $\mathbf{C}_{n+1}$ on those in $\mathbf{C}_n$ and $\mathbf{I}_n$ amounts to a logical operation. Elementary instances have been discussed as binary operations in Sect. 2.1.1. Whatever the value of M, this map can be reduced to a combination of the 16 binary gates listed in Sect. 2.1.1. In order to analyze the implementation of computing on the level of binary logical operations, it is therefore sufficient to focus on single binary gates.

How they can be realized by combining simple switches off the electrician's shelf is illustrated in Sect. 2.1.2. In particular, all the 16 binary gates can be implemented as combinations of exclusively NOR or exclusively NAND gates. To replace finger pressure by a likewise binary electric input, substitute fingertip switches with relays, vacuum tubes, or transistors. All these active elements are amplifiers, that is, weak input controls strong output, without any fundamental restriction concerning the amplification factor, i.e., the ratio of output vs. input magnitudes (force, voltage, power, etc.). No process is involved that would necessitate any consumption of energy. According to this physical criterion, these operations could be invertible.

The balance of binary information, however, leads to a different result. In the preceding subsection, it became clear already that switching a bit from one state to the other does not require investing energy or entropy, while the definite loss of a bit in an erasure invariably increases the entropy and produces heat. This physical result is reflected directly in the logical reversibility of the operations, which can be generalized to maps such as (9.2.12). Comparing the total binary information contents $I(\mathbf{C}_n, \mathbf{I}_n)$ and $I(\mathbf{C}_{n+1})$ it is evident that: If $I(\mathbf{C}_n, \mathbf{I}_n) > I(\mathbf{C}_{n+1})$, $\mathbf{C}_n$ and $\mathbf{I}_n$ could not both be retrieved from $\mathbf{C}_{n+1}$ alone. The map contracts the binary state space (some states on entrance map onto the same state on exit) and is not reversible. The loss of information, in turn, results in dissipation.

If by contrast, information is conserved, $I(\mathbf{C}_n, \mathbf{I}_n) = I(\mathbf{C}_{n+1})$, the binary logic in itself does not require any dissipation. This refers to the total M-bit state of the system: If the process is deterministic and reversible, there must exist exactly 2^M separate histories or trajectories in the binary state space that the system has to follow in each time step, neither branching nor joining, if not forced by a contingent input to jump to another trajectory.

The irreversibility of the binary gates is a simple consequence of the fact that a single output bit is insufficient to determine the two input bits uniquely. This suggests rendering them logically reversible by adding a second output bit that conserves either one of the two inputs. However, this does not work for all of the binary gates: If three or four of its output bits coincide, even a second output is not sufficient to restore the input. For this strategy to work, the gate must be balanced, i.e., have the same number of 0's and 1's in the output. This applies only to XOR and XNOR, see Sect. 2.1.1, apart from the identity with either one of the inputs or their negation. With these conditions, exactly 24 reversible binary gates are possible (Table 9.2d). In all other cases, both input bits have to be kept besides the output proper to make the gate reversible.

The fact that reversible binary logic conserves information has a remarkable consequence: Reversible gates can be implemented as closed Hamiltonian systems. In

Table 9.1 Reversible operation of binary gates can be achieved in exceptional cases. In this example, none of the three three-bit gates shown in (**a**), $A(a, b) = (a{\equiv}b)$, $B(b, c) = (b{\equiv}c)$, $C(a, b, c) = (a{\neq}(b{\neq}c))$, is symmetric under time reversal in itself. However, combining their output values (**b**), all the eight inputs can be retrieved, the inverse operation can be identified and reduced to three binary gates again, $a(B, C) = (B{\equiv}C)$, $b(A, B, C) = (A{\neq}(B{\neq}C))$, and $c(A, C) = (A{\equiv}C)$. The *Fredkin* or *switch gate* is its own inverse. Depending on the value of the first input a, it sends input b to output B and c to C, or vice versa [FT82]. For a mechanical implementation, see Fig. 9.14b. The CONTROLLED-CONTROLLED-NOT (**d**) inverts the input c, conditioned on the two other inputs a and b being both on 1. It is also its own inverse. The quantum equivalent of this gate forms an important building block for quantum computation [Fey82, Fey86, NC00], cf. Sect. 10.1.2.

a

a	b	c	$A{=}(a{\equiv}b)$	$B{=}(b{\equiv}c)$	$C{=}(a{\neq}(b{\neq}c))$
0	0	0	1	1	0
0	0	1	1	0	1
0	1	0	0	0	1
0	1	1	0	1	0
1	0	0	0	1	1
1	0	1	0	0	0
1	1	0	1	0	0
1	1	1	1	1	1

b

a	b	c	$a{=}(B{\equiv}C)$	$b{=}(A{\equiv}(B{\neq}C))$	$c{=}(A{\equiv}C)$
0	0	0	1	0	1
0	0	1	0	1	0
0	1	0	0	1	1
0	1	1	1	0	0
1	0	0	1	1	0
1	0	1	0	0	1
1	1	0	0	0	0
1	1	1	1	1	1

c

a	b	c	$A{=}a$	$B{=}a{\wedge}b{\vee}\neg a{\wedge}c$	$C{=}\neg a{\wedge}b{\vee}a{\wedge}c$
0	0	0	0	0	0
0	0	1	0	1	0
0	1	0	0	0	1
0	1	1	0	1	1
1	0	0	1	0	0
1	0	1	1	0	1
1	1	0	1	1	0
1	1	1	1	1	1

d

a	b	c	$A{=}a$	$B{=}b$	$C{=}\,c{\wedge}a{\wedge}b{\neg}c$
0	0	0	0	0	0
0	0	1	0	0	1
0	1	0	0	1	0
0	1	1	0	1	1
1	0	0	1	0	0
1	0	1	1	0	1
1	1	0	1	1	1
1	1	1	1	1	0

Table 9.2 (**a**) In order to be reversible, binary gates must have at least two outputs, say A and B, if the inputs are a and b. The condition that the input must be uniquely retrievable from the output, as in the CONTROLLED-NOT gate shown in (**a**) and (**b**) with its inverse in c, requires that each of the four combinations 00, 01, 10, and 11 appear exactly once. All the 24 possible reversible binary gates are listed in (**d**), in terms of the underlying single-output binary gates. In all combinations, at least one of the output bits repeats an input bit or its negation. Besides these two options, the only alternatives for the other output bit are XNOR ($\equiv$) and XOR ($\neq$). Important special cases highlighted in the table are the identity (pink), the switching or exchange gate (green) and the CONTROLLED-NOT (blue), detailed in (**b**) and (**c**).

a

b

a	b	$A{=}a$	$B{=}(a{\neq}b)$
0	0	0	0
0	1	0	1
1	0	1	1
1	1	1	0

c

A	B	$a{=}A$	$b{=}(A{\neq}B)$
0	0	0	0
0	1	0	1
1	0	1	1
1	1	1	0

d

A	B	A	B	A	B	A	B	A	B	A	B
a	b	$\neg a$	b	b	a	$\neg b$	a	$a{\equiv}b$	a	$a{\neq}b$	a
a	$\neg b$	$\neg a$	$\neg b$	b	$\neg a$	$\neg b$	$\neg a$	$a{\equiv}b$	$\neg a$	$a{\neq}b$	$\neg a$
a	$a{\equiv}b$	$\neg a$	$a{\equiv}b$	b	$a{\equiv}b$	$\neg b$	$a{\equiv}b$	$a{\equiv}b$	b	$a{\neq}b$	b
a	$a{\neq}b$	$\neg a$	$a{\neq}b$	b	$a{\neq}b$	$\neg b$	$a{\neq}b$	$a{\equiv}b$	$\neg b$	$a{\neq}b$	$\neg b$

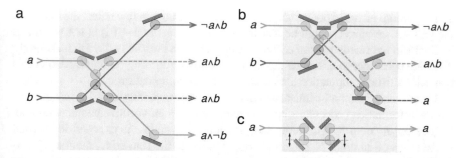

Fig. 9.15 Reversible binary logic can be implemented as ballistic motion of billiard balls, including elastic collisions between balls and specular reflection at hard walls. Presence of a ball at input or output represents "1", absence "0". An *interaction gate* with four exits (**a**) comprises binary operations, $\neg a \wedge b$, equivalent to $\neg (b \supset a)$, twice $a \wedge b$, and $a \wedge \neg b$, equivalent to $\neg (a \supset b)$. The *switch* or *Fredkin gate* (**b**, see Table 9.1c, d) sends input b to two alternative exits, $\neg a \wedge b$ or $a \wedge b$, controlled by the value of input a without altering it. Time delay elements (**c**), necessary to adjust collision points, are readily realized. Trajectories involving only reflection from fixed walls are shown as full lines, those including collisions between balls are dashed. Adopted from Refs. [FT82, Ben82].

more technical terms, it means that they can be realized employing exclusively frictionless ballistic motion. This idea has been proposed and worked out by Charles H. Bennett [Ben73, Ben82] and by Edward Fredkin and Tommaso Toffoli [Tof80, Tof81, FT82]. Their detailed constructions are not intended to be built in the lab, they are fictitious contraptions, to serve as convincing demonstrations that in principle, reversible computing is possible in classical devices. Moreover, they allow us to anticipate essential aspects of quantum computing, where reversibility is not only just barely achieved, as in the present context, but constitutes a fundamental boundary condition, to be accepted for better or worse.

In billiard-ball gates (Fig. 9.15), the presence or absence of a perfectly elastic ball at an entrance or exit of a gate represents the logical values "0" and "1". Inside the gate, to be considered as a black box, frictionless motion is controlled with fixed walls, where balls are reflected as light rays, and by collisions, according to the laws of elastic mechanical scattering. Summed over all outputs, the number of 1's must equal their number on entrance, that is, 4, since balls are neither lost nor gained during the process. Their identity, however, which could readily be recognized, for example by colour, is irrelevant here.

Given that in this manner, logically reversible gates can be accomplished, does that mean the entire computing process could be implemented in information conserving Hamiltonian mechanics, in equilibrium with a thermal environment? That would be an illusion. While certain intermediate steps can be performed in a reversible manner, they depend all the more on an irreversible reduction to nominal states, before and after the logical operations. For instance, billiard-ball gates are not protected against thermal and other noise. Errors in the initial conditions (position and momentum) of the balls propagate unabatedly along their trajectory. Collisions among balls are as unstable as, e.g., the Sinai billiard or a hard-sphere gas (Fig. 8.36), they even amplify initial errors exponentially. After passing through a few gates, the output of the ballistic computer would be lost in thermal fluctuations—if it is not stabilized by final and intermediate error correction steps, which invariably have to be strongly dissipative in order to eliminate unwanted entropy infiltrating en route. Also in this respect, ballistic computation anticipates quantum computation.

Genetic and other biochemical activity in the living cell, interpreted as instances of natural information processing, appears to constitute a provocative counterexample: It goes on at room temperature, and, at molecular scales, is exposed to thermal fluctuations of an order of magnitude comparable to the functional motion. Nevertheless, it works with astonishing accuracy and reliability. It is adopted as a model for another alternative implementation of computing, the Brownian computer [Ben82]. Brownian computation makes a virtue of the necessity of thermal noise. Instead of suppressing it, as in conventional computing, it is exploited to drive the computation in the intended direction, given a weak bias force, which requires the system to be out of equilibrium—but only marginally: It proceeds "two steps ahead, one step back", so that a trade-off between efficiency and energy consumption arises. Similar to ballistic computing, fictitious mechanical implementations of Brownian computers have been suggested to demonstrate and visualize their feasibility in principle. Figure 9.16 shows a proposed realization of an entire Turing machine (Sect. 9.3.1), including tape and read-write-equipment, made of rigid, frictionless, loosely fitting clockwork. Unlike their biological models, mechanical Brownian computers are protected by tight constraints against noise transverse to their programmed computation path. However, as the cutaway of a control element in Fig. 9.16b forebodes, they do remain very sensitive to external perturbations, particularly where computation paths bifurcate or cross.

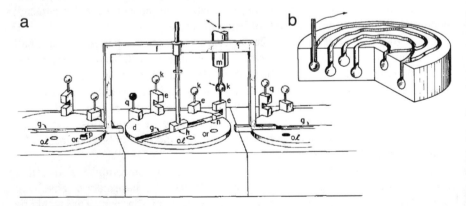

Fig. 9.16 Brownian Turing machine, realized as a mechanical device made of rigid, frictionless, loosely fitting clockwork. Tape advancing and read-write-equipment (**a**) operate on tape sections one by one, manipulating information-bearing elements according to transition rules governed by a control unit that is implemented as a master camshaft shown in (**b**). Reproduced with permission from [Ben82].

These examples of alternative realizations of digital computing, very different in their physical properties, provide convincing evidence that there exists no fundamental lower bound to the energy turnover per elementary digital operation—except for erasure. At the same time, separating the energy and time scales of nominal states and of their deterministic evolution from those of microscopic noise, which is indispensable to maintain the discrete nature of computation, does imply a tight trade-off between reliability and speed on the one hand and energy consumption on the other: The faster and more accurate the operation is required to be, the more energy it dissipates.

9.3 Global Structure of Classical Computing: Moving on Granular Spaces

Given the diversity of possible physical platforms for elementary memory and logical components, as exemplified above, the laws of physics do imply terms and conditions for digital as well as for any other type of computing, but they are indirect and leave plenty of space for variation and optimization. Conversely, we can summarize the general framework of digital computation in the following three conditions:

1. The set of admissible states is countable and finite, comprising a total of $N \in \mathbb{N}$ states.
2. The time evolution on this space advances in discrete steps.
3. Transitions from one state to the next follow deterministic rules.

They cannot be consequences of the physical laws underlying its implementation, as patent alternatives show (suffice it to mention analogue computing). Yet these principles delimit the behavioural repertoire of digital machines. They define fundamental laws governing all dynamical processes that can occur in the artificial universe created on a computer but are independent of the physics of the hardware. In this way, they constitute a category of dynamical systems of its own right and go through, up to even highest-level functions performed by such equipment.

9.3.1 Granular State Spaces

Principles equivalent to condition 1 above and their consequences have been contemplated independently of, and way before, the advent of digital computing. Notably in physics, a fundamental discretization of space–time, not immediately related to quantum mechanics, has been proposed from various sides, e.g., from cosmology (postulating the Planck length $l_P = \sqrt{\hbar G/c^3}$, with G, the gravitational constant, and c, the vacuum velocity of light, as smallest meaningful length scale) and from early computer science (considering the universe as a computational machine [Zus69]).

As an immediate corollary, a finite total number $N \in \mathbb{N}$ of nominal states implies an upper bound of

$$I_{max} = c \ln(N) \tag{9.3.1}$$

for the information that can be stored and processed in the device. Referring to the distinction between reversible (information conserving) and irreversible logic (discarding surplus information as waste), the deterministic dynamics on discrete state spaces (Fig. 9.17) can be subdivided into two main classes, reversible and dissipative time evolution.

Digital computing can be considered as a limiting case of Markov chains, introduced in Sect. 3.1.1, now restricting continuous transition probabilities to the alternative zero or one. Excluding contingent intervention through input devices as in Eq. (9.2.12), the transition from a state $\mathbf{C}_n$ of the total system to its state $\mathbf{C}_{n+1}$ amounts to a discrete mapping of the state space $\mathcal{L}$ onto itself,

$$\mathcal{L} \to \mathcal{L}, \mathbf{C}_n \mapsto \mathbf{C}_{n+1} = \mathbf{PC}_n. \tag{9.3.2}$$

In the reversible case, this map is bijective and represented by a permutation matrix $\mathbf{P}$ (with a single entry 1 in each row and each column). Examples of this kind of time evolution have been mentioned in the context of discretized chaotic maps, see Eqs. (5.5.15) and (8.5.10). In the irreversible case, it maps onto a smaller subspace, so that the matrix $\mathbf{P}$ is no longer of full rank.

The motion in state space induced by a permutation matrix generates periodic trajectories (Fig. 9.17a). They can eventually pass through the entire state space or,

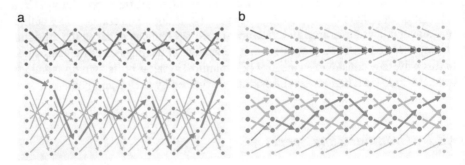

Fig. 9.17 Deterministic time evolution on finite discrete state spaces, reversible (**a**) or irreversible (**b**), amounts to mere permutations among the states. In the reversible case, the dynamics induces a partition of the total state space into disjoint subspaces, linking the elements of each one a single periodic trajectory (bold darker arrows in **a**), its period given by the cardinality of the subspace. In the irreversible case (**b**), the time evolution contracts these subspaces onto even smaller sets, comprising a single state, equivalent to a point attractor, or several ones, corresponding to a limit cycle. In panel (**b**), attracting subspaces are indicated by bold dots and arrows, their basins of attraction by coloured background.

if the permutation matrix decays into two or more blocks, subspaces of the size of these blocks. Their cardinality coincides with the period of the trajectories. In the example shown in Fig. 9.17a, the permutation matrix is block-diagonal. It consists of two blocks,

$$
\mathbf{P} = \begin{pmatrix} \mathbf{P}_1 & \mathbf{0}^{(4\times8)} \\ \mathbf{0}^{(8\times4)} & \mathbf{P}_2 \end{pmatrix}, \mathbf{P}_1 = \begin{pmatrix} 0\,0\,0\,1 \\ 0\,1\,0\,0 \\ 1\,0\,0\,0 \\ 0\,0\,1\,0 \end{pmatrix}, \mathbf{P}_2 = \begin{pmatrix} 0\,0\,0\,0\,0\,1\,0\,0 \\ 1\,0\,0\,0\,0\,0\,0\,0 \\ 0\,0\,0\,0\,1\,0\,0\,0 \\ 0\,0\,0\,0\,0\,0\,1\,0 \\ 0\,0\,0\,1\,0\,0\,0\,0 \\ 0\,0\,0\,0\,0\,0\,0\,1 \\ 0\,1\,0\,0\,0\,0\,0\,0 \\ 0\,0\,1\,0\,0\,0\,0\,0 \end{pmatrix}, \tag{9.3.3}
$$

generating trajectories of period 4 (magenta) and 8 (green), resp. In the irreversible case (Fig. 9.17b),

$$
\mathbf{P} = \begin{pmatrix} \mathbf{P}_1 & \mathbf{0}^{(4\times8)} \\ \mathbf{0}^{(8\times4)} & \mathbf{P}_2 \end{pmatrix}, \mathbf{P}_1 = \begin{pmatrix} 0\,0\,0\,0 \\ 1\,0\,0\,0 \\ 0\,1\,1\,1 \\ 0\,0\,0\,0 \end{pmatrix}, \mathbf{P}_2 = \begin{pmatrix} 0\,0\,0\,0\,0\,0\,0\,0 \\ 1\,0\,0\,0\,0\,0\,0\,0 \\ 0\,1\,0\,1\,0\,0\,0\,0 \\ 0\,0\,0\,0\,0\,1\,0\,0 \\ 0\,0\,1\,0\,0\,0\,1\,0 \\ 0\,0\,0\,0\,1\,0\,0\,0 \\ 0\,0\,0\,0\,0\,0\,0\,1 \\ 0\,0\,0\,0\,0\,0\,0\,0 \end{pmatrix}, \tag{9.3.4}
$$

$\mathbf{P}$ is not of full rank; there is an entry 1 in every column, but not in every row. The upper block $\mathbf{P}_1$ (magenta in Fig. 9.17b) here contracts to the identity in state 3, the lower block $\mathbf{P}_2$ (green) to a permutation $7 \to 9 \to 10 \to 8$, a limit cycle of period 4, both approached from states outside these attractors through transients of up to two steps.

The consequences of simulating dynamical systems on granular spaces comprising a finite set of discrete states has been amply discussed in the literature, see Refs. [CP82, HW85, WH86, BR87], and addressed above in the context of discretized chaotic systems, Sects. 5.5 and 8.5.2: The limited amount of total information in the system does not allow for a positive entropy production for an indefinite time, which condemns the time evolution to eventually repeat itself, at the latest after having visited all states once that the system disposes of. Chaotic motion is thus replaced by limit cycles. In the case of present-date digital equipment, the memory may reach tera- or petabytes, so that the corresponding finite recurrence time is of no practical relevance. In conclusion though, even with the vast versatility of digital computers, the degree of irregularity of chaotic dynamical systems remains out of their reach.

9.3.2 Navigation on Granular Spaces

Even if a permanent entropy production could be achieved in a digital computer, we would have to verify chaoticity in terms of an exponentially growing separation of trajectories, referring to a positive KS entropy, cf. Eq. (3.1.16). To measure distances, we have to define a metric on the discrete state space, where it less obvious to define than in a continuous space, a task analogous to measuring genetic distance in the space of nucleotide sequences, see Sect. 2.2.

For high-dimensional spaces of binary sequences such a metric does exist in the form of the *Hamming distance* [NC00], defined for a pair of binary sequences **a**, **b** of equal length as the number of digits where they do not coincide, or equivalently, the minimum number of bit flips necessary to reach one sequence from the other (Fig. 9.18). It meets the conditions defining a metric, i.e., for all sequences **a**, **b**, **c** ∈ $\{0, 1\}^M$, the Hamming distance satisfies.

- non-negativity, $d(\mathbf{a}, \mathbf{b}) \geq 0$,
- symmetry, $d(\mathbf{a}, \mathbf{b}) = d(\mathbf{b}, \mathbf{a})$,
- Identity of indiscernibles, $d(\mathbf{a}, \mathbf{b}) = 0 \Leftrightarrow \mathbf{a} = \mathbf{b}$,
- subaddivity (validity of the triangle inequality), $d(\mathbf{a}, \mathbf{b}) + d(\mathbf{b}, \mathbf{c}) \geq d(\mathbf{a}, \mathbf{c})$.

Comparing the Hamming distance to usual metrics on continuous spaces $\mathbb{R}^D$, it can be considered as a discrete version of a metric induced by the *p*-norm of order $p = 1$,

$$d(\mathbf{a}, \mathbf{b}) = \left(\sum_{n=1}^{D} |b_n - a_n|^p \right)^{1/p}, \qquad (9.3.5)$$

(a_n, b_n, denoting components of **a** and **b**), that is, it adds distances component by component.

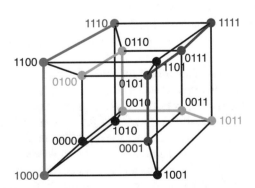

Fig. 9.18 A binary hyper-cube exemplifies Hamming distance as a metric in spaces of 2^M binary sequences, here $M = 4$. It measures distance between two sequences as the number of bits where they differ, i.e., the number of bit flips required to reach one sequence from the other. The figure highlights three examples in colour, $d(1110,1000) = 2$ (blue), $d(1111,1000) = 3$ (red), $d(1011,0100) = 4$ (green).

In the space of binary sequences, where each bit constitutes a dimension of its own, the number of states per dimension is minimal, two, the number of dimensions reaches a maximum. Grouping m bits to larger units, notably sets of $m = 8$ bits forming a byte, the resolution increases to 2^m points, while the total number of dimensions reduces by the same factor m. An important instance of the same relation is the genetic code (Sect. 2.2.1), with $4^3 = 64$ points per codon (to encode only some 20 amino acids plus a stop codon), that is, per dimension in genome space.

Computing amounts to controlled walks in this high-dimensional discrete space. In the above global view of motion in this space (Fig. 9.17), states were considered as closed entities, not resolving any internal structure, in particular ignoring that they are formed by ordered binary sequences. Switching from one state to another therefore could involve an unlimited number and an arbitrary distribution of positions along a symbol sequence. Computers in fact do not work like that. They access bits one by one, operating *locally* in sequence space. That is, each step only depends on, and only affects, a limited number of neighbouring symbols, it never encompasses, for instance, the entire memory content simultaneously. While Hamming distance defines a metric on discrete spaces by counting dimensions, it however does not allow talking of closeness, of neighbourhood, among dimensions. Defining local operation therefore requires to be more specific concerning the geometry of binary walks performed on computers.

In classical mechanics, for example, a general framework for dynamics is established by Hamilton's equations of motion. Their general form defines the state space—phase space—and imposes a basic structure on it, symplectic geometry with continuous coordinates (see Sect. 5.1). Cast in the form of partial differential equations, they are local in space and time, but the update of the system state is simultaneous all over phase space. As precise and rigorous, hence restrictive, the principles of Hamiltonian mechanics are (strict determinism, time-reversal invariance, conservation of phase-space volume, etc.), as vast is nevertheless the freedom of permissible flows on phase space that symplectic geometry allows, defined by the specific Hamiltonian and ranging from simple laminar flows (regular motion) to turbulence (fully developed chaos).

An instance of complex dynamics on a discrete space is genetic evolution: It is not deterministic but on the contrary, amounts to a random walk in genome space. However, this walk is guided and conditioned in turn by the concomitant variation of the position in a "fitness landscape" (Sect. 3.1.5, Fig. 3.14) to which it is coupled through the phenotype.

Also logical inference can be viewed as motion on a discrete space, the space of propositions. From a set of premises, representing the initial condition, inference rules guide a line of reasoning through the space of propositions to a set of logically valid conclusions. A intimately related concept already comes very close to computation: An *algorithm* (for a realistic example, see, e.g., the fast Fourier transform expounded in Sect. 2.3.4) receives data as input, mostly restricted to elements of some countably infinite set, e.g., the integers, and follows a sequence of precise operations to produce an output, numbers or a truth values.

9.3.3 Models of Classical Computing: The Turing Machine

Localizing computers in this repertoire of motion on discrete spaces, if only to assess where they can go and where not, which kind of process is accessible and which is out of reach of digital computation, is a formidable mathematical task. Computer science has produced important insights, but questions concerning computability, decidability, completeness remain an open frontier of research. It was Alan Turing who contributed what is arguably the first, strongest, and most influential attempt to construct a nucleus of a theory of computation: the Turing machine [Hop84, NC00].

Turing's construction anticipates the distinction of two basic components of a computer, memory and CPU, in rudimentary form (Fig. 9.19). It incorporates local operation on its total state space by delegating the entire time evolution to a *control unit* (its CPU), a subsystem comprising only a finite number of internal states (typically considered small, down to two), between which it switches according to a set of deterministic transition rules, its program. Its internal states include in particular an initial state ("start") without precursor and a final state ("halt") without successor. The Turing machine navigates in the remaining dimensions of the total state space by operating locally on a *tape* (its memory). It is a sequence of cells (not restricted in number, according to Turing's original concept), each one containing a single symbol from a finite alphabet, which the control unit can read as input and rewrite as its output, moving along the tape cell by cell. The Turing machine thus performs a walk on its infinite-dimensional discrete state space in bit-by-bit steps, steered in

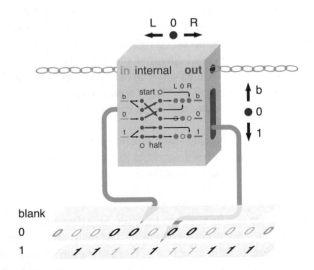

Fig. 9.19 The Turing machine navigates and operates locally on its discrete state space, represented by an infinite tape (bottom) with one symbol (here 0, 1, and blank) per cell, and the internal states of its control unit (top). The input read from the tape at the present position and the current internal state determine the output through a set of program lines: a symbol to be written on the tape and possibly a shift of the read–write head by one step to the left or right along the tape. Special internal states are the initial state "start", and "halt" if the present state does not match any programmed rule.

turn by the deterministic motion within the much smaller subspace spanned by the control unit. The number of the cell along the tape where the control unit currently operates defines the direction in state space along which the Turing machine moves, the value written to the cell defines the position in that dimension.

As elementary and austere as this construction appears—it could be realized combining off-the-shelf mechatronic components or even with mere clockwork, see, e.g., the fictitious mechanical computer depicted in Fig. 9.16—as comprehensive is its scope. The full computational capacity of any digital computer, however large, is within reach of the Turing machine, obviously at the expense of performance. This claim is substantiated by a complementary idea, the *universal Turing machine*: Instead of pre-wiring the program in the control unit, all individual features of a specific Turing machine can be transferred to the tape. Employing a procedure similar to Gödel numbering (Sect. 4.3.2, Table 4.1), the set of transition rules can be encoded as a sequence of natural numbers, the *Turing number* n_T of a specific Turing machine, to be written to the leading part of the tape. The control unit of the universal Turing machine then reads all possible combinations of input symbols and internal states as well as the current input from the tape and reaches an output again following the stored transition rules, that is, emulating the desired particular Turing machine.

Exactly how comprehensive is the concept of Turing machines? The *Church-Turing thesis* provides a succinct answer:

Every effectively calculable function is computable, and vice versa.

In this formulation, "effectively calculable" is to be interpreted as "calculable by a recursive algorithm" and "computable" as "to be computed by a Turing machine". "Recursive algorithms", in turn, are algorithms that may contain references to themselves (such as a subroutine calling itself in the course of execution). In effect, the Church-Turing thesis claims equivalence between a concept taken from mathematical logic and another from computer science, which thus define the same class of dynamical systems on discrete spaces. Only a few general attributes of the Turing machine are decisive for assigning it to this class: The unlimited size of the tape makes it qualitatively more powerful than a real computer with bounded memory (a "finite-state machine", cf. Eq. (9.3.1)). The fact that it only comprises a single control unit (and not two or more working in parallel), in turn, restricts its possibilities as compared to other, even more powerful classes of computing systems.

A remarkable limitation of the Turing machine and all equivalent computing devices is revealed by the *halting problem*. Unlike dynamical systems in physics, it includes the "halt" state, terminating the time evolution definitely. It is to be distinguished from steady states or limit cycles, which in the context of computing correspond to infinite loops. Viewing computers as tools, it is of evident interest whether a given task can be accomplished in finite time, that is, with the Turing machine arriving at "halt" and containing the required output on the tape. More precisely, the halting problem can be stated as follows:

Initiated with the input number n_I on the tape, does the Turing machine with Turing number n_T halt?

It can be proven that the halting problem, in turn, *cannot* be solved by a Turing machine (or an algorithm).

The halting problem resembles Gödel's Incompleteness Theorem. Indeed, both questions can be mapped to one another, and the proof of the undecidability of the halting problem follows the same strategy as Gödel's proof. The Turing number already resembles Gödel numbering. Introduce self-reference, as in Gödel's proposition (invoking the diagonal lemma), by giving the input number n_I the same value as the Turing number n_T (the "diagonal halting problem"):

Initiated with its own Turing number n_T on the tape, does the Turing machine with Turing number n_T halt?

Then disprove by contradiction that this can be decided by an algorithm. Formalize the answer to the diagonal halting problem by casting it in a function *halt*: $\mathbb{N} \to \{f, t\}, \{f, t\}$,

$$halt\,(n) := \begin{cases} f \text{ if Turing machine } n \text{ does not halt upon input of } n, \\ t \text{ if Turing machine } n \text{ halts upon input of } n. \end{cases} \quad (9.3.6)$$

Note that this program stops only if the subroutine HALT(N) finds that the Turing machine does *not* halt. Finally, the negative feedback crucial for Gödel's proof is implemented by supposing the existence of an algorithm HALTING−PROBLEM, alleged to produce the sought answer, which implements *halt(n)* as a fictitious subroutine HALT(N),

```
       PROGRAM HALTING - PROBLEM
       INTEGER*8 N
       LOGICAL HALTS
   10  HALTS = HALT(N)
           IF (HALTS = .FALSE.)
           THEN STOP
           ELSE GO TO 10
       END                                              (9.3.7)
```

If the Turing machine n_T *does* halt upon input of n_T, the subroutine HALT(N) produces the result TRUE and HALTING−PROBLEM is caught indefinitely in its loop 10, that is, never halts. In other words, the inability of the Turing machine to escape from an unending search penetrates to any algorithm programmed to solve the halting problem.

The proof reveals the close analogy between Gödel's Incompleteness Theorem and the undecidability of the halting problem. It indicates that axiomatization, proposed by Hilbert but questioned by Gödel as remedy for the foundational problems of mathematics, suffers from the same limitations as Turing machines and algorithms, hence belongs to the same class of formal systems.

In terms of the construction of the Turing machine, the two attributes addressed above are constitutive features of this class:

- a control unit that operates locally and sequentially, only on one cell at a time,
- an infinite tape that provides an unbounded storage capacity.

This is substantiated by considering other models of computation that differ from Turing machines precisely in these respects. The following two subsections will present two such models, cellular automata and Conway's game of life.

9.3.4 Cellular Automata: Parallel Computing on Granular Spaces

In Turing machines, all information processing goes on in a single small subsystem, the control unit. Altering even this seemingly evident condition leads to a substantially different class of discrete information processing systems with new properties, unattainable in Turing machines: *Cellular automata* [Wol83, Wol84] operate locally on an infinite-dimensional discrete state space, spanned by an ordered set of spatial sites, similar to the tape of the Turing machine. It does so applying a few deterministic transition rules, as in a Turing machine, but *simultaneously at every site*. That is, their state is updated in parallel, not sequentially. Cellular automata emerged in the heydays of cybernetics, with pioneering contributions by great minds like von Neumann, Wiener, and Zuse, and remarkably not as a comprehensive model of computation but as a guideline to explore the new world of discrete systems as a source of models for physical and even biological phenomena.

The current state of the system is represented by symbols a_i from an alphabet of k elements, assigned to sites i in an infinite one-dimensional ordered chain (for cellular automata on two- and more-dimensional spaces, see below). Time t proceeds in unit steps. The state of the system at cell i and time t is determined by the states of the same and $2r$ (the *coordination number*) neighbouring cells at the preceding time $t-1$ [Wol84],

$$a_i^{(t)} = F_N\left(\mathbf{a}_i^{(t-1)}\right),$$

(9.3.8)

where $F_N(\mathbf{a})$ is a discrete map from a finite symmetric neighbourhood of site i, from $i-r$ to $i+r$ ($2r+1$ sites), represented as a $(2r+1)$-dimensional vector

$$\mathbf{a}_i^{(t-1)} = \left(a_{i-r}^{(t-1)}, a_{i-r+1}^{(t-1)}, \ldots, a_{i-1}^{(t-1)}, a_i^{(t-1)}, a_{i+1}^{(t-1)}, \ldots, a_{i+r-1}^{(t-1)}, a_{i+r}^{(t-1)}\right),$$

(9.3.9)

to the updated state $a_i^{(t)}$ at site i (Fig. 9.20a). With a domain comprising k^{2r+1} distinct states (k possible values each at $2r+1$ sites) and k distinct possible outputs, there

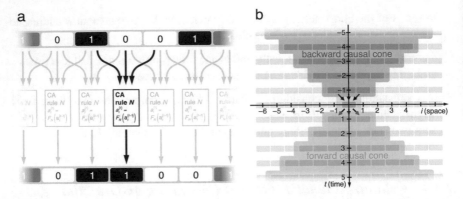

Fig. 9.20 Cellular automata "read" their input from a few discrete sites along a one-dimensional infinite sequence, determine their output from a simple deterministic rule and "write" it back as updated value to the "tape", as do Turing machines (**a**). They differ, however, in updating all sites simultaneously, not one at a time. If the input depends on the site operated on, plus r neighbours on both sides, a range of $2r + 1$ direct precursor sites results, increasing linearly by the same amount per time step backwards ("backward causal cone", pink in **b**), cf. Fig. 3.12. The same linearly increasing range of dependence on an initial site arises in the forward direction ("forward causal cone", green in **b**).

exists a total of $k^{(k^{2r+1})}$ such functions, also called rules, for cellular automata with these parameters. They can be enumerated with rule numbers N, $1 \leq N \leq k^{(k^{2r+1})}$, encoded uniquely as a k-nary number with $2r + 1$ digits (Fig. 9.21),

a

in		out		
000	0	0	0×2^0	0
001	1	1	1×2^1	2
010	2	1	1×2^2	4
011	3	1	1×2^3	8
100	4	1	1×2^4	16
101	5	0	0×2^5	0
110	6	0	0×2^6	0
111	7	0	0×2^7	0
		rule	**30**	

Fig. 9.21 Rule numbers encoding cellular automata with k symbols and $2r + 1$ neighbourhood sites, here $k = 2$, $r = 1$, are obtained associating one of k possible output symbols (third column in **a**) to each of the k^{2r+1} input configurations (first column), as in Eq. (9.3.9) and reading the sequence of k^{2r+1} symbols (here 01111000) as a k-nary number (columns 4 and 5), which gives the rule number for this cellular automaton (here 30), cf. Eq. (9.3.10). The cellular automaton with this rule number is run by applying this mapping to a given initial state (**b**), cf. Fig. 9.20, with periodic boundary conditions if the total number of state space cells is finite (here 8).

$$N = \sum_{j=0}^{2r} c_j k^j =: (c_0, \ldots, c_{2r}), c_j := a_{i-r+j}. \qquad (9.3.10)$$

Alternatively, therefore, the update (9.3.8) can also be written as

$$a_i^{(t)} = g\left(\sum_{j=0}^{2r} a_{i-r+j}^{(t-1)} k^j\right), \qquad (9.3.11)$$

with a discrete function $g(a)$ of a single integer argument [Wol84]. If the output does not depend on the exact position of each input symbol within the neighbourhood around site i, but only on the total number of cells containing a given symbol, the factors k^j in Eqs. (9.3.10, 9.3.11) can be replaced by 1, a case referred to as *totalistic* cellular automaton [Wol84]. A plausible condition, often taken for granted, is furthermore that the state comprising only zeros (referred to as "ground state" in the following) be stable (no "spontaneous ignition").

Totalistic rules break time-reversal invariance: They lose the information on the order of input symbols inside their neighbourhood, so that the input cannot be uniquely reconstructed from the output, and are inevitably irreversible. In general, however, cellular automata do not necessarily contract their state space. To be sure, the number of k^{2r+1} different states on input for a cellular automaton site i reduces to only k different output symbols. At the same time, however, the input sequences $\mathbf{a}_j^{(t-1)}$, $j \neq i$, of another $2r$ neighbouring sites overlap partially with the same input $\mathbf{a}_i^{(t-1)}$, so that the cellular automaton as a whole can well conserve information. In other words, the number $2r + 1$ of input cells, a given output symbol depends on, coincides with the number of output cells affected by a given input symbol. Indeed, the identity transformation, which is trivially reversible (rule 204 for $k = 2$, $r = 1$ automata is a patent example), or rigid shifts of an initial pattern can readily be achieved with cellular automata with $r > 0$. In more involved cases, however, a time arrow enters in a subtler way. Even balanced rules (all k symbols appear with the same frequency on output) in general map distinct $(2r + 1)$-tuples on input to the same output tuples. Their behaviour is comparable to Hamiltonian mechanical systems that conserve information, contracting phase space in some directions but expanding it at the same total rate in others, see comment 9 in Sect. 5.3. An inverse mapping that undoes a given cellular automaton step by step does not exist for them. This is manifest, for example, in the asymmetry under inversion of time of characteristic patterns produced by chaotic cellular automata, such as the typical triangular clearings that appear in panels (b) of Figs. 9.24, 9.25 and 9.26 (Fig. 9.22).

The time evolution generated by a cellular automaton comes in fact close to familiar forms of evolution laws in analogue physics, such as partial differential equations: The increment of the site value $a_i^{(t)}$ on the l.h.s. of Eqs. (9.3.8) or (9.3.11) can be interpreted as a discrete first-order partial derivative in time. The r.h.s., in turn, is equivalent to an expansion in partial derivatives in discrete position. Since

a

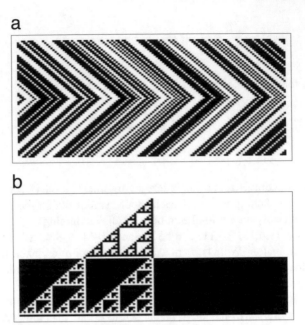

b

Fig. 9.22 Cellular automata evolve irreversibly in general, but in exceptional cases show reversible time evolution. Examples are rigid spatial shifts, as in $k = 2$, $r = 1$ automata 240 (spatial shift by $+1$ per time step) and 170 (shift by -1). (**a**) Starting from a random binary sequence, rule 240 has been run for 32 time steps, followed by another 32 steps with rule 170, resulting in exact recovery of the initial state. (**b**) Rules 102 and 153, in the same class of automata, perform a unit shift up or down, resp., in the space of binary triplets. Concatenated in the same way as in panel (**a**), they reproduce the localized initial state up to an interchange $0 \leftrightarrow 1$ and a single additional nonzero cell on the border of the automaton space.

expanding a continuous function $f(x)$ up to n^{th} order involves n independent coefficients, for example the function values at n different positions, the evolution of a $(2r + 1)$-site cellular automaton can be considered as a discrete analogue of a partial differential equation of first order in time and order $2r$ in position.

The update at site i depends on $2r + 1$ sites at the immediately preceding time. Following the dependence on precursor states further backward in time, the range of sites contributing to the state $a_i^{(t)}$ increases linearly with the time difference as

$$d(t'', t') = (2r + 1)(t'' - t') \text{ for } t'' > t', \tag{9.3.12}$$

defining the diameter of a one-dimensional "backward causal cone" (pink in Fig. 9.20b, in analogy to Fig. 3.12). Moving forward one step in time, all the sites from $a_{i-r}^{(t+1)}$ to $a_{i+r}^{(t+1)}$ depend on $a_i^{(t)}$, so that the range of sites codetermined by $a_i^{(t)}$ increases also as given by Eq. (9.3.12). They form the "forward causal cone" (green in Fig. 9.20b, examples are shown in Fig. 9.26). In the context of cellular automata, a linear increase of the range of cells in a "time-like" relation of a given site i implies

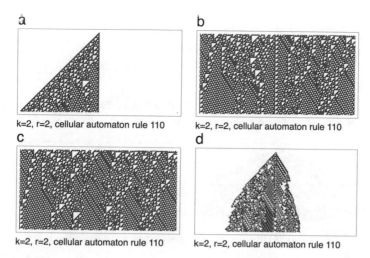

Fig. 9.23 Examples of the time evolution of the cellular automaton rule 110, Eq. (9.3.10), from a localized initial state with value 1 in a single cell on a constant background of zeros (**a**), and from two identical unbiased binary random sequences (**b**, **c**), up to the value of a single cell in the centre. Panel **d** shows the binary difference of panels **a** and **b**. Parameters are $k = 2$ (two symbols, 0 and 1) and $r = 1$ (neighbourhood of size 3). The first 64 time steps are shown. Data reproduced following Ref. [Coo04].

an exponential growth of the number of possible states of these cells. This means that the amount of information required to predict the evolution at this site over a time Δt also scales linearly with Δt and increases indefinitely for $\Delta t \to \infty$, indicating a close kinship to chaotic systems in continuous space and time.

It suggests itself to interpret the behaviour of cellular automata in similar terms as that of continuous dynamical systems, specifically for irreversible dynamics. Figures 9.23 to 9.26 show some examples of patterns generated by standard types of cellular automata, all with $k = 2$ (two symbols, 0 and 1). They are defined in Fig. 9.23 by rule 110, according to Eq. (9.3.10), with $r = 1$ (input comprising three adjacent cells), and in Figs. 9.24, 9.25, and 9.26 by totalistic rules (order of symbols in the input is irrelevant) with $r = 2$ (input comprising five adjacent cells). Patterns in Figs. 9.23a and 9.24 are generated from a localized initial condition, a few cells with a 1 in the centre, on a homogeneous background of zeros. In Figs. 9.23b, c and 9.25, initial conditions are unbiased binary random sequences extending over the full range of the figure. In order to visualize how information propagates in cellular automata, Figs. 9.23b–d and 9.26 compare pairs of patterns that evolved from random initial conditions as in Fig. 9.25, identical up to a single symbol at the centre, showing the difference mod 2 of the two patterns.

These simulations can also be considered as instances of *branching* in digital computation, exemplified in cellular automata. The notion originates in branching conditions in computer programs, such as "IF", "IF ... THEN ... ELSE", "WHILE ... DO", see, e.g., the FORTRAN code (9.3.7). In all these cases, even a single input bit suffices to steer the program into vastly distinct directions [ED19]. As Figs. 9.23d and

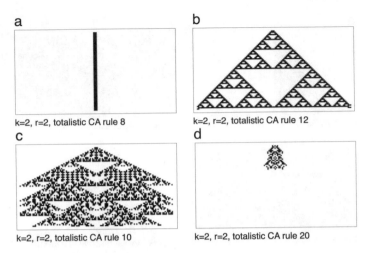

Fig. 9.24 Examples of the time evolution of cellular automata with totalistic rules, from localized initial states, sequences 111 (**a, b, c**) or 10101 (**d**) in the centre, on a constant background of zeros. Parameters are $k = 2$ (two symbols) and $r = 2$ (five cells). The rule numbers are 8 (**a**), 12 (**b**), 10 (**c**), and 20 (**d**). The first 64 time steps are shown. Data reproduced following Ref. [Wol84].

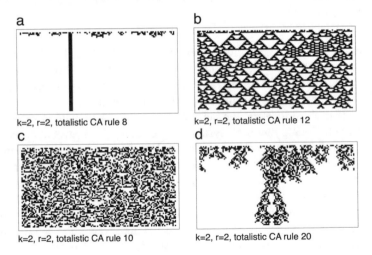

Fig. 9.25 Examples of the time evolution of cellular automata with totalistic rules, from initial states defined as unbiased binary random sequences. Parameters and rules as in Fig. 9.24. Data reproduced following Ref. [Wol84].

9.26b, c demonstrate, the difference between the respective outputs, measured, say, as Hamming distance, can increase up to linearly with time in certain cellular automata, thus mimicking logical branching and sensitive dependence on initial conditions.

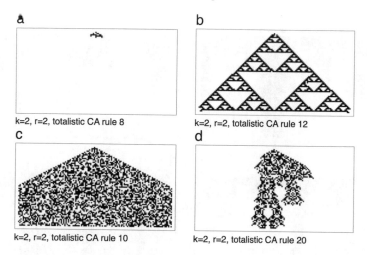

a

k=2, r=2, totalistic CA rule 8

b

k=2, r=2, totalistic CA rule 12

c

k=2, r=2, totalistic CA rule 10

d

k=2, r=2, totalistic CA rule 20

Fig. 9.26 Examples of the time evolution of cellular automata with totalistic rule, comparing pairs of initial conditions, prepared as unbiased binary random sequences, identical up to a single bit in the centre, as in Fig. 9.23d. Shown is the difference mod 2 of the respective patterns. Parameters and rules as in Fig. 9.24. Data reproduced following Ref. [Wol84].

In the examples shown, three qualitatively distinct types of behaviour can be identified. They represent the complete repertoire of time evolution in cellular automata, as has been confirmed in comprehensive numerical studies [Wol84]:

1. systems approaching point attractors or limit cycles (attractors that are periodic in time), including the homogeneous ground state as well as localized stable patterns: panels (a) in Figs. 9.24, 9.25 and 9.26,
2. chaotic systems producing spatio-temporal patterns without discernible regularity: panels (b) and (c) in Figs. 9.24, 9.25 and 9.26,
3. systems that, depending on the initial condition, exhibit all the three types, constant, periodic, or irregular patterns: Fig. 9.23, panels (d) in Figs. 9.24, 9.25 and 9.26, and Fig. 9.27.

While class 1 is analogous to dissipative dynamical systems contracting their state space in all directions, types 2 and 3 correspond to systems with positive Kolmogorov–Sinai entropy (Sect. 3.1.3). Their ability to produce information indefinitely is readily explained by the fact that the number of precursor states on which state $a_i^{(t)}$ depends increases by $2r + 1$ per time step, Eq. 9.3.12, injecting more information from the additional precursor sites at a constant rate. If the initial condition is an infinite random sequence, it can generate a correspondingly irregular output.

Simulations of cellular automata of class 3 provide astonishing evidence that they are capable of a more complex information processing than even chaotic systems, which pertain to class 2 above: Depending on the initial condition, the same system can produce irregular patterns of finite extension in space and time, patterns periodic in time of unlimited spatial complexity, and patterns moving in space with a constant offset per period in time (Fig. 9.27). This diversity suggests that cellular automata

a   b

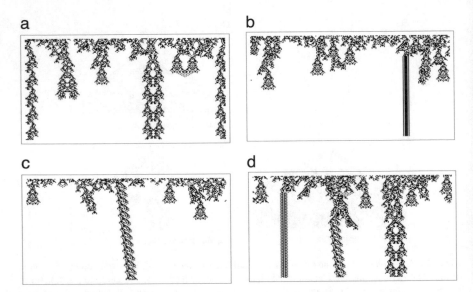

c   d

Fig. 9.27 Different patterns generated by a cellular automaton with totalistic rule 20 and parameters $k = 2$ and $r = 2$, as in panels **d** of Figs. 9.24, 9.25 and 9.26. Initiated with four different unbiased binary random sequences as in Fig. 9.25, they comprise irregular patterns local in space and time, symmetric as well as asymmetric (all panels), patterns periodic in time (panels **a, b, d**), patterns moving with constant velocity (**c, d**), and coexistence of all of these (**d**).

are capable of universal computing, as has been corroborated by an extensive study of the rule 110 cellular automaton [Coo04] (Fig. 9.23). Indeed, they are equivalent to universal Turing machines [Coo04].

9.3.5   Conway's Game of Life

These dynamical scenarios were discovered and have been studied most intensely in cellular automata in one dimension. This restriction, however, is by no means essential. Cellular automata in two and three dimensions are readily construed and lead to similar conclusions. A prominent example, which even gained some popularity outside academic circles, is *Conway's Game of Life* [Gar70].

Game of Life comprises the same components of a cellular automaton as introduced above. Implemented on a square grid, it defines the neighbourhood containing the input condition as the set of eight cells with coordinates 0 and ± 1 in both directions, relative to the central cell at the site (i, j) (termed *Moore neighbourhood*, Fig. 9.30a). The transition rule reads

$$a_{i,j}^{(t+1)} = \begin{cases} 0 & s_{i,j}^{(t)} \leq 1, \\ a_{i,j}^{(t)} & s_{i,j}^{(t)} = 2, \\ 1 & s_{i,j}^{(t)} = 3, \\ 0 & s_{i,j}^{(t)} \geq 4, \end{cases} \qquad s_{i,j}^{(t)} = \sum_{\substack{i'=i-1 \\ }}^{i+1} \sum_{\substack{j'=j-1 \\ |i'|+|j'|>0}}^{j+1} a_{i',j'}^{(t)}. \tag{9.3.13}$$

This is a totalistic rule, except for the central cell, which is evaluated separately. Accordingly, Game of Life is irreversible. The invariance of the rule (9.3.13) under rotations by integer multiples of $\pi/2$ and other redistributions within the neighbourhood further implies that in general, it does not factorize into a pair of rules taking only horizontal or only vertical neighbours into account: Game of Life cannot be reduced to a pair of independent one-dimensional cellular automata.

It exhibits a vast diversity of shapes evolving in space and time, which can all be associated to the three types of dynamical scenarios as listed above for one-dimensional cellular automata (stable and periodic patterns, chaotic attractors, complex switching between the foregoing forms), but with significant variations and extensions in detail. The following subclasses occur in the Game of Life:

1. the ground state: patterns die immediately or after a transient (Fig. 9.28a, b),
2. constant patterns (point attractors) (Fig. 9.28c, d),
3. periodic sequences of patterns (limit cycles) (Fig. 9.28e),
4. periodic sequences of patterns that undergo a rigid shift per period (generalized limit cycles) (Fig. 9.28g),
5. patterns that periodically spawn offshoot belonging to class 4 (generalized limit cycles),
6. patterns evolving without discernible regularity (chaotic attractors) (Fig. 9.29a),
7. patterns switching between the above types (Fig. 9.29b–d).

These phenomena resemble basic manifestations of life, birth, death, reproduction, budding, ... hence the poetic name of this cellular automaton. As in one dimension, the diversity of dynamical behaviours that can coexist in mutually uncoupled regions of the board and that interact in multiple ways when colliding lends substance to the suspicion that also Conway's Game of Life is capable of universal computation. For example, moving stable or periodic patterns can be compared to billiard balls in mechanical implementations of computing (see Fig. 9.15), and their different forms of interaction during collisions with fixed obstacles or with each other (reflection, transformation, annihilation, ..., Fig. 9.29b–d) invites to being interpreted as the processing of bits in gates. That would be sufficient to rank Conway's Game of Life in the same category of computing systems as the universal Turing Machine.

Conceiving cellular automata in two dimensions offers a host of variations. They concern not only the obvious freedom of defining transition rules, but even the very discretization of two-dimensional space. Dividing it into cells of identical shape amounts to a tessellation. Each regular tiling (e.g., square, triangular, and hexagonal, Fig. 9.30) opens different options to delimit neighbourhoods with different coordination numbers, which in turn lead to qualitatively distinct types of cellular automata.

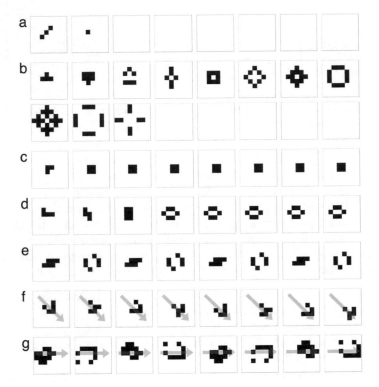

Fig. 9.28 Attractors approached after a short transient in the time evolution in Conway's Game of Life, Eq. (9.3.13). They comprise the void state (**a, b**), specific stable patterns forming point attractors (**c, d**), periodically repeating patterns, equivalent to limit cycles (**e**), and patterns that repeat periodically, but shifted by a constant integer vector (**f, g**). The pattern in (**f**) ("glider") moves by $(1, -1)$ (one cell towards the lower right) per period of four steps, the pattern in (**g**) ("light-weight space-ship", LWSS) by $(1, 0)$ per four time steps. Time advances from left to right.

9.4   The Hierarchical Structure of Computation

The previous subsections were dedicated to the physical basics of digital computing, its hardware, and barely scratched the most elementary forms of software, such as logical gates. As users, we are deliberately shielded from these inner workings and hardly ever interact directly with them. What distinguishes present-day computers, however, what is responsible for their deep impact on our daily lives, is owed to the topmost levels of the software, even far above the operating system. In between extends a multitude of intermediate levels, from machine code up to consumer software "in the cloud".

Such a high degree of organization is hardly found in any other material human product. It is characterized by a steep hierarchy of integration levels that localizes computers closer to biological systems than to the kind of systems typically considered in physics. It becomes manifest in the number of tiers that arise if their

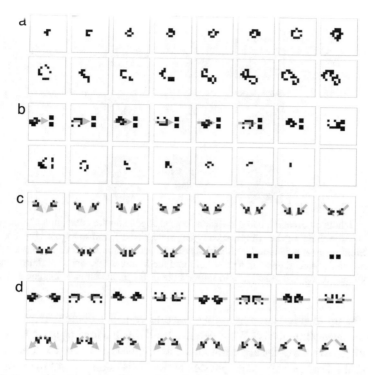

Fig. 9.29 Patterns evolving indefinitely without discernible regularity, corresponding to chaotic attractors, are within the repertoire of Game of Life. Even richer phenomena arise if moving patterns collide with fixed obstacles, such as an LWSS (see Fig. 9.28g) with a stable double block, resulting in their annihilation (**b**), two gliders with one another, ending in a stable double block (**c**), or two LWSS with one another, which converts them into a pair of diverging gliders (see Fig. 9.28f). Patterns evolve from left to right, top to bottom.

structure is organized as a tree, following Nicolai Hartmann's concept of "levels of reality" [Har40], cf. Fig. 1.6. On the range of intermediate scales between molecules (the highest common layer above which physics and biology diverge) and planetary scales, to be specific, physics usually considers two or three structural layers (e.g., clusters, crystals, macroscopic samples), while biology comprises some eight or more (macromolecules, cell organelles, cells, tissues, organs, organisms, superorganisms, ecosystems, ...).

In computers, this hierarchy does not primarily manifest itself in the complexity of their spatial structure, a feature they have in common with the human brain. The motherboard of a PC does not tell us anything about whether it runs Windows, Linux, or Macintosh operating systems. Rather, it is the organization of their software that competes with biology. At the same time, their input and output, in particular at the highest level of application software, may take a very tangible physical form, such as in robots, in self-driving cars, or in autopilots.

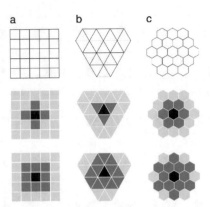

Fig. 9.30 Cellular automata in two-dimensional space depend on the tesselation chosen to define cells. There exist three regular tilings, square (**a**), triangular (**b**), and hexagonal (**c**). For each of them, different types of neighbourhoods with different coordination numbers can be defined, such as von Neumann's (second panel in **a**) with four and the Moore neighbourhood (third panel) with eight neighbours in the square lattice, and similar options for triangular and hexagonal tilings.

How far up in this hierarchy do we have to consider computers as legitimate objects of physical analysis? What determines the behaviour of the computer, the programmed advance of the algorithms executed at the top level or the time evolution following the laws of solid-state physics at the basis? Answers to these questions are by no means obvious. They are the object of active research at the crossroads of physics, computer science, and philosophy. The present subsection is intended to provide some orientation in this open debate.

9.4.1 Structured Organization of Computers: An Overview

Taking only the principal tiers into account of what in fact is even more complex, a standard computing device of our days exhibits about eight levels of a structural hierarchy, from Boolean logic on the bottom to networks of computers forming the (present) top of the pyramid [TA13], as sketched in Table 9.3. Legitimate subject of computer science, it is however by no means closed but continues upward into psychology and sociology and downward into physics. Gates processing binary input according to Boolean logic represent the border region between hardware and the lowest software level. The physical hardware, in turn, consisting of transistors, diodes, etc., in integrated circuits, decomposes further into charge carriers (in essence, electrons) moving in the crystalline lattice formed by doped semiconductors, continuing down into its atomic and subatomic components. Towards the top, computers, as man-made tools, receive input from and send output to their human users, who in turn are not isolated individuals but participate in families and companies, in social groups and societies, which strongly determine their behaviour. In this

Table 9.3 The structural and functional hierarchy of present-day computation comprises several layers, ranging from binary memory and Boolean logic on the lowest to computer networks on the highest level [TA13]. From its bottom, it continues further down into structural levels of solid-state, atomic and sub-atomic physics, and from its top further up into the psychology and sociology of its users.

Physics (hardware)	Computer science (software)	Psychology, sociology (users)
		⋮
	social networks, "the cloud"	agenices, companies, social groups
	internet ⟷	user groups
	application software ⟷	individual user
	operating system, programming language code	
	assembly language code	
	machine code	
integrated circuits ⟷	Instruction Set Architecture (ISA)	
electronic circuits ⟷	gates (Boolean logic)	
bistable states in ⟷ circuits and in ferro- magnetic/electric matter	binary memory	
charge carriers in struc- tured metal or semicon- ductor crystal lattices		
leptons, hadrons		
⋮		

way, computers link their own high level of complexity in inanimate matter with a likewise high level of human, thus basically biological, organization.

In this context, the distinction between hardware and software seems to play a central role. However, it is not as clear-cut as it may appear. In neural information processing, for instance, these concepts hardly apply at all. There, every reprogramming, otherwise involving a change of "software", is invariably achieved by a modification on the "hardware" level, for example by long-term potentiation or depression of synapses [BCP16]. Instead, the ironic term "wetware" has been coined to include both aspects. In electronic computing, "hardware" might refer to all those features of a computer that are invariant under any input condition, while "software" is everything stored in memory and modifiable through input channels. Still, the division is not sharp. In particular, the top structural level on the hardware side, the hard-wired organisation of gates on a chip, overlaps with the bottom level on the side of software, the Instruction Set Architecture (ISA) [ED19].

Towards the upper end of the hierarchy, distinguishing software from user decisions is not unique, either. Not only does user input control the software. Consumer software, such as browsers or social networks, also manipulates the user, in particular if it tracks user behaviour online. The border between autonomous evolution determined by the local software and the computer as a strongly interacting node in a complex man–machine network is therefore diffuse.

The pronounced asymmetry between bottom and top of the computing hierarchy, manifest in relations such as "is composed of" or "calls" between adjacent layers, is also reflected in an asymmetry in time, historical as well as operational. Historically, it is evident that the development advanced from bottom to top. The first assembly languages appeared in the late 1940s. Programming languages such as FORTRAN were generally available in the 1960s. Unix was developed in the 1970s, Microsoft Office, as a pioneering consumer software, was on the market in the 1980s, the first social network, Facebook, in the 2000s. Cloud computing emerged in the same decade, and further integration steps are ahead.

A similar asymmetry can be observed in the way these services are built up each time a computer is started, the *boot*, and are dismantled in the inverse process, the *shut down*. To initiate the computer, switch on power by hitting the "ON" button, not unlike that of an electric kettle. The computer will then boot its software hierarchy, layer by layer, bottom up in the order of their functional dependence, till the top level is reached. Conversely, to undo the process, an "OFF" button is not available, and simply pulling the plug may wreak havoc. Instead, choose "Shut down" from the main menu on the top level and confirm "Are you sure you want to shut down your computer now?", whereupon the levels of the hierarchy will disintegrate top down in inverse order. Finally, it is the last layer still in function, not the user, that switches off the power autonomously. This similarity between operational and historical development resembles Ernst Haeckel's recapitulation theory [Hae66], relating ontogeny to phylogeny [Gou77], and bears on analogous reasons.

While the hierarchy of computing is rooted in hardware, hence is evidently amenable to a description in physical terms, its crown reaches far above the proper realm of physics. If indeed it is possible to integrate computers as a legitimate subject into the physics of dynamical systems, which concepts could provide an adequate understanding, could help us locating them in this larger framework? A key concept, adopted from the theory of science, is *emergence*, to be sketched in the sequel.

9.4.2   Emergence in the Hierarchy of Computing

The structural and functional succession of layers we encounter in computers, see Table 9.3, reminds us of comparable tree structures in diverse contexts in science and culture, as sketched in Fig. 1.6. Phenomena further up in the tree and the concepts and theories describing them are related to those further down in a very special manner that is not captured appropriately with notions such as deduction, derivation, or causal explanation. In physics and biology, in particular, they are characterized instead as emergent phenomena [Gel94, Lau95, GHL19]. The hierarchy of computing not only resembles the organization of physics and biology, it even forms a bridge between the two (Table 9.3) and thus belongs to both. Yet it differs in many crucial aspects from these analogous models and invites us to take a fresh look on emergence.

Emergent features have been proposed to resolve the old controversy about reductionism vs. holism. Reductionists insist on fundamental laws having universal validity, hence should explain every observable phenomenon, if only applied comprehensively. Holists object that each level of reality creates phenomena of its own, following laws independent of and underivable from those holding on lower levels: Quarks are irrelevant for psychoanalysis. In physics, various open-minded thinkers have expressed the experience that unexpected phenomena appear as solutions of fundamental evolution laws, if only a large number of strongly interacting subsystems is involved, and worked out its far-reaching consequences [And72, Leg92, Gel94, Lau95]. The idea, often paraphrased as "the whole is more then the sum of its parts", found committed approval from other sciences, notably biology [Lor78, Wil98]. Termed "fulguration" by Konrad Lorenz [Lor78] as early as in the 1970s, it is now generally referred to as emergence. In biology, it is invoked frequently to interpret phenomena that appear unexplainable in the standard framework of evolutionary genetics, cellular biology, and neural information processing, such as consciousness and culture.

Tree structures as sketched in Fig. 1.6 are implemented, for example, in the file organisation under Unix and related operating systems. We encounter them in numerous organization schemes in daily life, from texts in literature (Fig. 1.6) through corporate structures in business and political administration: Every unit at level n of the hierarchy decomposes into a number of subunits at level $n-1$, which decompose further into subunits at level $n-2$, etc. Like this, the construction is self-similar, at least in a statistical sense. As a consequence, information contents at each level just add, as argued in Sect. 1.5.

Emergence is much more than this structural skeleton: It implies strong relations between the subunits of a system. In physics, it is forces between its parts, often described by interaction potentials, that couple their dynamics and become manifest in statistical correlations. Already with a small number of degrees of freedom, interaction can drastically change the behaviour of the total system: Systems with a single degree of freedom are integrable, from two freedoms upward they can be chaotic. If the interaction is too weak, the whole decays into subsystems that are nearly autonomous. If it is too strong, in turn, parts can no longer be told apart, they disappear as distinguishable subunits, and the total system remains as a structureless single unit. The maximum of complexity is found at an intermediate degree of coupling [BP97]. This conclusion generalizes the extreme cases of subdivisions addressed in the context of correlations and mutual information (Sect. 1.7): A set of independent, noninteracting subsystems reduces to the sum of its parts. If, on the contrary, subsystems are so strongly coupled that they lose their identity, they merge into a rigid unstructured whole.

Given an interaction in the relevant intermediate regime, the total number N of participating subsystems becomes a decisive parameter. A surprising insight gained in the field of many-body physics is that with N taking thermodynamic values (of the order of 10^{20}), a new simplicity evolves that can by no means be anticipated, inferring from individual features of the subsystems. Prototypical examples are.

- **Phase transitions**: Crossing a critical temperature from the hot to the cold side, the system settles into a spatial structure characterized by higher order and lower symmetry than on the hot side. Crystals, for example, replace homogeneity and isotropy of the liquid by periodic order and fixed orientation of symmetry axes on the solid side, a phenomenon known as *spontaneous symmetry breaking* [Gol92].
- **Laser**: Instead of emitting photons spontaneously, with random timing and direction, the atoms of the gain medium excite each other mutually, resulting in synchronous emission of photons with identical frequency and phase. Natural light with its statistical distribution of phases is replaced by coherent radiation. Inspired by the laser, the concept of *synergetics* has been coined for a class of self-organized non-equilibrium phenomena [Hak83], all pertaining to the larger category of emergence.
- **Quantum Hall effect**: In a metal strip exposed to crossed magnetic and electric fields, conduction electron states are quantized, forming a sequence of equidistant Landau levels in energy. At sufficiently low temperature, these states combine coherently to macroscopic quantum states, each one characterized by likewise quantized values of the ratio of current along and voltage across the strip and the electric field, with the dimension of a resistance. The quantized resistance depends exclusively on fundamental constants (Planck constant h and elementary charge e) as $R_{H,n} = h/ne^2, n \in \mathbb{N}$, and is independent of specific parameters of the sample. Even as a function of the magnetic field, it is constant over large intervals of the field strength, forming a staircase of well-defined plateaus. Macroscopic quantization as in the quantum Hall effect has been featured as a prime example of macroscopic simplicity emerging from contingent microscopic conditions [Lau95].

Summarizing these and numerous similar phenomena, emergence in many-body physics is characterized by the following properties:

- **Thermodynamic limit**: Emergent order arises in systems comprising a very large number of interacting identical particles and is more pronounced the larger the number of subsystems. At the same time, it is independent of details of the subsystems, as long as they satisfy a few necessary conditions.
- **Contraction of state space**: The total space of possible configurations of the many-body system reduces to a small discrete set of stable states, robust against perturbations and parameter changes within wide limits. They form *collective modes* of the many-particle system, described by parameters and obeying laws that cannot even be defined in terms of single-particle properties, but must be compatible with the fundamental microscopic laws. The eigenmodes of coupled chains of harmonic oscillators sketched in Sect. 5.6.2 provide an elementary example.
- **Spontaneous symmetry breaking**: These stable states exhibit a high degree of order, replacing continuous symmetries such as homogeneity and isotropy

by discrete symmetries (e.g. periodicity in space and/or time) or more complex structures, with correspondingly low values of entropy.

The interface between continuous physical systems and discrete nominal states at the bottom of the computation hierarchy (Sect. 9.2.1) belongs to this same category, as discussed above: Bi- or multistability requires large systems to achieve the dissipation stabilizing discrete states, and each of these attractors breaks the symmetry of the continuous space it is embedded in.

Higher levels of the computation hierarchy, however, do no longer fit well in this scheme. A step from layer $n-1$ to layer n always connects software units to other, more encompassing software units, differing only in their degree of integration. Units at the lower level already show a specific functionality, they are not identical as in many-body physics, and the higher level does not just add a large number of subunits, but integrates a well-defined number of distinct *modules* into a highly organized network. In the software hierarchy, the characteristics of emergent phenomena listed above therefore come in a new guise [ED19]:

– **Interface specificity**: Instead of an interaction potential, the relation between modules is characterized by specific codes and channels of information exchange, forming an interface between modules and towards higher levels of integration. Different types of modules can be involved, each one complying a specific function.
– **Shielding**: The only relevant condition modules have to fulfill is their function for the higher level. Their inner working is irrelevant (e.g., gates can be defined in terms of input and output voltages of an electronic circuit or in terms of billiard balls entering and leaving a scattering process, see Fig. 9.6), what matters is only the functionality achieved on the higher level.
– **Emergent functionality**: Connected to form a network with specific topology, comprising nodes and links, the modules accomplish a function that does not exist and cannot even be defined on the level of single modules (e.g., binary gates combine to an adder), it is **platform independent**.

The increase in structural and functional complexity with each tier in the software hierarchy can be expressed in terms of the order relation "*A emulates (simulates) B*" (Sect. 9.1.2), applied to different levels: Higher levels can emulate lower ones, not vice versa.

Emergent simplicity on the top level as in many-body phenomena, reflected in invariance properties, appears also in the present context. This becomes particularly evident in a capability shared by all computing equipment: universal computation. Before attempting a more detailed explication, universality can be characterized by a global symmetry: Every universal machine can emulate every other, i.e., for two universal computers A and B, both $A \succ B$ and $B \succ A$ are valid. Every possible algorithm can be implemented on every universal computer. That is, the diversity of different computing capacities on lower levels converges against a single comprehensive quality at the tip of the pyramid: universality. For example, the universal

Turing machine can perform all the more restricted tasks mastered by special-purpose machines.

The construction of the universal Turing machine also provides a hint how this quality is achieved. It attains its universality by reading the instruction set of a particular Turing machine as an input from its tape and applying it by its own control unit. In this manner, the algorithm to be performed becomes part of the total input, the initial condition of the machine. The way the "input proper" is to be processed is encoded itself in the head of the input. The distinction between "program" and "input" dissolves.

It is instructive to contrast this ambiguity with a related distinction in physics. Applying the analogy "natural systems as information processors" directly to physical systems and computers suggests associating their principal components as follows (Table 9.4):

For example, in classical mechanics, the input would be given by an initial set of coordinates in phase space, the program by Hamilton's equations of motion, the output by a final phase-space point. To be compared with universal computation, though, the definition of the Hamiltonian that determines the equations of motion would have to be integrated in the initial condition as a contingent feature of the system. Different regions of the total state space would then be reached by choosing different parameter values of a "universal Hamiltonian", leading to distinct dynamical behaviour (e.g., harmonic oscillator, two-body system with central force, billiard, ...). If the physical system referred to happens to be itself a computer, the right column in Table 9.4 is not just analogous to the left column, it is a high-level description of the *same system*. It pretends the same validity as a very low-level account in terms of a microscopic model Hamiltonian and thus competes with it.

As an example how to attain the high-level description from the physics that goes on on the lower level, consider a standard quantum mechanical Hamilton operator for the electron–ion system underlying a solid-state device such as a MOSFET (Fig. 9.12). In general, it comprises the following terms [ED19]

$$\hat{H}(\hat{\mathbf{p}}, \hat{\mathbf{x}}, \hat{\mathbf{P}}, \hat{\mathbf{X}}) = \hat{\mathbf{T}}_e(\hat{\mathbf{p}}) + \hat{\mathbf{T}}_i(\hat{\mathbf{P}}) + \hat{\mathbf{V}}_{ee}(\hat{\mathbf{x}}) + \hat{\mathbf{V}}_{ei}(\hat{\mathbf{x}}, \hat{\mathbf{X}}) + \hat{\mathbf{V}}_{ii}(\hat{\mathbf{X}}) + \hat{\mathbf{V}}_{input}(\hat{\mathbf{x}}, \hat{\mathbf{X}}, t).$$
$$(9.4.1)$$

Table 9.4 A plausible scheme how to associate elements of time evolution in physical systems with essential steps of information processing in computers.

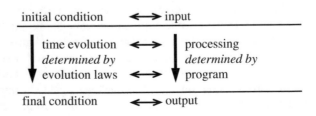

Momentum and position operators for N electrons and K ions are denoted as $\hat{\mathbf{p}} = (\hat{\mathbf{p}}_1, \ldots, \hat{\mathbf{p}}_N)$, $\hat{\mathbf{x}} = (\hat{\mathbf{x}}_1, \ldots, \hat{\mathbf{x}}_N)$, $\hat{\mathbf{P}} = (\hat{\mathbf{P}}_1, \ldots, \hat{\mathbf{P}}_K)$, $\hat{\mathbf{X}} = (\hat{\mathbf{X}}_1, \ldots, \hat{\mathbf{X}}_K)$, resp., and

$$
\hat{T}_e(\hat{\mathbf{p}}) = \sum_{\text{electrons } n=1}^{N} \frac{\hat{\mathbf{p}}_n^2}{2m_e}, \quad \hat{V}_{ee}(\hat{\mathbf{x}}) = \sum_{\text{electrons } n,n'=1}^{N} \frac{e^2}{4\pi\varepsilon_0 |\hat{\mathbf{x}}_{n'} - \hat{\mathbf{x}}_n|},
$$

$$
\hat{T}_i(\hat{\mathbf{P}}) = \sum_{\text{ions } k=1}^{K} \frac{\hat{\mathbf{P}}_k^2}{2M_k}, \quad \hat{V}_{ii}(\hat{\mathbf{X}}) = \sum_{\text{ions } k,k'=1}^{K} \frac{Z_k Z_{k'} e^2}{4\pi\varepsilon_0 |\hat{\mathbf{X}}_{k'} - \hat{\mathbf{X}}_k|},
$$

$$
\hat{V}_{ei}(\hat{\mathbf{x}}, \hat{\mathbf{X}}) = \sum_{\text{ions } k=1}^{K} \sum_{\text{electrons } n=1}^{N} \frac{Z_k e^2}{4\pi\varepsilon_0 |\hat{\mathbf{X}}_k - \hat{\mathbf{x}}_n|}. \tag{9.4.2}
$$

Here ε_0 denotes the vacuum permittivity, e and m_e charge and mass of the electron, Z_k and M_k proton number and mass of ion k, resp. The terms refer to kinetic energies, $\hat{T}_e(\hat{\mathbf{p}})$ and $\hat{T}_i(\hat{\mathbf{P}})$, of electrons and ions, and internal Coulomb interaction potentials, $\hat{V}_{ee}(\hat{\mathbf{x}})$, $\hat{V}_{ii}(\hat{\mathbf{X}})$, and $\hat{V}_{ei}(\hat{\mathbf{x}}, \hat{\mathbf{X}})$ within and between the two subsystems. The time-dependent potential $\hat{V}_{input}(\hat{\mathbf{x}}, \hat{\mathbf{X}}, t)$ represents the contingent user input to the computer, in the form of gate voltages at transistors and other externally controlled potentials. Already at this stage, disregarding all subatomic structures and interactions that become relevant at even lower levels and smaller scales, this Hamiltonian represents by no means the most fundamental account (which would address, e.g., the internal structure of atomic nuclei). Turning towards macroscopic variables and concepts, various assumptions, approximations, collective phenomena come into play that are indispensable to ascend beyond the microscopic level. In order to advance, for example, from the Hamiltonian (9.4.1) to the discretized response of a single transistor and thus to a description in terms of Boolean logic (Fig. 9.12), the following major steps are required:

- The **Born–Oppenheimer approximation** allows to separate the dynamics of electrons from that of the ions, neglecting correlations implied by quantum, based on a separation of characteristic time scales of the two subsystems.
- **Bloch theory** introduces energy bands as a feature of quantum systems with an idealized perfect discrete translation symmetry [AM76].
- Combined with **Fermi–Dirac quantum statistics** [Rei65] applied to electrons in semiconductors, it leads in particular to postulating the existence of **conduction and valence bands**.
- Electrons as elementary charge carriers are replaced by generalized particles: So-called quasiparticles combine, e.g., "naked" electrons with comoving excitations of the ion lattice (phonons) to form **dressed electrons**. Unoccupied electronic states in the valence band are considered as positive charge carriers of their own right, called **holes**.

A crucial insight is that these concepts and the theories they are embedded in cannot be derived by mere logical and mathematical inference from fundamental theories [Leg92, ED19]. In particular, the conception of suitable models, a key step in the construction of these theories, is a creative act that cannot be reduced to analytical deduction.

Even if the relevant spatial dimensions of the structures on a silicon wafer shrink to the order of a few nanometres, the numbers of electrons, N, and of ions, K, to be included in Eq. (9.4.1) are huge and may amount to several millions in a single gate. Already calculating the quantum mechanical ground state of a homogeneous sample of such a material, let alone excited states involving a non-trivial time dependence, is a formidable task that requires sophisticated analytical and numerical techniques such as **density functional theory**. In addition, in a device such as a MOSFET, this many-particle system is living in an extremely complex background potential, with p- or n-doped semiconductors combined with normal metal conductors in carefully designed and crafted geometries. At the microscopic level, coupled to the rest of the device acting as a heat bath, it has to be treated as an open thermodynamical system. Moreover, it is subject to a macroscopic time-dependent external driving, represented by the term $\hat{V}_{input}(\hat{\mathbf{x}}, \hat{\mathbf{X}}, t)$ in the Hamiltonian (9.4.1, 9.4.2), which contains all the complexity and contingency of the users' actions and other sources of input.

The repertoire of solutions for microscopic states and their time evolution, available for this open time-dependent dynamical system, is immense. Among them, the discrete macroscopic nominal states obeying Boolean logic stand out as a tiny minority. Still, the repertoire of these solutions is sufficient to allow for a behaviour as complex as universal computing.

9.4.3   Emergent Dynamics: Vertical Information Flow and Downward Causation

As the preceding discussion shows, the relationship between a microscopic physical description of the hardware and an account of the software in the conceptual framework of computer science cannot be grasped in simple rigorous terms. The notion of emergence just barely allows us comprehending the logical nature of this relation. However, what about physical dependencies between these levels? The question refers in particular to the exchange of information, maybe the only quantity that applies to both levels. The clear scheme developed in Sects. 5–7, of vertical top-down (macro–micro) and bottom-up (micro–macro) flows, linked by a horizontal interchange among microscopic freedoms, no longer captures the complexity of computers. For example, the phenomenon of shielding impedes ascending "vertical" information currents across the interface separating continuous physical from discrete logical processes, while the very functioning of computers requires macroscopic input to directly take effect on the microscopic level, as manifest in the term $\hat{V}_{input}(\hat{\mathbf{x}}, \hat{\mathbf{X}}, t)$ in the Hamiltonian (9.4.1).

The question concerns in particular the directed information flow involved in causal relations between the levels. It may appear obvious how to include a computer in a causal chain: A gust hits an airplane, the navigation system detects a deviation from the cruising altitude, the autopilot running on the on-board computer actuates the elevator, the airplane returns to cruising altitude. Both cause and effect are tangible macroscopic physical events, but mediated by a computing device controlled by an elaborate software (and without any direct human intervention). How exactly does the information propagate within the on-board computer?

The problem becomes more concrete as we take the hierarchy of computing levels (Table 9.3) into account. The input from the altimeter arrives, the command to the elevator leaves at a high level, coded as digital signals within the on-board electronics. Being a physical system, the processing inside the computer is determined by the quantum dynamics epitomized by a Hamiltonian such as (9.4.1, 9.4.2). At the same time, it follows the algorithms of the auto flight system. This suggests to conclude that the information enters at a top level, to be transmitted down to the level of electron motion determined by microscopic dynamical laws, a process referred to as *downward causation* [ED19]. From there, it ascends again to trigger the output at the top level, a process referred to accordingly as *upward causation*. It is downward causation, in particular, that gives reason for concern: Can the result of an "IF" condition in a program line alter the dynamics of electrons in a MOSFET transistor, which is already determined by the quantum mechanics of this semiconductor system?

We encounter comparable situations in all complex many-body systems that support macroscopic collective modes as emergent phenomena. Information entering as an external forcing is delegated to different levels of the structural hierarchy according to its spatial and temporal characteristics. Typically, large wavelengths and/or periods couple to low modes, high frequencies to fast modes, and trigger response at the same level. An instructive model is the chain of coupled harmonic oscillators, described in Sect. 5.6.2 in terms of its normal modes. For long chains, they range from "microscopic" length and time scales, corresponding to individual oscillators, up to "macroscopic" scales involving all or a large fraction of them. A "macroscopic input" driving the chain at a low frequency or long wavelength will excite predominantly the slow collective modes and, as they are defined to be separable, produce a likewise "macroscopic output" at similar scales.

As a more tangible example, imagine pushing a block of ice over the ground. The force you are applying moves the entire crystal as a whole. The causal relation between force and motion is neatly restricted to the macroscopic level, single electrons, O^- or H^+ ions forming the ice crystal are tagged along but do not participate as individual subsystems in the physical process. Talking here of downward or upward causation appears futile, since the information does not propagate between scales. Looking more closely, we observe that friction with the ground will heat the lower surface of the block and lead to partial melting, which in turn can reduce the friction coefficient and alter in turn the response of the whole block to the external force. Microscopic degrees of freedom are now more directly involved. Causal information transfer occurs within as well as among structural levels, depending on physical properties of the signal and the system.

Vertical information flow, such as switching between input and output at the software level and processing at the microscopic level in computers, would be unproblematic if the relation between the respective state spaces were one-to-one. In the case of the oscillator chains, for example, going from individual oscillators to normal modes is a mere reversible change of representation. The computing hierarchy, however, is situated at the opposite extreme. Descending from application software down to electron motion, the number of possible implementations of a higher-level operation on the next lower level multiplies at each step, ending with a vast variety of electronic processes that could correspond to the execution of a single program line.

In a generic thermodynamical system, see the discussion in Sect. 7.2, the existence of a multiplicity of microstates compatible with a given macrostate is reflected in fluctuations and noise. Causal relations between macroscopic states are no longer deterministic, they attain a probabilistic character. In precise physical terms, this applies also to the causal relationship between input and output. That the computer follows faithfully the deterministic evolution dictated by some algorithm, unaffected by any microscopic noise, is the successful result of highly sophisticated technology. It is true only within the narrow corridor of operating conditions of the device at hand and only with extremely high probability, not with absolute certainty. With low battery, at 50°C temperature or higher, or after hitting the computer with a sledgehammer, the system returns to quantum statistical mechanics as the only applicable law.

Even taking exceptional direct effects of microscopic processes into account, the label "causation", downward or upward, is not quite adequate. In causality, time order is a crucial condition (see Sect. 3.1). However, low-level electronic processes do not follow high-level logical operations with any time delay implied by a cause-effect relation. Rather, they are manifestations, not consequences, of these operations, described in a different language but occurring strictly simultaneously with them.

In computers, the relevant high-level modes are not static entities, they form dynamical systems of their own right, and of extreme complexity, culminating in universal computation. Nevertheless, they have causal power in the sense outlined above. Computing does determine part of what is going on at the microscopic level. However, this is only possible as far as it complies with the fundamental physical laws valid on that level.

Summarizing, information flows involved in computing are organized in two strictly separated layers (Fig. 9.31): Processing of digital information, related to the computer's function, occurs in its discrete nominal state space. It receives digital input from keyboard, touchscreen, mouse, accessories, and numerous other channels, transforms it on the various levels of the functional hierarchy outlined above, from application software down to individual logical gates implemented in integrated circuits, to result in a likewise digital output through the screen, through loudspeakers, printers, and other channels including the network. Simultaneously, the computer hardware is exposed to other physical input, contingent in the sense that it is independent of the digital nominal input, thus of the intentions of the user. It comprises countless sources, heat from the surrounding atmosphere, fluctuations of the supply voltage or the internal battery power, mechanical shocks, e.g., during

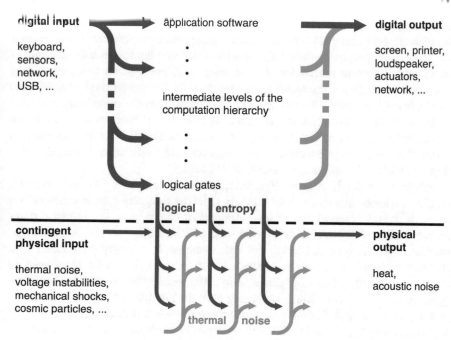

Fig. 9.31 Principal information flows in computers. Shielding results in a sharp separation between information currents related to the processing of digital information, distributed over various tiers of the functional hierarchy, from application software down to logical gates (Table 9.3), and entropy flows generated by all other physical processes occurring in the hardware. The interface between the two levels (bold horizontal line) is semipermeable: It allows logical information generated by unavoidable erasure in the digital information processing (and other entropy produced by dissipative electron motion in metals and semiconductors) to be absorbed as heat in the hardware and the environment, but protects logical operations against perturbation by thermal noise and quantum indeterminacy.

transport, energetic cosmic particles triggering bit flips in a memory, etc. In the same vein, the computer produces physical output, heat and infrared radiation produced in the CPU and other power-consuming components, the noise of ventilator and hard disk, ... The computer is designed so as to protect the digital information processing against perturbations, reducing a bottom-up flow of this entropy to a negligible minimum. That's shielding. At the same time, however, digital computing itself produces extra entropy, different from its output proper: Irreversible logical operations inevitably release entropy (Sect. 9.2) that merges as heat, $k_B T \ln(2)$ per erased bit, with that generated by dissipative electronic motion in the computer's circuits and other sources. That is, the interface between the two layers is semipermeable, it allows top-down entropy currents to enter the hardware and the surrounding atmosphere as global heat sinks, but blocks stochastic bottom-up perturbations by thermal noise.

From a physical point of view, information processing in computers of course bears on quantum theory. The Hamiltonian operator given in Eqs. (9.4.1, 9.4.2) as a microscopic model for the physics of chips and memory modules is a patent example. Semiconductor technology is based on solid-state physics; it relies in essential aspects on material properties that can only be understood and harnessed with the help of many-body quantum mechanics. At the same time, emblematic quantum effects, such as uncertainty and nonlocality, addressed in Chap. 8, are incompatible with the principles of classical digital information processing. They are undesirable and must be blocked out, an objective that is increasingly hard to achieve as miniaturization advances into the few nanometres range. Quantum computation, in turn, attempts to utilize these same quantum effects as a source for computing instead of suppressing them, as will be discussed in the subsequent Chapter.

While at the lower edge of Fig. 9.31, quantum mechanics has been omitted, another pertinent extension is missing on top. The integration of computers in a network appears there as one of several input/output channels, but in fact it plays an outstanding role. Notably the "world wide web" constitutes a new level of organization with the potential to lead to new qualities. It not only goes along with the simultaneous global interconnectedness of its users, it can be considered as the decisive motor promoting globalization. We should be prepared for unknown phenomena, for better or worse, emerging from the interplay of global connectivity of computers and global communication among users in a similar way as individual intelligence emerges from the complex connectivity of neurons. Are we on the verge of a new quality of collective thinking, of "swarm intelligence" [Lev99, Szu01]? In conclusion, computers, brainchildren of human ingenuity themselves, complement and augment the role of human intelligence as a catalyzer of entropy conversion on planet Earth.

The particular structure of information flows and causal relations we find in the computing hierarchy could serve as a model to understand other emergent phenomena, in particular in biological information processing. The brain as a natural computer: Since its advent, digital computing has stimulated the neurosciences, as an inspiring analogue of consciousness, emotion, conscience, and other high-level features of neural information processing. The emergent determinism of algorithms running on electronic platforms, far above the underlying physics with its proper rigorous laws, could even provide a fresh approach to the mind–body problem.

However, there are fundamental differences between the principles reigning in the two realms that cannot be ignored [ED19]:

– In biological systems, there is no such clear separation of two layers of information flows as outlined above. Their structural hierarchy appears rather as a nearly continuous sequence of levels. In particular, hardware and software merge gradually in the brain. The discharge of a neuron is an all-or-nothing event and thus of binary nature. This discreteness, however, is already sacrificed at the next level where not individual spikes but their frequency determines the response of a downstream neuron. Conscious decision making, in turn, is nearly independent of the spiking of individual neurons, but it does depend, for instance, on the concentrations of hormones and neurotransmitters, thus involving chemistry down to the molecular level. These and many other signalling channels integrate it in a living organism with its individual needs, and the individual in a society of conspecifics and in a more comprehensive ecosystem. As a consequence, vertical information flows, coupling neural information processing to higher and lower levels of biological organization, are vastly more important and more versatile than in computers.

- Both ontogeny and phylogeny are totally different. While a computer is *assembled* from component parts according to a pre-existing plan ad configured before it could function for the first time, a nervous system *develops*. To be sure, the genome controls most of its crucial features, from the molecular level to the global organization, but much of its structure arises by self-organization. Function and construction go hand in hand, the timescales of functional dynamics and restructuring of the network overlap. Machine learning does not call this distinction into question; it only works on platforms that are already functioning and cannot modify their hardware, while continuous learning in the brain does exactly that.

- As a product of evolution, the nervous system, an organ dedicated specifically to the communication and evaluation of information, formed on a timescale of millions of years. Even the human brain took some 100,000 years to reach its present state. In comparison, the several decades of history of digital computer technology shrink to an instant. The brain evolved and still evolves by mutation and selection in the "struggle for survival" in a natural environment. To be sure, also technological development involves trial and error. Yet the instantaneous appearance of the computer on evolutionary timescales can only be explained by its targeted design, based on scientific knowledge. They integrate in evolution, not as one of its direct products but as a tool invented by a particular species. Computers accurately serve specific purposes set externally and determining their design and function. Brains do not.

- Structures and dynamical parameters of computers can readily be observed and modified on all levels of hardware and software. By contrast, sciences of brain and mind confront a formidable challenge concerning the empirical test of their hypotheses. Direct experimental intervention is highly problematic. On the level of single neurons and neuron assemblies, recording allows some experimental observation and control, as do techniques like neuroimaging and EEG on larger scales. On the subjective side, self-observation and conversation open access to the mind, but they are notoriously unreliable and hardly allow any correlation with neurological data.

For these reasons, conclusions by analogy from computers to brains, from software to minds and vice versa, are inspiring but should be handled with care.

Chapter 10
Quantum Computation

Classical computers create a new world of its own, anchored in physical reality but hovering far above the laws of solid-state physics and electronics. Digital computing, in particular, reduces the continuous state space of classical physics to an artificial high-dimensional space of discrete nominal states, following their own rigorously deterministic evolution laws. Yet this space is versatile enough to simulate just about every phenomenon in the real world it is embedded in. Quantum computation owes itself to a completely different strategy. It aspires to utilize directly the discreteness of physics at atomic scales as a resource for computing. However, as elucidated in Chap. 8, quantization is not discretization. Hilbert space, the geometrical framework of quantum mechanics, is continuous. Discreteness comes in only indirectly, through the finite information capacity of small isolated systems. While in classical computing, the principles of the digital world are nearly independent of those of the underlying physics and are subject of a discipline of its own, computer science, it is quantum mechanics itself that dictates the rules of quantum computing. Quantum computation *is* applied quantum mechanics.

According to a common view, long-term tendencies in the development of computer hardware, epitomized in Moore's law, lead directly into quantum computing. As soon as structures created on a silicon wafer reach nanometer scales, quantum effects start posing a real threat to conventional computing. At the same time, quantum physics offers us discretization—viz. quantization—for free! So why not embrace this gift, drive miniaturization to the ultimate consequence, and write a bit on a single atom to do computation the quantum way from the very beginning?

This is often seen as the rationale behind quantum computation, repeated in countless variations in books [Llo06], articles, talks, PowerPoint presentations all over the world. We should not forget, however, that "conventional" digital computing already depends heavily on advanced quantum mechanics. Without the quantum theory of semiconductors, the concepts of valence and conduction bands, of electrons and holes, and the construction of the transistor (see Sect. 9.2.1) would not have been possible. Reducing features on a chip to the size of few atoms approaches the limits of validity of bulk solid-state theory. It requires a new approach, focussing on individual small systems.

Quantum computation rests on a distinct idea even of information in the quantum realm and requires an entirely new approach to computation. Attempting to take maximum advantage of the peculiarities of quantum mechanics, quantum computation highlights their consequences and forces us to understand them at all depth. Trying to overcome established strategies of classical computation, this effort also sheds light on what is disposable on that side, and which principles are of even deeper significance than expected.

At a stage where the initial quantum computation hype is waning and gives way to routine development, it cannot be the purpose of this Chapter to join in the praise. Rather, it is intended to reconsider information processing in quantum systems from a new angle and provide a sufficiently clear view of the underlying principles to enable a critical reflection on quantum computation. It is definitely not intended as a general introduction to quantum computing. In fact, the market offers a broad range of references from introductory works on a popular level [Llo06, Sei06] through comprehensive accounts including all technical details [NC00, BCS04].

10.1   What's so Special About *Quantum* Computing?

As a first, modest step from conventional towards quantum computation, one might think of hybrid solutions that exploit some particular quantum effect as a complementary resource. A prominent example is spintronics: Based on standard semiconductor technology, it attempts to include the electron spin as an internal degree of freedom, representing one bit, in addition to the usual voltage levels. This requires to generate and control currents with a significant polarization, stable on technologically relevant timescales, and to refine components such as transistors and memory elements so that they become sensitive to polarized currents. Evidently, magnetic fields and magnetic materials play a key role in such a technology.

Quantum computation goes much further. It attempts to unconditionally adopt the way nature is processing information at its very bottom, rethinking all the concepts established in conventional computation: Above all, what exactly is it that protons, neutrons, electrons, and all the other microscopic degrees of freedom are processing? If anything to be called "quantum information" can be meaningfully defined, see Chap. 8, it is not tied to facts and distinct from what we understand as information in the classical world. Can we interpret quantum mechanical time evolution as logical operations? Inhowfar can we compare them with logical gates? If so, which novel algorithms, different from all those running on classical computers, can we construct from them? Can quantum mechanics be harnessed so as to faithfully follow a predefined sequential program?

Every classical computer is basically quantum. Here, it is the quantum physics of many-body systems (Sects. 8.2.3, 8.3.3, and 8.6) together with a sophisticated technology that keeps quantum effects out of emergent classical facts. Then, if quantum computing makes positive use of precisely those coherence effects suppressed in classical computers, does every chunk of matter comprising a sufficient number of degrees of freedom qualify as a quantum computer? In fact, quantum computing also singles out a special narrow segment, now of quantum mechanics, as the privileged framework of its nominal states and processes, namely the unitary quantum mechanics of small closed systems (Sects. 8.3.1 and 8.3.2). It replaces the high-dimensional Cartesian discrete lattice of classical computing by Hilbert spaces with the topology of continuous hyperspheres in likewise high dimensions. In this sense, quantum computing is much less rigidly discrete than classical computation!

In particular, the conservation of entropy in unitary transformations forces quantum computation to abandon a basic principle of conventional computation. There, besides all common loss effects, non-invertible logical operations generate an irreducible amount of excess entropy that has to be disposed of. Conventional computing maintains a steep entropy gradient towards the environment to extract this waste and dump it as heat into microscopic freedoms. Quantum computing already works on the lowest level, there is no more sewer under the floor to dump trash to. It therefore has to restrict itself to invertible logical operations, see Sect. 9.2.2. At the same time, relying on coherence as a condition *sine qua non*, quantum computing has to block out all the other microscopic physics that unrelentingly pushes the computer towards classical behaviour and now goes on at the same level as the logical operations proper. A decisive question therefore is: Is it possible to isolate that microscopic piece of matter we want to use for quantum computation sufficiently from its environment, to prevent it from participating in the process?

Employing unitary quantum mechanics as the computing medium, another fundamental problem arises, which appears similarly in quantum mechanics in general: We can access and control a quantum computer only from our elevated position as macroscopic objects, preferably through a classical computer serving as interface. Since even the very representation of the relevant information is incompatible between the two levels, we have to create interfaces that translate the classical (e.g., binary) alternatives into whatever is processed in a quantum computer, and back. Their task is comparable to the preparation of quantum states, for the input, and to their measurement, for the output; they inevitably have to penetrate the splendid isolation required for the processing proper, at least once at the beginning and once at the end of each calculation.

The previous section introduced an order relation that allows to classify information processing systems in a hierarchy, including Turing machines and culminating in universal computing. How does quantum computing, the new kid on the block, integrate in this hierarchy? Classical computation goes on in quantum systems satisfying very special conditions, but it cannot be meaningfully considered as classical limit of quantum computing. The structural levels of the two technologies, the conceptual frameworks describing them, are just too different. A more appropriate comparison is suggested by the general criterion proposed to classify information processing systems: Can every quantum computer be simulated by a classical computer? Or every classical by a quantum computer? Or neither one of these relations? Or both?

In fact, such considerations were at the cradle of quantum computation. As early as 1982, Richard Feynman contemplated about quantum computers, at that time a completely fictitious concept, as an alternative solution employing two-state systems to simulate arbitrary other quantum systems [Fey82, Fey86], not yet as an option for general-purpose computing. His idea experienced a major impulse by a purely classical advance, the insight that reversible logic can be implemented [Ben73, Tof80, Tof81, FT82, Ben82] and allows for computation with arbitrarily low consumption of free energy (Sect. 9.2.2). Reversible classical gates then served as models to elaborate quantum gates performing reversible unitary transformations. They proved sufficient for universal computation. David Deutsch even conjectured that a stronger version of the Church-Turing thesis, cf. Sect. 9.3.1, referring explicitly to the physical implementation of algorithms, applies to a universal quantum computer, but not to the classical universal Turing machine [Deu85].

Stimulated by this theoretical progress, quantum algorithms were worked out that do not only match conventional classical computation, but are superior to it in efficiency or even cope with problems that cannot be solved efficiently on a classical computer. Successes such as Peter Shor's algorithm for prime factorization and Lov Grover's search algorithm gained particular fame. They employ techniques, e.g., quantum parallelism, based on quantum superposition and hence lacking a classical counterpart, to gain an uncatchable lead on classical information processing.

In parallel, substantial progress has been reached in the development of suitable quantum hardware. Challenges are on the one hand, to protect quantum coherence against perturbing impacts from the environment, and on the other, to allow for a limited but efficient access from the classical level to prepare the input and extract the output of an otherwise quantum algorithm. While the design of gates, circuits, and algorithms bears on the unitary quantum mechanics of small closed systems, these problems direct attention back to the dirty physics of quantum many-body systems with interactions that can at best be described statistically. Various options have by now been developed how to implement quantum computers in real systems that offer a reasonable performance in both respects, low decoherence and easy control. Since coherence can only be sustained for a finite time, typically insufficient for the desired minimum number of operations, other complementary strategies have to be found to reduce the unavoidable errors induced by incoherent processes to a tolerable level. Under the label of quantum error correction, various strategies have been proposed to protect the intended progression of quantum algorithms.

In what follows, this development is recapitulated from bottom up. Beginning in Sect. 10.2 with elementary units, quantum bits and basic reversible gates that are equivalent their classical counterparts, Sect. 10.3 advances into tasks where quantum computation gains a lead, presenting particular algorithms that take advantage of ingenious artifices such as quantum dense coding and quantum parallelism. Section 10.4 addresses the central threat to quantum computing, decoherence, and different strategies to reduce noise and correct errors where they cannot be avoided. Finally, Sect. 10.5 returns to harsh real life, presenting and comparing various alternatives to implement quantum computing in physical platforms and indicates the state of the art.

10.1   Tools for Quantum Computation: Qubits and Quantum Gates

10.2.1   The Qubit

If Shannon information is applied to classical physics, the bit figures as one possible unit among others to measure entropy, for example thermodynamic units. In quantum mechanics, replacing Shannon information with von Neumann entropy S_{vN} (Sects. 8.2.1 and 8.2.2), it attains a fundamental meaning. As the maximum of S_{vN}, reached in case of total ignorance about the state of a quantum system, is related to its Hilbert space dimension $D_{\mathcal{H}}$ by $S_{vN} = \mathrm{lb}(D_{\mathcal{H}})$, the absolute lower bound of this information capacity is defined by two-state systems (Sect. 8.1.2) with $D_{\mathcal{H}} = 2$, hence $S_{vN} = \mathrm{lb}(2) = 1$.

For quantum mechanics, therefore, the bit is more than just a unit, it refers to an information atom, the smallest possible and in this sense elementary quantum system. In view of this fundamental role, the term "qubit" has been introduced. Its meaning is somewhat ambiguous, though: Considered as a mere unit of information, applied to quantum systems, it is equivalent to the bit. If it refers to two-state systems with all the related quantum physics, it is a synonym for this known concept. According to Occam's razor, it is then dispensable. To be sure, the term is useful to remind us of the fundamental differences between binary logic and Boolean algebra, resting on the *tertium non datur* (Sect. 2.1.1) the bit alludes to, and the quantum mechanics of two-state systems associated to the qubit.

The principle of linear superposition, so crucial for quantum mechanics, implies that there is a continuous transition between the two basis states, say $|0\rangle$ and $|1\rangle$, of a two state system (cf. Eq. (8.1.8),

$$|\psi\rangle = \alpha_0|0\rangle + \alpha_1|1\rangle, \alpha_0, \alpha_1 \in \mathbb{C}, |\alpha_0|^2 + |\alpha_1|^2 = 1. \tag{10.2.1}$$

For a classical alternative, states a and $\neg a$, a continuous transition could be construed as well by introducing complementary probabilities $p(a)$ and $p(\neg a) = 1 - p(a)$, $0 \le p(a), p(\neg a) \le 1$, for the two states,

$$A(p(a)) = a \times p(a) + \neg a \times p(\neg a), p(a), p(\neg a) \in \mathbb{R}^+, p(a) + p(\neg a) = 1, \tag{10.2.2}$$

the " $+$" interpolating between logical conjunction and disjunction as in fuzzy logic. It suggests identifying $p = |\alpha_0|^2, q = |\alpha_1|^2$. That is misleading, however—the superposition (10.2.1) has a fundamentally different meaning. In particular, while classically the alternatives a and $\neg a$ are uniquely defined as extremes of the range connecting them, the basis states $|0\rangle$ and $|1\rangle$ can be readily replaced by other equivalent bases, for example rotating them by $\pi/4$ in their two-dimensional Hilbert space $\mathcal{H}^{(2)} \subset \mathbb{C}^2$,

$$|g\rangle = \frac{1}{\sqrt{2}}(|0\rangle + |1\rangle), |e\rangle = \frac{1}{\sqrt{2}}(|0\rangle - |1\rangle) \Leftrightarrow$$

$$|0\rangle = \frac{1}{\sqrt{2}}(|g\rangle + |e\rangle), |1\rangle = \frac{1}{\sqrt{2}}(|g\rangle - |e\rangle). \tag{10.2.3}$$

Rotations such as this one (it will be featured below as the Hadamard gate) represent unitary transformations and are central in the construction of gates for computation with two-state systems (Sect. 10.2.2). More generally, these rotations form a continuous family parameterized by an angle $\theta, 0 \le \theta < 2\pi$, specifying the direction in the two-dimensional Hilbert space,

$$|\psi\rangle = \exp(i\alpha)\Big(\cos(\theta)|0\rangle + \sin(\theta)|1\rangle \Big), \tag{10.2.4}$$

The global phase α has no physical meaning as long as the state $|\psi\rangle$ is considered alone. As soon as interference with other states is involved, however, it gains fundamental significance for quantum computation.

Classical probability does come into play in the broader framework of mixed states, represented by the density operator (Sect. 8.2.1). While they are not required to describe logical operations proper on qubits, they become indispensable to take account of all those processes that deteriorate the desired advance of a quantum computation. For a two-state system, in the representation in the basis $\{|0\rangle, |1\rangle\}$, it takes the form of a 2×2-matrix,

$$\hat{\rho} = \begin{pmatrix} \rho_{00} & \rho_{01} \\ \rho_{10} & \rho_{11} \end{pmatrix}, \rho_{ij} := \langle i|\hat{\rho}|j\rangle, i, j \in \{0, 1\}. \tag{10.2.5}$$

Restricting the density matrix to be normalized, $\mathrm{tr}[\hat{\rho}] = 1$, Eq. (8.2.6), and Hermitean, $\hat{\rho}^\dagger = \hat{\rho}$, Eq. (8.2.7), the number of independent real parameters reduces to three. They can be defined as components of a three-dimensional vector,

$$\mathbf{a} = (a_1, a_2, a_3), a_1 := \rho_{01} + \rho_{10}, a_2 := i(\rho_{01} - \rho_{10}), a_3 := \rho_{00} - \rho_{11}, \tag{10.2.6}$$

so that

$$\hat{\rho} = \frac{1}{2}(\hat{I} + \mathbf{a} \cdot \hat{\boldsymbol{\sigma}}), \tag{10.2.7}$$

$\hat{\boldsymbol{\sigma}}$ denoting the vector formed by the three Pauli matrices, Eq. (8.1.9),

$$\hat{\boldsymbol{\sigma}} = (\sigma_1, \sigma_2, \sigma_3), \sigma_1 = \begin{pmatrix} 0 & 1 \\ 1 & 0 \end{pmatrix}, \sigma_2 = \begin{pmatrix} 0 & -i \\ i & 0 \end{pmatrix}, \sigma_3 = \begin{pmatrix} 1 & 0 \\ 0 & -1 \end{pmatrix}, \tag{10.2.8}$$

and thus

$$a_n = \text{tr}\left[\hat{\rho}\hat{\sigma}_n\right] = \langle\hat{\sigma}_n\rangle, \, n = 1, 2, 3. \tag{10.2.9}$$

The fact that it is three matrices plus $\hat{\sigma}_0 := \hat{I}$ that form a basis of the space of Hermitean (2×2)-matrices should not be interpreted as a manifestation of three-dimensional physical space, as the alternative notation $\hat{\sigma} = (\sigma_x, \sigma_y, \sigma_z)$ insinuates. In the present context of (2×2)-density matrices, the three components $\hat{\sigma}_n$ are rather related to different symmetry properties of the mixed state, e.g., a_1 and a_2 measure coherence, a_3 measures the degree of polarization, see also Sect. 8.3.3.

The condition $\text{tr}\left[\hat{\rho}^2\right] \leq 1$ implies that $|\mathbf{a}|^2 \leq 1$, i.e., the tip of $\mathbf{a}$ is confined to lie on or inside the surface of a unit sphere, called the *Bloch sphere*, around the origin of the space $\mathbf{a} \in \mathbb{R}^3$ (Fig. 10.1). Specifically, if $\hat{\rho}$ represents a pure state, $\hat{\rho} = |\psi\rangle\langle\psi|$, Eq. (8.2.10), $|\mathbf{a}|^2 = 1$. If the system is completely depolarized, $\rho_{01} = \rho_{10} = 1/2$, then $|\mathbf{a}|^2 = 0$. In spherical coordinates, these relations read,

$$\mathbf{a} = a(\sin(\theta)\cos(\phi), \sin(\theta)\sin(\phi), \cos(\theta)), \, a := |\mathbf{a}|,$$

$$\hat{\rho} = \frac{1}{2}\begin{pmatrix} 1 + a\cos(\theta) & a\sin(\theta)\exp(-i\phi) \\ a\sin(\theta)\exp(i\phi) & 1 - a\cos(\theta) \end{pmatrix}. \tag{10.2.10}$$

In compliance with classical binary logics, pairs, triples, quadruples ... of qubits play a prominent role among higher-dimensional Hilbert spaces. The Hilbert space of a system comprising N qubits can be written as a Cartesian product,

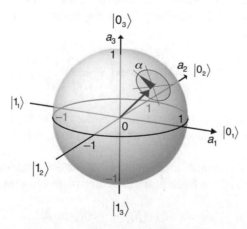

Fig. 10.1 The Bloch sphere represents the density operator for a two-state system as a point on the surface or inside of a unit sphere in three dimensions, Eqs. (10.2.6 and 10.2.7). The Bloch vector $\mathbf{a}$ = (a_1, a_2, a_3) (blue) contains the expectation values of the Pauli matrices, $\mathbf{a} = \langle\hat{\sigma}\rangle$. It reaches the surface if the system is in a pure state, $|\mathbf{a}| = 1$, otherwise ($|\mathbf{a}| < 1$, mixed state) it remains inside the sphere. The Bloch sphere represents the projective Hilbert space, hence is blind for global phases α, cf. Equation (10.2.4), of pure states, indicated here as a pink flag.

$$\left(\mathcal{H}^{(2)}\right)^N = \underbrace{\mathcal{H}_1^{(2)} \otimes \mathcal{H}_2^{(2)} \otimes \mathcal{H}_3^{(2)} \otimes \cdots \otimes \mathcal{H}_N^{(2)}}_{N \text{ factors}} = \bigotimes_{n=1}^{N} \mathcal{H}_n^{(2)}, \tag{10.2.11}$$

denoting the two-state Hilbert space as $\mathcal{H}_n^{(2)}$, $n = 1, \ldots, N$, and has dimension $D_{\mathcal{H}} = 2^N$. Other finite Hilbert space dimensions, not given by a power of 2, are of course possible but of minor relevance in quantum computation.

The canonical basis for an N-qubit system is the product basis $\{|0\rangle, \langle 1|\}^N$. For $N = 2$, for example, it reads

$$|0\rangle = \begin{pmatrix} 1 \\ 0 \\ 0 \\ 0 \end{pmatrix} = |0_1, 0_2\rangle, |1\rangle = \begin{pmatrix} 0 \\ 1 \\ 0 \\ 0 \end{pmatrix} = |0_1, 1_2\rangle, |2\rangle = \begin{pmatrix} 0 \\ 0 \\ 1 \\ 0 \end{pmatrix}$$

$$= |1_1, 0_2\rangle, |3\rangle = \begin{pmatrix} 0 \\ 0 \\ 0 \\ 1 \end{pmatrix} |1_1, 1_2\rangle \tag{10.2.12}$$

For special purposes, bases different from the canonical product basis may be more appropriate. Systems composed of N qubits are to be compared with classical state spaces comprising 2^N discrete states. However, the geometry, even the topology, and in particular the metric, in these spaces are fundamentally distinct. While in classical discrete spaces, distance is measured as Hamming distance, cf. Sect. (9.3.1), a natural number ranging from 0 to N, proximity between quantum states is quantified as their normalized overlap,

$$\theta(|a\rangle, |b\rangle) = \arccos\left(\frac{|\langle a|b\rangle|}{\sqrt{\langle a|a\rangle \langle b|b\rangle}}\right) \tag{10.2.13}$$

an angle between 0 and $\pi/2$. Its minimum is reached for orthogonal states, $\theta = \pi/2$. Quantum states cannot be more different than that, irrespective of the dimension of their Hilbert space.

The information assigned to a qubit is of obvious relevance for quantum computation. In view of von Neumann's definition of quantum entropy, Eq. (8.2.20),

$$S_{vN} = \langle -c \ln(\hat{\rho})\rangle = -c\text{tr}[\hat{\rho} \ln(\hat{\rho})], \tag{10.2.14}$$

the present context suggests setting $c = 1/\ln(2)$, hence $S_{vN} = -\text{tr}[\hat{\rho}\text{lb}(\hat{\rho})]$ (lb(x) denoting the binary logarithm of x). Representing the density matrix of a two-state system as a Bloch vector, Eq. (10.2.6), makes the definition (10.2.14) more explicit,

$$S_{vN} = -c\text{tr}\left[\frac{1}{2}(\hat{I} + \mathbf{a} \cdot \hat{\boldsymbol{\sigma}}) \ln\left(\frac{1}{2}(\hat{I} + \mathbf{a} \cdot \hat{\boldsymbol{\sigma}})\right)\right] = c(\ln(2) - \text{tr}[\ln(\hat{I} + \mathbf{a} \cdot \hat{\boldsymbol{\sigma}})]).$$

$$(10.2.15)$$

Expanding the logarithm in a Taylor series, $\ln(1 + x) = \sum_{n=1}^{\infty} (-1)^{n-1} x^n / n$, and using a corollary of the features of the Pauli matrices,

$$(\mathbf{a} \cdot \hat{\boldsymbol{\sigma}})^n = \begin{cases} a^n & n \text{ even,} \\ a^n \mathbf{e}_a \cdot \hat{\boldsymbol{\sigma}} & n \text{ odd,} \end{cases} \quad n \in \mathbb{N} \cup \{0\}, \qquad (10.2.16)$$

with $\mathbf{e}_a := \mathbf{a}/a$, allows to write the logarithm as

$$\ln(\hat{I} + \mathbf{a} \cdot \hat{\boldsymbol{\sigma}}) = \frac{1}{2}\left[\left(\hat{I} + \mathbf{e}_a \cdot \hat{\boldsymbol{\sigma}}\right) \ln(1 + a) + \left(\hat{I} - \mathbf{e}_a \cdot \hat{\boldsymbol{\sigma}}\right) \ln(1 - a)\right], \quad (10.2.17)$$

and thus, in units of bits,

$$S_{vN} = 1 - \frac{1}{2}[(1 + a)\text{lb}(1 + a) + (1 - a)\text{lb}(1 - a)]. \qquad (10.2.18)$$

That is, the von Neumann entropy depends only on the length a of the Bloch vector, not on its direction. In particular, for pure states, $a = 1$, it reaches its lower bound $S_{vN} = 0$, while for a completely depolarized system, $\hat{\rho} = \hat{I}/2$ and $a = 0$, the entropy attains its upper bound of one bit, justifying the term "qubit". An alternative quantity measuring the degree of coherence in a quantum system is the purity (8.2.16). Like the entropy of a two-state system, it depends on the system state only through a. With Eqs. (10.2.7) and (10.2.16), it reads

$$\xi_\rho = \text{tr}[\hat{\rho}^2] = \frac{1}{2}(1 + a^2). \qquad (10.2.19)$$

10.2.2 Unitary Operators, Reversible Computation, and Quantum Gates

One of the pivotal developments that stimulated quantum computation was the insight that logical gates can be implemented in physical systems without any fundamental lower bound of the energy dissipated per operation. Since the only step in computation that inevitably consumes energy and produces excess entropy is erasure, see Sect. 9.2.1, this obliges to work exclusively with reversible gates. This idea transfers directly to quantum computing: Here, it is not so much the avoidance of energy losses, but the requirement to suppress decoherent processes as far as possible that makes reversible operation obligatory for quantum computation.

Running a program on a classical digital computer, reversible or irreversible, means following a trajectory in a discrete state space, composed of jumps from one discrete state to another, performed in discrete time. It has the flavour of chess moves. In this respect, time evolution in quantum systems coincides rather with that in continuous classical physics: It is radically different, there are no preferred points within a continuous state space, and there is no natural clock giving rise to a global discretization of time. Even "quantum jumps", in particular the collapse of the wavefunction during measurement, see Sects. 8.3.3 and 8.4.2, upon careful analysis turn out to take place continuously in time [Zur84, Zeh93].

A closed quantum system evolves deterministically, following the Schrödinger equation, Eq. (8.3.1). Integrated over a finite time span, the time evolution takes the form of a unitary transformation,

$$\left|\psi\left(t'\right)\right\rangle \rightarrow \left|\psi\left(t''\right)\right\rangle = \widehat{U}\left(t'',t'\right)\left|\psi\left(t'\right)\right\rangle, \tag{10.2.20}$$

defining the time-evolution operator

$$\widehat{U}\left(t'',t'\right) = \widehat{T}\exp\left(-\frac{i}{\hbar}\int_{t'}^{t''} dt\,\widehat{H}(t)\right). \tag{10.2.21}$$

The explicit integration over time in the exponent is necessary if the Hamiltonian $\widehat{H}(t)$ is not constant but depends on time, as is inevitable in a programmable computer. The operator $\widehat{T}$ effects an ordering of the infinitesimal time steps in the integral from the earliest one $(t = t')$ rightmost to the latest $(t = t'')$ leftmost, see Eq. (8.3.4). Resuming its most important properties, unitary time evolution

1. is reversible, transforming under time reversal as

$$\widehat{U}\left(t',t''\right) = \left(\widehat{U}\left(t'',t'\right)\right)^{-1} = \widehat{U}^{\dagger}\left(t'',t'\right), \tag{10.2.22}$$

2. constitutes a dynamical group under concatenation,

$$\widehat{U}\left(t''',t'\right) = \widehat{U}\left(t''',t''\right)\widehat{U}\left(t'',t'\right), \tag{10.2.23}$$

3. conserves scalar products and the norm,

$$(\widehat{U}|a\rangle)^{\dagger}\widehat{U}|b\rangle = \langle a|\widehat{U}^{\dagger}\widehat{U}|b\rangle = \langle a|b\rangle,$$
$$\left|\widehat{U}a\right|^{2} = \langle a|\widehat{U}^{\dagger}\widehat{U}|a\rangle = \langle a|a\rangle = |a|^{2}. \tag{10.2.24}$$

4. conserves quantum entropy,

$$S_{vN}(t'') = -\text{tr}\left[\widehat{U}(t'',t')\hat{\rho}(t')\widehat{U}^\dagger(t'',t')\text{lb}\left(\widehat{U}(t'',t')\hat{\rho}(t')\widehat{U}^\dagger(t'',t')\right)\right]$$
$$= -\text{tr}\left[\hat{\rho}(t')\text{lb}(\hat{\rho}(t'))\right] = S_{vN}(t'). \tag{10.2.25}$$

For two-state systems, Eq. (10.2.25) means that surfaces of constant entropy in the Bloch-sphere representation are concentric spheres with radius a, $0 \le a \le 1$, around the origin.

In geometrical terms, unitary transformations of quantum systems are rotations of the unit sphere in the space $\mathbb{C}^{D_\mathcal{H}}$, with $D_\mathcal{H}$, the Hilbert-space dimension. For qubits, $D_\mathcal{H} = 2$, they are given by unitary (2×2)-matrices. A standard parameterization of this class of matrices, comprising four angles, is

$$\widehat{U}(\alpha, \phi_1, \phi_2, \theta) = \exp(i\alpha)\begin{pmatrix} \exp(i\phi_1)\cos(\theta) & \exp(i\phi_2)\sin(\theta) \\ -\exp(-i\phi_2)\sin(\theta) & \exp(-i\phi_1)\cos(\theta) \end{pmatrix}. \tag{10.2.26}$$

Its action on a density operator, $\hat{\rho} \to \hat{\rho}' = \widehat{U}\hat{\rho}\widehat{U}^\dagger$ (Eq. (8.3.14)), can be represented clearly by *vectorizing* the density matrix, that is, by arranging the four matrix elements as a four-vector, for example

$$\hat{\rho} = \begin{pmatrix} \rho_{00} & \rho_{01} \\ \rho_{10} & \rho_{11} \end{pmatrix} \leftrightarrow \hat{\boldsymbol{\rho}} = \begin{pmatrix} \rho_{00} \\ \rho_{11} \\ \rho_{01} \\ \rho_{10} \end{pmatrix}. \tag{10.2.27}$$

Vectorizing (2×2)-density matrices allows to represent unitary transformations of these matrices as (4×4)-matrices, so-called *unitary maps*,

$$\hat{\rho} \mapsto \hat{\rho}' = \widehat{U}\hat{\rho}\widehat{U}^\dagger \leftrightarrow \hat{\boldsymbol{\rho}} \mapsto \hat{\boldsymbol{\rho}}' = \mathcal{U}\hat{\boldsymbol{\rho}},$$

$$\mathcal{U} = \begin{pmatrix} (\cos\theta)^2 & (\sin\theta)^2 & \begin{pmatrix} e^{i(\phi_1-\phi_2)} & e^{i(\phi_2-\phi_1)} \\ -e^{i(\phi_1-\phi_2)} & -e^{i(\phi_2-\phi_1)} \end{pmatrix}\cos\theta\sin\theta \\ (\sin\theta)^2 & (\cos\theta)^2 & \\ \begin{pmatrix} -e^{i(\phi_1+\phi_2)} & e^{i(\phi_1+\phi_2)} \\ -e^{-i(\phi_1+\phi_2)} & e^{-i(\phi_1+\phi_2)} \end{pmatrix}\cos\theta\sin\theta & \begin{pmatrix} e^{2i\phi_1}(\cos\theta)^2 & -e^{2i\phi_2}(\sin\theta)^2 \\ -e^{-2i\phi_2}(\sin\theta)^2 & e^{-2i\phi_1}(\cos\theta)^2 \end{pmatrix} \end{pmatrix}. \tag{10.2.28}$$

With its four (2×2) blocks, the transformation exposes the different roles of the matrix elements involved: While the upper left block contains classical, real and positive, transition probabilities for the transitions $|0\rangle \to |0\rangle$, $|0\rangle \to |1\rangle$, $|1\rangle \to |0\rangle$, and $|1\rangle \to |1\rangle$, as in a Markov chain (see Sect. 3.1.1 and Fig. 3.2a), the other blocks include phase factors and are responsible for quantum coherence effects.

A more compact representation than the (4×4)-matrices of Eq. (10.2.28) is achieved by interpreting unitary operations on qubits as rotations of the Bloch sphere, that is, in three dimensions and parameterized by Euler angles [GPS01]. This relationship is a manifestation of the homomorphism from SU(2), the group of unitary (2×2)-matrices, onto the group SO(3) of orthogonal transformations with unit determinant in $\mathbb{R}^3$ [JS98, GPS01]. In addition, unitary operations include shifts of the global phase. Again, as in Eq. (10.2.26), this amounts to four angles, but now defined by (adopting the standard convention for spin rotations),

$$\widehat{U}(\alpha, \beta, \gamma, \delta) = \exp(i\alpha)\widehat{R}_3(\beta)\widehat{R}_2(\gamma)\widehat{R}_3(\delta), \qquad (10.2.29)$$

denoting rotations by an angle θ around the three axes of the Bloch sphere as

$$\widehat{R}_1(\theta) = \exp\left(-i\frac{\theta}{2}\hat{\sigma}_1\right) = \begin{pmatrix} \cos(\theta/2) & -i\sin(\theta/2) \\ -i\sin(\theta/2) & \cos(\theta/2) \end{pmatrix},$$

$$\widehat{R}_2(\theta) = \exp\left(-i\frac{\theta}{2}\hat{\sigma}_2\right) = \begin{pmatrix} \cos(\theta/2) & -\sin(\theta/2) \\ \sin(\theta/2) & \cos(\theta/2) \end{pmatrix},$$

$$\widehat{R}_3(\theta) = \exp\left(-i\frac{\theta}{2}\hat{\sigma}_3\right) = \begin{pmatrix} \exp(-i\theta/2) & 0 \\ 0 & \exp(i\theta/2) \end{pmatrix}. \qquad (10.2.30)$$

According to Euler's rotation theorem [JS98, GPS01], a sequence of three rotations around fixed axes as in Eq. (10.2.29) is equivalent to a single rotation by an effective angle θ around an effective axis $\mathbf{n}(\beta, \gamma, \delta) = (n_1(\beta, \gamma, \delta), n_2(\beta, \gamma, \delta), n_3(\beta, \gamma, \delta))$,

$$\widehat{U}(\alpha, \beta, \gamma, \delta) = \exp(i\alpha)\widehat{R}_{\mathbf{n}(\beta,\gamma,\delta)}(\theta),$$

$$\widehat{R}_{\mathbf{n}(\beta,\gamma,\delta)}(\theta) = \exp\left(-i\frac{\theta}{2}\mathbf{n}(\beta, \gamma, \delta)\cdot\hat{\sigma}\right) = \cos\left(\frac{\theta}{2}\right)\hat{I} - i\sin\left(\frac{\theta}{2}\right)\mathbf{n}(\beta, \gamma, \delta)\cdot\hat{\sigma}.$$

$$(10.2.31)$$

In SU(2)-matrix form, the corresponding unitary transformation reads

$$\widehat{U}(\alpha, \beta, \gamma, \delta) = \exp(i\alpha)\times$$

$$\begin{pmatrix} \cos(\theta/2) - n_3\sin(\theta/2) & -(in_1 + n_2)\sin(\theta/2) \\ (-in_1 + n_2)\sin(\theta/2) & \cos(\theta/2) + n_3\sin(\theta/2) \end{pmatrix}. \qquad (10.2.32)$$

In the geometric representation of the density matrix $\hat{\rho}$ as a Bloch vector $\mathbf{a}$, the transformation $\hat{\rho} \mapsto \hat{\rho}' = \widehat{U}(\alpha, \beta, \gamma, \delta)\hat{\rho}\widehat{U}^\dagger(\alpha, \beta, \gamma, \delta)$ translates to a rigid rotation $\mathbf{a} \mapsto \mathbf{a}' = R_{\mathbf{n}(\beta,\gamma,\delta)}(\theta)\mathbf{a}$ of the Bloch sphere, given by the SO(3)-matrix

$$R_{\mathbf{n}(\beta,\gamma,\delta)}(\theta) =$$

$$\begin{pmatrix} \cos(\theta) + n_1^2(1 - \cos(\theta)) & n_1 n_2(1 - \cos(\theta)) - n_3\sin(\theta) & n_3 n_1(1 - \cos(\theta)) + n_2\sin(\theta) \\ n_1 n_2(1 - \cos(\theta)) + n_3\sin(\theta) & \cos(\theta) + n_2^2(1 - \cos(\theta)) & n_2 n_3(1 - \cos(\theta)) - n_1\sin(\theta) \\ n_3 n_1(1 - \cos(\theta)) - n_2\sin(\theta) & n_2 n_3(1 - \cos(\theta)) + n_1\sin(\theta) & \cos(\theta) + n_3^2(1 - \cos(\theta)) \end{pmatrix}.$$

$$(10.2.33)$$

The components of the unit vector $\mathbf{n}(\beta, \gamma, \delta)$ and the effective rotation angle θ can be found by from characteristics of $R_{\mathbf{n}(\beta,\gamma,\delta)}(\theta)$ in whatever basis, e.g., $\theta = \arccos\left(\frac{1}{2}\left(\text{tr}\left[R_{\mathbf{n}(\beta,\gamma,\delta)}\right] - 1\right)\right)$.

A few special rotations of particular practical relevance have received proper names:

− **NOT gate:** $\alpha = \pi/2, \theta = \pi, n_1 = n_3 = 0, n_2 = 1$ in Eq. (10.2.32)

$$\widehat{N} = \begin{pmatrix} 0 & 1 \\ 1 & 0 \end{pmatrix}, \tag{10.2.34}$$

− **Hadamard gate:** $\alpha = \pi/2, \theta = \pi/2, n_1 = n_3 = 1/\sqrt{2}, n_2 = 0$ in Eq. (10.2.32)

$$\widehat{H} = \frac{1}{\sqrt{2}} \begin{pmatrix} 1 & 1 \\ 1 & -1 \end{pmatrix}, \tag{10.2.35}$$

− **$\pi/4$-phase shift:** $\alpha = \pi/4, \theta = \pi/2, n_1 = n_2 = 0, n_3 = 1$ in Eq. (10.2.32)

$$\widehat{S} = \begin{pmatrix} 1 & 0 \\ 0 & i \end{pmatrix}, \tag{10.2.36}$$

− **$\pi/8$-phase shift:** $\alpha = \pi/8, \theta = \pi/4, n_1 = n_2 = 0, n_3 = 1$ in Eq. (10.2.32)

$$\widehat{T} = \begin{pmatrix} 1 & 0 \\ 0 & \exp(i\pi/4) \end{pmatrix}. \tag{10.2.37}$$

All these gates involve at most two of the three basic rotations $\hat{\sigma}_1, \hat{\sigma}_2, \hat{\sigma}_3$. A particularly instructive case is a gate that permutes the three axes of the Bloch sphere among themselves, equivalent to a rotation by $2\pi/3$ around the space diagonal. It corresponds to the parameters $\theta = 2\pi/3, n_1 = n_2 = n_3 = 1/\sqrt{3}$ in Eq. (10.2.32) and is given by the transformation

$$\widehat{P} = \frac{i-1}{2} \begin{pmatrix} 1 & -i \\ 1 & i \end{pmatrix}, \tag{10.2.38}$$

equivalent to transforming the Bloch sphere by the SO(3) permutation matrix, cf. Eq. (10.2.33),

$$R_{(1,1,1)/\sqrt{3}}\left(\frac{2\pi}{3}\right) = \begin{pmatrix} 0 & 0 & 1 \\ 1 & 0 & 0 \\ 0 & 1 & 0 \end{pmatrix}. \tag{10.2.39}$$

For example,

$$\widehat{P}|0_1\rangle = \exp\left(i\frac{3\pi}{2}\right)|0_2\rangle, \ \widehat{P}|0_2\rangle = \exp\left(i\frac{\pi}{2}\right)|0_3\rangle, \ \widehat{P}|0_3\rangle = \exp\left(i\frac{3\pi}{2}\right)|0_1\rangle,$$

$$(10.2.40)$$

that is, up to phase factors, $\widehat{P}$ permutes eigenstates $|0_1\rangle \to |0_2\rangle \to |0_3\rangle \to |0_1\rangle$ and likewise $|1_1\rangle \to |1_2\rangle \to |1_3\rangle \to |1_1\rangle$.

The transformations considered till here conserve pure states of the qubit, but they also apply to mixed states. If we assume from the outset that it is in a pure state, the description simplifies considerably. It is then possible to interpret single-qubit transformations as elastic scattering processes. The states of the qubit before and after the action of the gate thus take the form of asymptotic states, composed of a right-going and a left-going wave. For example, on input, $|\psi_{in}\rangle = \psi_{right}|\psi_{in,r}\rangle + \psi_{left}|\psi_{in,l}\rangle$ with $\psi_{right}(x,t) = \exp(i(kx - \omega t))$, $\psi_{left}(x,t) = \exp(i(kx + \omega t))$. Passing an interaction zone, they interfere and leave the scattering region as output $|\psi_{out}\rangle = \psi_{right}|\psi_{out,r}\rangle + \psi_{left}|\psi_{out,l}\rangle$ (Fig. 10.2). Incoming and outgoing wave are related by a scattering matrix $\hat{S}$,

$$|\psi_{in}\rangle = \begin{pmatrix} \psi_{in,l} \\ \psi_{in,r} \end{pmatrix} \mapsto |\psi_{out}\rangle = \begin{pmatrix} \psi_{out,l} \\ \psi_{out,r} \end{pmatrix} = \hat{S}\begin{pmatrix} \psi_{in,l} \\ \psi_{in,r} \end{pmatrix} = \hat{S}|\psi_{in}\rangle, \ \hat{S} = \begin{pmatrix} t_{ll} & r_{lr} \\ r_{rl} & t_{rr} \end{pmatrix}.$$

$$(10.2.41)$$

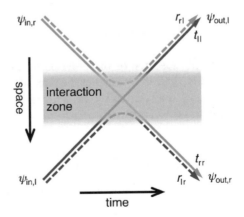

Fig. 10.2 Single qubit gates modeled as quantum scattering. The asymptotic incoming state $|\psi_{in}\rangle$ for $t \to -\infty$ comprises a right-going wave $\sim \psi_{in,r}$ and a left-going one $\sim \psi_{in,l}$. On exit, for $t \to \infty$, the outgoing wave $|\psi_{out}\rangle$ again contains right- and left-going components $\psi_{out,r}$ and $\psi_{out,l}$, resp. They are related by transmission and reflection amplitudes t_{ll}, t_{rr}, r_{rl}, and r_{lr}, the elements of the scattering matrix, cf. Eq. (10.2.41). Unitarity of the time evolution implies restricting conditions for these coefficients, see Eq. (10.2.42).

Its four elements $(t_{ll}, r_{rl}, r_{lr}, t_{rr})$ define scattering amplitudes for transmission and reflection into the left-going and reflection and transmission into the right-going waves, resp. The scattering matrix relates the asymptotic state for $t \to \infty$ to that for $t \to -\infty$. It represents a time-evolution operator and as such must be unitary,

$$\begin{pmatrix} 1 & 0 \\ 0 & 1 \end{pmatrix} = \hat{S}^\dagger \hat{S} = \begin{pmatrix} |t_{ll}|^2 + |r_{rl}|^2 & t_{ll}^* r_{rl} + r_{rl}^* t_{rr} \\ r_{lr}^* t_{ll} + t_{rr}^* r_{rl} & |t_{rr}|^2 + |r_{lr}|^2 \end{pmatrix} \Rightarrow \begin{array}{l} |t_{ll}|^2 + |r_{rl}|^2 = 1 = |t_{rr}|^2 + |r_{lr}|^2, \\ r_{lr}^* t_{ll} + t_{rr}^* r_{rl} = 0 = t_{ll}^* r_{lr} + r_{rl}^* t_{rr}. \end{array}$$

$$(10.2.42)$$

The pair of identities in Eq. (10.2.42) concerning diagonal elements of the unit matrix express the conservation of probability in the scattering process. This can be made explicit by a suitable parameterization, involving classical transition probabilities

$$\hat{S} = \begin{pmatrix} \sqrt{T_{ll}}\,\exp(i\phi_{ll}) & \sqrt{R_{lr}}\,\exp(i\phi_{lr}) \\ \sqrt{R_{rl}}\,\exp(i\phi_{rl}) & \sqrt{T_{rr}}\,\exp(i\phi_{rr}) \end{pmatrix}, \quad \begin{array}{l} T_{ll} = |t_{ll}|^2, \quad R_{lr} = |r_{lr}|^2, \\ R_{rl} = |r_{rl}|^2, \quad T_{rr} = |t_{rr}|^2, \end{array} \quad (10.2.43)$$

so that unitarity amounts to $T_{ll} + R_{rl} = 1 = T_{rr} + R_{lr}$.

Following quantum optics, a few standard situations, reflected in corresponding transformations, enjoy particular interest: For perfect transmission, $t_{ll} = t_{rr} = 1$, that is, without any interaction of the two beams, the transformation reduces to the identity. Assuming total reflection, by contrast, $|r_{rl}|^2 = |r_{lr}|^2 = 1$, an operation equivalent to the logical NOT, Eq. (10.2.34), is obtained. Intermediate between these extremes is the equal-probability beam splitter with $T_{ll} = R_{lr} = R_{rl} = T_{rr} = 0.5$. Choosing the phases as $\phi_{ll} = \phi_{rr} = 0$ (no phase shift for transmission) and $\phi_{rl} = \phi_{lr} = \pi/2$ (phase jump due to reflection at a hard wall) results in the scattering matrix

$$\hat{R} = \frac{1}{\sqrt{2}} \begin{pmatrix} 1 & i \\ i & 1 \end{pmatrix} = \hat{R}_1\left(\frac{3\pi}{2}\right), \qquad (10.2.44)$$

equivalent to a $3\pi/2$-rotation around the 1-axis. The values $\phi_{ll} = \phi_{lr} = \phi_{rl} = 0$ and $\phi_{rr} = \pi$, in turn, correspond to a Hadamard gate.

Another useful representation of single-qubit transformations is in terms of ladder (raising and lowering or creation and annihilation, resp.) operators, introduced in Eq. (8.3.42) for harmonic oscillators and in Eq. (8.5.66) for arbitrary angular momenta. For the z-component of spin-$\frac{1}{2}$ systems, they are defined as

$$\hat{a}_3 \equiv \hat{\sigma}_3^+ := \frac{1}{2}(\hat{\sigma}_1 + i\hat{\sigma}_2) = \begin{pmatrix} 0 & 1 \\ 0 & 0 \end{pmatrix}, \hat{a}_3^\dagger \equiv \hat{\sigma}_3^- := \frac{1}{2}(\hat{\sigma}_1 - i\hat{\sigma}_2) = \begin{pmatrix} 0 & 0 \\ 1 & 0 \end{pmatrix} \Leftrightarrow$$

$$\hat{\sigma}_1 = \hat{a}_3 + \hat{a}_3^\dagger, \hat{\sigma}_2 = i\left(\hat{a}_3^\dagger - \hat{a}_3\right), \hat{\sigma}_3 = \left[\hat{a}_3, \hat{a}_3^\dagger\right]. \qquad (10.2.45)$$

Some useful identities, implied directly by these definitions, are

$$\hat{a}_3^\dagger \hat{a}_3 = \begin{pmatrix} 0 & 0 \\ 0 & 1 \end{pmatrix} = |1\rangle\langle 1|, \, \hat{a}_3 \hat{a}_3^\dagger = \begin{pmatrix} 1 & 0 \\ 0 & 0 \end{pmatrix} = |0\rangle\langle 0|,$$

$$\hat{I}^{(2)} = \begin{pmatrix} 1 & 0 \\ 0 & 1 \end{pmatrix} = \hat{a}_3^\dagger \hat{a}_3 + \hat{a}_3 \hat{a}_3^\dagger = \left[\hat{a}_3^\dagger, \hat{a}_3\right]_+. \tag{10.2.46}$$

For the other components, permuting the subscripts circularly, they read

$$\hat{a}_1 := \frac{1}{2}(\hat{\sigma}_2 + i\hat{\sigma}_3) = \frac{1}{2}\begin{pmatrix} i & -i \\ i & -i \end{pmatrix}, \, \hat{a}_1^\dagger := \frac{1}{2}(\hat{\sigma}_2 - i\hat{\sigma}_3) = \frac{1}{2}\begin{pmatrix} -i & -i \\ i & i \end{pmatrix}, \tag{10.2.47}$$

etc. The subscript 3, for example, indicates the special role $\hat{\sigma}_3$ plays for the ladder operators, $\left[\hat{a}_3, \hat{a}_3^\dagger\right] = \hat{\sigma}_3$ (it will be omitted in what follows). Evidently, they are not themselves Hermitean but adjoints of one another.

The ladder operators allow to rewrite standard gates in a simple form, for instance,

− **NOT gate,** Eq. (10.2.34):

$$\hat{N} = \hat{a}_3 + \hat{a}_3^\dagger, \tag{10.2.48}$$

− **Hadamard gate,** Eq. (10.2.35):

$$\hat{H} = \left(\left[\hat{a}_3, \hat{a}_3^\dagger\right] + \hat{a}_3 + \hat{a}_3^\dagger\right)/\sqrt{2}, \tag{10.2.49}$$

− **Beam splitter,** Eq. (10.2.44):

$$\hat{R} = \left(\hat{I} + i\left(\hat{a}_3 + \hat{a}_3^\dagger\right)\right)/\sqrt{2}. \tag{10.2.50}$$

Computation involves at least two qubits. It is transformations in a four-dimensional Hilbert space, i.e., rotations of the unit sphere in $\mathbb{C}^4$, that have to be compared with binary gates of classical computation. To begin with, however, their properties are completely different from those of operations in Boolean algebra. Unitary transformations

1. are reversible (property 1 above),
2. conserve quantum entropy (property 4) at its minimum, $S_{vN} = 0$, and
3. are parameterized by continuous quantities (angles), Eq. (10.2.29), without any intrinsic preference for a discrete set of nominal states.

Comparing with classical computation, this suggests two striking conclusions: Besides the discretization of Hilbert space dimensions, with a lower bound of 2, there is no discretization inherent in quantum mechanics, it is totally analogue! Quantum computation therefore compares to classical computing as playing a violin to playing piano. Indeed, one of the challenges is to hit a subset of discrete states within Hilbert space that are preferred for the computation (the scale in use) and to prevent the process from leaving this subset. Moreover, keeping the entropy at the constant value zero, closed quantum systems do not redistribute any information, they just transform it between different representations.

A first obvious objective of quantum computation is devising quantum operations hat come as close as possible to binary classical gates. It suggests itself seeking inspiration from a closely related construction, proposed for classical computation even before the advent of quantum computation: reversible gates, see Sect. 9.2.2 and Table 9.1. In fact, it was the proposal by Fredkin, Toffoli, and Bennett of gates for classical computation without any lower bound for their energy consumption, thus avoiding the production of excess entropy [Ben73, Tof80, Tof81, FT82, Ben82], which triggered pioneering speculations about quantum computers [Fey82, Fey86]. The only reversible binary gates (besides the identity) are the XOR or CONTROLLED NOT as well as its negation and the exchange gate. Representing states in a basis such as the canonical product basis (10.2.11), N-qubit gates take the form of unitary $(2^N \times 2^N)$-matrices. As long as only real numbers are involved, unitarity reduces to orthogonality, $\hat{U}^\dagger \hat{U} = (\hat{U}^*)^{\mathrm{T}} \hat{U} = \hat{U}^{\mathrm{T}} \hat{U} = \hat{I}$.

To begin with, consider operations on a pair of two-state systems, $N = 2$. Similar to the case of single-qubit operations, the corresponding rotations of the unit sphere in $\mathbb{C}^4$ can be broken down to sequences of rotations within one of the single-qubit subspaces, i.e., operations on one of the two qubits at a time without affecting the other. Such gates are obtained in turn by combining, e.g., Pauli matrices, Eq. (10.2.8), or ladder operators, Eq. (10.2.45), for one qubit with the identity for the other. Denoting ladder operators $\hat{a}_3, \hat{a}_3^\dagger$ for the first qubit as $\hat{a}, \hat{a}^\dagger$, resp. (omitting the subscript 3), for the second one as $\hat{b}, \hat{b}^\dagger$, and Pauli matrices operating on qubits a and b as $\hat{\sigma}_{a,i}, \hat{\sigma}_{b,i}, i = 1,2,3$, resp., besides the two identities $\hat{I}_a$ and $\hat{I}_b$, one finds

$$\hat{a} = \begin{pmatrix} 0&0&1&0 \\ 0&0&0&1 \\ 0&0&0&0 \\ 0&0&0&0 \end{pmatrix}, \hat{a}^\dagger = \begin{pmatrix} 0&0&0&0 \\ 0&0&0&0 \\ 1&0&0&0 \\ 0&1&0&0 \end{pmatrix}, \hat{b} = \begin{pmatrix} 0&1&0&0 \\ 0&0&0&0 \\ 0&0&0&1 \\ 0&0&0&0 \end{pmatrix}, \hat{b}^\dagger = \begin{pmatrix} 0&0&0&0 \\ 1&0&0&0 \\ 0&0&0&0 \\ 0&0&1&0 \end{pmatrix}$$

$$\hat{\sigma}_{1a} = \begin{pmatrix} 0&0&1&0 \\ 0&0&0&1 \\ 1&0&0&0 \\ 0&1&0&0 \end{pmatrix}, \hat{\sigma}_{2a} = \begin{pmatrix} 0&0&-i&0 \\ 0&0&0&-i \\ i&0&0&0 \\ 0&i&0&0 \end{pmatrix}, \hat{\sigma}_{3a} = \begin{pmatrix} 1&0&0&0 \\ 0&1&0&0 \\ 0&0&-1&0 \\ 0&0&0&-1 \end{pmatrix},$$

$$\hat{\sigma}_{1b} = \begin{pmatrix} 0&1&0&0 \\ 1&0&0&0 \\ 0&0&0&1 \\ 0&0&1&0 \end{pmatrix}, \hat{\sigma}_{2b} = \begin{pmatrix} 0&-i&0&0 \\ i&0&0&0 \\ 0&0&0&-i \\ 0&0&i&0 \end{pmatrix}, \hat{\sigma}_{3b} = \begin{pmatrix} 1&0&0&0 \\ 0&-1&0&0 \\ 0&0&1&0 \\ 0&0&0&-1 \end{pmatrix}. \quad (10.2.51)$$

Two-qubit gates can be constructed by comparing directly with those matrices that represent the corresponding reversible operations in classical binary logic, that is, with permutation matrices, see, e.g., Eq. (9.3.3).

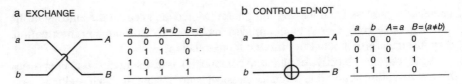

Fig. 10.3 Reversible quantum gates. The switching or EXCHANGE gate (**a**) only swaps its two input qubits without affecting their value. It is symmetric under interchange of the qubits and moreover is its own inverse. The CONTROLLED-NOT (**b**) performs a NOT on the second qubit, conditioned on the first one being in the 1-state. It is its own inverse but not invariant under interchange of the two inputs.

Arguably the simplest non-trivial reversible two-bit operation is the switching or EXCHANGE gate (Fig. 10.3a, cf. also Table 9.2d). It interchanges the two input bits without modifying them individually and amounts to the transformation

$$\hat{U}_{\text{XCH}} = \begin{pmatrix} 1 & 0 & 0 & 0 \\ 0 & 0 & 1 & 0 \\ 0 & 1 & 0 & 0 \\ 0 & 0 & 0 & 1 \end{pmatrix}, \tag{10.2.52}$$

expressed in terms of single-qubit ladder operators as

$$\hat{U}_{\text{XCH}} = \hat{a}^{\dagger}\left(\hat{I} + \hat{a}\hat{b}^{\dagger}\right)\hat{b} + \hat{a}\left(\hat{I} + \hat{a}^{\dagger}\hat{b}\right)\hat{b}^{\dagger}. \tag{10.2.53}$$

The switching gate is symmetric under an exchange of the two qubits and moreover under time reversal, i.e., it is its own inverse, $\hat{U}_{\text{XCH}} = \hat{U}_{\text{XCH}}^{\text{T}} = \hat{U}_{\text{XCH}}^{-1}$.

The other important operator of this type, the CONTROLLED-NOT (Fig. 10.3b, cf. also Table 9.2d), conditions the NOT operation on the second qubit on $\hat{a}^{\dagger}\hat{a}$ for the first one. As a matrix, it takes the form

$$\hat{U}_{\text{CNOT}} = \begin{pmatrix} 1 & 0 & 0 & 0 \\ 0 & 1 & 0 & 0 \\ 0 & 0 & 0 & 1 \\ 0 & 0 & 1 & 0 \end{pmatrix}. \tag{10.2.54}$$

In terms of ladder operators, it reads [Fey86]

$$\hat{U}_{\text{CNOT}} = \hat{a}\hat{a}^{\dagger}\hat{I}_b + \hat{a}^{\dagger}\hat{a}\left(\hat{b} + \hat{b}^{\dagger}\right) = \hat{I} + \hat{a}^{\dagger}\hat{a}\left(\hat{b} + \hat{b}^{\dagger} - \hat{I}_b\right), \tag{10.2.55}$$

Table 10.1 Dependence of the required number of bits or qubits of classical and quantum simulators on the physical size of the system to be simulated. Classically, it depends logarithmically on the size of the state space, hence linearly on the spatial dimension and / or the number of degrees of freedom. In quantum mechanics , the crucial parameter is the dimension of Hilbert space. It increases exponentially with the length, area, volume, ..., of the system, so that the number of qubits needed grows linearly with its size and exponentially with its dimension.

	classical		quantum	
resolution (cell size)	arbitrary, fixed by design		fixed by nature, $(2\pi\hbar)^f$ for f degrees of freedom	
dependence on	number of states	information content	Hilbert space dimension	information content
system size (spatial size, phase-space area, ...)	linear	logarithmic	exponential	linear
spatial dimension, number of degrees of freedom	exponential	linear	super-exponential	exponential

with $\hat{I}_b = \hat{b}^\dagger\hat{b} + \hat{b}\hat{b}^\dagger$, the identity for the second bit. As is the case for the switching gate, the matrix representation (10.2.54) of the CONTROLLED-NOT is symmetric, therefore it is its own inverse. In view of the special role of the control, however, it is not invariant under swapping the two qubits, as is reflected in the symbol used for it (Table 10.1b). With the help of the identities (10.2.45, 47) relating ladder to Pauli operators, these gates can be expressed as rotations in $\mathbb{C}^4$,

$$\hat{U}_{\mathrm{XCH}} = \frac{1}{2}\left(\hat{I} + \hat{\sigma}_{1a}\hat{\sigma}_{1b} + \hat{\sigma}_{2a}\hat{\sigma}_{2b} + \hat{\sigma}_{3a}\hat{\sigma}_{3b}\right),$$

$$\hat{U}_{\mathrm{CNOT}} = \frac{1}{2}\left(\hat{I}_a - \hat{\sigma}_{3a}\right)\hat{I}_b + \frac{1}{2}\left(\hat{I}_a + \hat{\sigma}_{3a}\right)\hat{\sigma}_{3b}. \tag{10.2.56}$$

A more sophisticated task is a single gate switching through all the canonical basis states (10.2.12) of the two-qubit system, for instance $|0_a, 0_b\rangle \to |0_a, 1_b\rangle \to |1_a, 0_b\rangle \to |1_a, 1_b\rangle \to |0_a, 0_b\rangle \to \dots$. Composing this circular permutation one by one of its four single-qubit steps,

$$\hat{U}_{0123} = \hat{U}_{01} + \hat{U}_{12} + \hat{U}_{23} + \hat{U}_{30}$$

$$= \begin{pmatrix} 0&0&0&0 \\ 1&0&0&0 \\ 0&0&0&0 \\ 0&0&0&0 \end{pmatrix} + \begin{pmatrix} 0&0&0&0 \\ 0&0&0&0 \\ 0&1&0&0 \\ 0&0&0&0 \end{pmatrix} + \begin{pmatrix} 0&0&0&0 \\ 0&0&0&0 \\ 0&0&0&0 \\ 0&0&1&0 \end{pmatrix} + \begin{pmatrix} 0&0&0&1 \\ 0&0&0&0 \\ 0&0&0&0 \\ 0&0&0&0 \end{pmatrix}$$

$$= \begin{pmatrix} 0&0&0&1 \\ 1&0&0&0 \\ 0&1&0&0 \\ 0&0&1&0 \end{pmatrix}, \tag{10.2.57}$$

it can in turn be expressed in terms of ladder operators or Pauli matrices,

$$\hat{U}_{0123} = \hat{a}\hat{b} + \hat{a}^\dagger\hat{a}\hat{b}^\dagger + \hat{a}^\dagger\hat{b} + \hat{a}\hat{a}^\dagger\hat{b}^\dagger = (\hat{a} + \hat{a}^\dagger)\hat{b} + \hat{I}_a\hat{b}^\dagger$$

$$= \frac{1}{2}\hat{\sigma}_{1a}(\hat{\sigma}_{1b} + i\hat{\sigma}_{1b}) + \hat{I}_a(\hat{\sigma}_{1b} - i\hat{\sigma}_{1b}). \tag{10.2.58}$$

Applying this gate, a single rotation of the four-dimensional Hilbert space, repeatedly to the ground state $|0_a, 0_b\rangle$ generates the other three basis states,

$$|0_a, 0_b\rangle = \hat{U}^0_{0123}|0_a, 0_b\rangle, \; |0_a, 1_b\rangle = \hat{U}^1_{0123}|0_a, 0_b\rangle,$$

$$|1_a, 0_b\rangle = \hat{U}^2_{0123}|0_a, 0_b\rangle, \; |1_a, 1_b\rangle = \hat{U}^3_{0123}|0_a, 0_b\rangle. \tag{10.2.59}$$

The gates thus emerge as a small subclass of the total space of operations, composed exclusively of rotations by integer multiples of π around the principal axes of the Hilbert spaces. This assures that, represented in the canonical basis $\{|0\rangle, |1\rangle\}^N$, they can be mapped directly to operators in binary logic or Boolean algebra.

More sophisticated things can be done with three qubits, and it is here where the differences in strategy between classical and quantum computing start to become apparent. A simple step from two- to three-qubit gates is conditioning the action of a two-qubit gate $\hat{G}_b$ on the third input a, resulting in a CONTROLLED-G. Equation (10.2.55) can be generalized to a recipe how to add a control to operator $\hat{G}_b$: Interpreting there the negation $\hat{b} + \hat{b}^\dagger$ on qubit b as the operator to be controlled by a, Eq. (10.2.55) can be written as

$$\hat{U}_{CG} = \hat{I} + \hat{a}^\dagger\hat{a}(\hat{G}_b - \hat{I}_b), \tag{10.2.60}$$

or as a $(2N \times 2N)$-matrix, if $\hat{G}$ is $(N \times N)$,

$$\hat{U}_{CG} = \begin{pmatrix} \hat{I}^{(N)} & \hat{0}^{(N)} \\ \hat{0}^{(N)} & \hat{G}_b \end{pmatrix}, \tag{10.2.61}$$

where $0^{(N)}$, $\hat{I}^{(N)}$ are the $(N \times N)$-zero and unit matrices, resp. Replacing $\hat{G}$ by the already controlled NOT operating on c and assigning the control to a and b, Eq. (10.2.60) results in the CONTROLLED-CONTROLLED-NOT [Fey86],

$$\hat{U}_{CCNOT} = \hat{I} + \hat{a}^\dagger\hat{a}(\hat{I}_{bc} + \hat{b}^\dagger\hat{b}(\hat{c} + \hat{c}^\dagger - \hat{I}_c) - \hat{I}_{bc})$$

$$= \hat{I} + \hat{a}^\dagger\hat{a}\hat{b}^\dagger\hat{b}(\hat{c} + \hat{c}^\dagger - \hat{I}_c), \tag{10.2.62}$$

and in matrix form

$$\hat{U}_{\text{CCNOT}} = \begin{pmatrix} I^{(N)} & 0^{(N)} \\ 0^{(N)} & \hat{U}_{\text{CNOT}} \end{pmatrix}. \tag{10.2.63}$$

In the same way, the EXCHANGE (Fig. 10.3a, Eq. (10.2.52)) becomes a CONTROLLED-EXCHANGE, equivalent to the classical switch or Fredkin gate (Table 9.1d). It interchanges inputs b and c on exit if and only if input $a = 1$. As a unitary matrix, it is composed of four (4×4)-blocks, using Eq. (10.2.61),

$$\hat{U}_{\text{CXCH}} = \begin{pmatrix} \hat{I}^{(4)} & 0^{(4)} \\ 0^{(4)} & \hat{U}_{\text{XCH}} \end{pmatrix}. \tag{10.2.64}$$

In terms of creation and annihilation operators, it reads, cf. Eq. (10.2.53),

$$\hat{U}_{\text{CXCH}} = \hat{I} + \hat{a}^\dagger \hat{a} \left(\hat{U}_{\text{XCH}} - \hat{I} \right) = \hat{I} + \hat{a}^\dagger \hat{a} \left(\hat{b}^\dagger \hat{c} \left(\hat{I}_{bc} - \hat{b} \hat{c}^\dagger \right) + \hat{b} \hat{c}^\dagger \left(\hat{I}_{bc} - \hat{b}^\dagger \hat{c} \right) \right). \tag{10.2.65}$$

These elements already allow constructing a gate of patent practical use, a single-digit module of a binary adder (Fig. 10.4). It comprises three input bits, the two summands a and b as well as the carry c transferred from the next lower digit, and three outputs, including one of the inputs, say a, the sum, and the carry to be consigned to the next higher digit. The Boolean sum $a + b$ (mod 2) of two bits is equivalent to the XOR operation, see Eq. (2.1.6). Qubit c is permanently kept at 0 on entrance. On exit, it contains the carry, the bit that takes the value 1 if both summands are equal to one so that the Boolean sum is set to zero by the mod 2, hence it is given by the Boolean product $a \times b$ or the AND gate $a \wedge b$ applied to the summands, see Eq. (2.1.6).

However, to have a fully scalable element of a binary adder of arbitrary size, a fourth qubit, say d, has to be included, to receive a possible non-zero carry from the next lower digit. Besides containing the total sum on exit, $D = a + b + d$ (mod

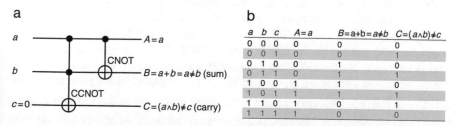

a

a ——————————— $A = a$

CNOT

b ——————————— $B = a + b = a \neq b$ (sum)

CCNOT

c = 0 ——————————— $C = (a \wedge b) \neq c$ (carry)

b

a	b	c	A=a	B=a+b=a≠b	C=(a∧b)≠c
0	0	0	0	0	0
0	0	1	0	0	1
0	1	0	0	1	0
0	1	1	0	1	1
1	0	0	1	1	0
1	0	1	1	1	1
1	1	0	1	0	1
1	1	1	1	0	0

Fig. 10.4 A single digit block of a binary adder. It can be composed (**a**) by combining a CONTROLLED-NOT, Eq. (10.2.54), with a CONTROLLED-CONTROLLED-NOT, Eq. (10.2.63), so that input qubit a is conserved on exit, qubit b contains the second summand on input and the Boolean sum $a + b$ on output, and qubit c, kept $c = 0$ on input, is set to the carry, $c = a \times b$, on output. The full truth table of this three-qubit gate is shown in panel (**b**). Lines corresponding to the fictitious value $c = 0$ are dimmed in grey.

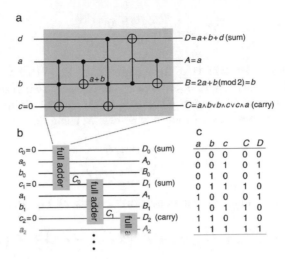

Fig. 10.5 A binary adder (**b**) is composed by concatenating single-digit blocks (**a**), one for each binary digit $a.n$, $b.n$ of the summands a and b, in such a way that the carry output of each block enters the next one as third summand. This requires to complement the adder as in Fig. 10.5a by a fourth input and output bit, resulting in the full adder (**a**). The binary operations for carry $C = (a \wedge b) \vee (a \wedge c) \vee (b \wedge c)$ and total sum $D = a + b + d (\text{mod } 2)$ are shown in panel (**c**).

2), it has to be taken into account in the next carry, $C = (a \wedge b) \vee (a \wedge c) \vee (b \wedge c)$, see Fig. 10.5a.

This full adder can then be concatenated digit by digit, so that the output d_n of block n is fed as input d_{n+1} into the next block $n + 1$. On final output, the nth digit of the sum D_n appears with the conserved nth digits of the summands as $A_n = a_n$ and $B_n = b_n$. If the adder comprises N input digits, the carry C_N of the last block becomes the $(N + 1)$th digit of the total sum on output.

All operations of classical Boolean logic can be composed of binary gates, which in turn can be reduced to combinations containing only a single type of gate, namely NAND or NOR (Sect. 2.1.1). In this sense, these two gates are universal. A similar reduction is possible in quantum computation: It can be shown that every unitary operation on a Hilbert space of arbitrary dimension $D_{\mathcal{H}}$ can be written as a product of single-qubit gates with the CNOT gate [NC00]. They can therefore be considered as universal. Single-qubit gates are rotations with unit determinant around the canonical axes of Hilbert space, the CNOT can be compared with a reflection, since $\det\left(\hat{U}_{\text{CNOT}}\right) = -1$. This universality is therefore analogues to the fact that general rotations in real space can be expressed as rotations around a set of non-colinear axes with unit determinant plus reflection.

10.3 Strategies for Quantum Computation: Quantum Algorithms

The examples presented in the foregoing subsection, culminating in a quantum version of a scalable adder, should have evidenced that quantum computation can compete with classical computers. This conclusion can be summarized and focussed in what might be called the *quantum Church-Turing thesis* [Deu85]:

Every finitely realizable physical system can be simulated by a universal quantum computer.

It differs from the original Church Turing thesis, see Sect. 9.3.3, in two crucial points: Firstly, it replaces the mathematical notion of *recursive algorithms*, a class of dynamical systems, by the physical concept of *finitely realizable physical system*, thus shifting the focus from the realm of formal mathematics to empirical science. Secondly, by referring to a "universal *quantum* computer", it invokes quantum mechanics as the more fundamental framework to discuss the computation capacities of dynamical systems in the spirit of the Turing machine.

Evidence that quantum computing is merely equivalent to classical computation would not yet justify dedicating so much effort to its study and implementation. We would rather expect to see that it is even superior by a decisive margin in part or all of the computation tasks: "Everything you can do, I can do better". In the sequel, a number of exemplary cases will be presented where this advantage becomes particularly manifest.

10.3.1 Quantum Dense Coding

In classical information processing, transforming a given bit string $\mathbf{a} \in \{0, 1\}^N$ to another bit string $\mathbf{a}' \in \{0, 1\}^N$ requires a zigzag walk of at least $d(\mathbf{a}', \mathbf{a})$ unit steps in the N-dimensional binary lattice, where $d(\mathbf{a}', \mathbf{a})$ is the Hamming distance between $\mathbf{a}$ and $\mathbf{a}'$ (see Sect. 9.3.1), i.e., the number of bits where they do not coincide. In quantum mechanics, an equivalent unitary transformation in a $2N$-dimensional Hilbert space amounts to a smooth rotation of the unit sphere in $\mathbb{C}^{2N}$. This allows for substantial shortcuts by rotating about skew axes, axes that may have components in all the dimensions of Hilbert space and need not coincide with those of the elementary rotations $\hat{\sigma}_{1m}, \hat{\sigma}_{2m}, \hat{\sigma}_{3m}$ around the canonical axes for each individual qubit m.

An elementary instance is *quantum dense coding*. It refers to the setting discussed in the context of the Einstein–Podolsky–Rosen (EPR) experiment, Sect. 8.4.3: The central concept are EPR or Bell pairs, entangled states of a pair of two-state systems that cannot be associated to anyone of the two subsystems involved but are shared, in the extreme case equally, between them, so that the information they carry is delocalized. For a pair of qubits a and b, there exist four Bell states, cf. Eq. (8.2.61). In the canonical basis (10.2.12), they are

$$|\psi_1\rangle = \frac{|00\rangle + |11\rangle}{\sqrt{2}} = \frac{1}{\sqrt{2}}\begin{pmatrix} 1 \\ 0 \\ 0 \\ 1 \end{pmatrix}, \quad |\psi_2\rangle = \frac{|01\rangle + |10\rangle}{\sqrt{2}} = \frac{1}{\sqrt{2}}\begin{pmatrix} 0 \\ 1 \\ 1 \\ 0 \end{pmatrix},$$

$$|\psi_3\rangle = \frac{|01\rangle - |10\rangle}{\sqrt{2}} = \frac{1}{\sqrt{2}}\begin{pmatrix} 0 \\ 1 \\ -1 \\ 0 \end{pmatrix}, \quad |\psi_4\rangle = \frac{|00\rangle - |11\rangle}{\sqrt{2}} = \frac{1}{\sqrt{2}}\begin{pmatrix} 1 \\ 0 \\ 0 \\ -1 \end{pmatrix}. \quad (10.3.1)$$

The four Bell states form an alternative orthonormal basis for their four-dimensional Hilbert space, thus define principal axes for that space inclined by 45° with respect to those corresponding to the canonical basis $\{|00\rangle, |01\rangle, |01\rangle, |11\rangle\}$. The clue for quantum dense coding is that they can be transformed into each other by rotations around the canonical axes, that is, by operating only on one of the two qubits, say the first one: Applying the Pauli matrices for qubit a, see Eq. (10.2.51), to $|\psi_1\rangle$, one obtains

$$\hat{I}_a|\psi_1\rangle = \frac{1}{\sqrt{2}}\begin{pmatrix} 1 & 0 & 0 & 0 \\ 0 & 1 & 0 & 0 \\ 0 & 0 & 1 & 0 \\ 0 & 0 & 0 & 1 \end{pmatrix}\begin{pmatrix} 1 \\ 0 \\ 0 \\ 1 \end{pmatrix} = \frac{1}{\sqrt{2}}\begin{pmatrix} 1 \\ 0 \\ 0 \\ 1 \end{pmatrix} = |\psi_1\rangle,$$

$$\hat{\sigma}_{1a}|\psi_1\rangle = \frac{1}{\sqrt{2}}\begin{pmatrix} 0 & 0 & 1 & 0 \\ 0 & 0 & 0 & 1 \\ 1 & 0 & 0 & 0 \\ 0 & 1 & 0 & 0 \end{pmatrix}\begin{pmatrix} 1 \\ 0 \\ 0 \\ 1 \end{pmatrix} = \frac{1}{\sqrt{2}}\begin{pmatrix} 0 \\ 1 \\ 1 \\ 0 \end{pmatrix} = |\psi_2\rangle,$$

$$\hat{\sigma}_{2a}|\psi_1\rangle = \frac{1}{\sqrt{2}}\begin{pmatrix} 0 & 0 & -i & 0 \\ 0 & 0 & 0 & -i \\ i & 0 & 0 & 0 \\ 0 & i & 0 & 0 \end{pmatrix}\begin{pmatrix} 1 \\ 0 \\ 0 \\ 1 \end{pmatrix} = \frac{1}{\sqrt{2}}\begin{pmatrix} 0 \\ -i \\ i \\ 0 \end{pmatrix} = -i|\psi_3\rangle,$$

$$\hat{\sigma}_{3a}|\psi_1\rangle = \frac{1}{\sqrt{2}}\begin{pmatrix} 1 & 0 & 0 & 0 \\ 0 & 1 & 0 & 0 \\ 0 & 0 & -1 & 0 \\ 0 & 0 & 0 & -1 \end{pmatrix}\begin{pmatrix} 1 \\ 0 \\ 0 \\ 1 \end{pmatrix} = \frac{1}{\sqrt{2}}\begin{pmatrix} 1 \\ 0 \\ 0 \\ -1 \end{pmatrix} = |\psi_4\rangle, \quad (10.3.2)$$

while manipulating only the second qubit b yields,

$$\hat{I}_b|\psi_1\rangle = |\psi_1\rangle, \, \hat{\sigma}_{1b}|\psi_1\rangle = |\psi_2\rangle, \, \hat{\sigma}_{2b}|\psi_1\rangle = i|\psi_3\rangle, \, \hat{\sigma}_{3b}|\psi_1\rangle = |\psi_4\rangle. \quad (10.3.3)$$

In conclusion, preparing a pair of qubits in Bell states, entanglement of the two subsystems allows to generate all the four basis states from a single initial state, manipulating only one of the qubits.

In this example, so far, classically unattainable economy in computation has been achieved by rotating entangled states, which themselves define skew directions in the original Hilbert space, around the canonical axes using the operators $\hat{I}_a, \hat{\sigma}_{ia}$ or $\hat{I}_b, \hat{\sigma}_{ib}, i \in \{1, 2, 3\}$. In fact, the same idea can be applied inversely to switch between the canonical factorizing (hence not entangled) states, rotating them in turn around skew axes. Consider the four Bell states (10.3.1) as canonical basis states for the Hilbert space spanned by a fictitious pair of "Bell qubits" $|\alpha\rangle_B, |\beta\rangle_B \in \{|0\rangle_B, |1\rangle_B\}$,

$$|\psi_1\rangle =: |0\rangle_B|0\rangle_B, |\psi_2\rangle =: |0\rangle_B|1\rangle_B, |\psi_3\rangle =: |1\rangle_B|0\rangle_B, |\psi_4\rangle =: |1\rangle_B|1\rangle_B,$$
$$(10.3.4)$$

where they appear as factorizing states. On these states, define rotation operators, say $\hat{I}_\alpha$ and $\hat{\tau}_{\alpha,i}, i \in \{1, 2, 3\}$, analogous to the Pauli matrices in the canonical basis, operating on the first of the two Bell qubits, and $\hat{I}_\beta, \hat{\tau}_{\beta,i}, i \in \{1, 2, 3\}$ operating on the second. In the original basis, the first set takes the form

$$\hat{I}_\alpha = \begin{pmatrix} 1 & 0 & 0 & 0 \\ 0 & 1 & 0 & 0 \\ 0 & 0 & 1 & 0 \\ 0 & 0 & 0 & 1 \end{pmatrix}, \hat{\tau}_{1\alpha} = \begin{pmatrix} 0 & 1 & 0 & 0 \\ 1 & 0 & 0 & 0 \\ 0 & 0 & 0 & -1 \\ 0 & 0 & -1 & 0 \end{pmatrix},$$

$$\hat{\tau}_{2\alpha} = \begin{pmatrix} 0 & 0 & i & 0 \\ 0 & 0 & 0 & i \\ -i & 0 & 0 & 0 \\ 0 & -i & 0 & 0 \end{pmatrix}, \hat{\tau}_{3\alpha} = \begin{pmatrix} 0 & 0 & 0 & 1 \\ 0 & 0 & 1 & 0 \\ 0 & 1 & 0 & 0 \\ 1 & 0 & 0 & 0 \end{pmatrix}. \quad (10.3.5)$$

Applied, for example, to the original basis state $|0_a, 0_b\rangle$ (both qubits in their ground state), they generate the three other elements of the canonical basis *by a single rotation*,

$$\hat{I}_\alpha|0_a, 0_b\rangle = \begin{pmatrix} 1 & 0 & 0 & 0 \\ 0 & 1 & 0 & 0 \\ 0 & 0 & 1 & 0 \\ 0 & 0 & 0 & 1 \end{pmatrix}\begin{pmatrix} 1 \\ 0 \\ 0 \\ 0 \end{pmatrix} = \begin{pmatrix} 1 \\ 0 \\ 0 \\ 0 \end{pmatrix} = |0_a, 0_b\rangle,$$

$$\hat{\tau}_{1\alpha}|0_a, 0_b\rangle = \begin{pmatrix} 0 & 1 & 0 & 0 \\ 1 & 0 & 0 & 0 \\ 0 & 0 & 0 & -1 \\ 0 & 0 & -1 & 0 \end{pmatrix}\begin{pmatrix} 1 \\ 0 \\ 0 \\ 0 \end{pmatrix} = \begin{pmatrix} 0 \\ 1 \\ 0 \\ 0 \end{pmatrix} = |0_a, 1_b\rangle,$$

$$\hat{\tau}_{2\alpha}|0_a, 0_b\rangle = \begin{pmatrix} 0 & 0 & i & 0 \\ 0 & 0 & 0 & i \\ -i & 0 & 0 & 0 \\ 0 & -i & 0 & 0 \end{pmatrix}\begin{pmatrix} 1 \\ 0 \\ 0 \\ 0 \end{pmatrix} = \begin{pmatrix} 0 \\ 0 \\ -i \\ 0 \end{pmatrix} = -i|1_a, 0_b\rangle,$$

$$\hat{\tau}_{3\alpha} |0_a, 0_b\rangle = \begin{pmatrix} 0 & 0 & 0 & 1 \\ 0 & 0 & 1 & 0 \\ 0 & 1 & 0 & 0 \\ 1 & 0 & 0 & 0 \end{pmatrix} \begin{pmatrix} 1 \\ 0 \\ 0 \\ 0 \end{pmatrix} = \begin{pmatrix} 0 \\ 0 \\ 0 \\ 1 \end{pmatrix} = |0_a, 1_b\rangle. \tag{10.3.6}$$

The same is true if they are applied to the other basis states, or if instead the second qubit is manipulated with the operators $\hat{\tau}_{\beta,i}$. These operators therefore achieve the same as the iterated permutation operator $\hat{U}_{0123}^n$, see Eq. (10.2.59), but operating only on one of the fictitious "Bell qubits" of Eq. (10.3.4).

Quantum dense coding has been introduced as a means that takes advantage of entanglement to effectively transmit two bits of information by manipulating only a single qubit. This is somewhat illusionary, however, as it depends on the reference qubit, the one that is left alone, to have been successfully prepared beforehand in a Bell state by some third party, not identical with the sender nor the receiver of the intended transmission. It appears more appropriate to see quantum dense coding as a shortcut, using the more compact topology of Hilbert space, as compared to the classical lattice $\{0, 1\}^M$, to efficiently navigate between basis states.

The increase in information density in quantum as compared to classical systems appears even more drastic if the task is storing bits, not transmitting them. As a fair benchmark, consider a computing device, able to simulate a spatially extended physical system, say of length L in one dimension. Classically, this system would be represented, discretizing the position with a resolution Δx, as a single integer n in the range $0 \leq n \leq N - 1$, if $N = \mathrm{int}(\Delta x / L) + 1$ is the number of bins to cover the spatial range L, or by a set of N complex Fourier coefficients, if the position is only specified by probabilities p_n to be at the site n. In any case, the required number of states of the simulator increases linearly with the size L of the system. In f dimensions or with f subsystems, it depends on this number as $N \sim L^f$, increasing exponentially with f. The information content, for example the required memory space, grows as $f \ln(N)$, logarithmically with the spatial system size and linearly with its dimension.

A suitable model for quantum mechanical simulation of an extended system would associate a two-state system to each point of the discretized space, representing an empty bin by the state $|0\rangle$ and an occupied one by $|1\rangle$ [Fey82]. Each bin could, for example, correspond to a phase-space cell of size $(2\pi\hbar)^f$ for f degrees of freedom. If every degree of freedom is represented by a finite number of M cells and every cell by a two-dimensional Hilbert space, the total Hilbert-space dimension for $N \sim M^f$ cells is then $2^{(M^f)}$. It is equivalent to a quantum memory size of M^f qubits, rapidly outnumbering the classical value. This comparison is summarized in Table 10.1.

In conclusion, omitting a host of details [Fey82], a classical computer, epitomized by a universal Turing machine, is insufficient to simulate a quantum system, if only for reasons of memory capacity. This seems to contradict the fact that, according to classical mechanics, the information density in phase space is indefinite, while quantum mechanics imposes an upper bound to it. A classical *computer*, however, is based from the outset on fixed finite number of states, organized in a particular rigid geometry.

10.3.2 Quantum Parallelism

As a bottom line of the previous subsection, quantum mechanics replaces zigzagging in an N-dimensional discrete lattice by transformations belonging to a continuous family of rotations in $\mathbb{C}^{2N}$, a space of much higher dimension. If quantum computation is applied to process classical information encoded in N bits, this has drastic consequences for the interfaces from the classical to the quantum level and back. The initial preparation of a quantum state representing a classical bit string ascends from a low dimensional to a high-dimensional space and thus is not unique; it leaves an enormous freedom of choice that can be exploited to optimize the efficiency of the computation. Conversely, going back from Hilbert space to the discrete classical lattice amounts to a projection, equivalent to quantum measurement (Sect. 8.4), part of the diversity in the higher-dimensional space has to be sacrificed. Programming these two steps in a clever way is at the heart of quantum computation; the solution is known as (*massive*) *quantum parallelism* (Fig. 10.6).

As indicated above, choosing rotation axes appropriately allows to perform unitary transformations of Hilbert space that affect all its dimensions simultaneously—such as, for example, a rotation about the space diagonal in three dimensions. This effect can be used in quantum computation to realize the same sequence of operations on a coherent superposition of a large number of independent quantum initial states in parallel, limited only by the total number of qubits available in the device. Terminating

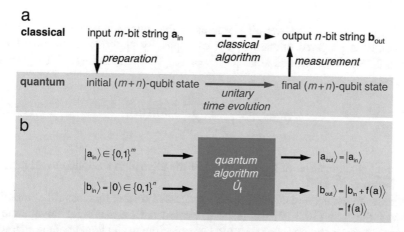

Fig. 10.6 Quantum parallelism. (**a**) In order to employ quantum computation to solve a classical problem, represented by an algorithm that maps an input string $\mathbf{a}_{in} \in \{0, 1\}^m$ to an output string $\mathbf{b}_{out} = \mathbf{f}(\mathbf{a}_{in}) \in \{0, 1\}^n$, a corresponding initial state composed of $(m + n)$ qubits has to be prepared. Processed in a suitable unitary time evolution, the output is read out by measurement. (**b**) If the quantum algorithm employed is designed such that the total output state includes the initial data $|a_{in}\rangle \in \{0, 1\}^m$ besides the output proper, it contains simultaneously the results for an entire 2^m-dimensional Hilbert space of input states, to be extracted by choosing the direction of projection by the final measurement appropriately, Eq. (10.3.11). See text.

the computation, however, only the result for a single initial condition can be read out by projection.

Consider a classical algorithm that maps an m-bit input string $\mathbf{a} \in \{0, 1\}^m$, $m \in \mathbb{N}$, to an n-bit output $\mathbf{f}(\mathbf{a}) \in \{0, 1\}^n$, $n \leq m$. To implement a quantum parallel evaluation, prepare an m-qubit data register for the input and an n-qubit target register for the output. According to the quantum Church-Turing thesis, the transformation

$$|\psi_{\text{in}}\rangle = |\mathbf{a}_{\text{in}}\rangle |\mathbf{b}_{\text{in}}\rangle \mapsto |\psi_{\text{out}}\rangle = |\mathbf{a}_{\text{out}}\rangle |\mathbf{b}_{\text{out}}\rangle = \hat{U}_{\mathbf{f}}|\psi_{\text{in}}\rangle = |\mathbf{a}_{\text{in}}\rangle |\mathbf{b}_{\text{in}} + \mathbf{f}(\mathbf{a}_{\text{in}})\rangle,$$
$$(10.3.7)$$

" $+$ " denoting binary addition, can be implemented on a quantum computer as a unitary transformation $\hat{U}_{\mathbf{f}}$ (the case that strictly $n < m$ appears to violate unitarity and requires special treatment). Choosing $|\mathbf{b}_{\text{in}}\rangle = |0_1\rangle \otimes \cdots \otimes |0_n\rangle =: |\mathbf{0}^{(n)}\rangle \in \{0, 1\}^n$, this transformation yields

$$|\mathbf{a}_{\text{in}}\rangle |\mathbf{0}^{(n)}\rangle \mapsto \hat{U}_{\mathbf{f}}|\mathbf{a}_{\text{in}}\rangle |\mathbf{0}^{(n)}\rangle = |\mathbf{a}_{\text{in}}\rangle |\mathbf{0}^{(n)} + \mathbf{f}(\mathbf{a}_{\text{in}})\rangle = |\mathbf{a}_{\text{in}}\rangle |\mathbf{f}(\mathbf{a}_{\text{in}})\rangle. \quad (10.3.8)$$

A suitable initial state $|\mathbf{a}\rangle$ to produce a coherent superposition of all outcomes $|\mathbf{f}(\mathbf{a})\rangle$ is product of Schrödinger-cat states for each of the input bits,

$$|\mathbf{a}_{\text{in}}\rangle = 2^{-m/2} \bigotimes_{j=1}^{m} \left(|0_j\rangle + |1_j\rangle \right) = 2^{-m/2} \sum_{\mathbf{a} \in \{0,1\}^m} |\mathbf{a}\rangle. \quad (10.3.9)$$

that is, a normalized sum over all elements of the canonical basis of $\{0, 1\}^m$. This state could be prepared efficiently, for example, by applying a Hadamard gate, Eq. (10.2.35), to the ground state $|\mathbf{0}^{(m)}\rangle = \otimes_{j=1}^{m} |0_j\rangle$. Operating with $\hat{U}_{\mathbf{f}}$ on this state gives

$$|\psi_{\text{out}}\rangle = 2^{-m/2} \sum_{\mathbf{a} \in \{0,1\}^m} |\mathbf{a}\rangle |\mathbf{f}(\mathbf{a})\rangle, \quad (10.3.10)$$

a coherent superposition of the outcomes of the algorithm $\mathbf{f}$ applied to all possible bit strings $\mathbf{a} \in \{0, 1\}^m$!

In order to extract a result that can be processed further by classical computing, however, the state (10.3.10) has to be projected onto a particular target state by measurement. An obvious choice would be projecting to an element of the canonical basis

$$\langle \mathbf{a}_{\text{out}}|\psi_{\text{out}}\rangle = 2^{-m/2} \sum_{\mathbf{a} \in \{0,1\}^m} \langle \mathbf{a}_{\text{out}}|\mathbf{a}\rangle |\mathbf{f}(\mathbf{a})\rangle = 2^{-m/2} |\mathbf{f}(\mathbf{a}_{\text{out}})\rangle. \quad (10.3.11)$$

So far, not much seems to be gained: In the end, all the prospects of having the results of 2^m calculations simultaneously in the output state has to be sacrificed to read out only a single one of them. That is too pessimistic, though. The superposition

(10.3.10) also allows to project onto more complex states which extract more global features of the result. The art of quantum programming is then to find such global aspects, sufficiently relevant also classically to lead to a real advantage for quantum computing.

10.3.3 The Deutsch and Deutsch-Jozsa Algorithms

One of the first quantum algorithms proposed to demonstrate the superiority of quantum computation in some tasks is an example for the clever use of the strategy alluded to above, extracting global features of the quantum mechanical result that classically would require a larger number of calculations from various initial conditions. The Deutsch algorithm and its extended version, the Deutsch-Jozsa algorithm, are of no immediate practical use but serve as models how to benefit from the special virtues of quantum computing.

The minimal condition for quantum "parallelism" is a device consisting of a pair of qubits, $n = m = 1$ in Eqs. (10.3.9–10.3.11), so that the algorithm $\mathbf{f}(\mathbf{a})$ reduces to a map $f(a)$, $\{0,1\} \rightarrow \{0,1\}$, that is, a classical unary logical gate such as the identity or NOT. Among the global features of these gates, there is one that can be checked in a single run of a quantum algorithm: the alternative "balanced" ($f(0) \neq f(1)$, identity or NOT) or "constant" ($f(0) = f(1)$, same output, 0 or 1, irrespective of input). Classically, this requires two evaluations of the algorithm. In order to solve the task with quantum parallelism, prepare both inputs as Schrödinger cat states, $|a_{in}\rangle = 2^{-1/2}(|0\rangle + |1\rangle)$, $|b_{in}\rangle = 2^{-1/2}(|0\rangle - |1\rangle)$, for example by passing initial values $|a\rangle = |0\rangle$, $|b\rangle = |1\rangle$, through a Hadamard gate, Eq. (10.2.35). A unitary quantum gate $\hat{U}_f$ as in Eq. (10.3.7) then generates the output

$$\hat{U}_f|a\rangle \in \mathbb{C}^2 \tag{10.3.12}$$

With $|a_{in}\rangle = 2^{-1/2}(|0\rangle + |1\rangle)$, the output depends on the map $f(a_{in})$ as follows,

$$\hat{U}_f|a_{in}\rangle|b_{in}\rangle = \frac{(-1)^{f(a_{in})}}{\sqrt{2}} \begin{cases} |0\rangle + |1\rangle \text{ if } f(0) = f(1) \\ |0\rangle - |1\rangle \text{ if } f(0) \neq f(1) \end{cases} \otimes \frac{1}{\sqrt{2}}(|0\rangle - |1\rangle), \tag{10.3.13}$$

Here the two alternatives do already not depend on the individual values of $f(a_{in})$ but on their coincidence "constant" versus "balanced".

The final step consists in rotating the two Schrödinger cats $|a\rangle = 2^{-1/2}(|0\rangle \pm |1\rangle)$ back to the canonical states $|0\rangle$, $|1\rangle$, by another Hadamard gate. The output is

$$|\psi_{out}\rangle = (-1)^{f(a_{in})} \begin{cases} |0\rangle \text{ if } f(0) = f(1) \\ |1\rangle \text{ if } f(0) \neq f(1) \end{cases} \otimes \frac{1}{\sqrt{2}}(|0\rangle - |1\rangle), \tag{10.3.14}$$

and can be read out by projection onto $|0\rangle$ or $|1\rangle$.

The gain by a factor 2 of quantum over classical computation may still appear marginal. However, it is straightforward to extend the Deutsch algorithm to scalar binary functions $f(\mathbf{a}_{in})$ of m-bit strings $\mathbf{a}_{in}$, corresponding to m-qubit states $|\mathbf{a}_{in}\rangle$. They include in particular the case $m = 2$, $n = 1$ of classical binary logical gates (Sect. 2.1.1).

On entrance, Hadamard gates operate individually on every qubit, giving

$$\left(\overset{m+1}{\underset{j=1}{\otimes}}\hat{H}\right)|\mathbf{a}_{in}\rangle|b_{in}\rangle = \left(\overset{m}{\underset{j=1}{\otimes}}\hat{H}\right)|\mathbf{0}^{(m)}\rangle\hat{H}|1\rangle = \frac{1}{\sqrt{2^m}}\overset{m}{\underset{j=1}{\otimes}}\left(|0_j\rangle + |1_j\rangle\right)\frac{|0\rangle - |1\rangle}{\sqrt{2^m}}$$

$$= \frac{1}{\sqrt{2^{m+1}}}\sum_{\mathbf{a}\in\{0,1\}^m}|\mathbf{a}\rangle(|0\rangle - |1\rangle). \tag{10.3.15}$$

As above in Eq. (10.3.12), the quantum gate $\hat{U}_f$ transforms this state to

$$\hat{U}_f\frac{1}{\sqrt{2^{m+1}}}\sum_{\mathbf{a}\in\{0,1\}^m}|\mathbf{a}\rangle(|0\rangle - |1\rangle) = \frac{1}{\sqrt{2^{m+1}}}\sum_{\mathbf{a}\in\{0,1\}^m}(-1)^{f(\mathbf{a})}|\mathbf{a}\rangle(|0\rangle - |1\rangle),$$

$$\tag{10.3.16}$$

a state that contains the results of the evaluation of $f(\mathbf{a}_{in})$ for every possible m-bit string as a coherent superposition. Finally, all the qubits of the data register $|\mathbf{a}\rangle$ are rotated back by another bit-by-bit Hadamard gate. Writing the action of this gate on a single qubit as

$$\hat{H}|a\rangle = \frac{1}{\sqrt{2}}\begin{Bmatrix}(|0\rangle + |1\rangle) \text{ if } a = 0\\(|0\rangle - |1\rangle) \text{ if } a = 1\end{Bmatrix} = \frac{1}{\sqrt{2}}\sum_{b\in\{0,1\}}(-1)^{ab}|b\rangle, \tag{10.3.17}$$

the action of $\hat{H}$ on the entire m-qubit state takes the form

$$\hat{H}|\mathbf{a}\rangle = \frac{1}{\sqrt{2}}\sum_{\mathbf{b}\in\{0,1\}^m}(-1)^{\mathbf{a}\cdot\mathbf{b}}|\mathbf{b}\rangle, \tag{10.3.18}$$

with the dot product to be understood as $\mathbf{a}\cdot\mathbf{b} = \sum_{j=1}^m a_jb_j$. Applied to the transformed state (10.3.16), this means

$$|\psi_{out}\rangle = \frac{1}{2^m}\sum_{\mathbf{a},\mathbf{b}\in\{0,1\}^m}(-1)^{\mathbf{a}\cdot\mathbf{b}+f(\mathbf{a})}|\mathbf{b}\rangle\frac{|0\rangle - |1\rangle}{\sqrt{2}}. \tag{10.3.19}$$

Equation (10.3.19) is readily evaluated in the two extreme cases of constant versus balanced function $f(\mathbf{a})$. Consider in particular the term $|\mathbf{b}\rangle = \otimes_{j=1}^m|0_j\rangle$ in the sum over $\mathbf{b} \in \{0, 1\}^m$ in Eq. (10.3.19), contributing the amplitude $2^{-m} \sum_{\mathbf{a} \in \{0,1\}^m} (-1)^{f(\mathbf{a})}$. If for all $\mathbf{a} \in \{0, 1\}^m$, $f(\mathbf{a}) = c$, $c \in \{0,1\}$, generalizing the upper option in Eq. (10.3.14) to m qubits, the sign $(-1)^{f(\mathbf{a})}$ factors out in the sum over $\mathbf{a} \in \{0, 1\}^m$ in Eq. (10.3.19). The term $|\mathbf{b}\rangle = |\mathbf{0}^{(m)}\rangle$ then exhausts already the full norm of $|\psi_{\text{out}}\rangle$, so that all other terms $|\mathbf{b}\rangle \neq |\mathbf{0}^{(m)}\rangle$ must vanish. In the opposite case of an equal number of function values $f(\mathbf{a}) = 0$ and $f(\mathbf{a}) = 1$, positive and negative terms $(-1)^{f(\mathbf{a})}$ in the sum over $\mathbf{a}$ cancel exactly, so that now the amplitude for $|\mathbf{b}\rangle = |\mathbf{0}^{(m)}\rangle$ vanishes and one of the remaining terms with one or more qubits $= |1\rangle$, must contribute with nonzero amplitude. Reading out all the m qubits in the data register $|\mathbf{a}\rangle$ therefore allows to exclude one of the two cases, constant or balanced, after just a single evaluation of the function f. By contrast, a classical computation would require evaluating it at least for half of all the possible input bit strings plus one, in order to exclude even the least favourable case for a balanced function, that is, at least $2^{m-1} + 1$ times.

10.3.4 Quantum Fourier Transform

The Deutsch and Deutsch-Josza algorithms are important and have a pedagogical value as paradigms of quantum parallel processing, yet are of limited practical use. A much more serious case, with a host of applications as essential ingredient of other quantum algorithms and also outside quantum computation, is quantum Fourier transform. The fundamental role Fourier transformations play in countless tasks of information-conserving data processing has been illustrated in Sect. 2.3.

Fourier transformation, as a unitary transformation, is very close to quantum mechanics. In particular, it relates the representations in position and in momentum space to one another and thus is essential for the wave-particle duality. Of special importance is the discrete Fourier transform: Restricting both position and momentum to arbitrarily large but finite spaces (as is inevitable in computer simulations), both coordinates become discrete at the same time, comprising the same number $N \in \mathbb{N}$ of coefficients per dimension. Denoting position and momentum amplitudes as $\psi_n = \langle n|\psi\rangle$ and $\tilde{\psi}_\lambda = \langle \lambda|\psi\rangle$, resp., they are related by

$$\psi_n = \sum_{\lambda=0}^{N-1} \left(\hat{F}_N\right)_{n\lambda} \tilde{\psi}_\lambda, \quad \tilde{\psi}_\lambda = \sum_{\lambda=0}^{N-1} \left(\hat{F}_N^{-1}\right)_{n\lambda} \psi_n, \quad \left(\hat{F}_N\right)_{n\lambda} = \frac{1}{\sqrt{N}} \exp\left(2\pi i \frac{n\lambda}{N}\right).$$

$$(10.3.20)$$

Constructing the matrix $\hat{F}_N$ is a central task in Fourier transformation and consumes the lion's part of the total amount of operations. Fast Fourier transform (FFT), specifically in the form of the Sande-Tukey algorithm (Sect. 2.3.4), provides a particularly efficient way to generate it, reducing the required number of floating-point operations from $O(N^2)$ to $O(N\ln(N))$.

It decomposes Fourier transformation into a self-similar sequence of $\ln(N)$ operations on the set of Fourier coefficients. Each one comprises a transformation to sum-and-difference coordinates with a phase shift on the difference term, applied to parts of the coefficient vector, which decrease at each step by 2 in size but increase in number by the same factor. Quantum Fourier transformation largely adopts this scheme but achieves further compactness and CPU time reduction by associating the nested bisection of the coefficient vector to a corresponding set of qubits, see Fig. 10.7. While the structure of 2^m sets of 2^{n-m} coefficients, if $N = 2^n$, induced by the Sande-Tukey algorithm, invites to being exploited in vectorizing or parallelizing computer architectures, quantum Fourier transform goes a radical step further: It associates each 2^{n-m}-dimensional coefficient vector to a single qubit, to undergo the same transformation. The way FFT handles this operation, namely as a linear vector transformation represented by a unitary matrix, is already very close to quantum algorithms, as is its recursive structure, proceeding step by step from (2×2)- to (4×4)-, ..., to $(2^n \times 2^n)$-blocks, see Eqs. (2.3.55–2.3.58).

In order to translate classical to quantum operations, it suggests itself to interpret the N-dimensional vector space of coefficient sets $\mathbf{f} = (f_0, ..., f_{N-1})$ as a Hilbert space of the same dimension, spanned by basis states $|l\rangle \in \{|0\rangle, |1\rangle\}^N$, so that, for a single basis state $|l\rangle$, Fourier transformation takes the form, cf. Eq. (2.3.55),

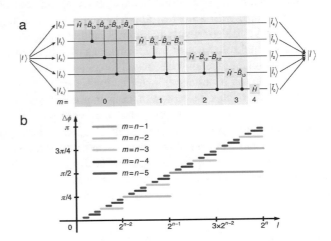

Fig. 10.7 Quantum Fourier transformation. **a** A quantum algorithm for discrete Fourier transformation, Eq. (10.3.22), is construed adopting the scheme of classical Fast Fourier Transformation (Sect. 2.3.4, Fig. 2.9a). It decomposes the transformation of an N-component coefficient vector, $N = 2^n$, into a sequence of $n = \mathrm{lb}(N)$, here $n = 5$, major steps $m = 1, ..., n$, each composed of a transformation to sum-and-difference coordinates, incorporated as a Hadamard gate $\hat{H}$ and $n-m$ controlled phase shifts $\hat{B}_{k,m}$, Eq. (10.3.32). They operate directly on qubits $|l_m\rangle$ representing the mth component of a binary decomposition of the input vector $|l\rangle$, $l = l_0 \times 2^0 + l_1 \times 2^1 + ...$ As in classical FFT, these binary components appear on exit in inverse order and have to be interchanged in the final output vector $|\vec{l}\rangle$, cf. Fig. 2.9b. **b** The accumulated phase shifts for each qubit $|l_m\rangle$ form a self-similar structure for $n \to \infty$. Note that owing to different conventions in classical and quantum Fourier transformation, the step index m counts in opposite order here, $m_{\mathrm{quantum}} = n + 1 - m_{\mathrm{classical}}$.

$$|l\rangle \mapsto |l'\rangle = \hat{F}_N|l\rangle = \frac{1}{\sqrt{N}} \sum_{k=0}^{N-1} \exp\left(2\pi i \frac{kl}{N}\right)|k\rangle, \tag{10.3.21}$$

and for an arbitrary Hilbert-space vector $|\mathbf{f}\rangle \in \mathbb{C}^{2N}$,

$$|\mathbf{f}\rangle = \sum_{l=0}^{N-1} f_l|l\rangle \mapsto |\tilde{\mathbf{f}}\rangle = \hat{F}_N|\mathbf{f}\rangle = \frac{1}{\sqrt{N}} \sum_{k,l=0}^{N-1} f_k \exp\left(2\pi i \frac{kl}{N}\right)|l\rangle. \tag{10.3.22}$$

In view of the nested structure of the algorithm, it is appropriate to represent the state index l in binary notation,

$$l = \sum_{\lambda=0}^{n-1} l_\lambda 2^{n-\lambda-1} = l_{n-1} + 2l_{n-2} + 4l_{n-3} + \cdots + 2^{n-1}l_1, \ l_\lambda \in \{0, 1\}, \tag{10.3.23}$$

which suggests interpreting the basis state $|l\rangle$ as composed of n qubits of decreasing significance from $|l_0\rangle$ down to $|l_{n-1}\rangle$.

Organized like this, the first of n subsequent similar operations counted by $m = 0, \ldots, n-1$ (coloured blocks in Fig. 10.7a), is equivalent to the Fourier transform for $n = 1$, that is, a single qubit. It reduces to the Hadamard gate applied to the least significant qubit $|l_{n-1}\rangle$, cf. Eq. (2.3.55) (lowermost horizontal line in Fig. 10.7a),

$$\hat{F}_2 = \frac{1}{\sqrt{2}} \begin{pmatrix} 1 & 1 \\ 1 & -1 \end{pmatrix} = \frac{1}{\sqrt{N}} \begin{pmatrix} 1 & 1 \\ 1 & \exp(2\pi i/2^1) \end{pmatrix}, \tag{10.3.24}$$

that means,

$$|l_n\rangle \mapsto \frac{1}{\sqrt{2^l}}\left(|0\rangle + \exp(2\pi i l_n 2^{-1})|1\rangle\right). \tag{10.3.25}$$

The step from $m = 1$ to $m = 2$, thus to a 2-qubit Fourier transform, already involves complex phase shifts $\exp(i\pi/2)$. The sum-and-difference transformation involved in each step of the Sande-Tukey algorithm, see Eq. (2.3.63), amounts to applying a Hadamard gate to both qubits. The additional phase shift in the difference component is accomplished by applying a $\pi/4$-phase shift, as in Eq. (10.2.37), controlled by the least significant qubit, to the second least significant qubit, $|l_{n-1}\rangle$,

$$\hat{R}_2 := \begin{pmatrix} 1 & 0 \\ 0 & \exp(2\pi i/2^2) \end{pmatrix}. \tag{10.3.26}$$

Together with the Hadamard gate applied to $|l_{n-1}\rangle$, this amounts to

$$|l_{n-2}\rangle|l_{n-1}\rangle \mapsto$$

$$\frac{1}{\sqrt{2^2}}\left[|0_{n-1}\rangle + \exp\left(2\pi i\frac{l_{n-1}}{2^1}\right)|1_{n-1}\rangle\right]\left[|0_{n-2}\rangle + \exp\left(2\pi i\left(\frac{l_{n-1}}{2^1} + \frac{l_{n-2}}{2^2}\right)\right)|1_{n-2}\rangle\right]. \quad (10.3.27)$$

Continuing in the same say through all the n qubits up to $|l_1\rangle$ results in

$$|l_0\rangle|l_1\rangle \ldots |l_{n-2}\rangle|l_{n-1}\rangle \mapsto \frac{1}{\sqrt{2^n}}\bigotimes_{m=0}^{n-1}\left[|0_m\rangle + \prod_{m'=m}^{n-1}\exp\left(2\pi i\frac{l_{m'}}{2^{n-m'}}\right)|1_m\rangle\right]$$

$$= \frac{1}{\sqrt{2^n}}\bigotimes_{m=0}^{n-1}\left[|0_m\rangle + \exp(2\pi i \times 0.l_{n-m-1}\ldots l_{n-1})|1_m\rangle\right],$$

$$(10.3.28)$$

employing binary notation $0.l_0\,l_1\,l_2\,\ldots\,l_{n-1} = \sum_{m=0}^{n-1}l_m 2^{-m-1}$, $l_m \in \{0, 1\}$.

As is the case in the Sande-Tukey algorithm, on output the qubits appear in inverse order, the least significant one in the leading position etc., and have to be reversed in a final swapping operation to give

$$|l_0\rangle|l_1\rangle \ldots |l_{n-2}\rangle|l_{n-1}\rangle \mapsto \frac{1}{\sqrt{2^n}}\left[|0_{n-1}\rangle + \exp(2\pi i \times 0.l_{n-1})|1_{n-1}\rangle\right]\otimes\cdots$$

$$\otimes\left[|0_0\rangle + \exp(2\pi i \times 0.l_0\ldots l_{n-1})|1_0\rangle\right]. \quad (10.3.29)$$

Equation (10.3.29) is equivalent to a discrete Fourier transformation. Indeed, using the binary decomposition of the basis states, $|l\rangle = |l_0\rangle \otimes \cdots \otimes |l_{n-1}\rangle$, $|l_m\rangle \in \{|0\rangle, |1\rangle\}$, $m = 0, \ldots, n - 1$,

$$|l\rangle \mapsto |\tilde{l}\rangle = \frac{1}{\sqrt{2^n}}\sum_{k=0}^{2^n-1}\exp\left(2\pi i\frac{kl}{2^n}\right)|k\rangle$$

$$= \frac{1}{\sqrt{2^n}}\sum_{k_1=0}^{1}\cdots\sum_{k_n=0}^{1}\exp\left(2\pi i l\sum_{m=1}^{n}2\pi i k_m/2^m\right)|k_1\rangle \otimes\cdots\otimes|k_n\rangle$$

$$= \frac{1}{\sqrt{2^n}}\sum_{k_1=0}^{1}\cdots\sum_{k_n=0}^{1}\bigotimes_{m=1}^{n}\exp\left(2\pi i k_m/2^m\right)|k_m\rangle. \quad (10.3.30)$$

Interchanging sums and product,

$$|\tilde{l}\rangle = \frac{1}{\sqrt{2^n}}\bigotimes_{m=1}^{n}\sum_{k_m=0}^{1}\exp(2\pi i l k_m/2^m)|k_m\rangle$$

$$= \frac{1}{\sqrt{2^n}}\bigotimes_{m=1}^{n}\left(|0_m\rangle + \exp\left(2\pi i l\sum_{m'=m}^{n}k_{m'}2^{-m'}\right)|1_m\rangle\right). \quad (10.3.31)$$

agreement with Eq. (10.3.29) is reached.

The sequence of factors in Eq. (10.3.31), counted by the index m, $1 \leq m \leq n = \mathrm{lb}(N)$, represents the quantum analogue of the n major steps in FFT, see Fig. 2.7. The difference is, however, that in the quantum algorithm, at each step m, the unitary transformation $\hat{U}_{\mathrm{QFT},m}$ operates only on a single qubit $|l_m\rangle$, (but controlled by $m-1$ other qubits). This operation can be expressed in turn as a sequence of $n-m$ elementary single qubit gates $\hat{B}_{k,m}$, $k = 2, \ldots, n-m+1$, controlled by qubit $|l_k\rangle$ and preceded by the Hadamard gate,

$$
\hat{U}_{\mathrm{QFT}} = \prod_{m=1}^{n} \hat{U}_{\mathrm{QFT},m}, \quad \hat{U}_{\mathrm{QFT},m} = \hat{B}_{2,m} \times \cdots \times \hat{B}_{n-m+1,m} \hat{H}_m,
$$

$$
\hat{B}_{k,m} = \begin{pmatrix} \hat{I}_k^{(2)} & \hat{0}^{(2)} \\ \hat{0}^{(2)} & \begin{pmatrix} 1 & 0 \\ 0 & \exp(2\pi i/2^k) \end{pmatrix} \end{pmatrix}_m, \tag{10.3.32}
$$

with $\hat{I}_k^{(2)}$ and $\hat{H}_k$ denoting the identity and the Hadamard gate acting on qubit k. In this way, the number of single qubit gates required sums up to $n + (n-1) + \cdots + 1 = n(n+1)/2$. Together with the final reordering of bits, a total of $O\big((\ln N)^2\big)$ operations results, to be compared with $O(N \ln N)$ for the classical FFT (Fig. 10.8). The gain for quantum Fourier transform is a consequence of the fact that at each step, only up to $\mathrm{lb}(N)$ gates are involved, while the corresponding number scales as N in the classical case.

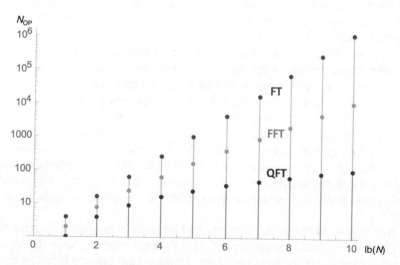

Fig. 10.8 Number of floating-point operations required for classical Fourier transform (FT, pink) versus Fast Fourier Transform implemented as Sande-Tukey algorithm (FFT, green), and number of elementary qubit operations required for the quantum Fourier transform (QFT, blue), as functions of the number of Fourier coefficients to be transformed.

Figure 10.7b depicts the phase shifts involved in quantum Fourier transformation in a schematic manner. It is evident that with growing number n of qubits and of transformation steps, relative phases $\Delta\phi$ involved in the transformation, specifically in the controlled phase-shift gates $\hat{B}_{k,m}$, Eq. (10.3.32), get increasingly smaller, comprising fractional values $\Delta\phi = 2\pi l/2^m$, $l = 1, \ldots, 2^n$, $m = 1, \ldots, n$. In this way, quantum Fourier transformation approaches continuous phases, challenging the precision of physical quantum processes required and threatening the discrete character of quantum computation.

This phase-sensitive character of quantum Fourier transformations is reflected also in its application as building block of more complex quantum algorithms. Notably algorithms for the estimation of eigenvalues $\exp(i\phi)$ of black-box gates, to identify the elementary period of periodic functions, and for the factoring of natural numbers employ quantum Fourier transformation as powerful tool to determine and manipulate phases. For details see, e.g., Ref. [NC00].

10.3.5   Quantum Search Algorithms

Quantum Fourier transformation with its arbitrarily small relative phases is a first indication that quantum computation tends to free itself from the narrow framework of strictly discrete combinations of $\pi/2$ rotations in Hilbert space. Quantum search algorithms go much further in this direction, in that they also work with arbitrary rotation angles. Above all, they do not satisfy the fundamental requirement that a classical algorithm provide a definite and unique answer to the question it is expected to solve. Instead, they generally achieve an answer that is only true with high probability, but not with certainty.

In essence, a search problem reduces to a pattern recognition task: In a search space of N items, find those items x, if any, that meet the criterion $F(x) = Y$. Quantum search algorithms emphatically are *not* intended to fulfil this task. They presuppose that some device exists that accomplishes the recognition proper, which may well be based on classical technology (e.g., a camera equipped with image recognition software). Once the positive items are registered as computer readable data, the job of the search algorithm is rather to identify and mark these positive items among a large number of negative ones. That is, they perform a very efficient quantum mechanical pattern matching, once the original feature profile has been encoded in qubits, transforming it to a subspace of the Hilbert space encompassing the full search space.

This suggests to introduce putatively two mutually orthogonal subspaces of the Hilbert space of a dimension corresponding to the entire search space, a Hilbert space $\mathcal{H}_+$ of solutions $|x\rangle$ of the search problem and the complementary Hilbert space $\mathcal{H}_-$ of states that are not solutions, with dimensions N_+ and N_-, resp., $N_+ + N_- = N$. Assume moreover a recognition algorithm to exist that marks solutions of the search problem with a single-bit flag,

$$f(|x\rangle) = \begin{cases} 1 \text{ if } |x\rangle \in \mathcal{H}_+ \Leftrightarrow F(x) = Y, \\ 0 \text{ if } |x\rangle \in \mathcal{H}_- \Leftrightarrow F(x) \neq Y. \end{cases} \tag{10.3.33}$$

It can be realized as a unitary operation that employs a signaling qubit $|q\rangle$, such that $|x\rangle|q\rangle \to |x\rangle|q \oplus f(x)\rangle$, denoting Boolean addition as $\oplus$. Within each subspace, representative states $|\psi_+\rangle$ and $|\psi_-\rangle$ can be defined as superpositions of all solution states $|x_+\rangle$ and all non-solution states $|x_-\rangle$, resp.,

$$|\psi_+\rangle = \frac{1}{\sqrt{N_+}} \sum_{|x_+\rangle \in \mathcal{H}_+} |x_+\rangle, \; |\psi_-\rangle = \frac{1}{\sqrt{N_-}} \sum_{|x_-\rangle \in \mathcal{H}_-} |x_-\rangle. \tag{10.3.34}$$

The key idea of the Grover [Gro97] search algorithm is to start from a neutral trial state $|\psi\rangle$ prepared within the two-dimensional subspace spanned by orthonormal basis states $|\psi_+\rangle$ and $|\psi_-\rangle$ (to be determined, e.g., by orthonormalization),

$$|\psi\rangle = \frac{1}{\sqrt{N}} \sum_{|x\rangle \in \mathcal{H}} |x\rangle = c_+|\psi_+\rangle + c_-|\psi_-\rangle,$$
$$c_+ = \sqrt{\frac{N_+}{N}} = \sin\left(\frac{\phi}{2}\right), \; c_- = \sqrt{\frac{N_-}{N}} = \cos\left(\frac{\phi}{2}\right), \; \phi := 2\arctan\left(\sqrt{\frac{N_+}{N_-}}\right). \tag{10.3.35}$$

and to rotate it within this subspace as close as possible towards $|\psi_+\rangle$. This amounts to eliminate as far as possible the components parallel to $|\psi_-\rangle$ from an unknown vector $|\psi\rangle$. The task can be solved with a geometrical trick (see Fig. 10.9): Flip $|\psi\rangle$ around the axis parallel to $|\psi_-\rangle$ (operation $\hat{O}$), flip the resulting vector back around the vector $|\psi\rangle$ (operation $\hat{G}$). This will reduce the $|\psi_-\rangle$ component in $|\psi\rangle$. This process is achieved by repeated application of a single operator that rotates its input state by a fixed angle in the desired direction, composed of the following elementary steps [NC00]:

1. Mark solution states $|x_+\rangle \in \mathcal{H}_+$ with a phase -1, applying the recognition (10.3.33), by multiplying them with a Schrödinger cat $|q\rangle =: I|q\rangle = (|0\rangle - |1\rangle)/\sqrt{2}$,

$$|x\rangle \frac{1}{\sqrt{2}}(|0\rangle - |1\rangle) \to (-1)^{f(x)}|x\rangle \frac{1}{\sqrt{2}}(|0\rangle - |1\rangle) \tag{10.3.36}$$

and discarding the auxiliary state again, so that $|x\rangle \to (-1)^{f(x)}|x\rangle =: \hat{O}|x\rangle$ remains, defining the operator $\hat{O}$ (referred to as "oracle").

2. Apply a Hadamard gate $\otimes_{n=1}^{N} \hat{H}_n$ to all components of $|\psi\rangle$.
3. Shift the phases of all states, except $|0\rangle$, by π : $|x\rangle \to (-1 + 2\delta_{x-0})|x\rangle$.
4. Undo the Hadamard operation by applying $\otimes_{n=1}^{N} \hat{H}_n$ anew.

The effects of the last three operations can be summed up to the operator $\hat{G}$.

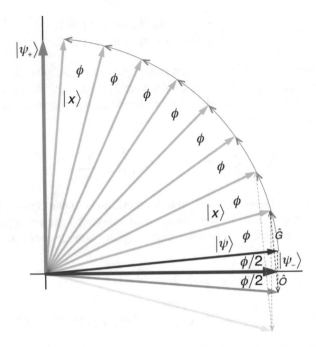

Fig. 10.9 The quantum search algorithm as repeated reflections and rotations. Within the two-dimensional Hilbert space spanned by balanced superpositions $|\psi_+\rangle$ of all solutions of the search (green vector) and of all blanks $|\psi_-\rangle$, the Grover search algorithm combines two reflections (dotted arrows), from the starting state $|\psi\rangle$ (blue) about $|\psi_-\rangle$ and then back again about $|\psi\rangle$, to a rotation by an angle ϕ towards $|\psi_+\rangle$. Iterating the procedure brings the system to within an angular distance $\leq \phi$ of the solution space $|\psi_+\rangle$, with a number of $O(\sqrt{N_+/N})$ of iterations, if N_+ is the dimension of the space of solutions and N the total number of items searched through.

$$|x\rangle \rightarrow \hat{G}|x\rangle = \left(\overset{N-1}{\underset{n=0}{\otimes}} \hat{H}_n\right)\left(2|0\rangle\langle 0| - \hat{I}_{\mathcal{H}}\right)\left(\overset{N-1}{\underset{n=0}{\otimes}} \hat{H}_n\right)|x\rangle = \left(2|\psi\rangle\langle\psi| - \hat{I}_{\mathcal{H}}\right)|x\rangle,$$

$$(10.3.37)$$

denoting the identity in $\mathcal{H}$ as $\hat{I}_{\mathcal{H}}$. Together with step 1, they indeed result in $|\psi\rangle$ approaching $|\psi_+\rangle$. It inverts the sign of all components of $|\psi\rangle$ parallel to $|\psi_+\rangle$, that is, reflects $|\psi\rangle$ about $|\psi_-\rangle$ (Fig. 10.9),

$$|\psi\rangle = \sin\left(\frac{\phi}{2}\right)|\psi_+\rangle + \cos\left(\frac{\phi}{2}\right)|\psi_-\rangle \mapsto -\sin\left(\frac{\phi}{2}\right)|\psi_+\rangle + \cos\left(\frac{\phi}{2}\right)|\psi_-\rangle,$$

$$(10.3.38)$$

performing a rotation by ϕ in the desired plane spanned by $|\psi_+\rangle$ and $|\psi_-\rangle$, but not necessarily towards $|\psi_+\rangle$. The remaining steps however, as is manifest in Eq. (10.3.37), amount to a reflection of the resulting state vector about $|\psi\rangle$ again, thus in a rotation

$$|\psi\rangle \mapsto \sin\left(\frac{3\phi}{2}\right)|\psi_+\rangle + \cos\left(\frac{3\phi}{2}\right)|\psi_-\rangle. \tag{10.3.39}$$

It increases the angle separating $|\psi\rangle$ from $|\psi_-\rangle$ by ϕ.

Iteration of the procedure k times leads to a total rotation by $k\phi$ with respect to the initial state $|\psi\rangle$,

$$|\psi\rangle \mapsto \sin\left(\frac{(2k+1)\phi}{2}\right)|\psi_+\rangle + \cos\left(\frac{(2k+1)\phi}{2}\right)|\psi_-\rangle. \tag{10.3.40}$$

The rationale of the algorithm is that in this way, the coefficient c_+ will eventually reach the value 1 or at least approach it so closely that a projection of the final state by measurement onto the canonical basis will reveal with high probability the binary code for the solution state $|\psi_+\rangle$ or the subspace of solutions of the search problem.

Only exceptionally will ϕ be a rational fraction of π, $\phi = p\pi/q$, $p, q \in \mathbb{Z}$, so that after a finite number of iterations of steps 1 to 4 above, the state $|\psi\rangle$ would coincide identically with $|\psi_+\rangle$, eliminating all negative items in the list. For example, if $N = 4$ and $N_+ = 1$, Eq. (10.3.35) gives $\phi = \pi/3$, so that, cf. Equation (10.3.39), the solution is reached exactly after only a single iteration with a single call of the recognition device. In all other cases, a small deviation from an exact hit will remain. In fact, the quantum search algorithm is intended in particular for the regime $N_+ \ll N_-, N$, of very rare positive search results among numerous blanks, that is, to find the proverbial "needle in a haystack" [Gro97]. In this case, the angle increment ϕ per iteration is small, the approach to a hit is creeping, and the process is close to a rotation in continuous time.

In this regime, the deviation in angle of the closest approximation from the exact hit $|\psi_+\rangle$ is no larger than $\phi/2 = \arcsin(\sqrt{N_+/N})$. This reduces the probability *not* to find $|\psi_+\rangle$, i.e., to obtain the output $|\psi_-\rangle$ instead, to N_+/N. Arriving there all the way from within the same distance from $|\psi_-\rangle$ requires

$$O\left(\frac{\arccos\sqrt{(N_+/N)}}{\phi}\right) = O\left(\frac{\arccos\sqrt{(N_+/N)}}{\arcsin\sqrt{(N_+/N)}}\right) = O\left(\sqrt{\frac{N}{N_+}}\right) \tag{10.3.41}$$

repetitions of the operation $\hat{G}$, while a classical search, browsing the list item by item, needs N steps.

The closeness of this procedure to a continuous rotation in Hilbert space can be substantiated by constructing a Hamiltonian that generates the Grover iteration as a unitary time evolution. Since the rotation takes place in the plane spanned by the solution state $|\psi_+\rangle$ and the trial state $|\psi\rangle$, it is plausible that the Hamiltonian required should contain projectors $|\psi_+\rangle\langle\psi_+|$ and $|\psi\rangle\langle\psi|$. Indeed, the simple choice

$$\hat{H} = |\psi\rangle\langle\psi| + |\psi_+\rangle\langle\psi_+| \tag{10.3.42}$$

does the job. From a physical point of view, it looks quite exotic, not resembling anything like kinetic or potential energies, but such interpretations lose their meaning in the context of two-state systems anyway. Applying an orthonormalization as above, $|\psi\rangle = c_+|\psi_+\rangle + c_-|\psi_-\rangle, |c_-|^2 + |c_+|^2 = 1$, and rotating $|\psi_+\rangle$ and $|\psi_-\rangle$ in the complex plane such that $c_-, c_+ \in \mathbb{R}^+$, allows to express $\hat{H}$ in terms of Pauli matrices,

$$
\hat{H} = \begin{pmatrix} 1 & 0 \\ 0 & 0 \end{pmatrix} + \begin{pmatrix} c_+^2 & c_+c_- \\ c_+c_- & c_-^2 \end{pmatrix} = \begin{pmatrix} 1 & 0 \\ 0 & 1 \end{pmatrix} + \begin{pmatrix} 0 & c_+^*c_- \\ c_+c_-^* & 0 \end{pmatrix} + \begin{pmatrix} c_+^2 & 0 \\ 0 & -c_+^2 \end{pmatrix}
$$

$$
= \hat{I} + c_+c_-\hat{\sigma}_1 + c_+^2\hat{\sigma}_3 = \hat{I} + c_+(c_-\hat{\sigma}_1 + c_+\hat{\sigma}_3). \tag{10.3.43}
$$

The time evolution of the trial state $|\psi\rangle$ then takes the form (see Eq. (10.2.28) for exponentials $\exp(-i\phi\mathbf{n}\cdot\hat{\boldsymbol{\sigma}})$ of Pauli matrices),

$$
\begin{aligned}
|\psi(t)\rangle &= \exp(-i\hat{H}t)|\psi\rangle \\
&= \exp(-it)\big(\cos(c_+t) - i\sin(c_+t)\big(c_-\hat{\sigma}_1 + c_+\hat{\sigma}_3\big)\big)|\psi\rangle \\
&= \exp(-it)(\cos(c_+t)|\psi\rangle - i\sin(c_+t)|\psi_+\rangle). \tag{10.3.44}
\end{aligned}
$$

Up to the physically irrelevant global phase $\exp(-it)$, the state $|\psi(t)\rangle$ rotates in the plane defined by $|\psi\rangle$ and $|\psi_+\rangle$ and passes through the target state $|\psi_+\rangle$ at times

$$
t_k = (2k+1)\frac{\pi}{2c_+}, \quad k \in \mathbb{Z}. \tag{10.3.45}
$$

In particular, if $|\psi\rangle$ is chosen, as in Eq. (10.3.35) with $N_+ = 1$, as a balanced superposition of all states of the canonical basis, so that $c_+ = 1/\sqrt{N}$, the first hit occurs at $t_1 = \pi\sqrt{N}/2$, independently of $|\psi_+\rangle$ and in agreement with Eq. (10.3.41).

With the Grover algorithm, quantum computation ultimately leaves the scheme of walks on a high-dimensional discrete lattice behind that underlies all classical computing. It ventures out into the vast space of arbitrary continuous rotations on a hypersphere, thereby establishing a new paradigm of computation, qualitatively different from all classical computation schemes.

10.4   Decoherence and Error Correction

In conventional classical computing, errors are not a serious concern. The reason is that computers are shielded against thermal and other noise sources, see Sect. 9.4.3, by a host of protection measures on the hardware as well as on the software level. For example, on the basic level of binary operations in semiconductor circuits, bits are protected by high potential walls forming wells and channels, much higher than the thermal energy k_BT, see Fig. 9.7, and manipulated by electromagnetic pulses with a rapidly abating response. They are therefore guided by a strongly structured energy landscape.

In quantum computation, there are no such guiding rails. The unitary character of its operations means that computing moves on a smooth energy landscape without any predetermined discrete paths. It is like steering a ball over a golf course, or rather over a billiard table where collisions with other balls are part of the dynamics. Errors are not damped out but accumulate or are even amplified. Quantum computation is therefore systematically more sensitive to errors than conventional classical computing, but closely resembles classical implementations of reversible computing, see Sect. 9.2.2, in their vulnerability by noise.

From the point of view of the deterministic computation process, errors become manifest as noise, irrespective of the specific source, be it thermal fluctuations, cosmic radiation, or imprecisions in the construction and handling of the apparatus. In quantum computing, it includes a phenomenon absent in classical computation: decoherence. The loss of phase correlations is a particularly critical threat of quantum computation. Noise, in turn, induces diffusion of the system state over its state space, see Sect. 7.1. While the computation process itself conserves information, in unitary quantum operations or frictionless classical computation, or even discards it, diffusion increases the entropy. As soon as it blurs the distinction of two states one bit apart, the computation is rendered useless. In the extreme, the computer thermalizes, thus loses the capacity for any meaningful computation.

Strategies to counteract diffusion of the quantum computer's state are an integral part of quantum computation. They include standard provisions also known in classical technology, such as in particular redundancy, as well as specifically quantum schemes, such as working in decoherence-free subspaces. The following subsections classify noise sources in 10.4.1, outline some error correction methods in 10.4.2, and indicate ways to systematically prevent errors from the outset in 10.4.3.

10.4.1 Sources, Types, and Effects of Noise

Taking errors into account, quantum computation is expelled from the paradise of coherence preserving unitary transformations. They can only be treated adequately in the framework of nonunitary time evolution of density matrices. A huge toolbox of mathematical formalisms is available for their description, originating in fields as diverse as quantum chemistry and solid-state physics, quantum optics, and quantum measurement. Quantum computation offers the opportunity to sketch the treatment of noise in a minimal case, two-state systems. How to represent unitary transformations of density matrices has been indicated above, in Sect. 10.2.2, in particular as rotations of the Bloch sphere. While they are characterized by conserving the length of the Bloch vector, nonunitary operations break this invariance and are generally reflected in contracting Bloch vectors. In the extreme case of a completely depolarized ensemble, $\hat{\rho} = \frac{1}{2}\hat{I}^{(2)}$, the Bloch sphere shrinks to a point at the origin. Noise in two-state systems becomes manifest as diffusion, not over any flat position or phase space but over Hilbert space with its spherical topology.

In Sect. 8.3.3, the phenomenon of decoherence has been elucidated on the basis of a microscopic model, comprising a spin, coupled to an environment composed of harmonic oscillators. It is a suitable starting point to discuss the effects of noise on a qubit. The Hamiltonian comprises three terms, $\hat{H}_S$, $\hat{H}_E$, and $\hat{H}_{SE}$, for spin, environment and coupling, resp.,

$$\hat{H} = \hat{H}_S + \hat{H}_{SE} + \hat{H}_E,$$

$$\hat{H}_S = \frac{1}{2}\hbar\omega_0\hat{\sigma}_3, \ \hat{H}_{SE} = g\hat{\sigma}_3 \sum_{n=1}^{N} (\hat{a}_n + \hat{a}_n^{\dagger}), \ \hat{H}_E = \sum_{n=1}^{N} \hbar\omega_n \left(\hat{a}_n^{\dagger}\hat{a}_n + \frac{1}{2}\right). \quad (10.4.1)$$

The time evolution it generates for any finite set of N discrete boson modes with frequencies ω_n is unitary, but in the limits $N \to \infty$ of a continuous spectrum and autocorrelations of the bath decaying instantaneously, the generated dynamics turns irreversible. Under these assumptions, the time evolution of the reduced density operator of the central system, $\hat{\rho}_S = tr_E[\hat{\rho}]$, can be written as a Markovian (memory-free) master equation,

$$\frac{d}{dt}\hat{\rho}_S(t) = \frac{-i}{\hbar}\left[\hat{H}_S, \hat{\rho}_S\right] + \frac{1}{4}\lambda\left[\hat{\sigma}_3, [\hat{\rho}_S, \hat{\sigma}_3]\right], \quad (10.4.2)$$

where the parameter $\lambda := 4g^2 T_E \left\langle(\hat{q}_E)^2\right\rangle$ measures the effective noise strength, depending on the temperature T_E, the coupling g to the spin, and the mean square displacement $\left\langle(\hat{q}_E)^2\right\rangle$ of the environment. In Eq. (10.4.2), the first, imaginary term stands for the unitary dynamics generated by the central system alone. The second, real term is responsible for all incoherent processes induced by the environment.

Decomposing the reduced density matrix in the basis of the three Pauli matrices, Eq. (10.2.7), the master equation becomes a coupled differential equation for the components of the Bloch vector,

$$2\dot{a}_0\hat{I}^{(2)} + \dot{a}_1\hat{\sigma}_1 + \dot{a}_2\hat{\sigma}_2 + \dot{a}_3\hat{\sigma}_3 = \omega_0(a_1\hat{\sigma}_2 - a_2\hat{\sigma}_1) - \lambda(a_1\hat{\sigma}_1 + a_2\hat{\sigma}_2), \quad (10.4.3)$$

or, separating the components,

$$\dot{a}_0 = 0, \ \dot{a}_1 = -\omega_0 a_2 - \lambda a_1, \ \dot{a}_2 = \omega_0 a_1 - \lambda a_2, \ \dot{a}_3 = 0. \quad (10.4.4)$$

The first member of Eq. (10.4.4) reflects the conservation of $tr[\hat{\rho}_S]$. The second and third members resemble the equations of motion for position and momentum of an underdamped harmonic oscillator, cf. Eqs. (6.2.1–6.2.4). The solution

$$a_1(t) = e^{-\lambda t}(a_1(0)\cos(\omega_0 t) + a_2(0)\sin(\omega_0 t)),$$
$$a_2(t) = e^{-\lambda t}(a_2(0)\cos(\omega_0 t) - a_1(0)\sin(\omega_0 t)), \quad (10.4.5)$$

combines rotation of the Bloch sphere about its 3-axis, generated by the first term on the r.h.s. of Eqs. (10.4.2 and 10.4.3) with exponential decay with the rate λ. The fourth member, $\dot{a}_3 = 0$, is solved by $a_3(t) = \text{const} = a_3(0)$: Bloch vectors spiral on planes $a_3 = \text{const}$ towards the 3-axis, the Bloch sphere as a whole shrinks to a vertical line (Fig. 10.10a).

Here, decoherence affects symmetrically the subspaces of $\hat{\sigma}_1$- and $\hat{\sigma}_2$-eigenvectors, and only them. The fact that the 3-component is left untouched is a consequence of the coupling $\sim \hat{\sigma}_3$: It is an instance of a decoherence-free subspace. Focusing on incoherent processes in the remaining Hilbert space, they can be viewed as classical stochastic processes. Random transitions between the eigenstates $|0_1\rangle$ and $|1_1\rangle$ of $\hat{\sigma}_1$ or $|0_2\rangle$ and $|1_2\rangle$ of $\hat{\sigma}_2$ can be described by rate equations for the probabilities to be in these states,

$$\dot{p}_{|0_n\rangle}(t) = \frac{d}{dt}\langle 0_n|\hat{\rho}_s(t)|0_n\rangle = \lambda\left(p_{|1_n\rangle}(t) - p_{|0_n\rangle}(t)\right), n = 1, 2, \qquad (10.4.6)$$

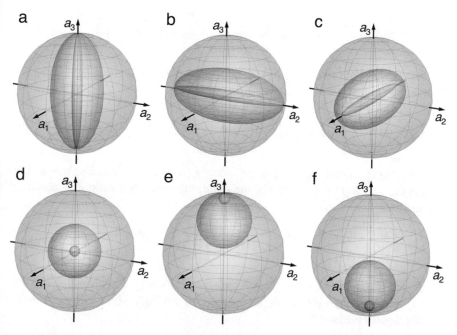

Fig. 10.10 Incoherent processes transforming a two-state system from a pure into a mixed state are reflected in a contraction of the Bloch sphere into the interior of the unit sphere that represents pure states. Depending on the type of incoherent transitions generated, the contraction may be nonisotropic, for example (**a**) towards the a_3-axis (phase flips), Eqs. (10.4.4–10.4.8), (**b**) towards the a_2-axis (bit-phase flips), Eqs. (10.4.13), or (**c**) towards the a_1-axis (bit flips), Eqs. (10.4.10–10.4.12), or isotropic, (**d**) contracting towards the centre (homogeneous depolarization), Eqs. (10.4.14 and 10.4.15), or eccentric, approaching a pure state such as (**e**) spin down, Eqs. (10.4.20 and 10.4.21), or (**f**) spin up, depending on the sign of the self-energy term in the Hamiltonian.

or, integrated from 0 to t, as a Markov chain (see Sect. 3.1.1),

$$|0_3\rangle \rightarrow |0_3\rangle, \quad |1_3\rangle \rightarrow \begin{cases} |1_3\rangle & \text{with probability } e^{-\lambda t}, \\ -|1_3\rangle & \text{with probability } 1 - e^{-\lambda t}. \end{cases} \qquad (10.4.7)$$

For the reduced density matrix, change of relative sign of $|0_3\rangle$ and $|1_3\rangle$ means

$$\hat{\rho}_S = \begin{pmatrix} \rho_{00} & \rho_{01} \\ \rho_{10} & \rho_{11} \end{pmatrix} \rightarrow \begin{cases} \hat{\rho}_S & \text{with prob. } e^{-\lambda t}, \\ \hat{\sigma}_3 \hat{\rho}_S \hat{\sigma}_3 = \begin{pmatrix} \rho_{00} & -\rho_{01} \\ -\rho_{10} & \rho_{11} \end{pmatrix} & \text{with prob. } 1 - e^{-\lambda t}. \end{cases} \qquad (10.4.8)$$

The sign change of its off-diagonal elements leads to a decay of the first and second components of the Bloch vector. For that reason, the incoherent transitions caused by an interaction $\sim \hat{\sigma}_3$ are called *phase flips*.

By a similar reasoning, it is evident that a Hamiltonian

$$\hat{H} = \frac{1}{2}\hbar\omega_0\hat{\sigma}_1 + g\hat{\sigma}_1 \sum_{n=1}^{N} (\hat{a}_n + \hat{a}_n^\dagger) + \hbar \sum_{n=1}^{N} \omega_n \left(\hat{a}_n^\dagger \hat{a}_n + \frac{1}{2} \right) \qquad (10.4.9)$$

(replacing $\hat{\sigma}_3$ in Eq. (10.4.1) by $\hat{\sigma}_1$), besides unitary rotation about the 1-axis, will generate incoherent transitions of the type

$$\dot{p}_{|0_n\rangle}(t) = \frac{d}{dt}\langle 0_n|\hat{\rho}_S(t)|0_n\rangle = \lambda\left(p_{|1_n\rangle}(t) - p_{|0_n\rangle}(t)\right), n = 2, 3, \qquad (10.4.10)$$

which now keep a_1 constant but reduce a_2 and a_3, so that the Bloch sphere shrinks towards the 1-axis (Fig. 10.10b). For the qubit, they become manifest in particular as transitions

$$|0_3\rangle \rightarrow \begin{cases} |0_3\rangle & \text{with prob. } e^{-\lambda t}, \\ |1_3\rangle & \text{with prob. } 1 - e^{-\lambda t}, \end{cases} |1_3\rangle \rightarrow \begin{cases} |1_3\rangle & \text{with prob. } e^{-\lambda t}, \\ |0_3\rangle & \text{with prob. } 1 - e^{-\lambda t}. \end{cases} \qquad (10.4.11)$$

This type of noise therefore interchanges the eigenstates of $\hat{\sigma}_3$, actual *bit flips*, and thus interchanges the diagonal elements of the reduced density matrix, without affecting the off-diagonal elements,

$$\hat{\rho}_S \rightarrow \begin{cases} \hat{\rho}_S & \text{with prob. } e^{-\lambda t}, \\ \hat{\sigma}_1 \hat{\rho}_S \hat{\sigma}_1 = \begin{pmatrix} \rho_{11} & \rho_{01} \\ \rho_{10} & \rho_{00} \end{pmatrix} & \text{with prob.} 1 - e^{-\lambda t}. \end{cases} \qquad (10.4.12)$$

Similarly, the effect of a coupling term $\sim \hat{\sigma}_2$ is found to be

$$\hat{\rho}_s \rightarrow \begin{cases} \hat{\rho}_s & \text{with prob. } e^{-\lambda t}, \\ \hat{\sigma}_2 \hat{\rho}_s \hat{\sigma}_2 = \begin{pmatrix} \rho_{11} & -\rho_{10} \\ -\rho_{01} & \rho_{00} \end{pmatrix} & \text{with prob. } 1 - e^{-\lambda t}, \end{cases} \qquad (10.4.13)$$

a combination of phase and bit flips, and is reflected in the Bloch sphere as a contraction towards the 2-axis (Fig. 10.10c).

In the three-dimensional space of the Bloch sphere, noise can be decomposed in three independent directions, given as in these examples for instance by the eigenvectors of the Pauli matrices. Another option is the set of eigendirections of spherical coordinates, $\{\mathbf{u}_r, \mathbf{u}_\theta, \mathbf{u}_\phi\}$. For example, the rate of change of the radial component $a(t)$ is

$$\dot{a}(t) = \frac{d}{dt}|\mathbf{a}(t)| = \frac{\mathbf{a}}{|\mathbf{a}|} \cdot \dot{\mathbf{a}}(t). \qquad (10.4.14)$$

An isotropic length reduction of the Bloch vector, $\dot{\mathbf{a}}(t) \| \mathbf{a}(t)$, is achieved by combining all the three noise types mentioned above, resulting in a stochastic process

$$\hat{\rho}_S \rightarrow \begin{cases} \hat{\rho}_S & \text{with prob. } e^{-\lambda t}, \\ \frac{1}{2}\hat{I}^{(2)} & \text{with prob. } 1 - e^{-\lambda t}, \end{cases} \qquad (10.4.15)$$

for the reduced density matrix. It leads asymptotically to a complete depolarization of the spin, shrinking the Bloch sphere into a point at the origin (Fig. 10.10d).

In all these processes, it is the entropy which is not conserved, and only it. Invoking Eq. (10.2.19), it can be expressed by the length $a = |\mathbf{a}|$ of the Bloch vector. For instance, in the case of phase flips, Eq. (10.4.5), in units of bits,

$$S_{vN}(t) = 1 - \frac{1}{2}((1 + a(t))\mathrm{lb}(1 + a(t)) + (1 - a(t))\mathrm{lb}(1 - a(t))) \text{ with}$$

$$a(t) = \sqrt{(a_3(0))^2 + e^{-2\lambda t}((a_1(0))^2 + (a_2(0))^2)}, \qquad (10.4.16)$$

so that for $t \rightarrow \infty$,

$$S_{vN}(t) \underset{t \to \infty}{\longrightarrow}$$

$$1 - \frac{1}{2}((1 + |a_3(0)|)\mathrm{lb}(1 + |a_3(0)|) + (1 - |a_3(0)|)\mathrm{lb}(1 - |a_3(0)|)), \qquad (10.4.17)$$

while in the case of isotropic noise, $a(t) = e^{-\lambda t}a(0)$,

$$S_{vN}(t) \underset{t \to \infty}{\longrightarrow}$$

$$c[\ln(2) + (1 - |a_3(0)|)\ln(1 - |a_3(0)|) - (1 + |a_3(0)|)\ln(1 + |a_3(0)|)],$$

$$(10.4.18)$$

Of course, a coupling to the environment can lead to energy losses of the central system, the qubit. Energy dissipation comes into play if the self-energy $\hat{H}_S = \hbar\omega_0\hat{\sigma}_n/2$ is no longer conserved. This requires spin operators other than $\hat{\sigma}_n$ to appear in the coupling $\hat{H}_{SE}$ to the environment, since ladder operators $\hat{\sigma}_n^\pm$ effecting the loss or gain of a quantum in $\hat{H}_S$ cannot be constructed from $\hat{\sigma}_n$ alone, see Eq. (10.2.45). For example, the Hamiltonian

$$\hat{H} = \frac{1}{2}\hbar\omega_0\hat{\sigma}_3 + g\hat{\sigma}_2\sum_{n=1}^{N}\left(\hat{a}_n + \hat{a}_n^\dagger\right) + \hbar\sum_{n=1}^{N}\omega_n\left(\hat{a}_n^\dagger\hat{a}_n + \frac{1}{2}\right) \qquad (10.4.19)$$

(the spin-boson model with N bosons [LC&87]) meets this condition. Along the same steps as above to eliminate the ambient degrees of freedom, equations of motion for the Bloch vector can be derived,

$$\dot{a}_0 = 0, \dot{a}_1 = -\omega_0 a_2 - \lambda a_1, \dot{a}_2 = \omega_0 a_1, \dot{a}_3 = \dot{a}_3(0) - \lambda a_3. \qquad (10.4.20)$$

They resemble Eq. (10.4.4), specifically concerning the rotation about the 3-axis, but differ from it in two points: Now the exponential decay affects a_1 and a_3, not a_2, and the equation for $\dot{a}_3$ contains a non-zero constant. As a consequence, $a_3(t)$ does not decay to zero, but approaches a residual polarization,

$$a_3(t)\underset{t\to\infty}{\longrightarrow}\frac{\dot{a}_3(0)}{\lambda}. \qquad (10.4.21)$$

If $\dot{a}_{3,0} = -\lambda$, $a_3(t)$ in the stationary state takes the value -1, that is, the two-state system decays towards its state of lowest energy, which is a pure state again (Fig. 10.10e). In that case, decoherence is dominated completely by dissipation. In the long-time limit, the entropy of the qubit, Eq. (10.4.17) even returns to zero, and its purity, Eq. (10.2.20), approaches unity (Fig. 10.11).

10.4.2   Error Protection and Correction

In classical data processing, digital computation proper is not significantly affected by noise, but in communication in general it is ubiquitous. The standard antidote is redundancy, see Sect. 1.7, in technical contexts as much as in everyday conversation and correspondence. In essence, redundancy means repetition, explicitly as a verbatim doubling of the signal to be transmitted, or implicitly by a high level of correlation among symbols and symbol strings. An instructive scheme is the checksum, a complementary code, in the simplest case the digit sum that depends sensitively on the original signal but is too short to permit its recovery, once it's corrupted. Instead,

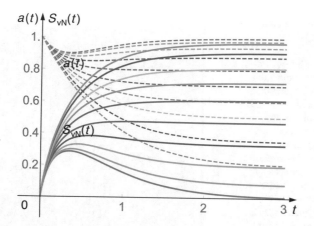

Fig. 10.11 Time dependence, caused by incoherent processes, of the size $a = |\mathbf{a}|$ of the Bloch vector $\mathbf{a}$ (dashed), representing a two-state system, and the von Neumann entropy $S_{vN}(t)$ (bold), related to $a(t)$ by Eq. (10.2.19). The set of curves ranges from the limit of dominating dissipation, $\lim_{t\to\infty} a(t) = 1$, to the limit of dominating decoherence, $\lim_{t\to\infty} a(t) = 0$, with values $a(\infty) = 1.0$, 0.98, 0.94, 0.88, 0.8, 0.7, 0.6, 0.5, 0.35, 0.2 (top to bottom for $a(t)$ and bottom to top for $S_{vN}(t)$).

it allows to detect (not localize) errors, so that a given transmission can be discarded and repeated. Evidently, the more digits are reserved for the checksum, the more reliable is the error detection.

This principle carries over unabatedly to quantum computation. How does redundancy reduce or prevent noise-induced diffusion? Its efficiency hinges on the assumption that noise is blind to the specific content of a signal and affects all symbols, all bits, indiscriminately. If that is true, the diffusion cloud, for example a Gaussian distribution, will spread symmetrically around the original signal, so that the signal can be recovered by averaging, as the centre of the cloud. Identical repetitions lift the signal into a higher-dimensional space where with the same noise strength, diffusion is slower: If the probability for a given bit to be corrupted is p, $0 \leq p \leq 1$, the probability to have the same bit flipped in n identical copies of the signal increases roughly as $p^n \leq p$. If the original signal is reconstructed by averaging, that is, by *majority vote* among the copies, the probability to recover the correct initial bit increases with the number of copies transmitted or processed in parallel.

Error protection in quantum computation follows largely this strategy, adapting it to the special conditions found in reversible unitary transformations. Beginning with the initial step, generating identical copies of the initial state, a first major obstacle has to be overcome. According to the no-cloning theorem (Sect. 8.3.1), copying of arbitrary quantum states from one Hilbert space to another contradicts unitarity. However, there is a gap in this verdict: copying is possible into a target state preset to a specific value. Indeed, the unitary CNOT gate (Sect. 10.2.2) copies the controlling input to the controlled one if the latter is preset to $|0\rangle$. Cascading CNOT gates therefore allows to generate as many copies of an arbitrary input state $|\psi\rangle$ as required. Copying a single qubit n times (pink box in Fig. 10.12),

Fig. 10.12 The Shor code protects, Eq. (10.4.26), against arbitrary combinations of bit flip and phase flip errors in quantum computation by generating a nested structure of N replicas of sets of N copies of the state $|\psi\rangle$ to be protected. A first N-fold replication combined with a Hadamard gate rotating all replicas by $\pi/4$, corresponding to the phase-flip code (Eq. (10.4.25), green box) is followed by another N-fold replication, without rotation, corresponding to the bit flip code (Eq. (10.4.22), pink box).

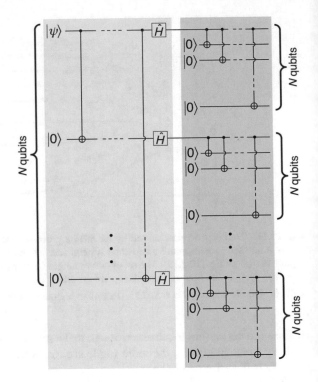

$$|0\rangle \rightarrow \underbrace{|00\ldots00\rangle}_{n \text{ bits}} =: |0\rangle^n, |1\rangle \rightarrow \underbrace{|11\ldots11\rangle}_{n \text{ bits}} =: |1\rangle^n, \qquad (10.4.22)$$

lifts it from its original two-state system into a Hilbert space of 2^n dimensions. Even if this is a unitary process, the potential increase of entropy from the subspace spanned by $|00\ldots00\rangle$ and $|11\ldots11\rangle$ to a random state in this 2^n-dimensional Hilbert space is much higher.

With the multiplicated qubit embedded as a two-dimensional subspace in a higher-dimensional Hilbert space, it is now possible to perform measurements projecting out components outside that subspace, to detect errors of transmission or computation. In order not to perturb the qubit, measurements are required that do not reveal any information on the state of the qubit, but only indicate where it possibly deviates from an n-fold identical copy. They can be constructed as follows:

$$\hat{P}_0 := |0\rangle^n \langle 0|^n + |1\rangle^n \langle 1|^n \qquad\qquad\qquad \text{no error}$$

$$\hat{P}_1 := \hat{\sigma}_{1,1}\big(|0\rangle^n \langle 0|^n + |1\rangle^n \langle 1|^n\big)\hat{\sigma}_{1,1}$$

$$= \left|\underbrace{1\ \ 0\ldots0}_{n-1\ \text{qubits}}\right\rangle\left\langle\underbrace{1\ \ 0\ldots0}_{n-1\ \text{qubits}}\right| + \left|\underbrace{0\ \ 1\ldots1}_{n-1\ \text{qubits}}\right\rangle\left\langle\underbrace{0\ \ 1\ldots1}_{n-1\ \text{qubits}}\right| \qquad \text{error in qubit 1}$$

$$\hat{P}_2 := \hat{\sigma}_{1,2}\big(|0\rangle^n \langle 0|^n + |1\rangle^n \langle 1|^n\big)\hat{\sigma}_{1,2} \qquad\qquad\qquad \text{error in qubit 2}$$

$$\vdots \qquad\qquad\qquad\qquad\qquad\qquad\qquad\qquad\qquad\qquad \vdots$$

$$\hat{P}_n = \hat{\sigma}_{1,n}\big(|0\rangle^n \langle 0|^n + |1\rangle^n \langle 1|^n\big)\hat{\sigma}_{1,n} \qquad\qquad\qquad \text{error in qubit } n$$

$$\text{(10.4.23)}$$

Indeed, for an arbitrary state $|\psi\rangle = c_0|0\rangle^n + c_1|1\rangle^n$ of the multiplicated qubit, corrupted, say, at position j, $|\psi'\rangle = \hat{\sigma}_{1,j}(c_0|0\rangle^n + c_1|1\rangle^n)$, all projections give zero, up to $\hat{P}_j := \hat{\sigma}_{3,j}\big(|0\rangle^n \langle 0|^n + |1\rangle^n \langle 1|^n\big)\hat{\sigma}_{3,j}$,

$$\hat{P}_j|\psi'\rangle\langle\psi'|\hat{P}_j = c_0\Big|0\Big\rangle^n \langle 0|^n c_0^* + c_1|1\rangle^n \langle 1|^n c_1^* = |c_0|^2 + |c_1|^2 = 1. \qquad \text{(10.4.24)}$$

According to the results of these measurements, correcting operations have to be applied: If for instance $\hat{P}_j|\psi'\rangle\langle\psi'|\hat{P}_j = 1$, the qubit at position j is corrupted and has to be flipped back by applying $\hat{\sigma}_{1,j}$ to $|\psi'\rangle$, which will recover $|\psi\rangle$. With this scheme, errors in at most a single qubit can be detected and fixed. If the probability for a single flip is p, hence for no flip is $q = 1 - p$, this occurs in a set of n copies with probability $q^n + npq^{n-1}$. In order to cover also multiple errors, i.e., double, triple, ..., flips, a corresponding number of additional projectors, 2^n-1 in all, would have to be included in the list (10.4.23).

This procedure works for errors caused by bit flips, that is, equivalent to operating with $\hat{\sigma}_1$ on the qubit. The two other error categories addressed in the previous subsection, corresponding to random actions of $\hat{\sigma}_2$ (combined bit and phase flips) or $\hat{\sigma}_3$ (phase flips), are not covered by the scheme as specified above. However, it is readily adapted to the other error types, by rotating the affected axis of the Bloch sphere into the 1-axis before the treatment and back thereafter. For example, operating with $\hat{\sigma}_3$ on the eigenstates $|0_1\rangle = 2^{-1/2}(|0_3\rangle + |1_3\rangle)$ and $|1_1\rangle = 2^{-1/2}(|0_3\rangle - |1_3\rangle)$ of $\hat{\sigma}_1$ interchanges them, that is, results in a bit flip in this basis: Transforming from the canonical basis, $\hat{\sigma}_3$- eigenstates, to $\hat{\sigma}_1$- eigenstates converts phase flips into bit flips which can then be detected and repaired as above (green box in Fig. 10.12). The gate accomplishing this transformation, in both directions, is the Hadamard gate (10.2.35).

It would be desirable to have an error protection scheme that automatically covers both types of error. This can be achieved by creating a hierarchy of two levels of replication and applying the Hadamard gate only to one of them. For example, following the transformation (10.4.25),

$$\underbrace{|0_3 \ldots 0_3\rangle}_{N \text{ qubits}} \to \overset{N}{\underset{n=1}{\otimes}} \hat{H}_N |0_3 \ldots 0_3\rangle = \underbrace{|0_1 \ldots 0_1\rangle}_{N \text{ qubits}} = 2^{-N/2} \overset{N}{\underset{n=1}{\otimes}} (|0_3\rangle_n + |1_3\rangle_n),$$

$$\underbrace{|1_3 \ldots 1_3\rangle}_{N \text{ qubits}} \to \overset{N}{\underset{n=1}{\otimes}} \hat{H}_N |1_3 \ldots 1_3\rangle = \underbrace{|1_1 \ldots 1_1\rangle}_{N \text{ qubits}} = 2^{-N/2} \overset{N}{\underset{n=1}{\otimes}} (|0_3\rangle_n - |1_3\rangle_n),$$

$$(10.4.25)$$

each qubit could be replaced again by an N-fold copy of itself, resulting in N^2 qubits in total, or a Hilbert space of 2^{N^2} dimensions,

$$\underbrace{|0_3 \ldots 0_3\rangle}_{N \text{ qubits}} \to 2^{-N/2} \overset{N}{\underset{n=1}{\otimes}} (|0_3\rangle_n + |1_3\rangle_n) \to 2^{-N/2} \overset{N}{\underset{n=1}{\otimes}} (\underbrace{|0_3 \ldots 0_3\rangle_n}_{N \text{ qubits}} + \underbrace{|1_3 \ldots 1_3\rangle_n}_{N \text{ qubits}}),$$

$$\underbrace{|1_3 \ldots 1_3\rangle}_{N \text{ qubits}} \to 2^{-N/2} \overset{N}{\underset{n=1}{\otimes}} (|0_3\rangle_n - |1_3\rangle_n) \to 2^{-N/2} \overset{N}{\underset{n=1}{\otimes}} (\underbrace{|0_3 \ldots 0_3\rangle_n}_{N \text{ qubits}} - \underbrace{|1_3 \ldots 1_3\rangle_n}_{N \text{ qubits}}),$$

$$(10.4.26)$$

Comprehensive error detection then requires to perform measurements of single qubits within each N-qubit block as well as of entire blocks. This procedure is known as the *Shor code* [Sho95], Fig. 10.12.

10.4.3   *Error Prevention: Computing In Decoherence-Free Subspaces*

Even as versatile as the Shor code, quantum error detection and correction schemes remain restricted to a relatively narrow range of errors. In particular, their construction is based on the assumption that errors occur in an uncorrelated manner in individual qubits, neglecting for example a residual unwanted interaction coupling spatially neighbouring two-state systems representing them. An all-round strategy preventing damages before they occur, instead of repairing them when they have already happened, by suppressing decoherence from the outset.

An obvious answer is cooling. Noise strengths and diffusion rates typically increase linearly with temperature, see, e.g., Eqs. (7.2.12) and (10.4.2). Cooling is therefore compulsory for quantum computation wherever it is possible. NMR computation with molecules in solution (Sect. 10.5.2) is a patent exception. In general, cooling is costly and severely restricts the accessibility of quantum computation. A much more promising approach is based on decoherence-free subspaces [MSE96, PSE96, LCW98, LW03, Lid13], alluded to in Sect. 8.3.3.

A central system interacting with a heat bath gets entangled with the degrees of freedom of the environment. However, in the total system, entanglement does not increase, it only changes its structure. The dimension of the total Hilbert space

$\mathcal{H} = \mathcal{H}_S \otimes \mathcal{H}_E$ is conserved. There always exists a set of subspaces that are not entangled, but straddle the boundary between the subsystems. Even in the presence of decoherence, these subspaces therefore remain neatly separated: Choosing qubits as states from different such subspaces immunizes them against decoherence.

Returning to the Eq. (10.4.2), the term representing incoherent processes contains the double commutator $[\hat{\sigma}, [\hat{\rho}, \hat{\sigma}]]$, a standard case of a diffusion term in a master equation for a reduced density operator. It almost reaches the most general form such terms can attain in a master equation for a Markovian dynamics in the limit of weak system-bath coupling, provided fundamental features of the density matrix such as positivity, Hermiticity, conservation of the trace are not violated. It can be shown to take the form of a linear superoperator operating on the density operator [Lin76],

$$\mathcal{L} : \hat{\rho} \mapsto \mathcal{L}\hat{\rho} = \frac{1}{2}\left(\left[\hat{L}, \hat{\rho}\hat{L}^{\dagger}\right] + \left[\hat{L}\hat{\rho}, \hat{L}^{\dagger}\right]\right) = \hat{L}\hat{\rho}\hat{L}^{\dagger} - \frac{1}{2}\left[\hat{\rho}, \hat{L}^{\dagger}\hat{L}\right]_{+}, \quad (10.4.27)$$

(where $[,]_+$ denotes the anticommutator, $[\hat{a}, \hat{b}]_+ := \hat{a}\hat{b} + \hat{b}\hat{a}$). The operators $\hat{L}$ and $\hat{L}^{\dagger}$, called Lindblad operators, induce excitation and deexcitation processes in the central system. In the special case that they are Hermitian, $\hat{L}^{\dagger} = \hat{L}$, Eq. (10.4.27) simplifies further to $\mathcal{L}\hat{\rho} = \frac{1}{2}[\hat{L}, [\hat{\rho}, \hat{L}]]$, recovering the form $[\hat{\sigma}, [\hat{\rho}, \hat{\sigma}]]$ with $\hat{L} = \hat{\sigma}$.

Decoherence-free subspaces are spanned by eigenstates $|l\rangle$ of $\hat{L}$ with eigenvalue zero,

$$\hat{L}|l\rangle = 0, l = 1, \ldots, N_{\text{dfs}}. \quad (10.4.28)$$

Within such a subspace $\mathcal{H}_{\text{dfs}} = \text{span}\{|l\rangle\}_{l=1}^{N_{\text{dfs}}}$, the density operator can be presented as

$$\hat{\rho}_{\text{dfs}} = \sum_{l,l'=1}^{N_{\text{dfs}}} \hat{\rho}_{\text{dfs},l,l'} |l\rangle\langle l'|. \quad (10.4.29)$$

Substituting it in Eq. (10.4.27), the conclusion

$$\mathcal{L}\hat{\rho}_{\text{dfs}} = \hat{L}\hat{\rho}_{\text{dfs}}\hat{L}^{\dagger} - \frac{1}{2}\left[\hat{\rho}_{\text{dfs}}, \hat{L}^{\dagger}\hat{L}\right]_{+} = 0 \quad (10.4.30)$$

is immediate.

In Hilbert spaces of N-fold replicas of a qubit, the relevant Lindblad operators are sums $\hat{S}_n = \sum_{j=1}^{N} \hat{\sigma}_{n,j}$ of spin operators $\hat{\sigma}_{n,j}$ operating on qubit j, the subscript n labeling $\hat{\sigma}_1, \hat{\sigma}_2, \hat{\sigma}_3$, or more appropriately in the present context, the triplet $\hat{\sigma}_3^+$, $\hat{S}^+|\psi\rangle = 2^{-1/2}(\hat{\sigma}_a^+ - \hat{\sigma}_b^+)$, and $\hat{\sigma}_3$. The rules of angular momentum addition in quantum mechanics imply that adding N spin-$\frac{1}{2}$ systems, the 3-component total angular momentum can take the values $-N/2, -N/2 + 1, \ldots, N/2-1, N/2$. The eigenvalue 0 therefore always exists for even N and is N_{dfs}-fold degenerate. The degree of degeneracy of this subspace is the dimension of the decoherence-free subspace. It depends on the total number N of qubits as [ZR97] as

$$N_{\text{dfs}}(N) = N! \bigg/ \left(\frac{N}{2}\right)! \left(\frac{N}{2}+1\right)! \tag{10.4.31}$$

(for even N) and increases monotonously with N. Asymptotically for $N \gg 1$, it approaches $N_{\text{dfs}}(N) \to N - (3/2)\text{lb}(N)$.

For example, for $N = 2$, the decoherence-free subspace is one-dimensional, spanned by the only eigenvector $|\psi^{(2)}\rangle = 2^{-1/2}(|01\rangle - |10\rangle)$. Indeed,

$$\hat{S}^+|\psi^{(2)}\rangle = 2^{-1/2}(\hat{\sigma}_a^+ + \hat{\sigma}_b^+)(|01\rangle - |10\rangle) = 2^{-1/2}(0 - |00\rangle + |00\rangle - 0) = 0,$$
$$\hat{S}^-|\psi^{(2)}\rangle = 2^{-1/2}(\hat{\sigma}_a^- + \hat{\sigma}_b^-)(|01\rangle - |10\rangle) = 2^{-1/2}(|11\rangle - 0 + 0 - |11\rangle) = 0,$$
$$\hat{S}_3|\psi^{(2)}\rangle = 2^{-1/2}(\hat{\sigma}_{3,a} + \hat{\sigma}_{3,a})(|01\rangle - |10\rangle)$$
$$= 2^{-1/2}(|01\rangle + |10\rangle - |01\rangle - |10\rangle) = 0. \tag{10.4.32}$$

For $N = 4$, Eq. (10.4.30) gives $N_{\text{dfs}}(4) = 2$, and a possible choice of degenerate eigenvectors with eigenvalue 0 is [ZR97]

$$\left|\psi_1^{(4)}\right\rangle = \frac{1}{2}(|1001\rangle - |0101\rangle + |0110\rangle - |1010\rangle),$$
$$\left|\psi_2^{(4)}\right\rangle = \frac{1}{2}(|1001\rangle - |0011\rangle + |0110\rangle - |1100\rangle). \tag{10.4.33}$$

(To form an orthonormal basis, these two eigenstates have to be orthonormalized.)

Preventing decoherence from the outset instead of repairing errors once they occurred, quantum computing in decoherence-free subspaces is an appealing option. However, its experimental realization and verification in the laboratory proves to be a challenge, even for a single qubit [KB&00, AH&04, FG&17]. In how far it plays any role in state-of-the-art commercial quantum computers, see Sect. 10.5.2, is not clear.

10.5 Physical Implementations

Quantum computation, above all, is a mathematical construction. It resembles classical computer science in that it reduces part of the real world to a precisely defined, but very limited set of nominal states and of canonical operations to be performed on them, in a radical idealization of the diversity and disorder of physical reality. Quantum computation, in turn, develops a new view and terminology of quantum time evolution in finite Hilbert spaces that consists of a set of basic elements such as gates, rotations, projections etc., organized along discrete computation paths. As such, it has grown into a theoretical body of its own right of enormous consistency and completeness that is even adopted in other fields of quantum mechanics, unrelated to computation.

At the same time, the pioneering spirit of quantum computation calls for an implementation in the laboratory, assembled from photons, spins, atoms, molecules in cavities and traps, and other elements of real physics. This is a formidable challenge, maybe more difficult even than the development of quantum algorithms that systematically outperform classical computation. The reason is that the idealizing assumptions implicit in quantum computation are plausible each one in itself, but as a whole combine almost incompatible conditions. Working at the lowest structural level of physics where the required nontrivial dynamics is possible, it lacks even lower levels where to dump trash information. This requires the most complete isolation of the qubits from any external noise and other perturbations. At the same time, a quantum computer, as any other information processing device, has to communicate, hence interact, with its environment, or else its function would go unnoticed in splendid isolation and be useless.

This subsection will sketch these challenges, beginning with a resume of the indispensable interactions between quantum computer and macroscopic world, specify the most important criteria for suitable systems, and present a brief list of specific options to implement quantum computation in the laboratory. However, the development in this field is so dynamic that this overview is inevitably incomplete and outdated the moment it is compiled. Readers are therefore referred to the current literature.

10.5.1 Peepholes: Communicating with a Quantum Computer

How a quantum computer can be integrated in an information processing protocol with classical front and end is sketched in Fig. 10.6a. Seen from this classical side, the quantum computer is a black box that receives a classical input, processes it faster than a classical device could do, to return classical output. This requires a classical-to-quantum interface on entrance to prepare the initial quantum state, a process control defining the sequence of operations, and a quantum-to-classical interface on exit to read out the result. These three operations will be sketched in the sequel. Both the preparation of an initial quantum state on input and the measurement of the final result are irreversible nonunitary processes that do not conserve entropy. The process control does, but it implies an inherent time arrow by the order in which noncommuting gates are applied.

Input: preparation

Quantum computation proper is unitary, that is, it conserves the purity of quantum states and requires a pure state as input. Which pure state it actually receives is of secondary relevance, since the intended input state can be generated from any pure initial state by applying quantum gates. As the quantum computer is a relatively small system embedded in a large environment, it will be in a mixed state, if this is not actively prevented. Disentangling it from its environment cannot be a unitary process, and there is no established method for it. Isolating the "quantum CPU" as the sanctum from the rest of the computer is one of the main tasks of implementing a quantum computer, as crucial as challenging. Complementary to preparation, measurement by definition should bring the measured system into a pure state upon leaving the apparatus, an eigenstate of the measured observable known as pointer state (see Sect. 8.4.2). However, which of these pointer states is reached is hard to control.

Program: process control

Were the time evolution of a quantum computer determined by a time-independent Hamiltonian, no particular states could be discerned as initial or final, not even a direction in time distinguishing input from output. Any non-trivial computation therefore requires some external control in the form of time-dependent potentials that initiate and terminate the computation and determine which gates operate in which order on the computer.

A rudimentary but instructive example is the same bistable system that has already served, in Sect. 9.2.1, as a model for a classical bit: the quartic double well, Eqs. (9.2.3 and 9.2.4). A direct quantization, replacing classical observables with operators, yields the Hamilton operator

$$\hat{H}(\hat{p}, \hat{x}) = \frac{\hat{p}^2}{2m} - \frac{m\Omega^2}{4}\hat{x}^2 + \frac{m^2\Omega^4}{64E_B}\hat{x}^4. \tag{10.5.1}$$

Here, Ω is the frequency of small oscillations around either one of the two minima at $x_\pm = \pm\sqrt{8E_B/m\Omega^2}$, E_B denotes the barrier height. Eigenstates of this Hamiltonian can be classified according to the same subdivision of phase space that arises for the classical system. For positive energies, above the barrier top, they are typical eigenstates of a strongly anharmonic oscillator. For negative energies, in particular close to the bottom of either well, they can well be approximated by the eigenfunctions of a harmonic oscillator with frequency Ω (Fig. 10.13a). However, their weak but finite coupling across the barrier lifts the degeneracy between harmonic oscillator eigenstates of the same quantum number in the left and the right well, so that they form pairs with a small energy splitting. They are alternatingly symmetric and antisymmetric with respect to spatial reflection $x \to -x$,

$$\psi_n(x) = \langle x|n\rangle = (-1)^n \psi_n(-x). \tag{10.5.2}$$

Deep inside each well, the energy splitting $I\Delta_n$ can be estimated using semiclassical approximations. For the ground state pair, it reads

$$\Delta_0 \approx \sqrt{\frac{128}{\pi}\hbar\Omega E_B} \exp\left(-\frac{16}{3}\frac{E_B}{\hbar\Omega}\right), \tag{10.5.3}$$

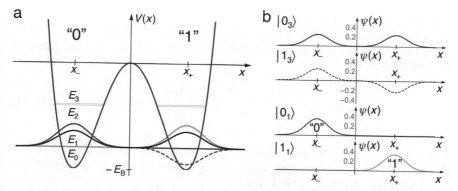

Fig. 10.13 Ground state pair of a quantum double well as a qubit. The eigenstates of a quartic double well potential (**a**), at energies close to the bottoms of the two wells, resemble harmonic oscillator eigenstates, but superposed as quasidegenerate doublets of symmetric and antisymmetric states, Eq. (10.5.2), separated in energy by small tunnel splittings Δ_n. The ground and first excited states are shown in panels (**a**) and (**b**) as bold and dashed black lines, resp. They can be recomposed to form states localized in the left or right well (pink and green, resp., in both panels). However, they are no eigenstates but tunnel between the wells with low frequency $\Delta_0/\hbar$, Eq. (10.5.3). This figure is a schematic rendering, shapes of wavefunctions are reproduced only qualitatively, energies are not drawn to scale.

and in this regime, $E_B \gg \hbar\Omega$, it is typically orders of magnitude smaller than the photon energy $\hbar\Omega$. The states coming closest to the nominal states "0" and "1" of the classical bit, localized in the left or the right well, resp., can be composed as superpositions of the ground and the first excited state of the double well (Fig. 10.13b),

$$|\text{left}_n\rangle = \frac{1}{\sqrt{2}}(|2n\rangle + |2n+1\rangle), \left|\text{right}_n\right\rangle = \frac{1}{\sqrt{2}}(|2n\rangle - |2n+1\rangle), n \in \mathbb{N}_0.$$

$$(10.5.4)$$

Since they are no eigenstates themselves but oscillate slowly with a frequency given by the splitting as $\Delta_n/\hbar$, giving rise to the phenomenon of *tunneling*.

In the context of quantum computation, it suggests itself to consider the ground state pair $|0\rangle$, $|1\rangle$ as a qubit. Adopting the notation introduced in Sect. 10.2.1, this pair is identified as eigenstates $|0_3\rangle$, $|1_3\rangle$ of the Pauli operator $\hat{\sigma}_3$, while the localized states of Eq. (10.5.4) become eigenstates of $\hat{\sigma}_1$

$$|0_1\rangle = |\text{left}_0\rangle = \frac{1}{\sqrt{2}}(|0_3\rangle + |1_3\rangle), |1_1\rangle = \left|\text{right}_0\right\rangle = \frac{1}{\sqrt{2}}(|0_3\rangle - |1_3\rangle). \quad (10.5.5)$$

Restricted to this pair, neglecting the participation of all higher excited states, the Hamiltonian reduces to $\hat{H} = \frac{1}{2}\Delta_0\hat{\sigma}_3$.

For classical bistable system interpreted as bits, a driving force, constant in space but depending on time, was introduced as an elementary control, Eq. (9.2.7). Applying an equivalent driving to the quantum double well amounts to augmenting the Hamiltonian (10.5.1),

$$\hat{H}(\hat{p}, \hat{x}, t) = \frac{\hat{p}^2}{2m} - \frac{m\Omega^2}{4}\hat{x}^2 + \frac{m^2\Omega^4}{64 E_B}\hat{x}^4 - F(t)\hat{x}. \tag{10.5.6}$$

with a time-dependent force $F(t)$, representing, for example, an electric field. Within the subspace of the ground state qubit, owing to the opposite parities of the two eigenstates, this term has matrix elements

$$\begin{aligned}
\langle 0_3 | F(t)\hat{x} | 0_3 \rangle &= \langle 1_3 | F(t)\hat{x} | 1_3 \rangle = 0, \\
\langle 0_3 | F(t)\hat{x} | 1_3 \rangle &= \langle 1_3 | F(t)\hat{x} | 0_3 \rangle = \chi F(t),
\end{aligned} \tag{10.5.7}$$

The constant $\chi := \langle 0_3 | \hat{x} | 1_3 \rangle = \langle 1_3 | \hat{x} | 0_3 \rangle$ measures a dipole moment. In this way, the effective Hamiltonian operating on this qubit takes the form

$$\hat{H}(t) = \Delta_0 \hat{\sigma}_3 - \chi F(t)\hat{\sigma}_1. \tag{10.5.8}$$

If the amplitude $F(t)$ of the driving is a complex number, e.g., an electromagnetic force $F(t) \sim \exp(i\omega t)$, the imaginary part would amount to adding a term $\sim \hat{\sigma}_2$ to the Hamilton operator. Controlling their time dependences appropriately then allows to navigate along arbitrary trajectories on the qubit's Bloch sphere to implement quantum gates.

It would be tempting to add more qubits to build Hilbert spaces of higher dimension $D_{\mathcal{H}} = 2^n$, by including also higher eigenstate pairs of the same double well, preferably in the semiclassical regime $E_B \gg \hbar\Omega$ where there are many eigenstate pairs, analogous to the ground state doublet, below the barrier top. An obvious problem with this option is, however, that the energy range to be covered by input and output and controls would grow as 2^n, while, for example, in a computer assembled from n separate two-state systems, it would only scale as n. This, together with unlimited scalability, is a strong point in favour of systems with a naturally finite Hilbert-space dimension, instead of systems in continuous space with a countably infinite dimension of Hilbert space such as the quartic double well or the harmonic oscillator.

Time-dependent Hamiltonians always indicate that the description is incomplete, reducing the part of the system that exerts the driving force to its action on the central system while ignoring the back action of the central system on the driving system. This is a legitimate approximation if, for example, the driving system has a high inertia and/or evolves only on much larger timescales than the central system.

Another systematic source of error, taking the orchestrated action of quantum gates for granted, is the precise calibration of their parameters. Being controlled from the classical level, amplitudes, phases, and time dependences are subject to fluctuations. Accumulated over time, they lead to decoherence, as does more generally the macroscopic nature of the laboratory equipment used to implement the driving. In this aspect, quantum computing suffers from the same instability as reversible classical implementations, such as in billiard ball gates, see Sect. 9.2.2.

Output: measurement

The final step in a quantum computation is reading out the result and converting it into classical information by a measurement. It is a crucial step, because in order to exploit the full potential of quantum computation, applying quantum parallelism, it is indispensable to run the quantum computer in a superposition state covering all single-qubit subspaces of its Hilbert space, cf. Sect. 10.3.2. The measurement therefore has two effects: It collapses the broad superposition to a single pure state, and it selects a single input from the $D_{\mathcal{H}}$ alternative ones that are contained in the output state before the measurement.

A typical laboratory experiment measures the expectation value of some observable $\hat{O}$, such as position $\hat{x}$ or momentum $\hat{p}$. If together with the mean $\langle \hat{O} \rangle$, all moments of the full probability distribution of this observable are to be determined, an entire set of operators, for example $\{|x\rangle\langle x| \, | \, x \in \mathbb{R}\}$, would have to be measured, resulting in a probability distribution

$$p(x) = \langle \psi | x \rangle \langle x | \psi \rangle = |\psi(x)|^2 \tag{10.5.9}$$

for an arbitrary state $|\psi\rangle$ of the measured system. In the standard situation for quantum computation, a broad superposition of qubits, for example, $|\psi\rangle = \sum_{m=0}^{2^n-1} c_m |m\rangle$ with $|c_m|^2 = 1/2^n$, this would result in

$$p(x) = \sum_{m,m'=0}^{2^n-1} c_m^* c_{m'} \langle m|x\rangle\langle x|m'\rangle = \sum_{m,m'=0}^{2^n-1} c_m^* c_{m'} \psi_m^*(x) \psi_{m'}(x), \tag{10.5.10}$$

that is, a global weighted average over all qubits involved. If only the state the system occupies within a discrete basis $\left\{ |m\rangle \, \middle| \, 0 \le m \le 2^n - 1 \right\}$ is to be measured, the distribution for the same example is

$$p_m = \sum_{m',m''=0}^{2^n-1} c_{m'}^* c_{m''} \langle m'|m\rangle\langle m|m''\rangle = |c_m|^2 = \frac{1}{2^n}, \tag{10.5.11}$$

and the measurement would select one of the basis states $|m\rangle$ at random, with equal probability in the range $0 \le m \le 2n{-}1$.

Much more in the spirit of quantum computation is it, however, to address individual qubits also on exit, in order to extract the intended output information after processing a superposition of input states. Individual qubit observations are part of the strategy, for example in the Deutsch-Jozsa algorithm, Sect. 10.3.3. A measurement of, say, the Pauli matrix $\hat{\sigma}_{1m_0}$ operating on qubit m_0, would amount to focussing on the local operator

$$\hat{O}_{m_0} = |0_{m_0}\rangle\langle 1_{m_0}| + |1_{m_0}\rangle\langle 0_{m_0}|. \tag{10.5.11}$$

In a system with continuous position space, such as the quartic double well, this will be a formidable technical task, and even in systems composed of two-state systems, addressing individual qubits in a final measurement is anything but trivial.

A simple model of quantum measurement has been discussed in Sect. 8.3.3 to demonstrate how it leads to decoherence and irreversibility. Observations in the course of the quantum algorithm, before it is completed, therefore have to be avoided, even if getting updated information on the state of the system at an intermediate state might be desired to optimize its control. Nevertheless, this can be achieved at least partially by reducing the coupling to the meter so far that decoherence can be neglected on timescales relevant for the measurement.

Resuming the various requirements implied by the three main phases of quantum computation, a good candidate system for the implementation of a quantum computer should meet the following criteria [Div00] as far as possible:

1. **Natural representation of qubits**, a Hilbert space of finite dimension $D_{\mathcal{H}} \in \mathbb{N}$, by the nature of the system or by efficient decoupling of higher excited states,
2. **Scalability**, the possibility to increase the number of qubits by adding identical elements,
3. **Complete control**, options to interact with the computing system without inducing decoherence and to implement a universal set of gates,
4. **High quality**, that is, a sufficiently long decoherence time to perform enough elementary operations before the system equilibrates.
5. **Individual readout** by addressing single qubits in measurements.

The simplicity of the basic principles and the rigour of the theoretical framework for quantum computation contrast with the diversity of systems and solutions that have been proposed for its implementation, ranging from quantum optics through solid-state physics through chemical physics. In the next subsection, four prototypical systems, considered as possible platforms for quantum computation, will be outlined and assessed according to these criteria.

10.5.2   Prototypical Platforms for Quantum Computing

The first and most basic decision in designing a quantum computer is how to realize the qubits required for the computation proper. As argued above, selecting two-state systems within the Hilbert space of countably infinite dimension of a bound system in continuous space, using some mechanism to project it out and isolate it from the rest of the system, is a viable possibility. The alternative is to use systems with a Hilbert space restricted naturally to two dimensions, or systems composed thereof. Accordingly, we can distinguish two major classes of physical implementations of qubits:

- **Spins** of massive fermions, in particular the individual spin of electrons or the collective spin of nuclei with an odd number of protons and/or neutrons.
- **Photons** representing two-state systems in various ways, such as Fock states in cavities or waveguide modes or circular polarization states, and atoms where two-level systems can be isolated by symmetry properties or resonance frequencies.

The following paragraphs proceed from quantum computation based on photons and/or atoms to realizations employing spins, passing also from the most volatile substrate, massless particles, to the most massive, entire molecules.

Photons only: quantum computation with the electromagnetic field

Qubits: Building qubits from quantum features of the electromagnetic field would amount to quantum computation without any material objects participating in the computation proper! Wherever discrete modes of the electromagnetic field can be identified, in particular in confined geometries, the scheme of second quantization [Hak76] introduces a quantum mechanical treatment by considering each mode as an independent harmonic oscillator, the most elementary model of a quantum system (Sect. 8.1.3). A photon is a component of the field that contains one quantum of energy, $\hbar\omega$.

However, the field is extended in continuous space, it is not obvious how to isolate such a component to harness it for computation. Cutting out one photon exactly is not possible, but approximate solutions do exist. A useful tool is the concept of coherent states (Sect. 8.1.3). Instead of representing them in position space, Eq. (8.1.33) or as a Wigner function, Eq. (8.1.34) [Sch01], they can be written as superpositions of energy eigenstates $|n\rangle$, $\hat{H}|n\rangle = E_n|n\rangle$, $E_n = \left(n + \frac{1}{2}\right)\hbar\omega$,

$$|\gamma\rangle = \exp\left(-\frac{|\gamma|^2}{2}\right) \sum_{n=0}^{\infty} \frac{\gamma^n}{\sqrt{n!}}. \tag{10.5.11}$$

The dimensionless complex parameter γ defines the centroid of the coherent state in phase space, therefore $|\gamma|^2 = \langle\hat{H}\rangle$ measures the mean energy, thus the degree of excitation of the field mode ω. Generating coherent states with a sufficiently small but non-zero value of $|\gamma|^2$ allows to restrict the photon number to 1 with high relative probability. However, the total probability for no excitation at all is much higher even, so that photons are generated at random at unpredictable times. A more controllable scheme to produce single photons is *parametric down-conversion* [Boy08], where in a nonlinear optical medium, an incoming photon of frequency ω is split into two photons of frequencies ω_1 and ω_2, with $\omega = \omega_1 + \omega_2$. If one of them is detected (and thus lost), the other is available for computation.

The pair of states $|0\rangle$ (no photon) and $|1\rangle$ (one photon) is not suitable for computation, if only for the marked physical asymmetry between them. Much more appropriate is distinguishing two modes, say a and b, with different frequencies, spatial location (e.g., two coupled cavities) etc., and to construct the qubit from the two states $|1_a 0_b\rangle =: |0\rangle$ and $|0_a 1_b\rangle =: |1\rangle$, a strategy referred to as *dual-rail representation*. If they refer to two separate light paths or to two distinct transverse modes travelling along a waveguide, the dual-rail representation translates time evolution into propagation along these channels.

Gates: Light interacts weakly with matter and almost not at all with itself. This strongly restricts options how to build gates transforming photon states. The standard toolbox is provided by traditional optics, specifically phase shifters and beam splitters, as introduced in Sect. 10.2.2.

Phase shifters are readily realized passing the pair of beams that represent the qubit through two distinct media with indices of refraction n_a and n_b, resp. In order to achieve a phase shift of $\Delta\phi$, this requires an optical path length

$$I = \frac{c}{\omega |n_a - n_b|} \Delta\phi, \qquad (10.5.12)$$

for light of frequency ω, where c is the vacuum velocity of light.

In addition to shifting phases, beam splitters interchange intensity between the two beams, cf. Eqs. (10.2.43, 44) and Fig. 10.2. In their simplest form, they do not require more than a half-silvered mirror. A somewhat more sophisticated version consists of a pair of 45° prisms separated by a metal layer. Combined with phase shifters, beam splitters comprise the full range of (2×2)-scattering matrices, for example parameterized as in Eq. (10.2.43), including in particular NOT and Hadamard gates, and thus are sufficient to implement all single-qubit gates.

The real challenge in quantum computing with photons is costructing two-(and more-)qubit gates, since they require light-light interaction. A linear treatment of the electromagnetic field, equivalent to neglecting all interactions between field modes, is ubiquitous in optics, Even in bulk material, nonlinear effects depending on the intensity are extremely scarce. A prominent exception is the *optical Kerr effect*, where the index of refraction depends on the intensity I,

$$n(I) = n(0) + \nu I, \qquad (10.5.13)$$

the coefficient ν measuring the strength of the nonlinearity. Combining a medium with stronger Kerr effect with another medium where it is negligible would allow to build a phase shifter, replacing one of the indices of refraction in Eq. (10.5.12) by $n(I)$. In this way, however, a trade-off arises between the phase difference desired, say π, and the attenuation along the required optical path. Even in the strongest Kerr media, the effect is so weak that a relative phase of π implies an unacceptable loss of intensity.

Decoherence: Conservation of intensity is a central challenge in optical technology. For example, optical fibres transmit light signals over distances of kilometres without intermediate amplification, and quality factors of optical cavities can reach values of the order of 10^{10}.

Readout: Photons provide the standard signal to detect elementary particles in experimental high-energy physics. Even neutrinos can be counted by amplifying the utterly weak scintillation flashes they trigger in a bulk medium. As a spin-off of this development, extremely efficient photomultiplier tubes are available that can be employed also in quantum computation for the detection of single photons.

Photons and atoms: quantum computation in cavities

Qubits: Quantum computation with light and only light suffers from the weak direct interaction between photons. A possible alternative is delegating the interaction proper to other particles that couple the photons indirectly. They should (i) interact strongly with photons, (ii) faithfully represent the information carried by the qubits involved, and (iii) provide sufficient internal interaction to implement two-qubit gates.

An attractive solution is atoms. Their structured charge distribution, manifest in their electric and magnetic dipole moments, interacts strongly with the electromagnetic field. Atoms typically have a complex energy level structure covering a broad range of excitation energies. Resonance energies, however, are so precisely defined, and symmetries such as parity so rich, that the frequency and polarization of photons employed as qubits is sufficient to select a single pair of atomic states as the two-state system representing specifically and exclusively an optical qubit.

Atoms interacting with light is an advanced topic referred to as quantum optics. The most common laboratory to study this interaction are optical cavities, metallic boxes with sufficiently high Q-factor for their internal field modes to warrant even a repeated swapping of qubits between the field modes and an atom in the cavity. Doing quantum computation in a high-Q cavity has an important general physical consequence: The perfect confinement defines a discrete set of eigenmodes of electromagnetic radiation with sharply defined natural frequencies and almost completely prevents irreversible decay processes.

Gates: For an atom of an overall size much smaller than the wavelength of the relevant light mode in the cavity, the light-matter interaction reduces to the coupling between the dipole moment of the atom, given by a spatial distance which in a two-state system is represented by $\hat{\sigma}_1$, and the field amplitude, represented by the boson operator $\hat{x} = \hat{b} + \hat{b}^\dagger$. It is described by the interaction Hamiltonian

$$\hat{H}_{AB} = g\hat{\sigma}_1\left(\hat{b} + \hat{b}^\dagger\right) = g\left(\hat{\sigma}_+ + \hat{\sigma}_-\right)\left(\hat{b} + \hat{b}^\dagger\right). \qquad (10.5.14)$$

Together with the Hamilton operators for atom and field alone, $\hat{H}_A = \frac{1}{2}\hbar\omega_0\hat{\sigma}_3$ and $\hat{H}_B = \hbar\omega_1\left(\hat{b}^\dagger\hat{b} + 1/2\right)$, resp., the total Hamiltonian takes the form

$$\hat{H} = \hat{H}_A + \hat{H}_{AB} + \hat{H}_B = \frac{1}{2}\hbar\omega_0\hat{\sigma}_3 + g(\hat{\sigma}_+ + \hat{\sigma}_-)(\hat{b} + \hat{b}^\dagger) + \hbar\omega_1\hat{b}^\dagger\hat{b}, \quad (10.5.15)$$

known as the spin-boson model with a single boson mode. It resembles the Hamiltonian for a spin interacting with an environment, modelled as a large number of harmonic oscillators, introduced in Sect. 8.3.3, Eqs. (8.3.30–8.3.33), with two exceptions: In the present spin-boson model, the spin interacts with the boson through its $\hat{\sigma}_1$ operator, and the number of bosons is reduced to one.

Of the atom–field operator products in the interaction term, $(\hat{\sigma}_+ + \hat{\sigma}_-)(\hat{b} + \hat{b}^\dagger) = \hat{\sigma}_+\hat{b} + \hat{\sigma}_+\hat{b}^\dagger + \hat{\sigma}_-\hat{b} + \hat{\sigma}_-\hat{b}^\dagger$, the first and the last one generate balanced processes, combining excitation of the atom with deexcitation of the field or vice versa. The second and third term correspond to simultaneous excitation or deexcitation of both subsystems. The latter two processes amount to strong decrease or increase of energy in both subsystems and are very infrequent. Neglecting them altogether (rotating-wave approximation) [Lou73] leads to the simplified model,

$$\hat{H} = \frac{1}{2}\hbar\omega_0\hat{\sigma}_3 + g(\hat{\sigma}_+\hat{b} + \hat{\sigma}_-\hat{b}^\dagger) + \hbar\omega_1\hat{b}^\dagger\hat{b}, \quad (10.5.16)$$

referred to as *Jaynes-Cummings Hamiltonian*, a workhorse of quantum optics.

The dominating effect on the atom of its interaction with the field is determined by the $\hat{\sigma}_1$ operator in the coupling and consists of rotations around the x_1 axis with a frequency $\Omega = \sqrt{g^2 + \Delta^2}$, with the *detuning* $\Delta = (\omega_0 - \omega_1)/2$. It rotates the atomic eigenstate $|0\rangle$ towards $|1\rangle$ and back, an effect known as *Rabi oscillation*, absorbing energy from the field but also inducing a shift in the relative phase of the photon states $|0\rangle$ and $|1\rangle$. It is this phase shift which can be used for quantum computation. Being nearly independent of the coupling, until very low values of g, it is much more efficient than the Kerr effect and readily allows to achieve the required relative phases of π. Together with beam splitters, it thus opens access to universal gates.

Swapping qubits between photons and atomic states, even more complex interactions come into reach. Given two photons with different frequencies ω_1 and ω_2, it is possible to interact simultaneously with a triplet of atomic states, with energies E_0, E_1, E_2. If the photon frequencies are tuned such that $\omega_1 = |E_1 - E_0|$ and $\omega_2 = |E_2 - E_0|$, transitions $E_1 \leftrightarrow E_0$ cannot occur if the atom is in its state $|2\rangle$ and vice versa. The logic of this interaction depends in particular on the order of the three energies; if $E_0 < E_1, E_2$, the triplet is in a V-, otherwise in a Λ-configuration.

With photons interacting with atoms in a cavity, the computation proper can therefore take place in the field, the atoms contributing only nonlinear interaction between photons, or in the atoms, the photons serving only for communication in space and time. In view of the latter option, one may ask, why involve photons in the first place? A compelling answer is that light opens a wealth of possibilities for very efficient communication, among the qubits forming the computer as well as with the outside world.

Decoherence and readout: Photons can bring their virtues to bear in particular concerning their superior stability against noise and other incoherent processes, in waveguides or cavities, and the advanced technology available to detect photons for the readout.

Atoms, ions, and lasers: quantum computation in traps and optical lattices

Quantum computation with trapped atoms or ions is arguably the implementation that comes closest to Feynman's vision of "an atom as a computer". Looking superficially at the physical apparatus, the difference between quantum optics in cavities and in traps may appear marginal, but conceptually it is substantial. Traps are not enclosed in any material confinement and thus strongly coupled to the surrounding electromagnetic field, which leads to irreversible decay processes. They do not prefer any specific internal modes but readily permit irradiation with, e.g., laser light of arbitrary frequency. At the same time, the confining potential does allow for bounded spatial motion of the trapped particles, for example vibrational modes.

Ion traps, invented by Wolfgang Paul, indirectly revolutionized quantum physics, by focussing attention on individual microscopic quantum systems instead of considering ensembles, the dominating perspective till then. It uses a trick to circumvent Earnshaw's theorem that no static electric or magnetic potential can confine a charge, a corollary of Gauss' law that does not permit the field pointing inwards over all of a closed surface in free space. However, a field oscillating periodically in time, on average over many cycles, can well have this effect. For example, vertical oscillations can stabilize a pendulum in its otherwise unstable inverted position [But01]. The Paul trap implements this effect either as a three-dimensional confinement, using hyperbolic ring and cap electrodes with cylindrical symmetry, Fig. 10.14, or as confinement in two dimensions generated by linear electrodes in a quadrupole

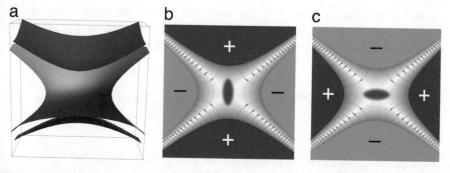

a   b   c

Fig. 10.14 Quadrupole ion traps employ three electrodes, two opposing caps of the same polarity and a ring electrode (cutaway rendering (**a**)) of opposite polarity to generate a quadrupole field of cylindrical symmetry. For static polarity, the equilibrium position in the centre of the trap is unstable. However, with periodically alternating polarity, generated by driving he electrodes with an AC voltage, charged particles (blue) can be stabilized on average in the equilibrium point. This effect allows to localize non-neutral ions without any contact with material walls. In panels (**b**) and (**c**), the instantaneous electric potential is represented by colour code (from red (negative) through white (zero) through blue (positive)), field lines are indicated by arrows.

configuration. While the three-dimensional trap, loaded with two or more particles, concentrates them into a cloud of spherical overall shape, the linear trap forces them into an approximately linear configuration, where repulsive electrostatic forces tend to induce a periodic order, called a Wigner crystal.

An alternative that does not require charged particles to be trapped but works with neutral atoms is optical lattices. They generate a potential, periodic in one spatial direction, as a standing wave resulting from two counterpropagating laser beams with identical wavelength. The Stark shift the oscillating field induces in neutral atoms converts either nodes or antinodes into effectively static potential wells, depending on the sign of the detuning between the laser frequency and that of the relevant atomic transition frequency. Unlike Wigner crystals, the configuration resulting for a larger number of trapped particles is exactly periodic in space, with a lattice constant given by half the wavelength of the confining laser, typically much smaller than the ion separation in a linear trap.

Qubits: By contrast to the previous realizations, quantum computation with trapped ions or atoms employs angular momentum states as qubits, taking advantage of the small finite dimensions of their Hilbert spaces. Working with atoms, angular momentum arises as electron spin, as orbital angular momentum of the electrons, and as nuclear spin. The interaction of these different spin systems, together with spatial symmetries of the atom, leads to a rich and complex level structure, ranging from orbital states down to the hyperfine structure on the smallest energy scales, the preferred subspace for quantum computation. This intricate web offers a host of alternatives to implement qubits, beyond the mere assembly of spin-$\frac{1}{2}$ state pairs.

A still elementary example is realizing a qubit pair with a single angular momentum 1 instead of two spins-$\frac{1}{2}$. In many systems, in particular most electron shells and nuclei, the relevant angular momentum operators, commuting with the Hamiltonian, are the square $\hat{J}^2$, with eigenvalues $j(j + 1)$, $j \in \left\{\frac{k}{2} | k \in \mathbb{N}\right\}$, and the component $\hat{j}_z$ parallel to the magnetic field, with eigenvalues $m_j \in \{-j, -j + 1, \ldots, j - 1, j\}$, called magnetic quantum numbers. For $j = 1$, their four simultaneous eigenstates $|j, m_j\rangle$ relate to the canonical two-qubit basis $\{|00\rangle, |01\rangle, |10\rangle, |11\rangle\}$ as

$$|0, 0\rangle = \frac{1}{\sqrt{2}}(|01\rangle - |10\rangle), |1, -1\rangle = |00\rangle, |1, 0\rangle = \frac{1}{\sqrt{2}}(|01\rangle + |10\rangle), |1, 1\rangle = |11\rangle.$$

$$(10.5.17)$$

Similarly, the eigenstates of these operators for $j = \frac{3}{2}$ can represent three qubits etc.

Scalability: Scaling up a trapped ion computer is straightforward by inserting more ions into the trap, or by filling more potential wells in an optical lattice, but is not easy to control and limited by the holding capacity of the device.

Gates: For an atom in a cavity, the dominating physics is the periodic, thus reversible, exchange of quanta with the electromagnetic field. In the present context of open traps, an ion in an excited state does not interchange energy with a specific field mode but decays irreversibly into the modes of the surrounding field. Spontaneous decays are described by the same spin-boson or Jaynes-Cummings Hamiltonians, Eq. (10.5.16), but now integrating over a continuum of bosonic degrees of freedom in infinite three-dimensional space. Excitations, in turn, are induced by irradiating the atom with laser pulses tuned to the frequency of the desired transition and of a duration tailored such as to achieve the desired rotation of the targeted two-state system in its Hilbert space. In this way, all single-qubit transformations are performed.

An interaction between qubits represented by individual ions or atoms is mediated by spatial vibration. In a chain of charged particles trapped in confining potentials, collective vibrational eigenmodes range from centre-of-mass motion down to alternating individual oscillation of neighbouring ions in counterphase, as in a classical linear chain of harmonic oscillators, see Sect. 5.6.2. They are quantized as well as harmonic oscillators, with energy quanta $\hbar\omega$, now called phonons. The interaction of ions with these modes is described again by the Jaynes-Cummings Hamiltonian. Collective modes extend over all ions in the trap, but with amplitudes varying from ion to ion. This allows to couple qubits selectively and thus to construct all two-qubit transformations, such as CONTROLLED-NOT and Hadamard gates.

Decoherence: The strong coupling of traps to the ambient field is a major threat to the coherence of stored qubits, as is manifest in particular in the spontaneous decay of excited atom states. Yet decay rates are so low that typical lifetimes of atomic states are sufficient to perform an appreciable number of gates on them. Lifetimes of vibration modes are much shorter, but they can be extended in effect by swapping information between motional and atomic states.

Readout: The stability and typical distances of ion configurations in traps are suitable for a targeted laser irradiation of individual ions and a specific detection of transmitted light, and thus for the readout of single qubits. Atoms trapped in optical lattices are localized even more precisely, facilitating a selective readout further.

Nuclear magnetic resonance: quantum computers in solution

Quantum computing in trapped ions and atoms goes to the extreme as concerns miniaturization of the physical structures involved. At the opposite extreme is an implementation of quantum computers that is largely based on a technology originally developed for applications in chemistry, biology, and even medicine: nuclear magnetic resonance (NMR). It has in common with ion trap technology that qubits are represented by nuclear spin states. However, it adopts a more traditional approach in that it addresses large ensembles of quantum systems and macroscopic manifestations of quantum phenomena.

The physical basis of NMR is the precession of nuclear spins around the direction of an external magnetic field with a frequency, the Larmor frequency, proportional to the magnitude of the field. The interaction of an additional external field, oscillating in resonance with the Larmor frequency, with the precession allows to induce transitions between angular momentum states and in particular to control the orientation of the spin relative to the static field. The field emitted owing to the precession of the spin, can be detected, in turn, with antennas.

The Larmor frequency $\omega_L = -\gamma B_0$ for a spin $\frac{1}{2}$ appears to be universal, depending only on the external field B_0 through the gyromagnetic ratio γ. However, the local field seen by the nucleus does not coincide exactly with the external field generated by some macroscopic coils. It is strongly modified by the electron shell of the atom, which tends to shield the nucleus against external field, and likewise by the entire environment of the nucleus on atomic and molecular scales. This explains the exquisite sensitivity of the Larmor frequency on the atomic and molecular structure, which not only allows to identify elements and isotopes, but also molecules or chemical groups the nuclei are embedded in. It is this specificity that opens a huge range of applications of NMR in chemistry and in the life sciences.

For quantum computation, this means that the natural units to be considered are not atoms or even two-state pairs, but entire molecules. While basic operations still proceed on the level of qubits, molecules play the role of quantum computers with a structure of communication between atoms implied by the configuration of the molecule. At the same time, it is not possible in NMR computation to address individual molecules. All operations affect simultaneously and in the same way all molecules of the same species in the sample serving for computation. It means that, as in quantum statistics, only ensembles can be prepared, manipulated, or measured. Liquid samples containing the molecules employed for computation are usually at room temperature, forming ensembles in thermal equilibrium with a broad energy distribution. This limitation makes it difficult to prepare qubits for example in their ground state and to read out specific dimensions of the molecules' Hilbert space.

Qubits: In NMR computation, qubits are represented by the spin-$\frac{1}{2}$ two-state systems of nuclear spins of the atoms present in the molecule (nuclei with proton and neutron numbers that are both even, so that both subsets form bosons, cannot be used). They are controlled by the combination of a strong static magnetic field (a typical magnitude is 12T) for the basic orientation of the spins, say in the direction $\mathbf{u}_3$, and a much weaker radio-frequency (RF) field for rotations in the transverse plane ($\mathbf{u}_1$, $\mathbf{u}_2$),

$$\mathbf{B}(t) = B_0 \mathbf{u}_3 + B_1 (\mathbf{u}_1 \cos(\omega t) + \mathbf{u}_2 \sin(\omega t)). \qquad (10.5.18)$$

The time evolution generated by the Hamiltonian

$$\hat{H}(t) = \frac{\omega_0}{2} \hat{\sigma}_3 + g(\hat{\sigma}_1 \cos(\omega t) + \hat{\sigma}_2 \sin(\omega t)), \qquad (10.5.19)$$

with $\omega_0 = -\gamma B_0$ and $g = -\gamma B_1$, consists of rotations of the spin around an effective axis $\mathbf{n} \sim \mathbf{u}_3 + (2g/(\omega_0 - \omega))\mathbf{u}_1$. By tuning the RF field into resonance with the Larmor frequency ω_0 and applying it only during pulses of defined duration, this allows for arbitrary rotations in the Hilbert space of the two-state system.

Scalability: The size of an NMR quantum computer is naturally limited by the size of the molecule employed. Trade-offs encountered choosing larger molecules are the increasing number of configurational degrees of freedoms and the complexity of couplings between qubits.

Gates: Single-qubit gates require mainly rotations by $\pi/2$ or π, which are common tools in NMR technology anyway to steer the nuclear spins in the sample into an optimal orientation for spectroscopy. Specifically, π-pulses around $\mathbf{u}_1$ are of special importance also in computation: They amount to a reflection $x_3 \rightarrow -x_3$ and therefore undo all previous precession around $\mathbf{u}_3$, a useful task to correct for unwanted inhomogeneities in the sample and as a reset for intended logical operations.

The spin–spin interaction necessary for two-qubit gates is provided by two channels, direct mutual coupling of the two magnetic dipoles and indirect interaction through the electrons shared among two nuclei as chemical bonds. While the first coupling is diminished on average over a large number of molecules, owing to the random orientation of the spins in thermal equilibrium, the coupling through electron clouds contains a scalar component, independent of the relative orientation of the two spins, and thus dominates the spin–spin interaction in thermal equilibrium. It is sufficient to devise two-qubit gates in NMR computation. Spin–spin couplings depend very sensitively on molecular structure, such as bond lengths and orientations. The specific configuration of the molecule used thus determines the "architecture" of the quantum computer built from it.

Decoherence: Working with liquid samples at ambient temperature, qubits in NMR technology are unrestrictedly exposed to thermalization, severely limiting their usable lifetimes. Other mechanisms that lead to a loss of coherence are inhomogeneities of the static magnetic field, leading to differences in precession frequencies and thus to a randomization of relative phases, as well as residual spin–spin couplings not used positively in the computation.

Readout: In nuclear magnetic resonance, it is standard technology to measure spin precession frequencies with antenna coils. However, individual qubits cannot be read out from a thermalized molecule. The result will be random and useless for computation. NMR computation thus prefers algorithms that only require the measurement of global properties of the qubit set.

Other implementations: Josephson junctions, quantum dots, and solid-state traps

The implementations outlined above have been the first ones proposed for quantum computing and have been most amply considered in theory and in the lab. Other options have come up in the meantime, based on novel ideas and employing different technologies. The enormous diversity of proposals indicates that the field is still in a very dynamic phase where no standard solutions have been established yet.

– Superconductivity is an attractive option for quantum computation, as it provides a quantum system of perfect stability. A discretization of states is achieved by inserting a weakly conducting barrier, forming a **Josephson junction**, which can be surmounted by Cooper pairs, the charge carriers in superconductivity, by tunnelling. Combining Josephson junctions, devices can be built where the number of Cooper pairs present on a superconducting island between two junctions, or the magnetic flux through a ring including two junctions, or the phases of oscillating superconducting currents across the junction, are available as control parameters. To maintain superconductivity, devices have to be cooled to extremely low temperatures (of the order of a few K). Quantum computation in superconducting circuits is explored mainly in commercial projects.

- **Quantum dots** are potential wells formed in semiconducting material by structured doping, of a size of the order of nanometres so that they can accommodate only a few bound electron states. They are the principal laboratory of mesoscopic physics, owing to their similarity to small atoms. Qubits are implemented in quantum dots as spin-$\frac{1}{2}$ states of electrons trapped in a single quantum dot, or as localized charge states in a double quantum dot. They can be controlled applying gate voltages or magnetic fields.

- Instead of trapping atoms or ions representing qubits in oscillating electromagnetic fields, they can be fixed as **donors or impurities in bulk solid-state material**, such as phosphorus donors in silicon or vacancies forming point defects in the diamond carbon lattice, or enclosed in fullerene nanospheres or in nanotubes. Qubits are then represented by internal states of these atoms or vacancies and manipulated by external fields as in traps or optical lattices (Table 10.2).

The latest and most advanced implementations of quantum computation (state of the art in 2020) have been presented by commercial companies, Google [AA&19] and IBM. The qubits employed in the Google device are based on macroscopic quantum states in superconducting circuits, a technology that could not yet be considered in the present section. Which kind of qubits are used in IBM computers is not clear. The processors of both products comprise around 100 qubits, a number rising with each new release. Companies claim that in benchmark tasks, their quantum computers are many orders of magnitude faster than state-of-the-art conventional supercomputers.

Table 10.2 The four physical implementations of quantum computation discussed in Sect. 10.5.2 are rated according to DiVincenzo's criteria [Div00].

Qubits	Scalability	Preparation	Control by gates	Decoherence	Readout
Photons	−	+	−	O	+
Photons and atoms in cavities	−	+	O	+	+
Ions in traps, atoms in optical lattices	+	−	+	−	+
NMR with molecules in solution	O	−	+	−	+

Chapter 11
Epilogue

Offering a synopsis of the dynamics of information, this book covers a huge range of scales, from atomic to astronomic, and touches a broad set of subjects, from quantum mechanics through logics through epistemology. This encompassing view owes itself to a central intention: to demonstrate that all over this diversity of fields, the concept of information not only plays a crucial role, but that it is the *same quantity*, physical entropy, that is behind the apparently divergent meanings and uses of the notion of information over the disciplines.

This idea may appear most daring in the intermediate scales, in life on Earth. To be sure, the significance of information in its synaptic for genetics is almost commonplace. That since its origin, life contributes increasingly to the conversion of low-entropy solar light to high-entropy thermal radiation of the Earth is an established fact. Less familiar is the view that even in its most complex manifestations where semantics is in the foreground, in human culture and communication, we are dealing with *physical entropy*. History and individual cognition are irreversible processes, they form a consequent step in the evolution of the universe through a sequence of non-equilibrium processes generating locally more and more order [Bre94]. On Earth, human cognition even increases the entropy production in a very direct sense: Purposeful human action guided by knowledge now modifies global parameters of the heat balance of the planet as an unintended side effect and will only be compensated, in turn, by further human insight. The particular question how the functioning of the nervous system can be interpreted in terms of information processing, from single neurons up to high-level brain functions, is an important research subject in the neurosciences. In this book, it could hardly be touched.

At the bottom of scales, dominated by the laws of quantum mechanics, we already have a quite a clear and consistent view of how information is transformed in quantum systems. In fact, at these scales, the familiar classical notion of information, based on hard facts, at most softened by probabilities, has to be abandoned altogether and gives way to a much more abstract and elusive measure. How classical facts can form on this substrate in macroscopic systems, in turn, is a long-lasting question quantum physics has begun to elucidate.

© Springer Nature Switzerland AG 2022
T. Dittrich, *Information Dynamics*, The Frontiers Collection,
https://doi.org/10.1007/978-3-030-96745-1_11

Much less is known about the opposite extreme, cosmic scales up to the size of the universe. Is it a closed system in the thermodynamical sense that conserves entropy? With which entropy content was it born at the Big Bang? How does it change with the growing universe? How do Special and General Relativity modify entropy transport? An apparently special detail is the time evolution of the entropy in black holes [Zeh10], resulting in particular in the so-called black hole information paradox [Mal20]. It is presently under active study and indeed already touches many fundamental questions, such as how to treat entropy under the extreme conditions in its core. They have much in common with the conditions during the big bang [Wei88] and therefore allow for more general conclusions regarding the role of entropy in cosmology [Mal20]. Again, even an outline of this field would have exceeded the scope of this book by far.

A coherent picture of information dynamics, bridging the borders between many disciplines, lends substance to the concept of an information age, discussed in the introduction. It is much more than just a label attached to a historical epoch. Now that it lasts already almost two centuries, are there any emerging ideas in sight that might replace information as a ubiquitous concept characterizing a new epoch? A number of promising candidates come to mind. My personal favourite is *networks*: Again originating as a precise technical term in applied mathematics, it rapidly conquered physics, from quantum mechanics to statistical mechanics, entered biology, in particular cell biology, and of course computer science, from where it massively invaded social science and daily life: Social networks have become a global phenomenon we cannot escape. "Networking" is almost obligatory for everyone striving for social or political impact. This development has reached a sufficiently mature stage to invite for a comprehensive appreciation.

Bibliography

[AA68] Vladimir I. Arnol'd, A. Avez: *"Ergodic Problems of Classical Mechanics"*, Benjamin (New York, 1968).

[AA&15] Carlos Abellán, Waldimar Amaya, Daniel Mitrani *et al.*, Phys. Rev. Lett. **115**, 250403 (2015).

[AA&19] Frank Arute, Kunal Arya, Ryan Babbush *et al.*, Nature **574**, 505 (2019).

[AB&03] Luis von Ahn, Manuel Blum, Nicholas J. Hopper, John Langford: *"CAPTCHA: Using Hard AI Problems for Security"*, in *"EUROCRYPT 2003: International Conference on the Theory and Applications of Cryptographic Techniques"* (2003).

[AC02] Andrew Adamatzky, Benjamin De Lacy Costello, Phys. Rev. E **66**, 046112 (2002).

[Ada09] Andrew Adamatzky, Phys. Lett. A **374**, 264 (2009).

[ADR82] Alain Aspect, Jean Dalibard, Gérard Roger, Phys. Rev. Lett. **49**, 1804 (1982).

[AFM15] Anthony Aguirre, Brendan Foster, Zeeya Merali (eds.): *"It from Bit or Bit from It? On Physics and Information"*, The Frontiers Collection, Springer (Cham, 2015).

[AGR81] Alain Aspect, Philippe Grangier, Gérard Roger, Phys. Rev. Lett. **47**, 460 (1981).

[AGR82] Alain Aspect, Philippe Grangier, Gérard Roger, Phys. Rev. Lett. **49**, 91 (1982).

[AH&04] J. B. Altepeter, P. G. Hadley, S. M. Wendelken, J. Berglund, P. G. Kwiat, Phys. Rev. Lett. **92**, 147901 (2004).

[AJR07] C. David Allis, Thomas Jenuwein, Danny Reinberg (eds.): *"Epigenetics"*, Cold Spring Harbor Laboratory Press (Cold Spring Harbor, 2007).

[AK07] Eduardo G. Altmann, Holger Kantz, Europhys. Lett. **78**, 10008 (2007).

[AK87] Zvia Agur, Michel Kerszberg, Am. Nat. **129**, 862 (1987).

[Alt18] Ehud Altman, Nat. Phy. **14**, 979 (2018).

[AM76] Neil W. Ashcroft, N. David Mermin: *"Solid State Physics"*, Holt-Saunders (Philadelphia, 1976).

[And58] Philip W. Anderson, Phys. Rev. **109**, 1492 (1958).

[And72] Philip W. Anderson: "More is different", Science **177**, 393 (1972).

[AP08] Nihat Ay, Daniel Polani, Advances in complex systems **11**, 1 (2008).

[AS93] Thomas P. Altenmüller, Axel Schenzle, Phys. Rev. A **48**, 70 (1993).

[Ash90] Robert B. Ash: *"Information Theory"*, Dover (New York, 1990).

[AV80] Yakir Aharonov, M. Vardi, Phys. Rev. D **21**, 2235 (1980).

[BA57] David Bohm, Yakir Aharonov, Phys. Rev. **108**, 1070 (1957).

[Bac96] Thomas Bäck: "Evolutionary Algorithms in Theory and Practice: Evolution Strategies, Evolutionary Programming, Genetic Algorithms", Oxford Univ. Press (Oxford, 1996).

[BAA06] D. M. Basko, I. L. Aleiner, B. L. Altshuler, Ann. Phys. (NY) **321**, 1126 (2006).

[BA&17] Manabendra Nath Bera, Antonio Acín, Marek Kuś, Morgan W. Mitchell, Maciej Lewenstein, Rep. Prog. Phys. **80**, 124001 (2017).

[Bak96] Per Bak: *"How Nature Works: The Science of Self-Organized Criticality"*, Copernicus (New York, 1996).

© Springer Nature Switzerland AG 2022
T. Dittrich, *Information Dynamics*, The Frontiers Collection,
https://doi.org/10.1007/978-3-030-96745-1

[Bal02] Tim Bale, Representation **39**, 15 (2002).

[Bar06] Luis Barreira: *"Poincaré recurrence: Old and new"*, in Jean-Claude Zambrini (ed.): *"XIVth International Congress on Mathematical Physics"*, pp. 415–422, World Scientific (New Jersey, 2006).

[Bat73] George Keith Batchelor: *"An introduction to fluid dynamics"*, Cambridge Univ. Press (Cambridge, UK, 1973).

[BB&71] V. Barkmann, P. Butera, L. Girardello, John R. Klauder, Rep. Math. Phys. **2**, 221 (1971).

[BB&91] R. Blümel, A. Buchleitner, R. Graham, L. Sirko, U. Smilansky, H. Walther, Phys. Rev. A **44**, 4521 (1991).

[BCP16] Mark F. Bear, Barry W. Connors, Michael A. Paradiso: *"Neuroscience: exploring the brain"*, 4th ed., Wolters Kluwer (Philadelphia, 2016).

[BC&00] Laurence Barker, Cağatay Candan, Tuğrul Hakioğlu, M. Alper Kutay, Haldun M. Ozaktas, J. Phys. A: Math. Gen. **33**, 2209 (2000).

[BCS04] Giuliano Benenti, Giulio Casati, Giuliano Strini: *"Principles of Quantum Computation and Information—Volume I: Basic Concepts"*, World Scientific (New Jersey, 2004).

[BE16] Philippe M. Binder, George F. Ellis, Phys. Scr. **91**, 064004 (2016).

[Bec67] Richard Becker: *"Theory of Heat"*, 2nd ed., revised by Günther Leibfried, Springer (Berlin, 1967).

[Bec75] William Beckner, Ann. Math. **102**, 159 (1975).

[Bel64] John S. Bell, Physics **1**, 195 (1964).

[Bel81] John S. Bell, J. Physique **42**, C2-41 (1981).

[Ben73] Charles H. Bennett, IBM J. Res. Dev. **17**, 525 (1973).

[Ben82] Charles H. Bennett, Int. J. Theor. Phys. **21**, 905 (1982).

[Ber71] John Desmond Bernal: *"Science in History. Volume 2: The Scientific and Industrial Revolution"*, MIT Press (Cambridge, MA, 1971).

[Ben03] Charles H. Bennett, Stud. Hist. Phil. Mod. Phys. **34**, 501 (2003).

[Ber77a] Michael V. Berry, Phil. Trans. R. Soc. (London) A **287**, 237 (1977).

[Ber77b] Michael V. Berry, J. Phys. A: Math. Gen. **10**, 2083 (1977).

[Ber78] Michael V. Berry: *"Regular and Irregular Motion"*, in S. Jorna (ed.): *"Topics in Nonlinear Mechanics"*, AIP Conf. Proceedings **46**, 16 (1978).

[Ber89] Michael V. Berry, Proc. R. Soc. (London) A **423**, 219 (1989).

[Ber91] Michael V. Berry: *"Some quantum-to-classical asymptotics"*, in Marie-Joya Giannoni, André Voros, Jean Zinn-Justin (eds.): *"Chaos and Quantum Physics"*, Les Houches Lectures LII 1989, NATO ASI, North-Holland (Amsterdam, 1991).

[BEZ00] Dirk Bouwmeester, Artur Ekert, Anton Zeilinger (eds.): *"The Physics of Quantum Information"*, Springer (Berlin, 2000).

[BF92] N. Brenner, S. Fishman, Nonlinearity **4**, 211 (1992).

[BGZ75] H. Bacry, A. Grossmann, J. Zak, Phys. Rev. B **12**, 1118 (1975).

[BG&89] R. Blümel, R. Graham, L. Sirko, U. Smilansky, H. Walther, K. Yamada, Phys. Rev. Lett. **62**, 341 (1989).

[BH20] Michael Bradie, William Harms, "Evolutionary Epistemology", The Stanford Encyclopedia of Philosophy (Spring 2020 Edition), Edward N. Zalta (ed.), https://plato.stanford.edu/archives/spr2020/entries/epistemology-evolutionary/.

[BI11] Heiko Bauke, Noya Ruth Itzhak, arXiv:1101.2683v1 [quant-ph] (2011).

[Bia06] Iwo Bialynicki-Birula, Phys. Rev. A **74**, 052101 (2006).

[Bia07] Iwo Bialynicki-Birula, AIP Conference Proceedings **889**, 52 (2007).

[BL85] Charles H. Bennett, Rolf Landauer, Sci. Am. **253**, 48 (1985).

[BM75] Iwo Białynicki-Birula, Jerzy Mycielski, Commun. Math. Phys. **44**, 129 (1975).

[Bol09] Ludwig Boltzmann: *"Wissenschaftliche Abhandlungen"*, vol. II (1875–1881), Fritz Hasenöhrl (ed.), Johann Ambrosius Barth (Leipzig, 1909).

[Boy08] Robert Boyd: *"Nonlinear Optics"*, 3rd ed., Academic Press (New York, 2008).

[BP97] Remo Badii, Antonio Politi: *"Complexity. Hierarchical Structures and Scaling in Physics"*, Cambridge Nonlinear Science Series 6, Cambridge Univ. Press (Cambridge, UK, 1997).

[BPM12] Jens H. Bardarson, Frank Pollmann, Joel E. Moore, Phys. Rev. Lett. **109**, 017202 (2012).
[BR56] J. Balatoni, A. Rényi: *"On the notion of entropy"* (in Hungarian), Publ. Math. Inst. Hungarian Acad. Sci. **1**, 9 (1956); English translation: *"Selected papers of Alfred Rényi"*, vol. I, pp. 558–584, Akademiat Kiado (Budapest, 1976).
[BR87] Christian Beck, Gert Roepstorff, Physica **25D**, 173 (1987).
[Bre94] Wilhelm Brenig: *"Erkenntnis und Irreversibilität"* (in German), in Klaus Mainzer, Walter Schirmacher (eds.): *"Quanten, Chaos und Dämonen. Erkenntnistheoretische Aspekte der modernen Physik"*, BI-Wissenschaftsverlag (Mannheim, 1994).
[Bri26] Leon Brillouin, Comptes rendus **183**, 24 (1926).
[Bri56] Leon Brillouin: *"Science and Information Theory"*, Academic (New York, 1956).
[Bri60] Leon Brillouin: *"Wave Propagation and Group Velocity"*, Pure and Applied Physics Series, Academic (New York, 1960).
[Bru14] Časlav Brukner, Nature Physics **10**, 259 (2014).
[BS88] James E. Bayfield, David W. Sokol, Phys. Rev. Lett. **61**, 2007 (1988).
[BS93] Christian Beck, Friedrich Schlögl: *"Thermodynamics of chaotic systems. An introduction"*, Cambridge Univ. Press (Cambridge, UK, 1993).
[BT12] John M. Beggs, Nicholas Timme, Fontiers in Physiology **3**, 163 (2012).
[BTW87] Per Bak, Chao Yang, Kurt Wiesenfeld, Phys. Rev. Lett. **59**, 381 (1987).
[BT&19] Jeremy M. Berg, John Tymoczko, Gregory Gatto, Lubert Stryer: *"Biochemistry"*, Ninth ed., W. H. Freeman (San Francisco, 2019).
[Bub05] Jeffrey Bub, Found. Phys. **35**, 541 (2005).
[But01] Eugene I. Butikov, Am. J. Phys. **69**, 1 (2001).
[BV87] Nandor L. Balazs, André Voros, Europhys. Lett. **4**, 1089 (1987).
[BV89] Nandor L. Balazs, André Voros, Ann. Phys. (NY), **190**, 1 (1989).
[BZ06] Robin Blume-Kohout, Wojciech H. Zurek, Phys. Rev. A **73**, 062310 (2006).
[CA97] Nicolas J. Cerf, Chris Adami, Phys. Rev. Lett **79**, 5194 (1997).
[CAP11] Giulio Chiribella, Giacomo Mauro D'Ariano, Paolo Perinotti, Phys. Rev. A **84**, 012311 (2011).
[Cav93] Carlton M. Caves, Phys. Rev. E **47**, 4010 (1993).
[CB06] Sam F. Cooke, Timothy V. P. Bliss, Brain **129**, 1659 (2006).
[CB99] Dante R. Chialvo, Per Bak, Neuroscience **90**, 1137 (1999).
[CBH03] Rob Clifton, Jeffrey Bub, Hans Halvorson, Found. Phys. **33**, 1561 (2003).
[CC&83] Giulio Casati, Boris V. Chirikov, Felix M. Izrailev, John Ford: *"Stochastic behavior of a quantum pendulum under a periodic perturbation"*, in Giulio Casati, John Ford (eds.): *"Stochastic Behavior in Classical and Quantum Hamiltonian Systems"*; Lecture Notes in Physics, vol. 93, p. 334, Springer (Berlin, 1983).
[CDL77] Claude Cohen-Tannoudji, Bernard Diu, Franck Laloë: *"Quantum Mechanics"*, vols. 1, 2, Wiley (New York 1977).
[CE80] Pierre Collet, Jean-Pierre Eckmann: *"Iterated Maps of the Interval as Dynamical Systems"*, Birkhäuser (Basel 1980).
[Cha75] Gregory J. Chaitin: *"Randomness and Mathematical Proof"*, Sci. Am. **232**, 47 (May 1975).
[Cha82] Gregory J. Chaitin, Int. J. Theor. Phys. **21**, 941 (1982).
[Chi10] Dante R. Chialvo, Nature Physics **6**, 744 (2010).
[Chi79] Boris V. Chirikov, Phys. Rep. **52**, 263 (1979).
[Chi93] Raymond Y. Chiao, Phys. Rev. A **48**, R34 (1993).
[CH&69] John F. Clauser, Michael A. Horne, Abner Shimony, Richard A. Holt, Phys. Rev. Lett. **23**, 880 (1969).
[Cla00] Gregory Claeys, J. History of Ideas **61**, 223 (2000).
[Coo84] Matthew Cook, Complex Sys. **15**, 1 (2004).
[Cop20] B. Jack Copeland: *"The Modern History of Computing"*, The Stanford Encyclopedia of Philosophy (Winter 2020 Edition), Edward N. Zalta (ed.), URL = <https://plato.stanford.edu/archives/win2020/entries/computing-history/>.
[CP82] James P. Crutchfield, Norman H. Packard, Int. J. Theor. Phys. **21**, 433 (1982).

[CL84]   Amir O. Caldeira, Anthony J. Leggett, Ann. Phys. (NY) **149**, 374; **153**, 445(E) (1984).

[CM10]   Dennis Coon, John O. Mitterer: *"Introduction to Psychology: Gateways to Mind and Behavior"*, 12th ed., Wadsworth, Cengage Learning (Belmont, CA, 2010).

[Cri58]   Francis H. C. Crick, Symp. Soc. Exp. Biol. **12**, 548 (1958).

[Cru84]   James P. Crutchfield, Physica **10D**, 229 (1984).

[CS90]   Dorothy L. Cheney, Robert M. Seyfarth: *"How Monkeys See the World"*, Chicago Univ. Press (Chicago, 1990).

[CSM07]   Shelley D. Copley, Eric Smith, Harold J. Morowitz, Bioorg. Chem. **35**, 430 (2007).

[CSS86]   Dorothy L. Cheney, Robert M. Seyfarth, Barbara Smuts, Science **234**, 1361 (1986).

[CT06]   Thomas A. Cover, Joy A. Thomas: *"Elements of Information Theory"*, 2nd ed., Wiley-Interscience (Hoboken, 2006).

[CX19]   Chris Chatfield, Haipeng Xing: *"The analysis of time series"*, 7th ed., Taylor and Francis, (Boca Raton, 2019).

[CZ&16]   Colin N. Waters *et al.*, Science **351**, 137 (2016).

[Dan09]   Antoine Danchin, FEMS Microbiol. Rev. **33**, 3 (2009).

[DC97]   Ken A. Dill, Hue Sun Chan, Nat. Struct. Biol. **4**, 10 (1997).

[Deu85]   David Deutsch, Proc. R. Soc. Lond. A **400**, 97 (1985).

[Deu91]   Josh M. Deutsch, Phys. Rev. A **43**, 2046 (1991).

[Dew84]   Alexander K. Dewdney, Sci. Am. **250**, 19 (1984).

[DG87]   Thomas Dittrich, Robert Graham, Europhys. Lett. **4**, 263 (1987).

[DG88]   Thomas Dittrich, Robert Graham, Europhys. Lett. **7**, 287 (1988).

[DG90a]   Thomas Dittrich, Robert Graham, Ann. Phys. (NY) **200**, 363 (1990).

[DG90b]   Thomas Dittrich, Robert Graham, Europhys. Lett. **11**, 589 (1990).

[DG90c]   Thomas Dittrich, Robert Graham, Phys. Rev. A **42**, 4647 (1990).

[DG91]   Thomas Dittrich, Robert Graham: *"Quantum Chaos in Open Systems"*, in Harald Atmanspacher, Herbert Scheingraber (eds.): *"Information Dynamics"*, NATO ASI Series B, vol. 256, Plenum (New York, 1991), pp. 289–302.

[DG92]   Thomas Dittrich, Robert Graham: *"Continuously Measured Chaotic Quantum Systems"*, in I. Percival, A. Wirzba (eds.): *"Quantum Chaos—Quantum Measurement"*, NATO ASI Series C: Mathematical and Physical Sciences, vol. 358, Springer (Berlin, 1992), pp. 219–229.

[DG&00]   Lu-Ming Duan, Geza Giedke, Juan Ignacio Cirac, Peter Zoller, Phys. Rev. Lett. **84**, 2722 (2000).

[DHM08]   Persi Diaconis, Susan Holmes, Richard Montgomery, SIAM Rev. **49**, 211 (2008).

[Dit14]   Thomas Dittrich, Eur. J. Phys. **36**, 015010 (2014).

[Dit19]   Thomas Dittrich, Entropy **21**, 286 (2019).

[Div00]   David P. DiVincenzo, Fortschritte der Physik **48**, 771 (2000).

[DJ92]   David Deutsch, Richard Jozsa, Proc. R. Soc. Lond. A **439**, 553 (1992).

[Dom05]   Gabor Domokos, Int. J. Bifurcation and Chaos **15**, 861 (2005).

[DO&08]   Ken A. Dill, Banu Ozkan, M. Scott Shell, Thomas R. Weikl, Ann. Rev. Biophys. **37**, 289 (2008).

[DRD08]   Mukeshwar Dhamala, Govindan Rangarajan, Mingzhou Ding, Neuroimage **41**, 354 (2008).

[Dur98]   Jean Durup, J. Mol. Struct. **424**, 157 (1998).

[Eck88]   Bruno Eckhardt, Phys. Rep. **163**, 255 (1988).

[ED19]   George Ellis, Barbara Drossel, Found. Phys. **49**, 1253 (2019).

[Eel91]   Ellery Eells: *"Probabilistic Causality"*, Cambridge Studies in Probability, Induction, and Decision Theory, Cambridge Univ. Press (Cambridge, UK, 1991).

[EFS01]   Andreas K. Engel, Pascal Fries, Wolf Singer, Nature Revs. Neurosci. **2**, 704 (2001).

[Ein05]   Albert Einstein, Ann. Phys. (Berlin) **322**, 549 (1905).

[EK95]   Artur Ekert, Peter L. Knight, Am. J. Phys. **63**, 415 (1995).

[Ell16]   George Ellis: *"How Can Physics Underlie The Mind? Top-Down Causation in the Human Context"*, The Frontiers Collection, Springer (Berlin, 2016).

[EPR35]   Albert Einstein, Boris Podolsky, Nathan Rosen, Phys. Rev. **47**, 777 (1935).

[ES79] Manfred Eigen, Peter Schuster: *"The Hypercycle—A Principle of Natural Self-Organization"*, Springer (Berlin, 1979).

[EW83] Manfred Eigen, Ruthild Winkler: *"The Laws of the Game: How The Principles of Nature Govern Chance"*, Princeton Univ. Press (Princeton, 1983).

[Fel68] William Feller: *"An Introduction to Probability Theory and Its Applications"*, vol. I, 3rd ed., Wiley Series in Probability and Mathematical Statistics, Wiley (Hoboken, 1968).

[Fey63] Richard P. Feynman: *"The Feynman Lectures on Physics"*, vol. 1, Addison-Wesley (Reading, MA, 1963), ch. 5–2.

[Fey82] Richard P. Feynman, Int. J. Theor. Phys. **21**, 467 (1982).

[Fey86] Richard P. Feynman, Found. Phys. **16**, 507 (1986).

[FF04] Gregory Falkovich, Alexander Fouxon, N. J. Phys. **6**, 50 (2004).

[FG04] A. V. Finkelstein, O. V. Galzitskaya, Phys. Life Rev. **1**, 23 (2004).

[FGP82] Shmuel Fishman, D.R. Grempel, Richard E. Prange, Phys. Rev. Lett. **49**, 509 (1982).

[FGP84] Shmuel Fishman, D.R. Grempel, Richard E. Prange, Phys. Rev. A **29**, 1639 (1984).

[FG&17] Mark Friesen, Joydip Ghosh, M.A. Eriksson, S.N. Coppersmith, Nat. Commun. **8**, 15923 (2017).

[FL85] Florencio, J. Jr., Howard, M. Lee, Phys. Rev. A **31**, 3231 (1985).

[Flo09] Luciano Floridi, Synthese **168**, 151 (2009).

[Flo10] Luciano Floridi: *"Information. A Very Short Introduction"*, Oxford Univ. Press (Oxford, 2010).

[Flo91] Hans Flohr, Theory and Psychology 1, **245** (1991).

[Flo95] Hans Flohr, Behavioural Brain Research **71**, 157 (1995).

[FMR91] Joseph Ford, Giorgio Mantica, Gerald H. Ristow, Physica D **50**, 493 (1991).

[Fre03] Edward Fredkin, Int. J. Theor. Phys. **42**, 189 (2003).

[FT82] Edward Fredkin, Tommaso Toffoli, Int. J. Theor. Phys. **21**, 219 (1963).

[FV63] Richard P. Feynman, Frank L. Vernon, Jr., Ann. Phys. (NY) **24**, 118 (1963).

[Gar70] Martin Gardner, Sci. Am. **223**, 120 (1970).

[Gas98] Pierre Gaspard: *"Chaos, Scattering, and Statistical Mechanics"*, Cambridge University Press (Cambridge, UK, 1998).

[GC95] Giulio Casati, Boris Chirikov: *"Quantum Chaos: Between Order and Disorder"*, Cambridge University Press (Cambridge, UK, 1995).

[Gel94] Murray Gell-Mann: *"The Quark and the Jaguar"*, Holt (New York, 1994).

[GH83] John Guckenheimer, Philip J. Holmes: *"Nonlinear Oscillations, Dynamical Systems, and Bifurcations of Vector Fields"*, Applied Mathematical Sciences, vol. 42, Springer (New York, 1983).

[GHL19] Sophie Gibb, Robin Findlay Hendry, Tom Lancaster (eds.): *"The Routledge Handbook of Emergence"*, Routledge Handbooks in Philosophy (Abingdon, 2019).

[Gib02] Willard Gibbs, J.: *"Elementary Principles in Statistical Mechanics"*, Charles Scribner's Sons (New York, 1902).

[Gib93] Willard Gibbs, J.: *"The Scientific Papers of J. Willard Gibbs"* (Vol. 1), Ox Bow Press (Woodbridge, 1993).

[Gie85] Alfred Gierer: *"Die Physik, das Leben und die Seele"* (in German), Piper (Munich, 1985).

[GK05] Christopher C. Gerry, Peter L. Knight: *"Introductory Quantum Optics"*, Cambridge Univ. Press (Cambridge, UK, 2005).

[GK&98] Rainer Goebel, Darius Khorram-Sefat, Lars Muckli, Hans Hacker, Wolf Singer, Eur. J. Neurosc. **10**, 1563 (1998).

[Gle11] James Gleick: *"The Information: A History, a Theory, a Flood"*, Pantheon (New York, 2011).

[GL&06] Sheldon Goldstein, Joel L. Lebowitz, Roderich Tumulka, Nino Zanghì, Phys. Rev. Lett. **96**, 050403 (2006).

[Goe49] Kurt Gödel, Rev. Mod. Phys. **21**, 447 (1949).

[Gol05] Rebecca Goldstein: *"Incompleteness: The Proof and Paradox of Kurt Gödel"*, W. W. Norton (New York, 2005).

[Gol92]   Nigel Goldenfeld: *"Lectures on Phase Transitions and the Renormalization Group"*, Taylor and Francis (Boca Raton, 1992).

[Goo90]   Jane Goodall: *"Through a Window. Thirty Years with the Chimpanzees of Gombe"*, Soko Publications (Eastbourne, 1990).

[Gou77]   Stephen Jay Gould: *"Ontogeny and Phylogeny"*, Belknap Press of Harvard Univ. Press (Cambridge, MA, 1977).

[GPS01]   Herbert Goldstein, Charles P. Poole, John L. Safko : *"Classical Mechanics"*, 3rd ed., Addison-Wesley (San Francisco, 2001).

[Gra69]   C. W. J. Granger, Econometrica **37**, 424 (1969).

[Gra90]   Robert Graves: *"The Greek Myths"*, Penguin (London, 1990).

[Gre98]   John R. Gregg: *"Ones and Zeros: Understanding Boolean Algebra, Digital Circuits, and the Logic of Sets"*, IEEE Press Understanding Science & Technology Series, IEEE Press / Wiley (Piscataway, NJ, 1998).

[Gro02]   Charles G. Gross, The Neuroscientist **8**, 512 (2002).

[Gro97]   Lov K. Grover, Phys. Rev. Lett. **79**, 325 (1997).

[Gru73]   Adolf Grünbaum: *"Philosophical Problems of Space and Time"*, 2nd ed., Boston Studies in the Philosophy of Science, Reidel (Dordrecht, 1973).

[GS&88]   E. J. Galvez, B. E. Sauer, L. Moorman, P. M. Koch, D. Richards, D. Phys. Rev. Lett. **61**, 2011 (1988).

[Gut90]   Martin C. Gutzwiller: *"Chaos in Classical and Quantum Mechanics"*, Interdisciplinary Applied Mathematics, vol. 1, Springer (New York, 1990).

[GVZ91]   Marie-Joya Giannoni, André Voros, Jean Zinn-Justin (eds.): *"Chaos and Quantum Physics"*, Les Houches Lectures LII 1989, NATO ASI, North-Holland (Amsterdam, 1991).

[GV&15]   Marissa Giustina, Marijn A. M. Versteegh, Sören Wengerowsky *et al.*, Phys. Rev. Lett. **115**, 250401 (2015).

[GZ09]   Crispin W. Gardiner, Peter Zoller: *"Quantum Noise: A Handbook of Markovian and Non-Markovian Quantum Stochastic Methods with Applications to Quantum Optics"*, 3rd ed., Springer Series in Synergetics, Springer (Berlin, 2004).

[Haa10]   Fritz Haake: *"Quantum Signatures of Chaos"*, 3rd ed., Springer Series in Synergetics, vol. 54, Springer (Berlin, 2010).

[Hae66]   Ernst Haeckel: *"Generelle Morphologie. I: Allgemeine Anatomie der Organismen. II: Allgemeine Entwickelungsgeschichte der Organismen"* (in German), Verlag von Georg Reimer (Berlin, 1866).

[Hae05]   Ernst Haeckel: *"Art Forms from the Ocean: The Radiolarian Atlas of 1862"*, Olaf Breidbach (ed.), Prestel (New York, 2005).

[Hak76]   Hermann Haken: *"Quantum Field Theory of Solids : An Introduction"*, North Holland, (Oxford, 1976).

[Hak83]   Hermann Haken: *"Synergetics, an Introduction: Nonequilibrium Phase Transitions and Self-Organization in Physics, Chemistry, and Biology"*, 3rd rev. enl. ed., Springer Series in Synergetics, Springer (New York, 1983).

[Hal06]   John L. Hall, Rev. Mod. Phys. **78**, 1279 (2006).

[Han06]   Theodor W. Hänsch, Rev. Mod. Phys. **78**, 1297 (2006).

[Har01]   Lucien Hardy, arXiv:0101012v4 [quant-ph] (2001).

[Har05]   James B. Hartle, Am. J. Phys. **73**, 101 (2005).

[Har40]   Nicolai Hartmann: *"Der Aufbau der realen Welt"* (in German), de Gruyter (Berlin, 1940).

[Haw17]   Stephen Hawking: *"A Brief History of Time: From the Big Bang to Black Holes"*, Bantam Books (New York, 2017).

[HB&15]   B. Hensen, H. Bernien, A. E. Dréau *et al.*, Nature **526**, 682 (2015).

[Hel84]   Eric J. Heller, Phys. Rev. Lett. **53**, 1515 (1984).

[HH82]   T. Hogg, Bernardo A. Huberman, Phys. Rev. Lett. **48**, 711 (1982).

[HH85]   T. Hogg, Bernardo A. Huberman, Phys. Rev. A **32**, 2338 (1985).

[HH94]   Ryszard Horodecki, Paweł Horodecki, Phys. Lett A **194**, 147 (1994).

[HH&96] Michał Horodecki, Paweł Horodecki, Ryszard Horodecki, Phys. Lett A **223**, 1 (1996).

[HH&09] Ryszard Horodecki, Paweł Horodecki, Michał Horodecki, Karol Horodecki, Rev. Mod. Phys. **81**, 865 (2009).

[Hir57] I. I. Hirschman, Jr., Am. J. Math. **79**, 152 (1957).

[Hit09] Peter Hitchens: *"The Broken Compass: How British Politics Lost its Way"*. Continuum International Publishing (London, 2009).

[HI&15] Masahito Hayashi, Satoshi Ishizaka, Akinori Kawachi, Gen Kimura, Tomohiro Ogawa: *"Introduction to Quantum Information Science"*, Springer (Berlin, 2015).

[HJ13] Roger A. Horn, Charles R. Johnson: *"Matrix analysis"*, 2nd ed., Cambridge Univ. Press (Cambridge, UK, 2013).

[HKO94] John H. Hannay, John P. Keating, Alfredo M. Ozorio de Almeida, Nonlinearity **7**, 1327 (1994).

[HM&88] William G. Hoover, Bill Moran, Carol G. Hoover, William J. Evans, Phys. Lett. A **133**, 114 (1988).

[Hof07] Douglas R. Hofstadter: *"I am a Strange Loop"*, Basic Books (New York, 2007).

[Hop84] John. E. Hopcroft, Sci. Am. **250**, 86 (1984).

[Hor97] Paweł Horodecki, Phys. Lett. A **232**, 333 (1997).

[HO84] John H. Hannay, Alfredo M. Ozorio de Almeida, J. Phys. A **17**, 3429 (1984).

[HO&84] M. Hillery, R. F. O'Connell, M. O. Scully, E. P. Wigner, Phys. Rep. **106**, 121 (1984).

[HP11] Daniel W. Hahs, Shawn D. Pethel, Phys. Rev. Lett. **107**, 128701 (2011).

[HP83] Hentschel, H.G.E. Itamar Procaccia, Physica **8D**, 435 (1983).

[HS13] Douglas R. Hofstadter, Emmanuel Sander: *"Surfaces and Essences. Analogy as the Fuel and Fire of Thinking"*, Basic Books (New York, 2013).

[Hug58] John R. Hughes, Physiol. Rev. **38**, 91 (1958).

[Hum90] David Hume: *"An Enquiry Concerning Human Understanding"*, Vintage Books (New York, 1990).

[HW08] Godfrey H. Hardy, Edward M. Wright: *"An Introduction to the Theory of Numbers"*, Roger Heath-Brown, Joseph Silverman, Andrew Wiles (eds.), 6th ed., Oxford Univ. Press (Oxford, 2008).

[HW85] Bernardo A. Huberman, W. F. Wolff, Phys. Rev. A **32**, 3768 (1985).

[IH&90] Wayne M. Itano, Daniel J. Heinzen, John J. Bollinger, David J. Wineland, Phys. Rev. A **41**, 2295 (1990).

[IPC21] Intergovernmental Panel on Climate Change: *"Climate Change 2021: The Physical Science Basis. Contribution of Working Group I to the Sixth Assessment Report of the Intergovernmental Panel on Climate Change"*, IPCC (Geneva, 2021).

[IS80] Felix M. Izrailev, Dima L. Shepelyanskii, Theor. Math. Phys. **43**, 553 (1980).

[Jac75] John David Jackson: *"Classical Electrodynamics"*, 2nd ed., Wiley (New York, 1975).

[Jac81] Donald Jackson: *"The Story of Writing"*, Shuckburgh Reynolds (London, 1981).

[Jay57] Edwin T. Jaynes, Phys. Rev. **106**, 4 (1957); *ibid*. **106**, 620 (1957); *ibid*. **108**, 171 (1957).

[Jay92] Edwin T. Jaynes: *"The Gibbs paradox"*, in C.R. Smith, G.J. Erickson, P.O. Neudorfer (eds.): *"Maximum Entropy and Bayesian Methods"*, Kluwer (Dordrecht, 1992).

[JA&00] Thomas Jennewein, Ulrich Achleitner, Gregor Weihs, Harald Weinfurter, Anton Zeilinger, Rev. Sci. Instrum. **71**, 1675 (2000).

[JC&11] M. Jofre, M. Curty, F. Steinlechner, G. Anzolin, J. P. Torres, M. W. Mitchell, V. Pruneri, Opt. Express **19**, 20665 (2011).

[Jer91] Harry Jerison: *"Brain Size and the Evolution of Mind"*, The Fifty-ninth James Arthur Lecture on the Human Brain, American Museum of Natural History (New York, 1991).

[Jev79] William Stanley Jevons: *"The Principles of Science: A Treatise on Logic and Scientific Method"*, 3rd ed., Macmillan (London, 1879).

[Joy02] Gerald F. Joyce, Nature **418**, 214 (2002).

[JS98] Jorge V. José, Eugene J. Saletan: *"Classical Dynamics. A Contemporary Approach"*, Cambridge Univ. Press (Cambridge, UK, 1998).

[JW&06] Vincent Jacques, E. Wu, Frédéric Grosshans *et al.*, arXiv:quant-ph/0610241 (2006).

[Kan04] Immanuel Kant: *"Prolegomena to Any Future Metaphysics That Will Be Able to Present Itself as a Science"* (German: *"Prolegomena zu einer jeden künftigen Metaphysik, die als Wissenschaft wird auftreten können"*), revised ed., Gary Hatfield (ed.), Cambridge Texts in the History of Philosophy, Cambridge Univ. Press (Cambridge, UK, 2004).

[Kan98] Immanuel Kant: *"Critique of pure reason"* (German: *"Kritik der reinen Vernunft"*), revised ed., Paul Guyer, Alan W. Wood (eds.), The Cambridge Edition of the Works of Immanuel Kant, Cambridge Univ. Press (Cambridge, UK, 1998).

[Kap03] Wilfred Kaplan: *"Advanced Calculus"*, 5th ed., Addison-Wesley (Reading, MA, 2003).

[Kar83] Charles F. F. Karney, Physica D **8**, 360 (1983).

[Kar97] Martin Karplus, Folding & Design 2, S69 (1997).

[Kau93] Stuart Kauffman: *"The Origins of Order: Self Organization and Selection in Evolution"*, Oxford Univ. Press (Oxford, 1993).

[Kau95] Stuart Kauffman: *"At Home in the Universe: The Search for Laws of Self-Organization and Complexity"*, Oxford Univ. Press (Oxford, 1995).

[KB07] Franz von Kutschera, Alfred Breitkopf: *"Einführung in die moderne Logik"* (in German), Alber (Freiburg, 2007).

[KB&00] Paul G. Kwiat, Andrew J. Berglund, Joseph B. Altepeter, Andrew G. White, Science **290**, 498 (2000).

[KCG11] Evan Kodra, Snigdhansu Chatterjee, Auroop R. Ganguly, Theor. Appl. Climatol. **104**, 325 (2011).

[KC&19] Peter M. Kappeler, Tim Clutton-Brock, Susanne Shultz, Dieter Lukas, Behav. Ecol. Sociobiol. **73**, 5 (2019).

[KL87] Stuart A. Kauffman, Simon Levin, J. Theor. Biol. **128**, 11 (1987).

[Kle10] Axel Kleidon, Phys. Life Rev. **7**, 242 (2010).

[Kle12] Axel Kleidon, Phil. Trans. R. Soc. A **370**, 1012 (2012).

[Kle16] Axel Kleidon: *"Thermodynamic Foundations of the Earth System"*, Cambridge Univ. Press (Cambridge, UK, 2016).

[Kli86] Stephen J. Kline: *"Similitude and Approximation Theory"*, 2nd ed., Springer (New York, 1986).

[Kol98] Andrey Kolmogorov, Theoretical Computer Science **207**, 387 (1998).

[KOS17] Dimitris Kakofengitis, Maxime Oliva, Ole Steuernagel, Phys. Rev. A **95**, 022127 (2017).

[KR08] Maya Kahana, Binyamin Gila, Rivka Adara, Ehud Shapiro, Physica D **237**, 1165 (2008).

[Kra81] Karl Kraus, Found. Phys. **11**, 547 (1981).

[KS04] Holger Kantz, Thomas Schreiber: *"Nonlinear Time Series Analysis"*, 2nd ed., Cambridge Univ. Press (Cambridge, UK, 2004).

[KT&16] Adam M. Kaufman, M. Eric Tai, Alexander Lukin, Matthew Rispoli, Robert Schittko, Philipp M. Preiss, Markus Greiner, Science **353**, 794 (2016).

[Kuh70] Thomas S. Kuhn: *"The Structure of Scientific Revolutions"*, 2nd ed., University of Chicago Press (Chicago, 1970).

[Kul68] Solomon Kulback: *"Information Theory and Statistics"*, Dover (New York, 1959).

[Kum17] Asutosh Kumar, arXiv:1504.07176v2 [quant-ph] (2017).

[Kup90] Bernd-Olaf Küppers: *"Information and the Origin of Life"*, MIT Press (Cambridge, MS, 1990).

[Lan61] Rolf Landauer, IBM J. Res. Dev. **5**, 183 (1961).

[Lan68] Serge Lang: *"Analysis I"*, Addison-Wesley (Reading, MS, 1968).

[Lan91] Rolf Landauer: *"Information is Physical"*, Physics Today **44**, 23 (May 1991).

[Lap51] Pierre-Simon de Laplace: *"A Philosophical Essay on Probabilities"*, translated into English from the original French 6th ed. by F. W. Truscott and F. L. Emory, Dover (New York, 1951).

[Lau05] Robert B. Laughlin: *"A Different Universe. Reinventing Physics From the Bottom Down"*, Perseus (New York, 2005).

[LB17] David J. Luitz, Yevgeny Bar Lev, Phys. Rev. B **96**, 020406(R), (2017).

[LCW98] Daniel A. Lidar, I. L. Chuang, K. Brigitta Whaley, Phys. Rev. Lett. **81**, 2594 (1998).

[LC&87]   A. J. Leggett, S. Chakravarty, A. T. Dorsey, M. P. A. Fisher, A. Garg, W. Zwerger, Rev. Mod. Phys. **59**, 1 (1987).

[Lea94]   Richard Leakey: *"The Origin Of Humankind"*, Science Masters Series, Perseus (New York, 1994).

[Leg92]   Anthony J. Leggett, Found. Phys. **22**, 221 (1992).

[Lev99]   Pierre Lévy: *"Collective Intelligence"*, Perseus (Cambridge, MA, 1999).

[Lid13]   Daniel A. Lidar, arXiv:1208.5791v3 [quant-ph] (2013).

[Lik82]   Konstantin K. Likharev, Int. J. Theor. Phys. **21**, 311 (1982).

[Lin76]   Göran Lindblad, Commun. Math. Phys. **48**, 119 (1976).

[LL92]   A. J. Lichtenberg, M. A. Lieberman, *"Regular and Chaotic Motion"*, 2nd ed., Applied Mathematical Sciences, vol. 38, Springer (New York, 1992).

[LLA15]   David J. Luitz, Nicolas Laflorencie, Fabien Alet, Phys. Rev. B **91**, 081103(R) (2015).

[Llo91]   Seth Lloyd: *"Causality and Information Flow"*, in Harald Atmanspacher, Herbert Scheingraber (eds.): *"Information Dynamics"*, NATO ASI Series B, vol. 256, Plenum (New York, 1991), pp. 131–142.

[Llo02]   Seth Lloyd, Phys. Rev. Lett. **88**, 237901 (2002).

[Llo06]   Seth Lloyd: *"Programming the Universe"*, Random House (New York, 2006).

[Lor63]   Edward N. Lorenz, J. Atmos. Sci. **20**, 130 (1963).

[Lor68]   Konrad Lorenz: *"Vom Weltbild des Verhaltensforschers. Drei Abhandlungen"* (in German), Deutscher Taschenbuch-Verlag (München, 1968).

[Lor78]   Konrad Lorenz: *"Behind the Mirror: A Search for a Natural History of Human Knowledge"*, Harcourt Brace Jovanovitch (New York, 1978).

[Lot88]   Hermann Lotze: *"Logic in Three Books. Of Thought, of Investigation and of Knowledge"*, Bernard Bosanquet (ed.), Clarendon Press (Oxford, 1888)

[Lou73]   William H. Louisell: *"Quantum Statistical Properties of Radiation"*, Wiley (New York, 1973).

[Lov00]   James Lovelock: *"Gaia: A New Look at Life on Earth"*, 3rd ed., Oxford Univ. Press (Oxford, 2000).

[Low63]   Per-Olov Löwdin, Rev. Mod. Phys. **35**, 724 (1963).

[LP00]   Robert B. Laughlin, David Pines, Proc. Nat. Acad. Sci. **97**, 28 (2000).

[LR85]   Patrick A. Lee, Tiruppattur V. Ramakrishnan, Rev. Mod. Phys. **57**, 287 (1985).

[LSV08]   S. Lievens, N. I. Stoilova, J. Van der Jeugt, J. Phys. Conf. Ser. **128**, 012028 (2008).

[Lud50]   Gerhart Lüders, Ann. Phys. Leipzig **443**, 322 (1950).

[LW03]   Daniel A. Lidar, K. Brigitta Whaley, arXiv:quant-ph/0301032v1 (2003).

[Mac03]   David J. C. MacKay: *"Information Theory, Inference and Learning Algorithms"*, Cambridge Univ. Press (Cambridge, UK, 2003).

[Mac82]   Ernst Mach: *"Die ökonomische Natur der physikalischen Forschung"*, K. k. Hof- und Staatsdruckerei (Wien, 1882).

[Mac97]   Ernst Mach: *"Contributions to the Analysis of Sensations"*, C. M. Williams (ed.), The Open Court Publishing Co. (Chicago, 1897).

[Mal20]   Juan Maldacena, Nature Revs. Phys. **2**, 123 (2020).

[Mar03]   Norman Margolus, Int. J. Theor. Phys. **42**, 309 (2003).

[May74]   Robert M. May: *"Stability and complexity in model ecosystems"*, 2nd ed., Princeton Univ. Press (Princeton, NJ, 1974).

[May76]   Robert M. May, Nature **261**, 459 (1976).

[Maz17]   James E. Mazur: *"Learning and Behaviour"*, 8th ed., Taylor & Francis (New York, 2017).

[MH&12]   Xiao-Song Ma, Thomas Herbst, Thomas Scheidl *et al.*, Nature **489**, 269 (2012).

[Mic10]   Karo Michaelian, Earth Syst. Dynam. Discuss. **1**, 1 (2010).

[MK79]   Steven W. McDonald, Allan N. Kaufman, Phys. Rev. Lett. **42**, 1189 (1979).

[MK&13]   Xiao-Song, Johannes Kofler, Angie Qarry *et al.*, Proc. Nat. Acad. Sci. **110**, 1221 (2013).

[MM&19]   Z. K. Minev, S. O. Mundhada, S. Shankar *et al.*, Nature **570**, 200 (2019).

[MOE09]   Nancey Murphy, Timothy O'Connor, George Ellis (eds.): *"Downward Causation and the Neurobiology of Free Will"*, Understanding Complex Systems, Springer (Berlin, 2009).

[Mon71] Jacques Monod: *"Chance and Necessity: An Essay on the Natural Philosophy of Modern Biology"*, Alfred A. Knopf (New York, 1971).

[MR&94] F. L. Moore, J. C. Robinson, C. Bharucha, P. E. Williams, M. G. Raizen, Phys. Rev. Lett. **73**, 2974 (1994).

[MR&95] F. L. Moore, J. C. Robinson, C. Bharucha, Bala Sundaram, M. G. Raizen, Phys. Rev. Lett. **75**, 4598 (1995).

[MS77] B. Misra, E. C. G. Sudarshan, J. Math. Phys. **18**, 756 (1977).

[MS86] Vitali D. Milman, Gideon Schechtman: *"Asymptotic Theory of Finite Dimensional Normed Spaces"*, Lecture Notes in Mathematics, vol. 1200, pp. 5–6 and 140–141, Springer (Berlin, 1986).

[MSE96] Massimo Palma, G. Kalle-Antti Suominen, Artur Ekert, Proc. R. Soc. Lond. A **452**, 567 (1996).

[Mul07] Scott J. Muller: *"Asymmetry: The Foundation of Information"*, Frontiers Collection, Springer (Berlin, 2007).

[MW70] John Mathews, R. L. Walker: *"Mathematical Methods of Physics"*, 2nd ed., Benjamin/Cummings (Menlo Park, 1970).

[Nac02] Werner Nachtigall: *"Bionik. Grundlagen und Beispiele für Ingenieure und Naturwissenschaftler"* (in German), 2nd ed., Springer (Berlin, 2002).

[Nas05] See, e.g., NASA's Deep Impact Mission that culminated in the partial destruction of comet Tempel I in July 4, 2005: http://www.jpl.nasa.gov/missions/deep-impact/.

[NC00] Michael A. Nielsen, Isaac L. Chuang: *"Quantum Computation and Quantum Information"*, Cambridge Univ. Press (Cambridge, UK, 2000).

[Neu18] John von Neumann: *"Mathematical Foundations of Quantum Mechanics"*, N.A. Wheeler (ed.), Princeton Univ. Press (Princeton, NJ, 2018).

[Neu29] John von Neumann, Z. Phys. **57**, 30 (1929).

[Neu66] John von Neumann: *"The Theory of Self-Reproducing Automata"*, Arthur W. Burks (ed.), Univ. of Illinois Press (Urbana, IL, 1966).

[NN56] Ernest Nagel, James R. Newman: *"Gödel's Proof"*, Sci. Am. **194**, 71 (June 1956).

[NN85] Ernest Nagel, James R. Newman: *"Gödel's Proof"*, New York Univ. Press (New York, 1985).

[NPM99] Vilém Novák, Irina Perfilieva, Jiří Močkoř: *"Mathematical principles of fuzzy logic"*, Kluwer (Dordrecht, 1999).

[NR&16] C. Neill, P. Roushan, M. Fang *et al.*, Nature Physics **12**, 1037 (2016).

[Nus04] Christiane Nüsslein-Volhard: *"Von Genen und Embryonen"* (in German), Reclam (Stuttgart, 2004).

[Nus06] Christiane Nüsslein-Volhard: *"Coming to Life: How Genes Drive Development"*, Florian Maderspacher (ed.), Kales Press (San Diego, 2006).

[OCB12] Ognyan Oreshkov, Fabio Costa, Časlav Brukner, Nature Comm. **3**, 1092 (2012).

[OGY90] Edward Ott, Celso Grebogi, James A. Yorke, Phys. Rev. Lett. **64**, 1196 (1990).

[Ott93] Edward Ott: *"Chaos in Dynamical Systems"*, Cambridge Univ. Press (Cambridge, UK, 1993).

[OZK01] Haldun M. Ozaktas, Zeev Zalevsky, M. Alper Kutay: *"The Fractional Fourier Transform with Applications in Optics and Signal Processing"*, Wiley (Chichester, 2001).

[Ozo88] Alfredo M. Ozorio de Almeida: *"Hamiltonian systems: Chaos and quantization"*, Cambridge Monographs on Mathematical Physics, Cambridge Univ. Press (Cambridge, UK, 1988).

[PA&10] Stefano Pironio, Antonio Acín, Serge Massar *et al.*, Nature **464**, 1021 (2010).

[Pea09] Judea Pearl, *"Causality: Models, Reasoning and Inference"*, 2nd ed., Cambridge Univ. Press (Cambridge, UK, 2009).

[Pei34] Charles Sanders Peirce: *"Collected Papers of Charles Sanders Peirce"*, Harvard Univ. Press (Cambridge, MA, 1934).

[Per71] A. M. Perelomov, Theor. Math. Phys. **6**,156 (1971) [Russian: Teor. Mat. Fiz **6**, 213 (1971)].

[Per95] Asher Peres, *Quantum Theory: Concepts and Methods*, Fundamental Theories of Physics, vol. 57, Kluwer Academic (Dordrecht, 1995).

[Per96] Asher Peres, Phys. Rev. Lett. **77**, 1413 (1996).

[PH10] Arijeet Pal, David A. Huse, Phys. Rev. B **82**, 174411 (2010).

[PH19] Nathalie Picqué, Theodor W. Hänsch, Nature Photonics **13**, 146 (2019).

[PH69] Paris, D.T. Hurd, F.K. *"Basic Electromagnetic Theory"*, McGraw-Hill (Boston, 1969).

[Pie80] John R. Pierce: *"An Introduction to Information Theory. Symbols, Signals, and Noise"*, 2nd ed., Dover (New York, 1980).

[Pla36] Max Planck: *"Vom Wesen der Willensfreiheit"* (in German), in *"Vorträge und Erinnerungen"*, 5th ed., S. Hirzel (Stuttgart, 1949), pp. 301–317.

[Pla43] Max Planck: *"Zur Geschichte der Auffindung des physikalischen Wirkungsquantums"* (in German), in *"Vorträge und Erinnerungen"*, 5th ed., S. Hirzel (Stuttgart, 1949), pp. 15–27.

[PN77] Ilya Prigogine, Gregoire Nicolis: *"Self-Organization in Non-Equilibrium Systems"*, Wiley (New York, 1977).

[Poi96] Henry Poincaré: *"Calcul des Probabilités"* (in French), George Carré (Paris, 1896).

[Pru98] Stanley B. Prusiner, Proc. Nat. Acad. Sci. **95**, 13363 (1998).

[PSE96] Massimo Palma, G. Kalle-Antti Suominen, Artur K. Ekert, Proc. Roy. Soc. Lond. A **452**, 567 (1996).

[PSW06] Sandu Popescu, Anthony J. Short, Andreas Winter, Nature Physics **2**, 754 (2006).

[PS&12] Alberto Peruzzo, Peter Shadbolt, Nicolas Brunner *et al.*, Science **338**, 634 (2012).

[PV07] Milan Paluš, Martin Vejmelka, Phys. Rev. E **75**, 056211 (2007).

[Raa15] Panu Raatikainen: *"The Diagonalization Lemma"*, in *"Stanford Encyclopedia of Philosophy"*, Edward N. Zalta (ed.), The Metaphysics Research Lab, Center for the Study of Language and Information, Stanford University (Stanford, 2015).

[RB&95] J. C. Robinson, C. Bharucha, F. L. Moore, R. Jahnke, G. A. Georgakis, Q. Niu, M. G. Raizen, Phys. Rev. Lett. **74**, 3963 (1995).

[RDO08] Marcos Rigol, Vanja Dunjko, Maxim Olshanii, Nature **452**, 854 (2008).

[Rec73] Ingo Rechenberg, *"Evolutionsstrategie. Optimierung technischer Systeme nach Prinzipien der biologischen Evolution"* (in German), problemata, vol. 15, Frommann Holzboog (Stuttgart, 1973).

[Rei56] Hans Reichenbach, *"The Direction of Time"*, Maria Reichenbach (ed.), Univ. of California Press (Berkeley, 1956).

[Rei65] Frederick Reif, *"Fundamentals of Statistical and Thermal Physics"*, McGraw-Hill (Boston, 1965).

[Rei92] Linda E. Reichl, *"The Transition to Chaos. In Conservative Classical Systems: Quantum Manifestations"*, Institute for Nonlinear Science, Springer (New York, 1992).

[RGL08] Kenneth J. Rothman, Sander Greenland, Timothy L. Lash, *"Modern Epidemiology"*, 3rd ed., Wolters Kluver (Philadelphia, 2008).

[Ris89] Hannes Risken, *"The Fokker-Planck Equation. Methods of Solution and Applications"*, 2nd ed., Springer Series in Synergetics, vol.18, Springer (Berlin, 1989).

[Roe00] Juan G. Roederer, *"Information and Its Role in Nature"*, Frontiers Collection, Springer (Berlin, 2000).

[Ros85] Robert Rosen, *"Anticipatory Systems: Philosophical, Mathematical and Methodological Foundations"*, Pergamon (Oxford, 1985).

[Ros91] Robert Rosen, *"Life itself. A Comprehensive Inquiry Into the Nature, Origin, and Fabrication of Life"*, Columbia Univ. Press (New York, 1991).

[Rot97] Gerhardt Roth, *"Das Gehirn und seine Wirklichkeit. Kognitive Neurobiologie und ihre philosophischen Konsequenzen"* (in German), Suhrkamp Taschenbuch Wissenschaft 1275 (Frankfurt am Main, 1997).

[RPK97] Michael G. Rosenblum, Arkady S. Pikovsky, Jürgen Kurths, Phys. Rev. Lett. **78**, 4193 (1997).

[RUR18] Josef Rammensee, Juan Diego Urbina, Klaus Richter, Phys. Rev. Lett. **121**, 124101 (2018).

[San21] Flavio Del Santo: *"Indeterminism, Causality and Information: Has Physics Ever Been Deterministic?"*, in Anthony Aguirre, Zeeya Merali, David Sloan (eds.): *"Undecidability, Uncomputability, and Unpredictability"*, The Frontiers Collection, Springer (Cham, 2021), p. 63.

[Sar90] Marcos Saraceno, Ann. Phys. (NY) **199**, 37 (1990).

[Sau77] Ferdinand de Saussure: *"Course in General Linguistics"*, Fontana/Collins (Glasgow, 1977).

[SB10] Josef Stoer, Roland Bulirsch: *"Introduction to Numerical Analysis"*, 3rd ed., Texts in Applied Mathematics, vol. 12, Springer (New York, 2010).

[SB95] Dimitris Stassinopoulos, Per Bak, Phys. Rev. E **50**, 5033 (1995).

[SBB15] Anil K. Seth, Adam B. Barrett, Lionel Barnett, J. Neurosci. **35**, 3293 (2015).

[Sch00] Thomas Schreiber, Phys. Rev. Lett. **85**, 461 (2000).

[Sch01] Wolfgang P. Schleich, *"Quantum Optics in Phase Space"*, Wiley-VCH (Berlin, 2001).

[Sch35] Erwin Schrödinger, Naturwissenschaften **23**, 807 (1935).

[Sch44] Erwin Schrödinger: *"What Is Life? The Physical Aspect of the Living Cell"*, Cambridge Univ. Press (Cambridge, UK, 1944).

[Sch84] Heinz Georg Schuster: *"Deterministic Chaos: an Introduction"*, Physik-Verlag (Weinheim, 1984).

[Seb76] Thomas A. Sebeok: *"Contributions to the Doctrine of Signs"*, Indiana Univ. Press (Bloomington, IN, 1976).

[Sei06] Charles Seife: *"Decoding the Universe"*, Penguin (New York, 2006).

[SG64] Leonard Susskind, Jonathan Glogower, Physics **1**, 49 (1964).

[Sha48] Claude E. Shannon, Bell Syst. Tech. J. **27**, 379 (1948).

[Sha80] Robert Shaw, Z. Naturforsch. A **36**, 80 (1980).

[Sho95] Peter W. Shor, Phys. Rev. A **52**, R2493 (1995).

[SH&15] Michael Schreiber, Sean S. Hodgman, Pranjal Bordia *et al.*, Science **349**, 842 (2015).

[Sim00] Rajiah Simon, Phys. Rev. Lett. **84**, 2726 (2000).

[Sin02] Wolf Singer: *"Der Beobachter im Gehirn. Essays zur Hirnforschung"* (in German), Suhrkamp (Frankfurt a. M., 2002).

[Sin70] Yakov G. Sinai, Russ. Math. Survey **25**, 137 (1970).

[SKR13] Ole Steuernagel, Dimitris Kakofengitis, Georg Ritter, Phys. Rev. Lett. **110**, 030401 (2013).

[Skl93] Lawrence Sklar: *"Physics and Chance. Philosophical Issues in the Foundations of Statistical Mechanics"*, Cambridge Univ. Press (Cambridge, UK, 1993).

[SL&16] J. Smith, A. Lee, P. Richerme *et al.*, Nature Physics **12**, 907 (2016).

[SM09] Dmitry A. Smirnov, Igor I. Mokhov, Phys. Rev. E **80**, 016208 (2009).

[SM95] Eörs Szathmáry, John Maynard Smith, Nature **374**, 227 (1995).

[Smi67] William Smith: *"Dictionary on Greek and Roman Biography and Mythology"*, vol. III Little, Brown, and Co. (Boston, 1867).

[SO82] Attila Szabo, Neil S. Ostlund: *"Modern Quantum Chemistry"*, Macmillan (London, 1982).

[Sob95] Dava Sobel: *"Longitude: The True Story of a Lone Genius Who Solved the Greatest Scientific Problem of His Time"*, Walker (New York, 1995).

[Sol64] Ray Solomonoff, Information and Control **7**, 1, *ibid.* **7**, 224 (1964).

[SS95] Eörs Szathmáry, John Maynard Smith, Nature **374**, 227 (1995).

[SS&15] Lynden K. Shalm, Evan Meyer-Scott, Bradley G. Christensen *et al.*, Phys. Rev. Lett. **115**, 250402 (2015).

[Sre94] Mark Srednicki, Phys. Rev. E **50**, 888 (1994).

[Sta90] William Stalling: *"Cryptography and Network Security: Principles and Practice"*, Prentice Hall (New Jersey, 1990).

[Ste45] Saul Steinberg: *"All in Line"*, Duell, Sloan & Pearce (New York, 1945).

[Ste83] Laurence Sterne: *"The Life and Opinions of Tristram Shandy, Gentleman"*, Oxford World's Classics, vol. IV, chap. XIII, Oxford Univ. Press (Oxford, 1983).

[Sto15]   James V. Stone: *"Information Theory: A Tutorial Introduction"*, Sebtel Press (Sheffield, 2015).

[Sto99]   Hans-Jürgen Stöckmann: *"Quantum Chaos: An Introduction"*, Cambridge Univ. Press (Cambridge, UK, 1999).

[Str15]   Steven H. Strogatz: *"Nonlinear Dynamics and Chaos: With Applications to Physics, Biology, Chemistry, and Engineering"*, 2nd ed., Westview Press (Boulder, 2015).

[ST&08]   Tetsu Saigusa, Atsushi Tero, Toshiyuki Nakagaki, Yoshiki Kuramoto, Phys. Rev. Lett. **100**, 018101 (2008).

[SU&10]   Thomas Scheidl, Rupert Ursin, Johannes Kofler *et al.*, Proc. Nat. Acad. Sci. **107**, 19708 (2010).

[SW49]   Claude E. Shannon, Warren Weaver: *"The Mathematical Theory of Communication"*, University of Illinois Press (Urbana, 1949).

[SW85]   G. Schmidt, B. W. Wang, Phys. Rev. A **32**, 2994 (1985).

[SY&09]   Woodrow L. Shew, Hondian Yang, Thomas Petermann, Rajarshi Roy,Thomas Plenz, J. Neurosci. **29**, 15595 (2009).

[Szi29]   Leo Szilard, Z. Phys. **53**, 840 (1929).

[Szu01]   Tadeusz M. Szuba: *"Computational Collective Intelligence"*, Wiley (New York, 2001).

[TA13]   Andrew S. Tanenbaum, Todd Austin: *"Structured Computer Organization"*, sixth ed., Pearson (Boston. MS, 2013).

[TB&98]   W. Tittel, J. Brendel, H. Zbinden, N. Gisin, Phys. Rev. Lett. **81**, 3563 (1998).

[Tof80]   Tommaso Toffoli, *Reversible computing*, MIT Report MIT/LCS/TM-151 (1980).

[Tof81]   Tommaso Toffoli, Math. Systems Theory **14**, 13 (1982).

[US&18]   Y. Ulrich, J. Saragosti, C. K. Tokita, C. E. Tarnita, D. J. C. Kronauer, Nature **560**, 635 (2018).

[Vle28]   John H. van Vleck, Proc. Natl. Acad. Sci. **14**, 178 (1928).

[Vol05]   Gerhard Vollmer: *"How is it that we can know this world? New arguments in evolutionary epistemology"*, in Vittorio Hösle, Christian Illies (eds.): *"Darwinism And Philosophy"*, University of Notre Dame Press, (Notre Dame, 2005), pp. 259–274.

[Wal16]   Mark Walport (Government Chief Scientific Adviser): *"Distributed Ledger Technology: beyond block chain"*, UK Government Office of Science Report (2016).

[Weh78]   Alfred Wehrl, Rev. Mod. Phys. **50**, 221 (1978).

[Weh79]   Alfred Wehrl, Rep. Math. Phys. **16**, 353 (1979).

[Weh91]   Alfred Wehrl: *"Information Theoretical Aspects of Quantum Mechanical Entropy"*, in Harald Atmanspacher, Herbert Scheingraber (eds.): *"Information Dynamics"*, NATO ASI Series B, vol. 256, Plenum (New York, 1991), pp. 267–278.

[Wei71]   Carl F. von Weizsäcker: *"Die Einheit der Natur"* (in German), Hanser (Munich, 1971).

[Wei85]   Carl F. von Weizsäcker: *"Aufbau der Physik"* (in German), Hanser (Munich, 1985).

[Wei88]   Steven Weinberg: *"The First Three Minutes"*, Basic Books (New York, 1988).

[Wey31]   Hemann Weyl: *"The Theory of Groups and Quantum Mechanics"* (English translation of *"Gruppentheorie und Quantenmechanik"*), Dover (New York, 1931).

[Wey52]   Hemann Weyl: *"Symmetry"*, Princeton Univ. Press (Princeton, NJ, 1952).

[WH86]   W. F. Wolff, Bernardo A. Huberman, Z. Phys. B **63**, 397 (1986).

[WH&87]   James D. Watson, Nancy H. Hopkins, Jeffrey W. Roberts, Joan Argetsinger Steitz, Alan M. Weiner: *"Molecular Biology of the Gene"*, fourth ed., Benjamin/Cummings (Menlo Park, 1987).

[Whe00]   Nicholas Wheeler: *"Phase Space Formulation of the Quantum Mechanical Particle-in-the-Box Problem"*, preprint, Reed College Physics Department (Portland, OR, 2000).

[Whe83]   John A. Wheeler: *"Law without Law"*, in J. A. Wheeler, W. H. Zurek (eds.): *"Quantum Theory and Measurement"*, Princeton Univ. Press (Princeton, 1983), pp. 182–213.

[Whe90a] John A. Wheeler: *"Information, Physics, Quantum: The Search for Links"*, in W.H. Zurek (ed.): *"Complexity, Entropy, and the Physics of Information"*, Santa Fe Institute Studies in the Sciences of Complexity, vol. VIII, Westview Press (Santa Fe, 1990), pp. 3–28.

[Whe90b] John A. Wheeler, in S. Kobayashi, H. Ezawa, Y. Murayama, S. Namura (eds.): *"Proceedings of the 3rd International Symposium on Foundations of Quantum Mechanics in the Light of New Technology"*, Physical Society of Japan (Tokyo, 1990), pp. 354–368.

[Whi61] J. Eldon Whitesitt: *"Boolean Algebra and Its Applications"*, Addison Wesley (Reading, Mass., 1961).

Wie61] Norbert Wiener: *"Cybernetics: Or Control and Communication in the Animal and the Machine"*, Cambridge Univ. Press (Cambridge, UK, 1961).

[Wil98] Edward Wilson *"Consilience. The Unity of Knowledge"*, Random House (New York, 1998).

[Wol83] Stephen Wolfram, Rev. Mod. Phys. **55**, 601 (1983).

[Wol84] Stephen Wolfram, Physica **10D**, 1 (1984).

[WR10] Alfred North Whitehead, Bertrand Russell: *"Principia Mathematica"*, vols. 1, 2, 3, Cambridge Univ. Press (Cambridge, UK, 1910, 1912, 1913).

[Wri71] Georg Henrik von Wright: *"Explanation and Understanding"*, Cornell Univ. Press (Ithaca, 1971).

[WZ&16] Colin N. Waters, Jan Zalasiewicz, Colin Summerhayes *et al.*, Science **351**, 137 (2016).

[YC&13] Juan Yin, Yuan Cao, Hai-Lin Yong *et al.*, Phys. Rev. Lett. **110**, 260407 (2013).

[YC&17] Juan Yin, Yuan Cao, Yu-Huai Li *et al.*, Science **356**, 1140 (2017).

[Zak89] J. Zak, J. Math. Phys. **30**, 1591 (1989).

[Zas79] George M. Zaslavsky, Phys. Lett. **69A**, 145 (1979).

[Zeh10] H. Dieter Zeh: *"The Physical Basis of the Direction of Time"*, 5th ed., The Frontiers Collection, Springer (Berlin, 2010).

[Zeh93] H. Dieter Zeh, Phys. Lett. **A 172**, 189 (1979).

[Zei99] Anton Zeilinger, Found. Phys. **29**, 631 (1999).

[Zim81] Erik Zimen: *"The Wolf. A Species in Danger"*, Delacorte Press (New York, 1981).

[ZGP83] Dongmei Zhang, László Györgyi, William R. Peltier, Chaos **3**, 723 (1983).

[ZP94] Wojciech H. Zurek, Juan Pablo Paz, Phys. Rev. Lett. **72**, 2508 (1994).

[ZPP08] Marko Žnidarič, Tomaž Prosen, Peter Prelovšek, Phys. Rev. B **77**, 064426 (2008).

[ZR97] Paolo Zanardi, Mario Rasetti, Phys. Rev. Lett. **79**, 3306 (1997).

[Zur81] Wojciech H. Zurek, Phys. Rev. D **24**, 1516 (1981).

[Zur84] Wojciech H. Zurek: *"Collapse of the wavepacket: how long does it take?"*, in Gerald T. Moore, Marlan O. Scully (eds.): *"Frontiers of Nonequilibrium Statistical Physics"*, NATO ASI Series B: Physics, vol. 135, Springer (Berlin, 1984), p. 145.

[Zur89] Wojciech H. Zurek, Phys. Rev. A **40**, 4731 (1989).

[Zur91] Wojciech H. Zurek, Physics Today **44**, 36 (1991).

[Zus69] Konrad Zuse: *"Rechnender Raum"*, Elektronische Datenverarbeitung **8**, 336 (1967, in German); *"Calculating Space"*, MIT Technical Translation AZT-70–164-GEMIT (1969).

Index

© Springer Nature Switzerland AG 2022
T. Dittrich, *Information Dynamics*, The Frontiers Collection,
https://doi.org/10.1007/978-3-030-96745-1

Printed in the United States
by Baker & Taylor Publisher Services